Treatise on Physiological Optics

Volume II

Hermann von Helmholtz

Edited by
James P. C. Southall
Formerly Professor of Physics in Columbia University

DOVER PUBLICATIONS, INC.
Mineola, New York

DOVER PHOENIX EDITIONS

Bibliographical Note

This Dover edition, first published in 1962 in one volume, and reprinted in 2005 as two separate volumes, is an unabridged and corrected republication of the English translation of *Handbuch der Physiologischen Optik,* originally published in 1924 in two volumes by The Optical Society of America, Rochester, New York. The color plates originally facing pages 352 and 353 have been moved to face pages 356 and 357; likewise the color plates originally facing pages 384 and 385 have been moved to face pages 388 and 389.

Library of Congress Cataloging-in-Publication Data

Helmholtz, Hermann von, 1821–1894.
[Handbuch der physiologischen Optik. English]
Treatise on physiological optics / Hermann von Helmholtz ; edited by James P. C. Southall—Dover ed.
p. cm.—(Dover phoenix editions)
"An unabridged and corrected republication of the English translation of Handbuch der physiologischen Optik, originally published in 1924 by the Optical Society of America, Rochester, New York."
Includes bibliographical references.
ISBN 0-486-44260-8 (v.1)—ISBN 0-486-44264-0 (v.2)—ISBN 0-486-44246-2 (v.3)
1. Physiological optics. I. Southall, James P. C. (James Powell Cocke), b. 1878. II. Title.

QP475.H48613 2005
612.8'4—dc22

2005042046

Manufactured in the United States of America
Dover Publications, Inc., 31 East 2nd Street, Mineola, N.Y. 11501

Treatise on Physiological Optics

Volume II

Handbuch

der

Physiologischen Optik

von

H. von Helmholtz.

Dritte Auflage

ergänzt und herausgegeben in Gemeinschaft mit

Prof. Dr. A. Gullstrand und **Prof. Dr. J. von Kries**

Upsala Freiburg

von

Professor Dr. W. Nagel (†)

Rostock

Zweiter Band

Mit 80 Abbildungen im Text und 3 Tafeln

Die Lehre von den Gesichtsempfindungen

herausgegeben von Prof. Dr. **W. Nagel** und Prof. Dr. **J. v. Kries**

Hamburg und Leipzig

Verlag von Leopold Voss

1911.

Table of Contents of Volume II

Part II

The Sensations of Vision

Appendix by W. Nagel

Adaptation, Twilight Vision, and the Duplicity Theory

Appendix by v. Kries

EDITOR'S NOTE

The new material in this volume comprises three Notes specially prepared for the English translation by Professor v. KRIES, a chapter at the end contributed by Dr. CHRISTINE LADD-FRANKLIN, and a partial bibliography of works relating to the sensations of vision, which have appeared in the interval since the publication of the third German edition. A Table of Corrigenda for Volume I has been appended. The coloured plates for this edition were made in Germany.

As stated in the Preface, in the preparation of Part II of this work, the Editor has received much assistance from Professors HENRY LAURENS (§§17, 18, 18A and Appendices of W. NAGEL and v. KRIES), M. DRESBACH (§§22, 23, 24 and 25), and L. T. TROLAND and E. J. WALL (§§19, 20, and 21). Miss TOWNSEND and Mr. TRELEAVEN have aided him in reading the proof.

JAMES P. C. SOUTHALL

Department of Physics,
Columbia University,
New York, N.Y.
October 1, 1924.

Part Second

The Theory of the Sensations of Vision

§17. Stimulation of the Organ of Vision

The nervous system of the body is acted on by external agents of various kinds, which produce changes in the state of the nerves. These changes may sometimes be detected by auxiliary apparatus, for example, by studying the electrical reactions; but they are also manifested by their actions on other parts of the body with which the nerves are connected. The change of state of the so-called *motor nerves* is accompanied by contractions of corresponding muscles. Under the same circumstances, other nerves, known as *sensory nerves*, excite sensations in the brain which is the organ of consciousness of the body. Now in the case of the motor nerves, no matter how diverse the external action may be—tearing, crushing, cutting, burning, eroding, shocking by electricity,—the invariable result is the contraction of the corresponding muscle, the only difference being one of degree. Therefore, apart from their qualitative differences, these various influences, so far as their relation to the motor nerves are concerned, are called *stimuli*. Quantitatively, we speak of a stimulus as being strong or weak according to the amount of twitching that is produced. The resulting alteration of the state of the nerve due to a stimulus is called *stimulation* or excitation. Similarly, the ability of the stimulated nerve to make the muscle contract is known as its *excitability*. The latter is affected by mortification and by external influences of many kinds.

The sensory nerves may be analyzed in the same way. External agencies, which acting on a motor nerve would cause contraction of the muscle, have another peculiar sort of effect on a sensory nerve and give rise to a sensation, provided the nerve is alive and not disconnected with the brain. But there is undoubtedly an essential difference here, because there are qualitative differences in the sensation corresponding to qualitative differences in the stimulus. But although different stimuli cause different sensations, still their effects are invariably sensations, that is, invariably actions of a kind that do not occur otherwise and are peculiar to the living body. Accordingly, the abstract conception of stimuli and stimulation as used first with reference to the motor nerves has been transferred likewise to the sensory nerves. Thus, the external agencies which acting on the sensory nerves excite sensations are also called *stimuli*, and the change itself that takes place in the nerve is said to be a *stimulation*.

The state of stimulation that may originate at any part of a nerve fibre through the action of stimuli is always conducted to all other parts of the nerve fibre. This is manifested partly by a difference in the electrical actions and also by its effect on other organic structures (muscles, brain, glands, etc.) with which the nerve is connected. What occurs is a contraction of the muscle, or a sensation, or increased glandular secretion, etc. Conduction of the stimulation is never impeded unless the nervous structure has been seriously damaged by mechanical or chemical actions or by coagulation of the nervous tissue in death. Thus, an uninjured nerve fibre possesses not only *excitability* or the capacity of being stimulated everywhere, but *conductivity* also. A separation of these two characteristics has not yet been conclusively demonstrated.[1] Moreover, thus far there are no known differences in the structure and function of the sensory and motor fibres, that might not be attributed to differences in their connection with other organic systems. The fibres themselves seem to be indifferent and to have no other office except to be conductors; transmitting the stimulation either to a muscle, in which case they are motor nerves, or to the sensitive parts of the brain, in which case they are sensory nerves.

According to their quality human sensations fall into five groups corresponding to the so-called *five senses*. The qualities of the sensations cannot be compared with each other unless they belong to the same group. For example, we can compare two different sensations of the sense of sight as to intensity and colour, but we cannot compare either of them with a sound or a smell.

As far as it has been possible to test it, physiological experience shows that *the only sensations that can be produced by stimulation of a single sensory nerve fibre are such as belong in the group of qualities of a single definite sense;* and that *every stimulus which is capable of exciting this nerve fibre at all arouses sensations of this particular sense.* A complete experimental proof of the statement is not possible except with nerve fibres that are collected together in special stems separate from all fibres of the other senses, as in the *nervus opticus* of the sense of sight, in the *nervus acusticus* of the sense of hearing, in the *nervus olfactorius* of the sense of smell, and in the posterior spinal roots of the sense of touch. If different kinds of stimuli act on these nerves different sensations arise, but the sensations are always such as belong to the group of qualities of that particular sense. On the other hand, in the case of fibres that run along the same nerve with those of another quality (for example, gustatory fibres mixed with tactile in

[1] ¶See E. D. ADRIAN, Conduction in peripheral nerve and in the central nervous system. *Brain*, **41**, (1918), 23–47. Idem, The recovery process of excitable tissues. *Jour. Physiol.*, **54**, 1920, 1–31; **55**, 1921, 193–225.—R. S. LILLIE, Transmission of physiological, influence in protoplasmic systems, especially nerve. *Physiol. Rev.*, **2**, (1922), 1–37.—K. LUCAS. *The conduction of the nervous impulse* (London and New York, 1917). (H. L.)

the tongue in the *nervus glossopharyngeus* and *nervus lingualis*) there is at least a probability of the same sort of thing, since we find that in morbid conditions there is sometimes an isolated paralysis of the gustatory sensations alone without paralysis of the tactile sensations, or *vice versa*; and because also no other tactile nerves have the faculty of exciting gustatory sensations.

Light sensations belong to the *sense of sight*. They can all be compared as to intensity and colour. That part of the nervous system where sensations of this nature can be excited is what J. MÜLLER called the *visual substance*, or, as it is also sometimes called, the *nervous mechanism of vision*. It comprises the retina, the optic nerve, and a part of the brain that is still not exactly defined where the radical fibres of the optic nerve lie. No other nervous mechanism in the body can produce a sensation of light, that is, a sensation of the same quality as that of the mechanism of vision, although the vibrations of the luminiferous aether may also be perceived by the tactile nerves. However the quality of the sensation of radiant heat is entirely different from the sensation of light. It is the same way with aerial vibrations which the auditory nerve perceives as sound, whereas at the same time they excite in the skin a tactile sensation of buzzing. Similarly, vinegar tastes sour on the tongue; but smarts when it touches a raw place on the skin or a delicate mucous membrane like the conjunctiva of the eye.

On the other hand, there are many other kinds of stimuli besides the vibrations of the luminiferous aether which may excite the organ of vision. Mechanical forces and electrical currents possess the power of stimulating all the nervous mechanisms of the body. But when these stimuli act on the optic nerve, they always excite sensations of vision, and never any other kinds of sensation like that of sound or of smell. If at the same time they excite tactile sensations, we must suppose that this is because there are likewise special tactile nerves in the eye and perhaps even in the optic nerve itself (as in all internal parts of the body). These tactile sensations due to pressure on the eye or electrical action are distinct from the simultaneous sensation of light in still another way also; because whereas the former are perceived at the place of the stimulation, the latter are misconstrued as bright objects in the field of view. This question will be considered again in connection with a more detailed description of the mechanical stimulation of the eye.[1]

[1] ¶A stimulus that can arouse a specific sensation is commonly said to be "adequate" or "inadequate," according as it does or does not excite the sensation under ordinary circumstances. Thus, for instance, objective light is an adequate stimulus for the eye, but pressure on the eyeball is an inadequate one. (J. P. C. S.)

As all the other organs of sense behave similarly, it may be said that the nature of a sensation depends primarily on the peculiar characteristics of the (receptor) nervous mechanism; the characteristics of the perceived object being only a secondary consideration. A sensation must belong to the group of qualities associated with a certain one of the senses; but what particular sense this is, does not depend at all on the external object, but simply on the nature of the nerve that is stimulated. But the quality of the sensation that is aroused does depend on the nature of the external object that excites it. Whether the sun's rays will be perceived as light or heat, is simply a question of whether they are perceived by the optic nerve or by the cutaneous nerves. But whether they will be perceived as light that is red or blue, and dim or bright, or as heat that is mild or intense, depends both on the nature of the radiation and on the condition of the nerve. The quality of the sensation is thus in no way identical with the quality of the object by which it is aroused. Physically, it is merely an *effect* of the external quality on a particular nervous apparatus. The quality of the sensation is, so to speak, merely a *symbol* for our imagination, a sort of earmark of objective quality.

The first and most important means of stimulating the optic nerve is by *objective light*; because this stimulus acts on the optic nerve far more frequently and continuously than any others. Thus the chief method of perception of external objects is through sensations of the visual mechanism that are aroused by objective light. Accordingly it is not necessary to assume a particular, specific relation or homogeneity between the objective light and the nervous agency of the optic nerve, as was generally supposed by earlier philosophers and physiologists. For the optic nerve is not the only nerve that may be stimulated by objective light (because this is true also of the skin nerves), nor is objective light the only stimulus for the optic nerve. The reason why it is the most common, and therefore the most important, is simply because the optic nerve and the retina are so situated at the back of the eye that while it is easy for light to penetrate to them, they are much more inaccessible to mechanical and electrical actions. This excessive frequency and importance of stimulation by objective light led people to give the name *light* to those aetherial vibrations that are capable of exciting the sensation of light. Properly speaking, the word should be used only in this latter sense, that is, to denote the sensation that is produced by this means. Solar radiation includes "*sunlight*" and "*sunheat*," depending on the different sensations it excites. As long as man did not ponder over the nature of his sensations, it was natural for him to transfer the qualities of his sensations directly to the external objects, and so to suppose that the rays of the sun were of two kinds corresponding to his two sensations. Besides, at first he knew nothing about the solar radiations except what his sensations told him. He noticed that some radiations, which, like the rays of the sun, contain a preponderance of waves of higher frequencies, affect the eye much more than they do the skin; while others containing a preponderance of waves of lower frequencies, act on the skin but hardly affect the eye at all. Naturally, the two agencies were considered as objectively separate. In very recent years careful investigation of the phenomena of radiation with respect to their properties that are independent of the nervous mechanism, has shown that the only difference between the so-called light rays and heat rays is in the frequency of the vibrations. And thus in this instance at least physics has succeeded in freeing itself from entanglement with the subjective sensations that were so long confused with the objective causes. The detailed description of objective light as a means of stimulation of the retina will be given in the next chapter.

The phenomena resulting from *mechanical stimulation* of the organ of vision differ according to the extent of the stimulus. In case of a sudden blow on the eye there is a sensation of light which appears and disappears with lightning speed, and which may be very bright and extend over the entire visual field. As opposed to old-fashioned incorrect views of this phenomenon, it may be pointed out here that when this happens in the dark, no trace of light in the injured eye can be seen by another person; no matter how strong the subjective flash may be. And it is impossible to discern any real object in the outside world by virtue of this subjective illumination of the dark field.[1]

The effect of local pressure is easier to investigate. If somewhere at the edge of the orbit a blunt point, like the finger nail, for example, is pressed against the eyeball, it produces a luminous effect, or so-called *pressure-image* or *phosphene*. It is seen in that part of the field that corresponds to the place affected on the retina. Thus when the pressure is exerted from above, the bright spot appears on the lower edge of the field; and when it is exerted at the external angle of the eye, it appears to be on the nasal side of the field. Similarly, when the pressure is exerted from below or at the inner angle, the light seems to be above or on the outside part of the field, respectively. If the object that exerts the pressure is not large, the phenomenon usually has a bright centre surrounded by a dark ring and by an outer bright one. To the

[1] Concerning a legal action in which it was alleged that a man standing at a window received a blow on the eye, and was able to recognize his assailant in the glow of light that was caused thereby, see J. MÜLLER, *Arch. f. Anat.*, 1834, page 140.

writer it is brightest when the pressure is exerted at or near the equator of the eye where the sclerotica is thinnest. The pressure-image appears then on the edge of the dark visual field as a bright arc, nearly semicircular in form. Under these conditions it is quite far from the point of fixation, that is, from the place in the field corresponding to foveal vision. It coincides, therefore, with the region where objects lie that are not seen distinctly when the eyes are open. However, with some practice in indirect vision, particularly when conspicuous bright objects happen to be at the apparent place of the pressure-image, it is possible to notice that figures in the vicinity of the pressure-image are distorted, due to the curved hollow form of the sclerotica and retina. Often too they are dark in spots. But the pressure-image can be brought nearer the point of fixation by turning the eye far inwards and at the same time pressing on it from the outside; or *vice versa*. The image then is somewhat fainter, because the posterior surface of the sclerotica offers more resistance to pressure. Certain individuals are able to bring the pressure-image to the place of direct vision simply by pressing at the outer angle. THOMAS YOUNG could do this; and although the writer cannot quite succeed at it, the pressure-image comes so near the point of fixation that images of external objects disappear at its centre. The pressure-image is represented in Fig. 1 of Plate I as it looks to the author when a sheet of white paper is placed against the face between the eye and nose, the eye turned inwards as far as possible, and pressure exerted with a blunt instrument on the outer edge of the orbit. The nasal side is at N, and the image consists of a dark spot traversed by a bright vertical band. When the pressure is exerted at the right level, there is a horizontal continuation of the dark spot, the tip of which reaches the point of fixation at a. Moreover, somewhere near the place where the optic nerve enters there is an indistinct shadow b. How the place where the optic nerve enters the eye may be recognized in the field of view, will be explained in §18. PURKINJE observed and depicted a system of fine parallel curved lines extending between the dark pressure-image and the point of fixation. The author sees them best (but not in the way they are represented in PURKINJE's drawing) when the corresponding place in the field of view is very bright.

On the other hand, in the dark visual field there is a bright yellowish circular area within which there is sometimes a dark spot or a dark ring. A dim light is also seen at the entrance of the optic nerve, so that the appearance is similar to that shown in Fig. 1, Plate I, provided the light and dark portions of the drawing are supposed to be transposed. But the author has not been able to detect in the dark field the continuation extending towards the yellow spot.

The phenomena are again different when a moderate uniform pressure is exerted on the eyeball for a longer space of time; for example, by pressing it from in front either with the soft part of the hand or by the tips of the fingers of one hand. In a short space very brilliant and variable luminous patterns will appear in the visual field, which execute curious and fantastic movements, frequently not unlike the most gorgeous kaleidoscopic figures that are shown nowadays by electric projection. PURKINJE has studied these phenomena very carefully, and accurately described and represented them. They seem to have had a high degree of regularity for him. The background generally consisted of fine quadrangles in regular array, on which there were either stars with eight rays, or dark or bright rhombs with vertical and horizontal diagonals; and the patterns were surrounded by alternately bright and dark bands. In the author's own experience there is no such regularity in the figures. The background of the visual field is usually finely patterned at first, but in the most manifold way and in very different colours. Frequently, it is as if the field were strewn with fine leaves or covered with moss; then presently they look like bright brownish-yellow quadrangles everywhere with fine line patterns; and at last they usually develop in the form of dark lines on a brownish-yellow background. Sometimes they assume very complex star-shaped figures, and sometimes they are in the form of an inextricable labyrinth or maze, which seems to be waving or flowing continually. There are often bright blue or red sparks in certain parts of the field which may last for a considerable time. If, when the phenomenon is at its maximum, the pressure is released, without letting extraneous light enter the eye, the play of figures proceeds for a long time still, gradually getting darker until it ceases entirely. But if the eye is opened as the pressure is released and directed towards a bright object, there is absolute darkness at first; and then gradually single bright objects shining brilliantly begin to be manifest in the middle of the field. For instance, in the writer's own case, separate sheets of white paper appear in their true form but of dazzling brightness, and superposed on them are the remnants of the previous patterns, the dark places in them now showing bright. The abnormal brightness gradually fades away just as the pressure-images do before the eye when it is shut. But the eye on which the pressure was exerted is for a longer time still different from the other eye; because the field looks more violet to it, whereas it looks yellowish to the unpressed eye. VIERORDT and LAIBLIN maintain that with continuous pressure on the eye they have seen the ramifications of the blood-vessels on the retina, red on a dark ground; but the writer has tried in vain to obtain this effect. Moreover, VIERORDT frequently saw the retinal vessels in this way with a bright blue colouring. Both observers witnessed, as STEINBACH and PURKINJE had done before, a network of vessels with blood circulating in them. PURKINJE supposed they were the retinal arteries; but as the appearance was visible along with the previously mentioned blood-vessels of the retina, LAIBLIN concluded from his observations that the circulation which was perceived here must belong "to another layer of the retina, more to the outside, and containing more blood-vessels." In the pressure-images of the eye, except for occasional sudden flashes of the familiar vascular figure of the retina, neither MEISSNER nor the writer himself has ever succeeded in seeing anything similar to a network of vessels. The flowing movement of the labyrinthine system of lines during the last stages of the phenomenon has no similarity at all to a network of vessels.[1] As to the theory of these phenomena, it seems from DONDERS's investigations with the ophthalmoscope that the effect of pressure on the eye is undoubtedly to produce changes in the blood-vessels of the retina, so that the veins begin to pulsate and finally become entirely emptied of blood. This was seen in several cases. The restless and constantly shifting images produced by sustained pressure on the eye might be compared to the sensation of ants running over the skin, such as occurs in limbs that have "gone to sleep" when the nerves have been pressed on for some time. When pressure is exerted on the nerves in the thigh, the foot and lower leg very soon lose the capacity of feeling contact with external objects. Accompanying it there is an intense tingling sensation in the numbed parts of the skin; which in similar fashion soon arouses variable excitements of the sensitive nerve fibres, such as are manifested by the delicate moving figures in the visual field during corresponding states of the retina. On releasing the pressure, the ability of perceiving external objects returns, and the first movements of the foot are often painful; whereas in the case of the eye, the outside light is blinding in its power.

Another phenomenon apparently connected with mechanical stimulation of the retina, consists of certain spots of light that are visible to sensitive eyes in the dark field when they have just executed a quick movement. These are represented in Fig. 2, Plate I, as they look in the field of view of the writer's eyes, when they have been

[1] In my own case (says NAGEL) there is regularly a dense net-work of bright lines on a dark ground when I close one eye almost tight for at least 20 minutes, whether the eye is pressed or not. The bright lines exhibit a *rapid* flowing or flickering, which is very clear. For several minutes at first the phenomenon does not appear and then develops gradually. After a half hour or an hour it is so distinct that it is disturbing when one tries to read with the other eye. The flowing image is most conspicuous when both eyes are closed. Far from being *absent* in the fovea, this flowing is particularly clear there.—N.

moved to the left in the direction of the arrow. The spots marked L and R are the appearances in the left eye and right eye, respectively. The effect is less developed in the eye that turns inwards (the right eye in this case) than in the one that turns outwards. It occurs with the writer only in the morning; either on waking or as a result of indisposition; but other observers, for example, PURKINJE and CZERMAK[1] perceived these spots in the dark at any time of day as fiery rings or half-rings. Their distance from the point of fixation is such than an observer who is familiar with the phenomenon of the so-called blind spot (which will be described later) can infer that they are situated where the optic nerve enters. Therefore, a probable explanation of their origin is that, with sudden motions of the eye, the optic nerve being set in motion along with the eyeball is stretched at the place where it comes into the eye. When PURKINJE[2] turned his eye far inwards, he saw a steady ring of light where the optic nerve enters, surrounded by concentric bright bands towards the middle of the field; but in the writer's case the phenomena are never anything but momentary. If the experiment is tried with the eye open in front of a uniformly illuminated white surface, dark spots corresponding to the entrance of the optic nerve make their appearance when the eye is turned far to one side. They are produced more easily by turning the eye inwards, as was observed by CZERMAK, and have a regular circular form when the eye is turned outwards. In the reddish field produced by closing the eyelids and illuminating them from outside, these dark spots appear blue. In the writer's own case, the dark spots show traces of the same luminous appearances that are visible in the dark field; but CZERMAK insists that with him the latter phenomenon is not a negative reproduction of the former. Here also the stimulated nerve-fibres seem to lose their sensibility to external stimuli on account of the pull on them. The fibres which are here stimulated must be those whose ends are in the immediate neighbourhood of the optic nerve, because the place where the optic nerve enters is itself not sensitive to light, and hence it cannot be supposed that any fibres capable of light sensation end at that place and are responsible for a sensation of light at this very spot. And, finally, the *accommodation phosphene* seen by PURKINJE[3] and CZERMAK[4] has to be considered here. When a person looks out of a window with his eyes fixed on something very near, and then suddenly accommodates for distant vision, a fairly

[1] *Physiologische Studien.* Abteilung I. § 5. S. 42 u. Abt. II. S. 32. — *Wiener Sitzungsber.* XII. S. 322 u. XV. 454.

[2] *Beiträge zur Kenntnis des Sehens.* S. 78.

[3] *Zur Physiologie der Sinne.* Bd. I. 126. II. 115.

[4] *Wiener Sitzungsber.* XXVII. 78.

small luminous border will be seen near the periphery of the field of view, which, having the form of a closed ring, flashes out at the instant when accommodation is consciously relaxed. PURKINJE observed the phenomenon also when uniform pressure on the eye was suddenly released. The writer himself has never seen it. CZERMAK thinks the reason of it is because at the instant when the tension of the ciliary muscle ceases, the relaxed zonule is again stretched, while the lens is still shortened radially; which results in a sudden stretching of the outermost edge of the retina where it is attached to the zonule.

When the writer exerts his accommodation and looks towards an uniformly illuminated white surface, there is a shadowy spot at the point of fixation. It shades off brown at the edge, perhaps with brown or bright violet lines radiating from it in various directions. The field of view then usually gets dark rapidly, with net-like designs and parts of the blood-vessels appearing dark against a white background. Everything vanishes when the accommodation is relaxed. PURKINJE describes the brown spot, but says that its centre is white. In this same category belongs an elliptical and spotted luminous effect seen by PURKINJE[1] in the dark visual field when pressure on the eyelids was suddenly released. In order to produce this phenomenon, it was necessary to expose the eye to light a little while in advance. The writer himself cannot see it.

Dogs show no sign of pain when the exposed optic nerve is cut and pulled; but the same kinds of injury to a cutaneous nerve of equal size produces the most intense agony. The human eye sometimes has to be extirpated on account of cancer. In such cases when the optic nerve itself has not degenerated, large masses of light are said to be perceived at the moment the optic nerve is severed,[2] accompanied by somewhat greater pain than is caused by cutting the adjacent parts. It is hardly reasonable to suppose that the severance of the optic nerve would be entirely devoid of pain like that perceived by the tactile nerves. All the other large nerve trunks have their *nervi nervorum*, that is, particularly sensitive fibres which belong to them just as much as to all the rest of the internal parts of the body and which mediate their local sensibility.[3] It can be shown that such *nervi nervorum* are sent from the posterior sensory roots to the anterior roots of the spinal nerves, through which motor fibres alone leave the cord. If the ulnar nerve is struck at the elbow joint, there is a sensation

[1] *Zur Physiologie der Sinne.* II. 78.

[2] TOURTUAL in J. MÜLLER, *Handbuch der Physiologie.* Koblenz 1840. Bd. II. S. 259.

[3] ¶The various sensations mediated by the "tactile" nerves, of which HELMHOLTZ writes, have been divided into protopathic, epicritic and deep. See any standard textbook of Physiology. (H. L.)

of pain referred to the region of distribution of the nerve in the fourth and fifth finger, as well as another localized at the place struck, which is more unpleasant than that resulting when the skin alone is stimulated. This must be referred to the nerves of the nerve trunk. In the same way when the eyeball is pressed at the outer angle, the pain of the pressure is felt locally by means of the sensory nerves of this region, and the light that is seen is supposed to be in the region of the bridge of the nose. Something of a similar nature may happen when the optic nerve trunk is stimulated.

That the optic nerve and the retina, both capable of being stimulated by so delicate an agency as light, are tolerably insensitive to the roughest mechanical maltreatment, that is, have no sensation of pain, has seemed a remarkable paradox. The explanation, however, is simple, because the quality of all sensations of the optic nerve belongs to the group of light sensations. The sensibility is not lacking, but the *form* of the sensation is different from that usually associated with this particular kind of stimulus.

Light sensations due to *internal conditions* are very varied. There are a number of luminous phenomena, occurring in all diseased conditions of the eye or of the entire body, that may take up the whole field or may be localized in it. In the latter case they take sometimes the form of irregular spots and sometimes fantastic figures of men or animals, etc. Mechanical causes often participate in these effects, as, for example, increased blood-pressure in the vessels or humors of the eye. Thus, on releasing the eye from uniform pressure, parts of the vascular figure often flash out; and sometimes, after violent exertion separate pulsating parts, maybe larger portions, of the vascular figure are visible.[1] In other cases there may be a sort of chemical stimulation due to altered condition of the blood, for example, by narcotic poisoning. Finally, many of these phenomena also may be explained as due to a spread of a state of excitation within the central part from other parts of the nervous system to the origin of the optic nerve. When the state of excitation in a stimulated sensory nerve is imparted to another that is not acted on by the stimulus at all, we call it an associated sensation. For example, looking at large bright surfaces, such as sunlit snow, causes many persons to feel a simultaneous tickling in the nose. The sound of certain scraping or squeaking noises makes a cold chill run down the back. Apparently, such associated sensations may occur also in the visual apparatus when other sensory nerves are

[1] Purkinje, *Zur Physiologie der Sinne.* I. 134. II. 115. 118. — *Subjektive Erscheinungen nach Wirkung der Digitalis* II. 120.

stimulated, *e.g.*, by intestinal worms in children or by retained faeces, retarded circulation and other abnormal conditions in hypochondriacs. The origin of peculiar fantastic shapes or luminous images associated with the appearance of familiar external objects is due apparently to a similar transference of the state of excitation from the part of the brain that is active in the formation of ideas to the visual apparatus. These have been noted by many observers who state that while they were seeing them they were thoroughly aware of their subjective nature.[1] Certain individuals, for example, GOETHE and J. MÜLLER, could indeed see similar phenomena at any time by simply closing their eyes and remaining for a long time in darkness.

As a matter of fact, the field of vision of a healthy human being is never entirely free from appearances of this kind which have been called the *chaotic light* or *luminous dust of the dark visual field*.[2] It plays such an important part in many phenomena, like after-images, for example, that we shall call it the *self-light* or *intrinsic light* of the retina. When the eyes are closed, and the dark field is attentively examined, often at first after-images of external objects that were previously visible will still be perceived (as to their origin, see §§24 and 25 below). This effect is soon superseded by an irregular feebly illuminated field with numerous fluctuating spots of light, often similar in appearance to the small branches of the blood-vessels or to scattered stems of moss and leaves, which may be transformed into fantastic figures, as is reported by many observers. A quite common appearance seems to be what GOETHE describes as floating cloud-ribbons (*"wandelnde Nebelstreifen"*). PURKINJE speaks of them as "broad streamers, more or less curved, with black intervals between them, which sometimes move in concentric circles towards the centre of the field, to become lost there, or maybe to disintegrate into floating curls, or to revolve as curved radii of circles around this place; the movements being so sluggish that ordinarily it takes eight seconds for a streamer to complete its performance and vanish out of sight." The author's experience is that they generally look like two sets of circular waves gradually blending together towards their centre from both sides of the point of fixation. The position of this centre for each eye seems to correspond to the place of entrance of the optic nerve; and the movement is synchronous with the respiratory movements. One of PURKINJE's eyes being weaker

[1] Cases of this sort are summarized by J. MÜLLER, *Über phantastische Gesichtserscheinungen.* Koblenz, 1826, page 20.

[2] ¶"The completely dark-adapted eye when sheltered from all external stimuli gives a sensation which is variously described as the light chaos, the intrinsic light of the retina, and so on. HERING calls this sensation 'mean grey'." J. H. PARSONS, *An introduction to the study of colour vision.* 1915, p. 251. (J. P. C. S.)

than the other, he could not see these floating clouds except in his right eye. The background of the visual field, on which these phenomena are projected is never entirely black; and alternate fluctuations of bright and dark are visible there, frequently occurring in rhythm with the movements of respiration; as observed by both J. MULLER[1] and the writer. Moreover, with every movement of the eyes or eyelids, and with every change of accommodation, there are accompanying variations of this "luminous dust." The shapes that are assumed are very curious, especially when one happens to be in a strange place that is perfectly dark, as, for instance, in an unlighted hallway where it is necessary to grope one's way; because then these imaginary figures are apt to be mistaken for real objects. Under such circumstances PURKINJE noticed that every unexpected contact and every uncertain movement produced instantaneous oscillations of the eye which were accompanied by gossamer clouds of light and other luminous appearances, such as may easily have been the origin of many ghost stories.

After strenuous exercise and when the body is overheated, PURKINJE[2] noticed a faint glow of light glimmering in his dark field, like the last expiring flickers of a flame of alcohol burning on the top of a table. Upon closer examination he detected countless tiny little points of light darting to and fro and leaving little trails of light behind them. He got a similar effect when he closed his right eye and strained to see with his other weak eye.

Another important fact is, that after a person has lost one of his eyes, or in case the optic nerves and eyes have degenerated and cease to function, he may still have subjective sensations of light.[3] Such experiences show that not merely the retina, but the trunk and roots of the optic nerve in the brain as well, are capable of giving rise to sensations of light as a result of being stimulated.

Lastly, another powerful agency for stimulating not only the optic nerve but all the other nerves of the body is by a current of electricity. As a rule, the motor nerves do not produce twitching except at the instants when the current traversing them is suddenly increased or diminished; but sensations are excited in the sensory nerves not only by fluctuations in the current but by a steady flow; and the quality of the sensation depends on the direction of the current.

When the optic nerve is stimulated by *fluctuations in the strength of a current of electricity*, bright flashes of light are produced extending

[1] *Phantastische Gesichtserscheinungen.* S. 16.

[2] *Beobachtungen und Versuche, etc.* I. 63, 134. II. 115.

[3] See J. MÜLLER, *Phantastische Gesichtserscheinungen.* S. 30 — A. v. HUMBOLDT, *Gereizte Muskel- und Nervenfaser.* Tl. II. S. 444. — LINCKE, *de fungo medullari.* Lips. 1834.

over the entire visual field. These effects can be obtained by discharges of Leyden jars just as easily as from a galvanic pile, provided a strong enough portion of the current flows through the optic nerve as nearly parallel as possible to the direction of its fibres. A convenient way of doing this is to place one electrode on the forehead or on the closed eyelid and the other on the neck; or the latter may be held in the hand, if the electrical apparatus is so powerful that a great resistance does not matter. The electrodes should be in the form of plates or cylinders; and if they are covered with damp pasteboard and the parts of the body where they are attached thoroughly moistened beforehand, the pain on the skin can be diminished. Not many experiments of this sort have thus far been made by discharges of Leyden jars. On account of the proximity of the brain, it is well to be careful, because FRANKLIN and WILKE[1] noted that discharges through the head may result in unconsciousness. LE ROY[2] passed the discharge through a young man who was blind from cataract. His head and right leg were wound with a brass wire, and a Leyden jar discharged through its ends. At every discharge the patient thought he saw a flame pass rapidly downwards from above, accompanied by a noise as of heavy firing. When the shock was made to pass only through the blind man's head, by attaching metal plates above the eye and at the back of the head, and connecting them with a jar, the patient had sensations of fantastic figures, individual persons, crowds of people in lines, etc.

Experiments with galvanic currents are more numerous. In order to perceive simple flashes of light due to making or breaking the circuit, a few zinc-copper cells are sufficient, or even a single cell in case of excitable eyes. For example, when a piece of zinc is placed on the moistened lid of one eye and a piece of silver on that of the other, and the two pieces of metal brought into contact, a flash appears at the instant of contact, and again at the instant of separation. The experiment is more instructive when one metal is placed on one eye and the other taken in the mouth, because in this way the connection between the brightness of the flash and the direction of the current can be easily made out at the same time. According to PFAFF's observations, the flash is more striking when the circuit is closed, provided the positive metal (zinc) is placed on the eye and the negative electrode (silver) taken in the mouth; because then the positive electricity flows upwards through the optic nerve. The writer has never had any success with these experiments with a simple circuit, probably because his eye is not sensitive enough to such stimulation. But the flashes of light are very

[1] *Mém. de mathém. de l'Acad. de France.* 1755. pp. 86–92.

[2] FRANKLIN, *Briefe über Elektrizität.* Leipzig 1758. S. 312.

brilliant with a small galvanic pile of about a dozen elements. For example, when a battery of DANIELL cells is used that gives a constant current, the flash on closing the circuit is found to be greater when the current flows upwards; whereas the flash on breaking the circuit is greater when the current flows downwards. There are similar differences of effect in the case of the motor nerves depending on the direction of the current; but here these differences are due also to the strength of the current.

In order to perceive *the continuous action of a uniform current*, most eyes require a small galvanic pile, although RITTER perceived it even with a single cell. To avoid blinding the eyes by the flash of light and the unpleasant muscular twitching in opening and closing the circuit, the writer suggests placing two metallic cylinders on the edge of the table near which the patient is seated; the cylinders being wrapped with pasteboard that has been dipped in salt water, and connected with the two terminals of a DANIELL's battery of from a dozen to two dozen cells. The forehead is pressed firmly against one of the cylinders, and then the hand makes contact with the other. Thus, by gradually touching the electrode, the effects of fluctuations of the current are very slight, and the circuit may be easily made or broken at will. The direction of the current can be reversed by applying the other cylinder to the forehead. In this way pressure is not exerted on the eyes, which is something to be avoided.

When a weak ascending current is conducted through the optic nerves, the dark field of the closed eyes becomes brighter than before and assumes a faint violet colour. During the first moments the optic disc appears in the brightened field as a dark circular area. The brightness quickly diminishes in intensity and disappears completely when the current is interrupted. This effect can be produced without a flash of light by slowly letting go the cylinder in contact with the hand. Then as the field begins to darken, in contrast to the previous blueness, it takes on a reddish yellow tinge due to the intrinsic light of the retina.

On making the circuit in which the current flows in the opposite or descending direction, the striking result is that only that part of the visual field that is illuminated by the intrinsic light of the retina becomes darker than before, and has a somewhat reddish yellow colour. The optic disc is conspicuous on the dark background as a bright blue circular area, although frequently only the half of it towards the middle of the field is visible. When the circuit is broken, the field again becomes brighter and bluish white, and the optic disc appears dark.

The darkening of the field caused by the descending current indicates that in these experiments it is not primarily a question of an electrical stimulation, but that changes of excitability due to the passage of the current are also involved. PFLÜGER's experiments[1] tend to show that the excitability of the nerve is enhanced by a feeble current in the portion where the positive electricity enters it, and reduced where it leaves it. Accordingly, with an ascending current the excitability would be enhanced in the portion of the optic nerve towards the brain and diminished in the portion next the retina; exactly the reverse being the case when the current is a descending one. PFLÜGER's law affords an explanation of the decrease and increase of the intrinsic light of the eye, provided it is assumed that the internal stimuli that produce this effect act on the end of the optic nerve that is towards the brain. This being the case, the ascending current must result in augmenting, and the descending current in lowering, the intrinsic light. Whether the opposite illumination at the optic nerve is to be regarded as contrast or as internal stimulation near the place where it comes into the retina, is still a moot question. RITTER found that external objects were less clear while the current was descending, and more clear when the current was ascending; which is in conformity with the above explanation; because when the retina itself is stimulated, ascending currents must increase its sensibility. The writer can corroborate this for dimly illuminated objects. Moreover, PURKINJE's explanation of the decrease of clearness in objective vision as being due to the increase of the intrinsic light of the eye, which acts as a kind of mist, is in perfect harmony with the above. At all events this brightening and darkening of the visual field prevents one from being sure whether the light from an isolated object is perceived more strongly or more feebly.

PFLÜGER finds that when the steady current is interrupted there is increased sensibility in those parts of the nerve that had become less sensitive; as is shown in our case by the brightening of the visual field. On the other hand, for a short space (as long as ten seconds) there is at first reduced sensitivity in those parts of the nerve which were previously more sensitive; which is then succeeded by a slight increase of sensitivity again. In our case the darkening of the field when the ascending current is interrupted corresponds to the first state; and the only sign of the latter state is that the darkening seems to be soon succeeded by the normal condition.

With stronger currents obtained by using from 100 to 200 zinc-copper cells, RITTER observed a reversal of colouration, but the increase or decrease of brightness was the same as with weak currents.

[1] *Untersuchungen über die Physiologie des Elektrotonus.* Berlin 1859. On this subject, see § 25.

Thus strong ascending currents aroused in him a bright green sensation; which was bright red with still stronger currents. Strong descending currents gave a faint blue sensation. In the former case, when the circuit was broken, the sensation was blue at first, which quickly changed over into the red left behind by the weak current. On the other hand, on interrupting a strong descending current, the sensation was red at the first instant, rapidly changing to the customary blue. The writer's own experience with strong currents[1] is that they produce a wild interplay of colours in which no regularity can be discovered.

Another thing that RITTER reports is that external objects appear not only more indistinct but also smaller when the eye is traversed by an ascending current. This leads us to suspect that his eyes were accommodated for near vision. The current causes so much pain at the place where it enters that it is almost impossible to avoid stretching the adjacent muscles, wrinkling the forehead, and closing the eyelids tightly. Whenever the eye and its adjacent parts are strained, there is a tendency with most people to accommodate for near vision, and this has also a certain influence on the impression one gets as to the size of something seen. DU BOIS-REYMOND[2] calls attention to the fact that when an electric current flows through the eye, the pupil contracts; and, doubtless, there is likewise some change in the mechanism of accommodation. Conversely, in the case of descending currents, RITTER reports that objects appeared larger and more distinct.

Finally, PURKINJE describes other special forms of luminous phenomena produced by electrical stimulation, when the current is made to flow from a small pointed conductor either into the middle of the closed eyelids or in the vicinity of the eye. The effect of the current as described above was always most noticeable at the place where the axis of the eye meets the retina. Here there was a diamond-shaped spot surrounded by several alternately dark and bright diamond-shaped bands. On the other hand, the place where the optic nerve enters invariably exhibited the opposite phase of electrical action. For instance, when the current was ascending, the axial point of the eye was like a bright blue diamond immediately surrounded by a dark band, and the optic disc like a dark circle surrounded by a blue sheen. When the current was descending, the axial point appeared as a dark diamond surrounded by red-yellow bands, and the optic nerve

[1] The current of 24 DANIELL cells was conducted to forehead and neck by metal plates covered with moist pasteboard. The resistance in this circuit was very much less than in RITTER's arrangement. He used a battery of high resistance and had his arm in the circuit besides. Consequently, the connection between the current-strengths in the two experiments is not easy to be ascertained.

[2] *Untersuchungen über tierische Elektrizität.* Berlin 1848. Bd. I. S. 353.

as a bright luminous disc. As the current became steadier, the figures soon vanished; but when the current was more intermittent (which PURKINJE caused by moving the circuit about), the blue figure persisted, being brighter by far than the red-yellow figure.

These phenomena at the place where the optic nerve enters the eye, as described by PURKINJE, are usually seen by most persons; but instead of the diamond-shaped figures, the writer and others who have tried it at his request can see simply indefinite patches of light. As a result of pressure on the eye, PURKINJE saw entirely similar rhombic figures. So far as the writer is aware, these rhombs have never been seen by any other observer; and it is a question therefore whether their regular form was not due to idiosyncrasies of PURKINJE's eyes.

When the current was introduced near the eye through a small conductor, the appearances of light corresponding to the yellow spot and the optic disc were the same as before. But in addition a dark arc was noticeable on the edge of the field and parallel to it; which kept its apparent place during movements of the eye; whereas the phenomena dependent on the optic nerve and yellow spot seemed to follow the movements of the eye. This dark arc is in the upper part of the field when the electrode is placed below the eye; and on the right when the electrode is on the left, and *vice versa*. Hence it follows that those portions of the retina which are nearest the electrode perceive no light. In order to see this phenomenon distinctly, PURKINJE used chain conductors, so that with every movement of them, the current was interrupted.[1]

In old days, without any positive knowledge of the subject, the theory of the visual sensations was entirely a matter of philosophy. The first thing that had to be comprehended was that the sensations are nothing but the effects of external things on our bodies, and that perception is a result of sensation by means of psychical processes. This is the view of Greek philosophy.[2] It begins

[1] G. E. MÜLLER has corroborated the interesting fact that the threshold for galvanic light perception is practically the same for light adaptation and dark adaptation of the eye. This is also true for my eye (writes NAGEL), and is remarkable, because the sensibility for the adequate *light* stimulus increases very much when the eye is dark-adapted. I found too that the threshold of the pressure-phosphene (which, however, cannot be accurately determined) was not appreciably different in the light-adapted eye and the dark-adapted eye. The pressure-phosphene certainly is qualitatively changed at the beginning of dark adaptation. While pressure with a blunt instrument on the temporal side of the eyeball in the light-adapted eye causes a small clear yellowish ring to appear in the dark visual field, the ring is much larger and a brilliant bluish white when the eye has been dark-adapted for a half-hour. This makes the phenomenon more striking; but, as stated, it is impossible to find a threshold difference when the pressure stimuli are nicely regulated.—N.

G. E. MÜLLER, Über die galvanischen Gesichtsempfindungen. *Zft. f. Psych. u. Physiol. d. Sinnesorgane*, XIV. 329.

W. NAGEL, Einige Beobachtungen über die Wirkung des Druckes und des galvanischen Stromes auf das dunkeladaptierte Auge. Ibid. XXXV. 285.

[2] See WUNDT, Zur Geschichte der Theorie des Sehens in HENLE und PFEUFFERS *Zeitschrift für rationelle Medizin.* 1859.

with naïve suppositions as to how images of objects can possibly reach the mind. DEMOCRITUS and EPICURUS believed that the images were let loose from the objects and flew into the eye. EMPEDOCLES made the rays proceed to the object not only from the source, but from the eye also, and argued that the object was thus, so to speak, touched by the eye. PLATO's opinions vacillated. In the *Timaeus* he accepts the views of EMPEDOCLES: the rays issuing from the eye are like rays of light except that they are without heat, and the only way that vision occurs is when the internal light from the eye proceeds to the object and encounters the external light. On the other hand, in the *Theaetetus*, his reflections as to the spiritual basis of the perceptions lead him to entertain views that are not very far apart from the more mature standpoint of ARISTOTLE.

ARISTOTLE[1] made a delicate psychological analysis of the part played by the spiritual reality in the sense-perceptions. Physically and physiologically, sensation is clearly different from what it is psychically; and the perception of external objects does not depend on some kind of delicate tactile feelers emanating from the eye (such as EMPEDOCLES's nerves of vision), but is due to an act of judgment. Physically, indeed, his ideas are very undeveloped, but in the fundamental conceptions the germ of the undulatory theory can be traced. For according to ARISTOTLE, light is nothing corporeal, but an activity (ἐνέργεια) of the intervening transparent medium, which when at rest constitutes darkness. However, he still does not abandon the notion that the effect of light on the eye is not necessarily of the same nature as that of the luminous source by which it is excited. He tries rather to account for this correspondence between cause and effect by the fact that the eye also contains transparent substances, which may be put in the same state of activity as the external transparent medium.

ARISTOTLE's peculiar and striking contributions to the theory of vision passed without notice during the middle ages. FRANCIS BACON and his successors were the first to take up these threads again in their keen discussions of the connection between ideas and sensations: until KANT in his *Critique of Pure Reason* put an end to their theory.

At this same time natural philosophers were interested only on the physical side of the theory of vision, which had developed rapidly from the time of KEPLER. HALLER formulated the general theory of nerve excitability; and described quite clearly and correctly the relation between light and sensation and between sensation and perception.[2] But more exact knowledge concerning excitation of the eye by other stimuli was still lacking; or at least what was known was fragmentary and regarded as simply curious. To GOETHE belongs the credit of having brought the importance of this knowledge to the attention of German scientists; although he did not succeed in winning them over to a revised theory of the physics of light from the standpoint of the direct visual sensations, which was the real purpose of his famous treatise on Colour Theory. Soon after came the important observations of RITTER and other electrical workers concerning excitations of the sensory nerves; and above all, PURKINJE's observations; so that in 1826 J. MÜLLER could state the chief laws of the subject in his Theory of Specific Sense Energy as it was first published in his work on the Comparative Physiology of Vision, to which reference was made at the beginning of this chapter. This work and that of PURKINJE are closely related to GOETHE's Colour Theory, although J. MÜLLER subsequently abandoned its physical concepts. MÜLLER's law of specific energies was a step forward of the greatest importance for the whole

[1] *De sensibus, de anima* lib. II. c. 5–8 and *de coloribus*.

[2] *Elem. Physiolog.* Tom. V lib. 16, 17.

theory of sense perceptions, and it has since become the scientific basis of this theory. In a certain sense, it is the empirical fulfilment of KANT's theoretical concept of the nature of human reason.

Even ARISTOTLE was aware of the images produced by pressure on the eye. NEWTON[1] conjectured that mechanical disturbance of the retina produces a motion in it similar to that made by the impact of rays of light. He considered this motion as the cause of the sensation of light. The opinion that in the case of pressure-images, and in other cases also, objective light is developed in the eye has had its adherents until quite recently. An example of this view is the medico-legal case mentioned above in which the capable physician SEILER seemed to think it necessary to admit the possibility of such a contingency. But no one has ever been able to see the light thus developed in another person's eye. To strengthen this view, its adherents have cited the cases of persons like the Emperor TIBERIUS, CARDANUS and KASPAR HAUSER who were able to see in the dark, that is, with very little light. Another argument which they use is the so-called luminosity of animal eyes and of the eyes of albinos and certain other human beings whose eyes are malformed; which is due simply to the reflection of light. Finally, they instance distinct after-images which old people see in the evening after the light is extinguished, and which sometimes persist for a long time; as proving the possibility of development of light in the eye. Quite recently more accurate descriptions of pressure-images have been given by PURKINJE and SERRES D'UZÈS. How THOMAS YOUNG utilized these effects in his theory of accommodation has been mentioned in Vol. I, page 158.

VOLTA was aware of the flash of light when the current flowing through the eye was turned on or off. RITTER perceived the persistent luminous actions even with a simple cell; and subsequently they were minutely described, especially by PURKINJE.

Supplement by HELMHOLTZ *in the First Edition*

It was expressly stated above that the actions of steady currents of electricity on the visual apparatus were not to be considered as a stimulation (as they used to be regarded), but as changes of excitability due to the electrification. But the author's assumption that the continuous internal excitation of the fibres of the optic nerve, whose sensibility is thereby increased, takes place on the side of the nerve towards the brain, does not agree with the phenomena that occur when an electric current flows through a small electrode right into the eyeball itself. These phenomena as observed by PURKINJE were partly described on page 17. A more probable inference here would be that it is the electrified condition of the radial fibres of the retina that is responsible, and that their steady stimulation takes place on the posterior surface of the retina.

If the negative electrode is placed on the neck, and the positive electrode, consisting of a pointed cone-shaped piece of sponge soaked in salt water and fastened to a handle of metal, is applied to the moistened eyelids near the outer angle of the eye, the visual field

[1] *Optice*, at the end of Quaestio XVI.

appears dark on the nasal side, and bright on the temporal side; and the optic disc which is within the bright portion appears dark. When the eye is turned so that the point of fixation falls on the boundary between the bright and dark areas, a bright tuft of light seems to radiate out from it towards the dark portion and a dark tuft towards the bright portion of the field. These two oval tufts just about cover the area of the yellow spot.

If the direction of the current is reversed, the light and dark areas change places. Breaking the circuit has the same instantaneous effect as reversing the current.

All these phenomena may be simply explained as due to the electrified state of the radial nerve fibres of the retina, on the supposition that there is a permanent weak stimulation at their posterior ends as the result of internal causes; the presence of which seems to be indicated by the intrinsic light of the retina.

When positive electricity enters the eyeball from the outer side of the eye and returns from the inner and posterior side, the excitability of the posterior surface of the retina will be diminished where the current enters and increased where it leaves; and hence the inner half of the visual field, corresponding to the outer half of the retina, must appear dark, and the outer half bright. Probably the optic nerve acts as a poor conductor, so that the current is reduced near the place where this nerve enters the eye; which makes this place stand out through contrast. If the yellow spot is at the border of the portions of the retina through which the current is passing in opposite directions, the current flows through it along the surface of the retina. In the yellow spot, however, there are bundles of fibres which run also along the surface of the membrane. Accordingly, these fibres are traversed by positive electricity from the temporal towards the nasal side, that is, the current flows through the fibres on the temporal side of the fovea in the direction towards their ends that are connected with the cones, and on the nasal side it flows the opposite way. Thus, on the temporal side the excitation will be increased, and on the nasal side diminished; and this is why the bright tuft appears on the nasal side of the point of fixation in the field of view, and the dark tuft on the temporal side.

When the place is changed where the current enters, the entire phenomenon is correspondingly shifted.

Note by W. Nagel.—Neither with ascending nor with descending current has the editor succeeded in showing that there is any variation of the luminous threshold. N.

Mechanical Stimulation

1706. I. NEWTON, *Optice,* at the end of Quaestio XVI.
1774. EICHEL in *Collectan. soc. med. Havniensis* 1774.
1797. A. v. HUMBOLDT, *Versuche über die gereizte Muskel- und Nervenfaser.* II. 444.
1801. TH. YOUNG, On the mechanism of the eye. *Phil. Trans.* 1801. I. 23.
1819–25.* PURKINJE, *Beobachtungen und Versuche zur Physiologie der Sinne.* I. 78, 126, 136, II. 115.
1825. MAGENDIE, *Journal de Physiologie.* IV. 180. V. 189.
1826. J. MÜLLER, *Über die phantastischen Gesichtserscheinungen.* Koblenz. S. 30.
1832. D. BREWSTER in POGGENDORFFS *Ann.* XXVI. 156.—*Phil. Mag.* I. 56.
1833. SEILER in HENKES *Zeitschr. für gerichtl. Med.* 1833. 4. Quartal. S. 266.
1834. LINCKE, *De fungo medullari.* Lipsiae.
QUETELET, POGGENDORFFS *Ann.* XXXI. 494.
J. MÜLLER in *Archiv für Anat. und Physiol.* 1834. S. 140.
1840. TOURTUAL in J. MÜLLERS *Handbuch der Physiologie* II. 259.
1850. SERRES D'UZÈS, Du Phosphène. *C. R.* XXXI. 375–378.
1854–55.* CZERMAK, *Physiologische Studien.* Abt. I. § 5. S. 42 and Abt. II. S. 32.—*Wiener Sitzungsberichte* XII. 322 and XV. 454.
1856. A. E. LAIBLIN, *Die Wahrnehmung der Choroidealgefässe des eigenen Auges. Dissert.* Tübingen.
MEISSNER, Bericht über die Fortschritte der Physiologie im Jahre 1856, S. 568 in HENLES *Zeitschr. für ration. Medizin.*
1858. J. CZERMAK, Über das Akkommodationsphosphen. *Wiener Ber.* XXVII. 78–86.—*Archiv für Ophthalmologie.* VII, 1. pp. 147–154.

Electrical Stimulation

1755. LE ROY, *Mém. de Mathém. de l'Acad. de France.* 1755. pp. 86–92.
1794. PFAFF in GRENS *Journal der Physik* VIII. 252, 253.
1795. PFAFF, *Über tierische Elektrizität.* S. 142.
1798. RITTER, *Beweis, dass ein beständiger Galvanismus den Lebensprozess im Tierreiche begleite.* Weimar 1798. S. 127.
1800. VOLTA, *Colezione dell' Opere.* Tom. II, P. II. p. 124.
*RITTER, *Beiträge zur näheren Kenntnis des Galvanismus.* Bd. II. St. 3, 4. S. 159, 166. § 93.
1801 and 1805 RITTER in GILBERTS *Annalen.* VII. 448. XIX. 6–8.
1819. PURKINJE, *Beobachtungen und Versuche zur Physiologie der Sinne.* Bd. I. Prag 1819. S. 50. Bd. II. Berlin 1825. S. 31. KASTNERS *Archiv für die gesamte Naturlehre* 1825. V. 434.
1823. MOST, *Über die grossen Heilkräfte des in unseren Tagen mit Unrecht vernachlässigten Galvanismus.* Lüneburg 1823. S. 812.
1829. FECHNER, *Lehrbuch des Galvanismus und der Elektrochemie.* Kap. 39. S. 485 ff.
1830. HJORT, *De functione retinae nervosae.* Part. II. Christiania 1830. (Dissert.) p. 34 § 17.
1848. E. DU BOIS-REYMOND, *Untersuchungen über tierische Elektrizität.* I. 283–293; 338–358.
1863. R. SCHELSKE, Über Farbenempfindungen. *Archiv für Ophthalmol.* IX. (3) S. 39–62.
1864. AUBERT, *Physiologie der Netzhaut.* Breslau. S. 333–390.

Note to §17

On the Stimulation of the Visual Apparatus by RÖNTGEN *Rays and* BECQUEREL *Rays. By* W. NAGEL

In addition to HELMHOLTZ's account in §17 of the action of adequate and inadequate stimuli of the visual organ, mention should be made of the fact that the sensation of light in the eye can be aroused both by RÖNTGEN *rays* and by the emanations of the so-called *radio-active* substances. The sensation of light produced by the impact of X-rays in the eye was noticed first by BRANDES and DORN[1]. COWL and LEVY-DORN[2] supposed this optical effect could be traced to illusions produced chiefly by electrical action at a distance. But BRANDES and DORN, also RÖNTGEN,[3] HIMSTEDT and NAGEL,[4] and others showed that the luminous sensation was produced when such sources of error were guarded against. The vacuum tube can be enclosed in a light-proof box made of thin sheet aluminum opaque to light; and yet a powerful glow of light will be produced in the eye, provided the eye has been dark-adapted for fifteen minutes or more prior to the test.

On the other hand, if the rays are allowed to fall only on a certain area of the retina, the rest of it being shielded from the action of X-rays by a thick lead screen, the luminous effects are likewise sharply outlined. Thus, for example, when a diaphragm, consisting of a thick plate of lead with a hole in it 3 mm in diameter, is held on one side near the eye, so that the X-rays cross the eyeball from the temporal to the nasal side, the result is that the bundle of rays meets the retina twice; and, consequently, two bright circles are seen, which are projected out in the field exactly in the same way as the pressure-phosphenes.

When the radiation enters the eye from the temporal side, the spot projected on the nasal side is brighter than that on the opposite side. This is easily explained because the nasal glow is the result of stimulation of the temporal half of the retina and the glow projected on the temporal side is the result of stimulation of the nasal half; but ere the rays reach this part of the retina, they have been absorbed to some appreciable extent in the vitreous humor.

The contrast with the above is very striking when no diaphragm is used at all, and the rays are allowed to fall freely on the eye. The brightest sensation is always on the side from which the rays come; that is, usually on the temporal side, when the radiation is lateral. Radium emanation, as above stated, acts in the same way.

When a screen with a slit in it is moved to and fro between the eye and the RÖNTGEN tube, the phenomena are very instructive. The X-rays traverse the eye without being refracted; and, hence, the lines of intersection of the wedge-shaped bundle of rays with the nearly spherical retina appear projected in the field as curved lines depending

[1] WIEDEMANNS *Ann.* LX. 478, 1897; LXIV, 620, 1897; LXVI, 1171, 1898.

[2] *Arch. f. (Anat. u.) Physiol.* 1897.

[3] *Ber. d. preuss. Akad.* 1897. 576.

[4] *Ann. d. Physik.* IV F. 4, S. 537, 1901.

on the locus of the stimulation. The most conspicuous effects are obtained when the rays come through the diaphragm in the frontal direction from the temporal side of the eye, and when the aperture is in the form of a rectangular cross with vertical and horizontal beams. If the vertical slit is in the equatorial plane of the eye, two crosses are seen composed of approximately straight lines intersecting each other at right angles. If the diaphragm is shifted more towards the back of the eye, the crosses become much distorted and finally blend into their horizontal portions.

HIMSTEDT and NAGEL endeavoured to ascertain whether X-ray stimulation is a direct one, comparable with that of light; or whether fluorescence of the ocular media has a distinct part in it, as is the case in the perception of ultra-violet and BECQUEREL rays (see below). The mere fact that it is possible to stimulate precise parts of the retina by X-rays, indicated that it was extremely unlikely that there was any noteworthy fluorescence of the ocular media. And, as a matter of fact, not a single trace of fluorescence could be detected in these substances. But in the retina itself there was certainly some slight effect of this nature, much less, however, than that produced by ultra-violet rays. But whether the retina in the live eye does not fluoresce still more, is a question that has not been answered.

The eye must be dark-adapted in order to be sensitive to X-rays; and hence it seems likely that the same elements whose sensitivity to *light* rays increases so much in darkness must also be the perceptive agencies for the X-ray stimulation; that is, according to our assumption, the perceptive elements in this case must be the rods. Although the fluorescence of the retina under radiation is feeble, the layer that emits the fluorescent light and the layer that is sensitive to light must be exceedingly close together; in fact, they may partly coincide with each other. From this point of view, it may be that the perceptibility of X-rays has something to do with the fluorescence of the retina. But this explanation still does not account for the fact mentioned above, that the light sensation is greater on that side of the retina where the X-rays come to it through the vitreous humor.

It is often stated that the totally colour-blind are peculiarly able to see X-rays. But the only reason for this that can be suggested is, that these persons are accustomed by experience to shield their eyes as much as possible from bright light, and so when they enter the dark room, they are already more dark-adapted than other subjects with normal vision.

The writer put a bandage over both eyes of a totally colour-blind young girl. It was composed of many layers of black velvet, bound so tight that even after an hour in the lighted room she had not the slightest sensation of light. When the X-ray tube was started, the diffused light was visible to her a metre away. With her eyes blindfolded, she was able to tell when a lead plate was placed between her and the tube, and when it was removed. But the results are similar for normal individuals after an hour's dark adaptation.

Undoubtedly, the effect of radium emanation or Becquerel *rays* on the visual apparatus is due to the fluorescence of the transparent parts of the eye, including the retina. Himstedt and Nagel easily succeeded in demonstrating this fluorescence in the eyes of various animals. Since the entire contents of the eye, particularly the lens, are self-luminous under the influence of radium, there can be no question of a localization of the stimulus. The glow that is perceived is therefore likewise a diffused one, except that, just as with X-ray radiation, the strongest sensation of light is on the side where the radium is placed.—N.

§18. Stimulation by Light

What we have now to consider is the excitation of the organ of vision by means of objective light or aether vibrations. The vibrations of the aether are not included among the general agencies of stimulating nerves such as electricity and mechanical injury, which can disturb every part of any nerve fibre. It may be demonstrated that the fibres of the optic nerve within the trunk of this nerve and in the retina are no more stimulated by light than are the motor and sensory nerve fibres of other nerves. Some special apparatus in the retina at the end of the optic nerve fibres must be present, which is adapted to enable objective light to start a nervous impulse.

First of all, let us show that the nerve fibres in the trunk of the optic nerve are not stimulated by objective light. The bulk of these fibres are at the place where the optic nerve comes into the eye through the sclerotica. This nerve, lying exposed on the side towards the transparent ocular media, is not overlaid with any black pigment; and yet it is so translucent that light falling on it may penetrate it to an appreciable extent. This can be shown with an ophthalmoscope which often reveals ramifications of the central blood-vessels that are inside the optic nerve and covered completely by the mass of nerves. If such ramifications can be recognized in the interior of the nerve-substance, light must penetrate that far so as to return from there to the observer's eye. Thus, there is nothing to prevent the light falling on the eye from penetrating to a certain depth in the substance of the optic nerve. But *this light that falls on the place where the optic nerve enters the eye is not perceived.*

Holding the book, with the lines in the usual horizontal position, at a distance of about one foot away, close the left eye, and look at the white cross in Fig. 1 with the other eye. There is a certain adjustment for which the white circle vanishes entirely from the field, and no gap is to be seen in the black background. To succeed with this experiment, one must look steadily at the little cross and not look to one side. When the book is moved closer or farther away, the white disc

Fig. 1.

reappears, and is distinctly seen by indirect vision; and the same thing happens also when the book is slanted so as to throw the white disc a little higher or lower. All other objects, white, black or coloured, that are not larger than the disc, disappear in like manner when they are laid on the disc, and the experiment is conducted in the same way. We learn, therefore, that in the field of each eye there is a certain gap where nothing can be discerned; and that accordingly there is a corresponding place on the surface of the retina that is not conscious of an image when it falls there. This place is called the *blind spot*. As the blind place in the visual field of the right eye is to the right of the point of fixation, and in that of the left eye to the left of the point of fixation, the blind spot of the retina must be on the nasal side of the yellow spot; in the region where the optic nerve comes into the eye.

That the blind spot is actually identical with the place of entrance of the optic nerve had been shown previously by measurement of its apparent size and of its apparent distance from the point of fixation of the eye. A still more direct proof was given by DONDERS[1] with his ophthalmoscope. This instrument was used to reflect the light of a small flame some distance away into the patient's eye; and the latter was made then to turn his eye until the little image of the flame fell on the place where the optic nerve enters the eye. The image here was not sharply outlined, and at the same time the entire optic disc, which was at least twenty times greater than the little image, was rendered

[1] *Onderzoekingen gedaan in het Physiol. Labor. d. Utrechtsche Hoogeschool.* VI. 134.

quite luminous. This is due to the translucent nature of the nervous substance. On the retina itself, near the disc, he saw scarcely any trace of light that might be due to diffusion in the transparent media of the eye or that was reflected sideways from the brightly illuminated surface of the optic nerve. The patient was not conscious of any sensation of light as long as the little luminous image fell entirely on the disc. Some patients thought they perceived a faint glimmer of light; which possibly might be due to the feeble illumination of the retina mentioned above. By small movements of the mirror, the image could be made to move from one side to the other of the disc, but there was never any consciousness of light until a part of the image was distinctly across the boundary so as to fall on a place where the various layers of the retina were present. This means simply that the blind spot corresponds to the entire area where the optic nerve enters, and certainly is not just at those places where the blood-vessels enter the eye.

Subsequently, Coccius[1] showed how the same experiment could be performed by the observer in his own eye; which is still more instructive. For this purpose, a plane or convex mirror is used, with a hole in it, like the mirror of an ophthalmoscope. The observer holds it close in front of his eye, and allows the light from a lamp to come through the hole to his eye. If the eye is first turned right towards the edge of the hole, the little inverted red image of the flame on the retina of one's own eye can easily be seen. Now trying not to let the image go, turn the eye more and more inwards, until presently the image falls at the place where the optic nerve is; and then make the observations described above. Incidentally, the flame ought to be small or far away; otherwise, too much light will enter the eye. In this way the larger blood-vessels are seen, but the field, of course, is very small. When a larger flame is used, the eye is so blinded by it that it is impossible to see much. If the quantity of light falling on the place where the optic nerve enters the eye is large, a faint glimmer of light will certainly be perceived, but the explanation of it, according to these experiments, is that some of the light spreads out over adjacent parts of the retina. Sometimes too in experiments of this kind there is a red glow of light in the eye, perhaps when a blood-vessel on the surface of the optic nerve is highly illuminated and reflects light. This was seen by A. Fick and P. du Bois-Reymond by using for the object the image of the sun as made by a convex lens.

The form and apparent size of the blind spot in one's own visual field may be easily found as follows. Place the eye 8 or 12 inches above

[1] *Über Glaukom, Entzündung und die Autopsie mit dem Augenspiegel.* Leipzig 1859. S. 40, 52.

a sheet of white paper on which a little cross is marked to serve as point of fixation of the eye.

Take a white or at any rate brightly coloured pen, with some ink on its point, and move it over the paper into the projection of the blind spot until the point disappears. Then move it from this place, first in one direction, and then in another, towards the periphery of the spot where it again begins to be visible. Fig. 2 shows the blind spot of the author's right eye, as drawn in this way with respect to *a* as point of fixation. The length of the straight line *AB* is one-third of the distance betwen the eye and the paper for this particular figure. The spot has the form of an irregular ellipse, on which the writer can discern the beginnings of the larger blood-vessels, as seen also by HUECK. If a small black spot is made on the paper, and different points fixated one after another, the continuation of the vessels far in the field of the retina will be found to be blind places. The easiest way to do this is by finding first the direction of the blood-vessels in one's own eye by COCCIUS's method.

Fig. 2.

Let f denote the distance of the eye from the paper; F the distance of the second nodal point from the retina (which is about 15 mm on the average); d the diameter or any other linear dimension of the blind spot in the drawing; and D the corresponding magnitude on the retina; then

$$\frac{f}{F}=\frac{d}{D},$$

from which D can be calculated. In a measurement of this kind, the magnitude denoted by F can never be perfectly accurately determined for the individual eye; and so without using it, a better way is to measure the visual angle, that is, the angle between the direction-lines (see Vol. I, p. 96) corresponding to the different points of the drawing. If the lines of vision drawn to the point a in Fig. 2 may be assumed to be perpendicular to the plane of the figure, and if the distance ad is denoted by β, and the angle subtended by ad is denoted by α, then

$$\frac{\beta}{f}=\tan\alpha\ .$$

from which the angle a can be calculated. In the same way the visual angle between a and any other point in the drawing may be found. The following are results obtained in this way by different observers:

(1) Apparent distance of the point of fixation from the nearest part of the edge of the blind spot: LISTING[1] 12° 37.5′. HELMHOLTZ 12° 25′. THOS. YOUNG 12° 56′.

(2) Apparent distance of the most distant part of the edge: LISTING 18° 33.4′. HELMHOLTZ 18° 55′. THOMAS YOUNG 16° 1′.

(3) Apparent diameter of the blind spot in the horizontal direction: HANNOVER and THOMSEN[2] for 22 eyes 3° 39′ to 9° 47′; average of all measurements 6° 10′. LISTING 5° 55.9′. GRIFFIN,[3] largest value, 7° 31′. HELMHOLTZ 6° 56′. THOMAS YOUNG, who rather inconveniently used two lights for finding the boundary of the spot, 3° 5′.

(4) Actual diameter of the blind spot, using LISTING's value of 15 mm for F, in LISTING's eye, 1.55 mm. HELMHOLTZ 1.81 mm. HANNOVER and THOMSEN, average, 1.616 mm.

A measurement by E. H. WEBER of the diameter of the place of entrance of the optic nerve in two cadavers gave 2.10 mm and 1.72 mm. The distance from its centre to the centre of the yellow spot was in one eye 3.8 mm; whereas the same distance as calculated for LISTING's eye was 4.05 mm. The greatest and smallest diameters of the vessels in the middle of the nerve were 0.707 and 0.314 mm; the greatest in the other eye 0.633 mm.

Even before DONDERS's experiments it might have been inferred from these measurements that the entire optic disc was insensitive to light.

Another way of forming some idea of the apparent size of the blind spot in the field of view is to try to realize that eleven full moons can be placed side by side along its diameter. In it a human face 6 or 7 feet away will disappear.[4]

That the fibres in the trunk of the optic nerve cannot be stimulated by light, is shown by the phenomena of the blind spot that have been described. That the ramifications of this nerve which spread out from the disc all over the anterior surface of the retina are likewise insensitive to light, may be inferred from the fact that perfectly definite bright places in the field of view are seen also as actually definite places.

[1] *Berichte der Königl. sächs. Ges. der Wiss.* 1852. S. 149. E. H. WEBER's observations are also given here.

[2] A. HANNOVER, *Bidrag til Øjets Anatomie.* Kjöbenhavn. Cap. VI. S. 61.

[3] GRIFFIN, Contributions to the physiology of vision. London, *Medical Gazette.* 1838 May. p. 230.

[4] ¶It is related that that "merry monarch," King Charles II, got much amusement by making his courtiers see how they would look when their heads were off their shoulders. (J. P. C. S.)

When light falls on a place A on the retina, it meets here not only those fibres that end in A but also those which pass over A and end in the more peripheral parts of the retina. Now since the place at which a nerve fibre is stimulated is not discriminated in the sensation, so far as sensation is concerned, the result would be the same as if light had fallen on those peripheral parts of the retina. If this were the case, we should see a streak of light extending from every illuminated point to the borders of the visual field, which of course is not the case. In other words, the fibres of the optic nerve spreading out over the retina cannot be stimulated by objective light.

On the other hand, the evidence of the sensitiveness of the posterior layers of the retina to light is afforded by the ability of seeing the shadows of the retinal vessels (Vol. I, 211). The retinal vessels lie in the layer of the optic nerve fibres, some finer ones also in the layer immediately posterior to that of the ganglion cells (No. 6 in Fig. 14 of Vol. I) and in the fine granular layer (No. 5 of same diagram). From the movements of the shadow of these vessels when the source of light is moved, the inference is that the layer by which the shadow is perceived, the layer in which the light on the edge of the shadow gives rise to nerve excitation, must be a very little distance beyond the vessels. According to H. Müller's measurements (Vol. I, p. 220), the distance of the vessels from the surface which perceives their shadow must be between 0.17 and 0.36 mm. The distance of the vessels from the most posterior part of the retina where the rods and cones are, according to the same authority, is between 0.2 and 0.3 mm. Thus in any case the sensitive layer must be one of the most posterior layers of the retina; that is, the layer of rods and cones or the outer granular layer. At the place where vision is most distinct, which, according to Remak and Koelliker is in the central cavity of the yellow spot, there is nothing but ganglion cells and cones; consequently, the latter would seem to be the elements that are peculiarly sensitive.[1] H. Müller and Koelliker express the opinion that the

[1] In the second edition of this treatise, a briefer statement was substituted in place of the rest of this paragraph as given above, which follows the text of the first edition. With respect to the function of the rods, the new statement was different from the original view, and reads as follows (p. 255): "From the perfectly analogous anatomical structure of the rods it is extremely probable that they also have the same sort of capacity; which was the opinion of H. Müller and Koelliker. Nevertheless, they must play an entirely different rôle in the localization of sensations, because, in spite of their being finer and more numerous in the peripheral parts of the retina where they predominate, the power of discrimination between very nearly similar impressions is more imperfect in this region than it is in the fovea.

"Since the investigation of the delicacy of perception of different parts of the retina is essentially bound up with the question as to what elements of the retina are sensitive to light (that is, excite a sensation when light acts on them), and how they are connected with nerve fibres, let us consider this question first.

rods are also sensitive, because, like the cones, they too are connected with similar fibres traversing the retina perpendicularly. But as remarked by E. H. WEBER, this assumption seems to be opposed to the fact that there are nothing but cones at the place of most distinct vision; whereas out towards the periphery of the retina where more and more rods are found between the cones, acuity of vision becomes less and less perfect. If the rods were sensitive elements, it might be expected that the sensitivity and exactness of perception would necessarily be greater where the rods are more numerous, because there are more rods than cones in the same area. The connection with radial fibres is no proof of the nervous nature of the rods, because a large number of the radial fibres are attached to the *membrana limitans*, and it is therefore extremely probable that these are connective tissue, and not nerve fibres at all. In saying that the posterior layer of the retina and particularly the cones are the last elements of the nervous mechanism of vision that are sensitive to light, of course, what is meant is that external light stimulates changes in these structures that result in nervous excitation and, finally, in sensation, if this excitation is transmitted to the brain. In fact, the *light-sensitive* elements of the retina, as they may be called, just as we speak of a sensitive plate in photography, are functionally different from all other parts of the nervous system simply by virtue of their sensitivity to light, just as on the other hand they are in so many respects different in their anatomical structure. Another result also is that the action of light on the peculiar nervous substance of the retina and optic nerve is not an immediate one, as in the case of electricity and of mechanical disturbance, whereby in every nerve fibre at every place in it the molecular changes can be started that constitute the process of stimulation. The action of light is more indirect. It acts directly only on the special light-sensitive apparatus or the cones. We have no idea as to what kind of an effect this is and as to what, if any, similarity there is between it and nerve stimulation; whether a vibration is set up, as NEWTON,[1] MELLONI,[2] SEEBECK[3] and other physicists supposed;

"The part of the retina which is capable of the finest spatial discrimination is a perfectly regular mosaic of individual parts, that is, the cones. Each of these is connected with a nerve fibre, which is then connected with the cells of the retina. Accordingly, the assumption that every single cone has its own independent nerve conduction to the brain, and that, consequently, the sensation excited in it can be distinguished from qualitatively equal sensations in the adjacent cones, does not seem improbable."

For the more modern conceptions of the functional difference between rods and cones, see the Note at end of § 25.—N.

[1] *Optice*. Lib. III. Quaestio XVI.

[2] POGG. *Ann.* LVI. 574.

[3] Ibid. LXII. 571.

whether there is a shift in molecular arrangement, as E. du Bois-Reymond supposes to take place in the electromotor molecules of muscles and nerves; whether there is some heating effect, as Draper[1] thinks; or whether this light-sensitive layer of the retina is some part of a photo-chemical apparatus, as Moser[2] assumed. At any rate, stimulation of those nerve fibres connected with the cones that are acted on by light is merely a secondary result of these changes, whatever they are.

Acuity of visual perception is also connected with the size of the retinal element stimulated by light. The light that falls on a single sensitive element can produce nothing but a single light sensation. In such a sensation there is no way of telling whether some parts of the element are highly illuminated as compared with other parts. A luminous point can be perceived when its image on the retina is very much smaller than a single retinal element, provided the amount of light from it that falls on the eye is sufficient to affect the sensitive element appreciably. Thus, for example, the fixed stars, are perceived as objects of great brilliancy, although their apparent sizes are vanishingly small. Similarly, too, a dark object on a bright background may be perceived even when its image is smaller than a sensitive nerve element, provided the amount of light that falls on the element is perceptibly diminished by the dark image around it. Thus, suppose that with the given degree of illumination, the eye is capable of perceiving differences of two percent in light intensity; then a dark image whose area was two percent of that of a sensitive element might still be perceived. Obviously, on the other hand, two bright points cannot be distinguished as separate unless the distance between their images is greater than the diameter of a retinal element. Were the distance less than this, the two images would have to fall on the same element or on two adjacent elements. In the first case, both would excite simply a single sensation; and in the second case, there would indeed be two sensations, but in adjacent nerve elements, so that it would be impossible to tell whether there were two separate points of light or only one whose image happened to be on the border of both elements. The distance between the two bright images, at least between their centres, must be greater than the width of a single sensitive element if the two images are to fall on two different and non-contiguous elements, with another element in between them that is not stimulated by light at all, or at least is more feebly stimulated than the others.

[1] *Human physiology.* p. 392.
[2] Pogg. *Ann.* LVI. 177.

According to HOOKE's dictum,[1] two stars whose apparent distance apart is less than 30″ appear as one; and scarcely one person in a hundred can distinguish two stars when their apparent distance apart is less than 60″. Others, who have made their observations, not on stars, but on illuminated white lines or rectangles, have found the resolving power of the eye rather less than this. The best eye examined by E. H. WEBER was able to distinguish two white marks whose middle lines were 73″ apart. With higher illumination, and under the most favourable conditions, the author has been able to make out lines of this sort that were only 64″ apart. In LISTING's schematic eye a visual angle of 73″ corresponds to a distance on the retina of 0.00526 mm; an angle of 63″ to 0.00464 mm; and an angle of 60″ to 0.00438 mm. KOELLIKER's measurements give for the diameter of the cones in the yellow spot values between 0.0045 and 0.0054 mm. This agrees almost exactly with the above figures, and incidentally also tends to confirm the assumption that the cones are the last sensitive elements of the retina.[2]

At the same time, it is clear that the optical characteristics of a well constructed and correctly accommodated eye are quite sufficient for attaining the resolving power that the eye is actually known to have. As a matter of fact, with a pupil of 4 mm diameter, the blur circle on the retina due to chromatic aberration was found to be 0.0426 mm in diameter (see §13); that is, almost ten times greater than the width of a cone. But it was explained at the time why this blur circle, in spite of its size, did not sensibly disturb good vision. The aberrations due to asymmetry of the eye (see §14) are generally

[1] SMITH's System of Compleat Opticks, Vol. I, Book I, Chap. 3, 97.

[2] ¶See NAGEL's note at end of this section.

From the standpoints of both geometrical and physical optics, it is extremely improbable, even under the most favourable circumstances, that the image on the retina is ever actually so small as not to exceed the diameter of a single cone. It must certainly be larger than this if either diffraction-effects or aberrations are taken into account. It has been found recently that the apparent size of minute objects subtending angles as much as 2 or 3 minutes of arc depends essentially on the intensity of illumination.

The *aligning power* of the eye is essentially different from its *resolving power* in the sense of being able to distinguish the components of a double star. The perception of width, that is, of slight lateral differences of position, is distinctly sharper than that of vertical dimensions. The precision with which the eye can adjust a mark on a vernier scale to coincide with a division on the principal scale is extraordinarily great. Under the best conditions skilled observers succeed in making such settings with an average error of not more than 3″ of arc. In coincidence range-finders the two images can be aligned with an error that usually does not exceed 12″, and in some instances with an error of not more than 2″. —In daylight a dark line on a bright background whose length is not less than a few minutes of arc can be perceived provided its thickness is as much as 1.2 seconds. Under the same circumstances a bright line on a dark background must be at least 3.5 seconds wide to be recognized. (J. P. C. S.)

much smaller and have less bad effect on vision, unless one tries to see horizontal and vertical lines at the same time.

The resolving power is much less in the lateral parts of the retina than in the yellow spot; the decrease near the centre of the retina being less than it is farther away from it. The measurements made by AUBERT and FÖRSTER show that the decrease in different directions from the centre proceeds at different rates, being most rapid up and down, and slowest towards the outer side of the retina. However, individual differences in this respect are fairly large. Another noteworthy result from these measurements is that in accommodation for distance, the falling off of the resolving power towards the periphery of the retina seems to proceed more rapidly than it does in near vision. These observers found that a similar decrease of the power of discrimination of optical images out towards the periphery of the retina certainly does not occur in rabbits' eyes. This proves that the imperfection of vision in the peripheral parts of the retina depends simply on the peculiarity of the retina, and not at all on the quality of the optical images.

TOB. MAYER, and subsequently E. H. WEBER also, used parallel white lines separated by black ones of the same width as test-object for finding the smallest interval that can be discerned. VOLKMANN used spider filaments on a bright background. For convenience of illumination, the author found it better to use a grating of black wires separated by intervals equal to the diameter of the wire; which was set up against the background of the bright sky. TOB. MAYER also used white squares, partly separated by a black grating and partly arranged like a chess board.

In making these tests, the eye should be able to accommodate perfectly. When coarser objects are used, which have, therefore, to be set up farther away, an appropriate concave lens should be placed in front of the eye. The illumination must be strong, but not too dazzling. The author observed a striking change in the form of the bright and dark straight lines. The width of each bright and dark band in his grating was $\frac{13}{24} = 0.4167$ mm. At a distance between 110 and 120 cm the effect began to be apparent; the grating assuming an appearance something like that shown in Fig. 3*A*. The white lines seemed to be curved partly like waves, and partly like a string of pearls with places alternately thicker and thinner. In Fig. 3*B* the little hexagons are supposed to be sections of cones in the yellow spot; and *a*, *b*, and *c* represent optical images of three bars of the grating. Above *dd* these images are shown in their real form; but below *dd*, all the hexagons, that were predominantly black, are made entirely black, and all that were predominantly white, are made entirely white; the idea being that the predominant character is responsible for the sensation that is perceived. Thus in the lower portion of Fig. 3*B* a pattern is obtained that is similar to that in *A*. PURKINJE[1] saw something of the kind; and BERGMANN also noticed that sometimes just before the lines of a grating disappeared completely, the grating looked like a chess-board, the bands sometimes being seen at right angles to their actual direction; all of which may be explained in the same way as above.[2]

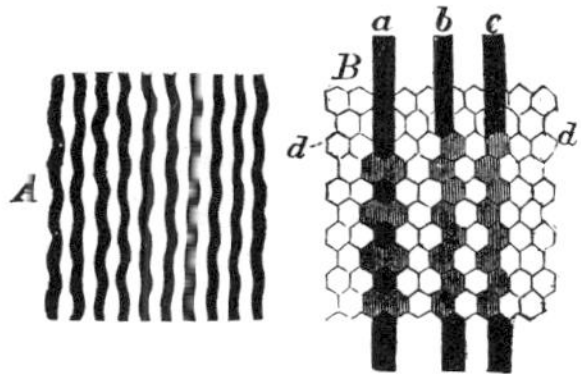

Fig. 3.

When the widths of two luminous objects used in the test are vanishingly small as compared with the interval between them, they cannot be seen as separate unless there is an unstimulated retinal element between the retinal elements on which their images fall. In other words, the diameter of such an element must certainly be less than the interval between the two images. However, if the width of the object is the same as the dark interval between each pair of them, it is not absolutely necessary that the retinal elements shall be smaller than the image of the dark band. A retinal element on which the image of the dark band falls, and which extends on its sides partly into the bright bands, is thus in a position to perceive less light than the adjacent elements; provided the total amount of light that falls on it is less than that that falls on its neighbours. In such cases, therefore, the most we can say is that the retinal elements are smaller than the interval between the middle lines of the bright bands. As a matter of fact, the results of TOB. MAYER's experiments, as given below, show that with parallel lines the resolving power remains the same as before when the width of the black or white bands is changed, without varying the total width of the two. This is the reason why the author has always used this sum of the two as the width of the object, that is, the interval between the middle lines of two adjacent objects (contrary to the usage of MAYER and WEBER); and this is the distance used in calculating the smallest visual angle or angle of distinctness.

The explanation of the greater resolving power of the author's eye as compared with that of other adults may be due to the brighter illumination that was possible with his grating. The keenest eye was that of a ten year old boy examined by BERGMANN. TOB. MAYER studied the influence of illumination. He found that systems of lines

[1] *Beobachtungen und Versuche.* I. 122.

[2] HENLE and PFEUFFER. *Zeitschrift für ration. Medizin.* (3.) II. 88.

could be seen better when illuminated by quite bright daylight, and that any higher illumination did no good. At night he obtained lesser degrees of illumination by placing a light at different distances from the paper. The farther away the light was, the nearer he had to come. When the distance of the light was gradually increased from 6 inches to 13 feet, the visual angle for white bands with an equally wide interval between them, increased (as above calculated) from 138 to 344″.

Observer	Object	Size of Object in mm	Distance from eye in mm	Distance divided by size of object	Visual angle in seconds
1. Hooke	Fixed stars	....	....	....	60
2. Tob. Mayer	(a) Parallel lines with intervals of same width	1.624	3573	2200	94
	(b) Parallel lines with wider and narrower intervals	1.354	3086	2280	90
3. Tob. Mayer	White squares separated by black grating	1.985	5035	2536	81
4. Tob. Mayer	Chess board	2.346	3898	1661	124
5. Volkmann	Spider webs	0.141	190	1346	153.2
6. N. N. by Volkmann	Spider webs	....	352	2500	80.4
7. Th. Weber by E. H. Weber	Parallel lines with intervals of same width	0.113	249	2210	93.3
8. N. N. by E. H. Weber	Parallel lines with intervals of same width	....	311	2760	73
9. N. N. by E. H. Weber	Parallel lines with intervals of same width	....	249	2210	90.6
10. Helmholtz	Rod grating	1.083	3500	3232	63.82
11. O. H. by Helmholtz	Rod grating	....	2400	2215	93
12. Bergmann	Parallel lines with intervals of same width	2	5500 to 8000	2750 4000	75 51.6

He found an empirical formula which corresponds fairly well to his measurements, namely $s = 158'' \sqrt[3]{a}$, where s denotes the visual angle and a denotes the distance of the light. As the brightness $h = \frac{1}{a^2}$, it follows that $s = \frac{158}{\sqrt[6]{h}}$.

Supplement by HELMHOLTZ (*in the first edition*)

A. VOLKMANN has published an account of some new experiments which lead him to think that the foveal cones are not fine enough to explain the actual visual acuity of the human eye. The principal experiments were made with a pair of fine wires stretched in front of a bright background, which, by means of a micrometer screw, could be brought so close together that the interval between them apparently vanished. VOLKMANN considered this interval as the smallest visible object, and subtracted from its actual value the irradiation fringe by which the width of the wires is apparently increased. Thus he found extraordinarily minute values for the smallest images, apparently very much smaller than the retinal cones. The author, however, is obliged to take exception to VOLKMANN's results, because, as was pointed out above, it is not right to conclude from such experiments that the perceptive elements of the retina are smaller than the image of the space between the wires, but merely that they are smaller than the distance from the middle of one dark band to the middle of the next. These latter distances in VOLKMANN's experiments are not very much smaller than those previously found by other observers.

The experiment with systems of parallel wires, as described above in the text, has been repeated by Dr. HIRSCHMANN, with many variations, in order to find the best conditions. He obtained values of the angle of distinctness in some cases as small as 50 seconds, which means on the retina a width of 0.00365 mm. But the most recent measurements of the diameter of foveal cones are as follows: M. SCHULTZE, 0.0020 to 0.0025 mm; H. MÜLLER, 0.0015 to 0.0020 mm; and WELCKER, 0.0031 to 0.0036 mm. Thus, the cones are minute enough to account for the actual resolving power of the eye.

In other experiments VOLKMANN used letters, figures and other forms of objects, and attempted to establish the fact that the number of cones on which the image of the object fell is not large enough to enable its form to be discerned. But it should be remembered that when the eye is moved, the image of a letter may be formed successively on different groups of cones, in relatively different positions on the single cones; and that differences which perhaps disappear in one position of the image may become clear in another.

The author does not believe, therefore, that we are forced to abandon the view that the retinal cones are the perceptive elements. But it is possible, judging from the most recent observations of M. SCHULTZE, that the rod-like ends of the cones in the yellow spot, turned towards the choroid and separated from each other by black pigment, which measure only 0.00066 mm, may be the only sensory elements, and not the entire cones.

Oculists usually measure visual acuity by means of letters of different size which the patient is made to view from a considerable distance, with spectacle glasses to aid the accommodation. The measure of the visual acuity of the eye is expressed by a fraction, whose numerator is the distance at which those letters are still legible, whereas the denominator is the distance at which they subtend an angle of 5′. These latter distances were used in SNELLEN's test-charts.

According to VROESOM DE HAAN, the visual acuity at ten years of age is 1.1; at forty years, 1.0; and at 80 years, 0.5; showing a gradual decrease with advancing age. But E. JAVAL finds that when astigmatism is corrected and the illumination is good (equal to 500 candles at a distance of one metre) the visual acuity is from 1/4 to 1/3 higher than that given by DE HAAN.

Note by W. NAGEL

In open daylight or with pretty good artificial illumination, the visual acuity as determined by the SNELLEN test above mentioned is found to be on the average higher than unity. Some letters can be recognized better than others; and therefore at present instead of letters it is common to use hook-shaped figures in the form of E or ⊏, and the test consists in saying which side of the figure is open. Just as in the case of the SNELLEN test-type, the visual acuity is put equal to unity when the hook-figure that can just be distinguished subtends a visual angle of 5′. The figures are constructed so that the width of each separate black line is one-fifth the total width of the figure.

Higher values of the visual acuity are obtained by tests with these hooks than with letters and numerals; for example, with good illumination or sky-light, the visual acuity of the normal eye measured in this way averages between 1.5 and 2.0. And it is not rare to find persons with visual acuity even as high as 4, particularly in the case of savages, some of whom, according to measurements of H. COHN, KOTELMANN and G. FRITSCH, certainly have a visual acuity of 6. COHN finds that the average visual acuity of savages is not strikingly greater than that of civilized people; but the more thorough investigations made by G. FRITSCH indicate a distinct, although no very great, superiority among savages in this respect.[1]

In LISTING's schematic eye, according to the data given above in the text, an angle of 60″ corresponds to a length of 0.00438 mm on the retina.

[1] ¶It is well known that the sense perceptions of savages are very extraordinarily developed in some respects, as shown by their astonishing quickness in shooting, hunting, etc. Quality of vision and high visual acuity depend largely on training in youth and practice. HUMBOLDT, describing his adventures in the Andes mountains, relates incidentally how the Indians in his party were able to discern his guide in a white cloak a long way off as a white point moving in front of black basaltic rock walls, before HUMBOLDT himself could make him out through his field glasses; although soon afterwards he also was able to see the guide with his naked eye. The natives were cleverer and quicker in perceiving this faint object, but their actual visual acuity was probably not much higher. Apparently, the visual acuity of mankind has not changed materially in several thousand years. We know from the records of antiquity that the seven stars of the Pleiades appeared the same to former generations as they do today. Stars of the seventh magnitude were invisible to the normal eye then as now. (J. P. C. S.)

With a visual acuity of 5, on the basis of SNELLEN's system of measurement, a letter or hook that could just be discerned would subtend an angle of 60″ or 1′.

Assuming, as HELMHOLTZ did on the basis of KOELLIKER's measurements, that the diameter of the smallest visual element of the retina is 0.0045 mm, the image of the tiniest test-figure that could be discerned by an eye whose visual acuity was equal to 5 would have to be formed on the surface of a single retinal cone. If that were the case, the power of discerning form would be incomprehensible. But, as HELMHOLTZ himself states in his supplement to this section, H. MÜLLER found that the foveal cones are much narrower, their diameters being from 0.0015 to 0.002 mm (which is in accordance with the most recent measurements made by G. FRITSCH). Thus on the assumption of fine cones of this kind, even the high values of the visual acuity recently obtained by COHN and FRITSCH are comprehensible at least for the simple E-figures. (N.)

The researches of AUBERT and FÖRSTER as to the precision of vision in the peripheral parts of the retina were carried out by two methods. In the first method, in order to secure the position of the eye and also to protect it from lateral glare, the observer looked through a firmly clamped black tube at an arc (2 feet wide and 5 feet long) on which various characters, letters and numerals, were marked at equal intervals apart. The contrivance could be moved on rollers, so that after each test the portion of it within sight of the observer could be quickly changed. The letters and numerals were in no regular series of any sort, and so the observer could never guess any numbers except those which he had actually seen. A Leyden jar in front of the arc was discharged from time to time, thereby illuminating the characters for an instant. In the intervals it was so dark that, while the observer was barely able to see the place where the letters were, he could not discern their forms. An assistant adjusted the arc with the letters on it in any position he pleased, and after each inspection the observer told what characters he had recognized. Four arcs were used with characters of different size; and the distance between the observer and the objects could be varied.

The angle subtended by the portion of the arc which contained characters that could be recognized, that is, double the angle between the visual axis and the extreme outside visible letter or figure, AUBERT called the *space-angle* (*Raumwinkel*); and the angle subtended by the longest dimensions of the visible characters he called the *number-angle* (*Zahlenwinkel*). In terms of this nomenclature, the result of the tests was found to be, that *with characters of the same actual size, the ratio between the number-angle and the space-angle was nearly constant*; provided the space-angles were not more than 30° or 40°, in which case the number-angles were rather larger than they should have been in order to obey this rule. On the other hand, *with characters of the same apparent size, small characters nearer the eye were easier to discern than larger characters farther away.* The following table gives the results that were obtained in this case for the ratio between the space-angle and the number-angle:

Actual size of characters in mm	Limiting value of space-angle	Ratio of space-angle to number-angle		
		Minimum	Maximum	Mean
26	25°	7	7.9	7.18
26	40	6	7.3	6.69
13	27	11	12	11.14
7	27	9.7	14.5	12.79

In the second column, under "limiting value of the space-angle," the extreme value is given for which the ratio of the two angles was found to be approximately constant. The last column shows that the ratio between the two angles increases when the actual size of the characters is diminished. The reason for this is difficult to explain. Could it be that the mechanism of accommodation somehow alters the peripheral parts of the retina? AUBERT supposes that in far vision the rods are obliquely disposed in the marginal portions of the retina, and consequently hamper the normal procedure of the rays of light.

In the second method the apparatus represented in Fig. 4 was used with ordinary daylight. *A* is a strip of white lacquered tin, 30 cm long and 5 cm wide, which can be turned around the axle *u* like the wing of a windmill. It can be raised and lowered on the upright *B* which is fastened to the baseboard *C*. The observer's eye is placed at the other end of the board, opposite the axle *u*; his other eye being screened by a piece of black paper *D*, which is fastened to a wooden upright *d* and capable of being turned to the right and left. The axle of the tin strip is 20 cm away from the middle of the line connecting the two eyes. The board *C* has a handle underneath.

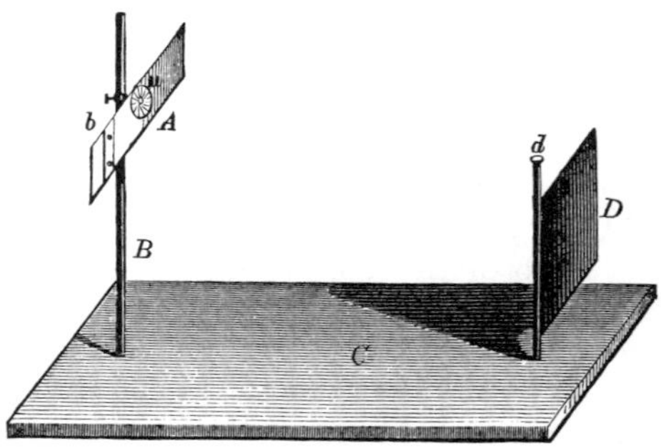

Fig. 4.

The observations were made by putting the nose on the wooden rod d, covering one eye with the screen, supporting the chin on the board in front of the screen and adjusting the axle of the tin strip at the same level with the eyes. Thus stationed, the observer gazes steadily at the middle of the strip (or the tip of the axle). There is a white card b which slides in grooves in the tin strip; and it has two points marked on it. He gradually moves it from one side towards the point of fixation; and as soon as he can distinguish the two points by their images in the periphery of the retina, the card is halted, and the distance between the two points and the point of fixation is read on a scale on the tin strip. The same measurements are repeated with the tin strip inclined to the horizontal at various angles. There were several white cards, each with two round black spots on them, of various sizes and at different distances apart; but the two spots were always symmetrically situated with respect to the axle u.

The results of these measurements for a pair of black spots 2.5 mm in diameter and at a distance of 14.5 mm apart are exhibited in Fig. 5. The continuous contour line is the diagram for AUBERT's eye, and the dotted line for FÖRSTER's eye. The *radii vectores* all intersect in the point corresponding to the point of fixation of the eye; and those that are drawn in the figure indicate the directions of the strip of tin for which actual measurements were made. The points designated by O and U are the upper and lower limits of the field in the vertical meridian, and the points designated by A and I are the temporal and nasal limits in the horizontal meridian. The length of the line ab shows the distance of the eye from the tin strip, which was 20 cm; all the linear dimensions in the diagram being one-fifth actual size.[1] Accordingly, these areas are, first of all, those parts of the field of view within which two dots of the given dimensions and interval apart may be distinguished. The corresponding areas on the retina are obtained by inverting these diagrams. The irregular oval form of the contour is quite different even for the two eyes of the same individual.

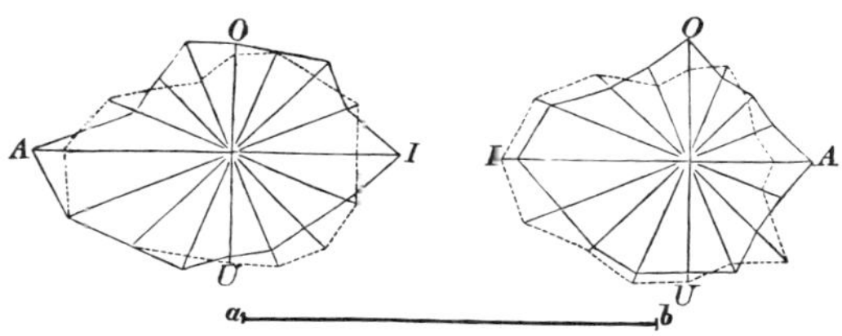

Fig. 5.

[1] AUBERT states that the dimensions in the diagram are reduced one-fourth; but this does not agree with his numerical data.

The average results of measurements with different pairs of dots are shown in Fig. 6; where *a* designates the point of fixation. With each pair of dots the measurements were made on four eyes in eight different meridians. The length of the line *ab* is the average distance between the eye and the tin strip for one pair of dots; similarly, *ac* is the average distance for another pair of dots, etc. The pair of dots at *c* is the pair for which Fig. 5 is drawn. Evidently, the farther away the eye is, the more rapidly the interval increases between the pair of dots. The average results as actually obtained were as follows (distances all given in millimetres):

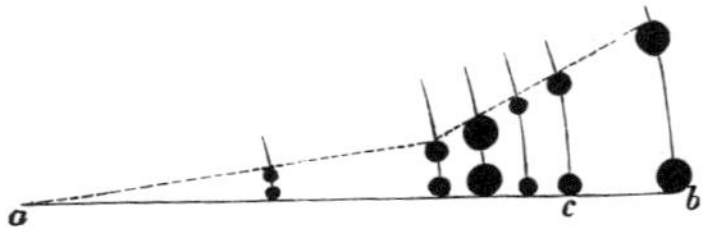

Fig. 6.

Interval between dots	Diameter of dot	Average distance from middle of tin strip
3.25	1.25	31
6.5	2.5	50
9.5	3.75	55
12.	1.25	60
14.5	2.5	65
20.5	3.75	77

Moreover, in these tests both AUBERT and FÖRSTER frequently discovered also places in the retina that were insensitive, small blind spots as it were, where one of the dots or both of them suddenly vanished. At some of these places the blindness was apparently only temporary; but there were others where it was more or less permanent and which could always be found again.

The phenomena of the blind spot were discovered by MARIOTTE who was interested in finding out what sort of vision there was at the place of entrance of the optic nerve. The experiment aroused so much interest that he performed it before King CHARLES II of England in 1668. PICARD modified the experiment in such a way that even with both eyes open, the observer could not see anything. The way he did it was to fasten a piece of paper on the wall, and stand off from it at a distance of about ten feet; then holding his finger right in front of his face, he converged both eyes on it so that its image fell on the blind spot of each eye and therefore was not seen at all. Otherwise, under the same circumstances it would have been seen double. MARIOTTE made still another improvement of the experiment by making two objects vanish with both eyes open. Two bits of paper were fastened at the same level on a wall, three feet apart. The observer stands about twelve or thirteen feet from the wall with his thumb held about eight inches from the eye, so that it hides the left-hand piece of paper from the right eye and the other piece from the left eye. Now if he looks at his thumb, both pieces of paper will disappear because their images fall on the blind spot of that eye from which neither of them is screened by his thumb. LE CAT tried also to calculate the size of the blind spot on the retina, but he found it far too small, namely, from 0.20 to 0.25 Paris line. DANIEL BERNOULLI drew its outline on the floor. The way he did it was to close one eye and hold the cord of a plummet close to the open eye; the plummet itself just missed the floor. Looking vertically down the cord, he tried to find the places on the floor where a coin would have to be put so as just to begin to disappear from view. The figure he got was nearly elliptical. But lacking sufficiently accurate data as to the optical constants of the eye, he got too high a value for the size of the blind spot. According to his calculation its diameter was about one-seventh of that of the eye.

MARIOTTE's discovery led to a lengthy discussion of a question, which, with the meagre knowledge of the nerve-functions at that time, was perhaps bound to arise; namely, as to whether it was really the retina that perceived light, as had been assumed by KEPLER and SCHEINER. MARIOTTE decided that it was the choroid, because this coating was absent in the blind spot, while the fibres of the retina are very numerous there. In fact, a whole line of men notable in optical science, as, for example, MERY, LE CAT, MICHELL and, among more recent ones, BREWSTER, espoused MARIOTTE's view of the matter. Thus, it was argued that, since the retina was transparent, it could not retain the light, and that it was too thick to give a sharp image. LE CAT even tried to show that the choroid was a continuation of the *Pia Mater* of the brain. The sensitivity of the retina to light was defended by PECQUET, DE LA HIRE, HALLER, PORTERFIELD, PERRAULT and ZINN. Curiously enough the chief argument they used to support this view was that the retina was the anatomical development of a very large nerve, whereas the choroid has only a few small nerves. The other arguments that were advanced to support their opinion and to offset the difficulties about MARIOTTE's experiment were not of much value. PORTERFIELD maintained that the optic nerve was still surrounded and permeated by the sinewy sheaths of the nerve at the place where it entered the eye, and was not soft and delicate enough to be sensitive to so nice an agency as light. HALLER also argued that there was no real retina at the entrance of the optic nerve but a white porous membrane which may be unsuitable for vision, without implying that the retina itself is unsuitable. Others, for example, RUDOLPHI, and COCCIUS too at first, believed that the non-sensitive place corresponded merely to the central vessels of the optic nerve; but as soon as the optical constants of the eye were more accurately known, this view was shown to be untenable by authorities like HANNOVER, E. H. WEBER, A. FICK, and P. DU BOIS-REYMOND. J. MÜLLER thought that the matter could be explained on the assumption that the MARIOTTE effect was analogous to the disappearance of the images of coloured objects when they are formed on a white background on the peripheral parts of the retina; which will be discussed later in §23. This is due to fatigue of the retina. But he supposed this occurred very much more rapidly and suddenly at the place of entrance of the optic nerve. The objection to this is that a bright object which emerges suddenly in the invisible gap in the field of view is not perceived at all, and so does not stimulate the visual substance at all; and hence there can be no question of fatigue.

The necessary conclusions from the facts as above set forth were formulated by the writer as long ago as 1851. He took the position then that objective light was incapable of affecting the fibres of the optic nerve as well as the fibres that spread over the anterior surface of the retina. At that time it had not been discovered that there was any anatomical connection between the layer of rods and the nervous elements of the retina; and hence the only assumption that could be made was that the retinal nerve cells or granules were the light-sensitive elements. Soon thereafter H. MÜLLER discovered the radial fibres of the retina connected with the elements by the cones and rods. KOELLIKER showed that they were in the human eye. Both of them conjectured that the elements of the layer of rods were the light-sensitive elements; and, finally, the physiological proof of it was produced by MÜLLER. It is true the same view had been previously put forth by TREVIRANUS, but without sufficient information as to the microscopical structure. He called the light-sensitive elements *nerve papillae*.

Precision of vision has been an object of research ever since the time when telescopes began to be made. HOOKE used the right principle before anybody else, by trying to find out what had to be the angular distance of the components of a double star in order for it to be recognized as such. But nearly all subsequent investigators have tried to ascertain the smallest size of a black dot that can still be made out by the eye; and, naturally, the results obtained have been very various. Among these may be mentioned HEVELIUS, SMITH, JURIN, TOB. MAYER, COURTIVON, MUNCKE, and TREVIRANUS. The influence of illumination in these tests was recognized by JURIN and MAYER. JURIN supposed that the reason why the visual angle has to be greater in order to distinguish two marks as separate from each other than it has to be in order to recognize each mark by itself, was because the eye trembled and hence the images formed on two rods merged together. VOLKMANN gave the reason why the only way to have a constant measure of visual acuity is by the method of separating two distinct objects. Measurements by this method were made by E. H. WEBER, BERGMANN and MARIÉ DAVY.

Blind Spot and Localization of the Light Sensation

1668. MARIOTTE, *Oeuvres*. p. 496–516; also in *Mém. de l'Acad. de Paris 1669 et 1682*. *Phil. Transact.* II. 668. *Acta Eruditorum* 1683. p. 68.
1670. PECQUET, *Phil. Transact.* XIII. 171.
PERRAULT, ibid. XIII. 265.
1694. DE LA HIRE, *Accidens de la vue.*
1704. MERY, *Hist. de l'Acad. de Paris.* 1704.
1709. DE LA HIRE, ibid. 1709. p. 119. 1711. p. 102.
1728. D. BERNOUILLI, *Comment. Petropol. vet.* T. 1. p. 314.
1738. SMITH, *Optics.* Cambridge 1738. Remarks. p. 6.
1740. LE CAT, *Traité des sens.* Rouen. p. 171. 176–180.
1755. ZINN, *Descriptio oculi humani.* p. 37.
1757. HALLER, *Physiologia.* T. V. p. 357, 474.
1759. PORTERFIELD, *On the eye.* II. 252, 254.
1772. MICHELL in PRIESTLEY, *Geschichte der Optik.* 4. Per. 5. Abt. 2. Kap. (German edition, S. 149.)
1819. PURKINJE, *Beobachtungen und Versuche.* I. 70 and 83.
1835. D. BREWSTER, POGGENDORFFS *Ann.* XXIX. 339.
G. R. TREVIRANUS, *Beiträge zur Aufklärung der Erscheinungen und Gesetze des organ. Lebens.* Bremen.
1838. GRIFFIN, Contributions to the physiology of vision. *London medical gazette.* 1838, May. p. 230.
1840. J. MÜLLER, *Handbuch der Physiologie.* II. 370.
1844. VALENTIN, *Lehrbuch der Physiologie.* 1. Ausgabe. II. 444.
1846. VOLKMANN, Art.: Sehen in WAGNERS *Handwörterbuch der Physiol.* III. 272.
1850. A. HANNOVER, *Bidrag til Øjets Anatomie, Physiologie og Pathologie.* Kjöbenhavn. Cap. VI. p. 61.
1851. HELMHOLTZ, *Beschreibung eines Augenspiegels.* Berlin. S. 39.

1852. E. H. Weber, Über den Raumsinn und die Empfindungskreise in der Haut und im Auge. *Verhandl. der Leipz. Gesellsch.* 1852. S. 138.
A. Koelliker, Zur Anatomie und Physiologie der Retina. *Verhandl. d. phys. med. Ges. zu Würzburg.* 3. July 1852.
Donders, *Onderzoekingen gedaan in het physiol. Labor. d. Utrechtsche Hoogeschool.* VI. 134.

1853. D. Brewster, Account of a case of vision without retina. *Report of the British Assoc. at Belfast.* p. 3.
A. Fick and P. du Bois-Reymond, Über die unempfindliche Stelle der Netzhaut im menschlichen Auge. J. Müllers *Archiv für Anat. und Physiol.* 1853. p. 396.
Coccius, *Die Anwendung des Augenspiegels.* Leipzig. S. 20.

1855. H. Müller, *Verhandl. d. phys.-med. Ges. zu Würzburg.* IV. 100. V. 411–446.

1856. Idem, Anatomisch-physiolog. Untersuchungen über die Retina bei Menschen und Tieren. Siebold *und* Köllikers *Zeitschr. für wissensch. Zoologie.* VIII. 1–122.

1857. Aubert and Förster, Über den blinden Fleck und die scharfsehende Stelle im Auge. *Berliner allg. med. Zentralzeitung.* 1857. No. 33. S. 259, 260.

1859. Coccius, *Über Glaukom, Entzündung und die Autopsie mit dem Augenspiegel.* Leipzig. S. 40 and 52.

Acuity of Vision

1705. Hooke, *Posthumous works.* p. 12. 97.

1738. Smith, *Optics.* I. 31.
Jurin, ibid. *Essay on distinct and indistinct vision.* p. 149.

1752. Courtivron, *Hist. de l'Acad. de Paris.* p. 200.

1754. Tob. Mayer, *Comment. Gotting.* IV. 97 and 135.

1759. Porterfield, *On the eye.* II. 58.

1824. Amici in: *Ferussac bull. sc. math.* 1824. p. 221.

1829. Lehot, ibid. XII. 417.

1830. Holke, *Disquis. de acie oculi dextri et sinistri.* Lipsiae.

1831. Ehrenberg in Poggendorffs *Ann.* XXIV. 36.

1840. Hueck in J. Müllers *Archiv für Anat. und Physiol.* 1840. S. 82.
J. Müller, *Handbuch der physiologie.* II. 82.

1841. Burow, *Beiträge zur Physiologie und Physik des menschlichen Auges.* Berlin. S. 38.

1846. Volkmann, art.: Sehen in Wagners *Handwörterbuch d. Physiol.* III. 331, 335.

1849. Marié Davy, *Institut.* No. 790. p. 59.

1850. W. Petrie, *Institut.* No. 886. p. 415.

1852. E. H. Weber, *Verhandl. der sächs. Ges.* 1852. S. 145.

1854. Bergmann, Zur Kenntnis des gelben Flecks der Netzhaut. Henle *und* Pfeuffer, *Zeitschr.* (2) 245–252.

1855. Budge, Beobachtungen über die blinde Stelle der Netzhaut. *Verhandl. des naturhist. Vereins d. Rheinlande.* 1855. S. XLI.

1857. Bergmann in Henle *und* Pfeuffer, *Zeitschr. für rat. Med.* (3) II. 88.
Aubert and Förster in Graefe, *Archiv für Ophthalmologie.* III. Abt. 2. S. 1.

1860. G. Braun, Notiz zur Anatomie der Stäbchenschicht der Netzhaut. *Wien. Ber.* XLII. S. 15–19.

— G. M. Cavallieri, Sul punto cieco dell' occhio. *Atti dell' Istituto Lombardo.* II. 89–91.

1861. H. Müller, Bemerkungen über die Zapfen am gelben Fleck des Menschen. *Würzburger Zeitschrift für Naturk.* II. 218–221.

1862. H. Snellen, *Letterproeven ter bepaling der gezigtsscherpte.* Utrecht.

— J. Vroesom de Haan, *Onderzoek naar den invloed van den leeftijd op de gezigtsscherpte.* Utrecht.

— A. W. Volkmann, *Physiologische Untersuchungen im Gebiete der Optik.* Leipzig. Heft 1, S. 65.

1863. Wittich, Studien über den blinden Fleck. *Archiv für Ophthalm.* IX. 3. S. 1–38.

— K. Vierordt, Über die Messung der Sehschärfe. Ibid. S. 219–223.

1864. AUBERT, *Physiologie der Netzhaut.* Breslau. S. 187–251.
— W. ZEHENDER, Historische Notiz zur Lehre vom blinden Fleck. *Archiv für Ophthalm.* X. 1. S. 152–155.
— O. FUNKE, Zur Lehre von den Empfindungskreisen der Netzhaut. *Bericht der naturforsch. Ges. zu Freiburg i. Br.* III. S. 89–116.
— DONDERS, *Anomalies of accommodation and refraction.* London. p. 177–203.

Following references (compiled by H. L.) are to some recent literature concerning Visual Acuity:

1914. H. LAURENS, Über die räumliche Unterscheidungsfähigkeit beim Dämmerungssehen. *Zeit. f. Sinnesphysiol.* **48**, 233–239.
1917. A. GLEICHEN, Beitrag zur Theorie der Sehenscharfe. GRAEFES *Arch. f. Ophth.*, **93**, 303–356.
1920. J. W. FRENCH, The unaided eye. III. *Trans. of the Optical Soc.* **21**, 127–147.
1920. J. P. C. SOUTHALL, Refraction and visual acuity of the human eye. *Amer. Jour. Physiol. Optics*, **1**, 277–316.
1921. C. SHEARD, Some factors affecting visual acuity. *Amer. Jour. Physiol. Optics*, **2**, 168–184.
1921. C. E. FERREE and G. RAND, An acuity lantern and the use of the illumination scale for the detection of small errors in refraction and their correction. *Proc. of the Amer. Psychol. Assoc. Psychol. Bull.* **17**, 46, 47. — An apparatus for determining acuity at low illuminations, for testing the light and colorsense and for detecting small errors in refraction and in their correction. *Journ. Exper. Psychol.*, **3**, 59–71. — Lantern and apparatus for testing the light sense and for determining acuity at low illuminations. *Amer. Jour. of Ophthalmol.*, **3**, 335–340. — An apparatus for testing light and color sense. *Amer. Jour. of Ophthalmol.*, **3**, 812–814. — Visual acuity at low illumination and the use of the illumination scale for the detection of small errors in refraction. *Amer. Jour, of Ophthalmol.*, **3**, 408–417.
1922. H. HARTRIDGE, Visual acuity and the resolving power of the eye. *Jour. Physiol.*, **57**, 52–67.
1922. P. W. COBB, Individual variations in retinal sensitivity and their correlation with ophthalmologic findings. *Jour. Exper. Psychol.*, **5**, 227–246.
1923. E. E. ANDERSEN and F. W. WEYMOUTH, Visual perception and the retinal mosaic. *Amer. Jour. Physiol.*, **64**, 561–594.

A Supplement by W. NAGEL

§18. A. Changes in the Retina due to Light

There are certain positive effects produced in the retina that are manifestly dependent on the action of light.

1. *Structural Changes*

When the retina of the eye is subjected to microscopical examination, numerous differences are found between preparations which have been kept in the dark before fixing and staining and those which have been exposed to light. In the latter case the nuclei in the various layers and the cone-ellipsoids will not take up the stain in the same way as they do when the retina has not been exposed to light. For example, after exposure to light acid dyes do not produce so marked a stain.[1]

Changes also have been found in the ganglion cells of the retina; particularly by BIRCH-HIRSCHFELD[2] in treatment of preparations by the NISSL method of staining. With exposure to greater illumination these changes may be even great enough to lead to the formation of vacuoles, etc. In such cases it is not certain whether the result is due to direct action of light on the ganglion cells and cone nuclei, or whether this excessive stimulation of light takes place simply in the cones, while the changes in the nerve cells are after-effects of conduction of powerful stimuli. It is more likely, perhaps, direct action of light.

ANGELUCCI[3] found the chemical reaction of the retina was alkaline in the dark and acid in the light; and subsequent researches have confirmed this result.[4]

The phenomena of the phototropic migration of pigment and contraction of the cones under the influence of light have been studied by many investigators. These effects are particularly interesting because the whole microscopical appearance of the retina may be essentially changed as a result of these processes.

F. BOLL[5] found that the retina of a frog's eye that had been kept in the dark for several hours may be easily removed from the eyeball as a translucent membrane; but that when the eye was exposed to light, the retina clung fast to it and usually would not come off except in pieces, which are deep black in appearance. In the latter case, the pigment epithelium remains attached to the layer of rods and cones, whereas in the former case it does not. BOLL,[5] CZERNY,[6] ANGELUCCI,[7]

[1] See BIRNBACHER, v. GRAEFES *Arch. f. Ophthalm.* **40**, 1894. Summary of the literature is given by S. GARTEN, GRAEFE-SÄMISCHS *Handbuch der gesamten Augenheilkunde.* I. Teil, III. Band, XII. Kapitel, Anhang.

¶See also brief review of literature on this subject as given by S. R. DETWILER, The effect of light on the retina of the tortoise and the lizard. *Jour. Exper. Zool.*, **20**, 1916, 165–191. (H. L.)

[2] v. GRAEFES *Archiv. f. Ophthalm.* **50**, 1900 and *Arch. f. Anat. u. Physiol.* 1878.

[3] *Encyclopédie française d'opht.* II. S. 108.

[4] ¶See R. DITTLER, Über die Zapfencontraktion an der isolierten Froschnetzhaut. PFLÜGERS *Archiv.*, **117**, 1907, 295–328. Idem, Über die chemische Reaktion der isolierten Froschnetzhaut. Ibid., **120**, 44–50. — S. GARTEN, Die Veränderungen der Netzhaut durch Licht. GRAEFE-SÄMISCH *Handb. d. ges. Augenheilkunde* I. Teil, III. Band, XII. Kap., Anhang. S. 1–130. — G. F. ROCHAT, Über die chemische Reaktion der Netzhaut. *Arch. f. Ophthalm.*, **59**, 1904, 170–188 (in which the position is taken that light does not cause any change in the reaction of the retina). — L. B. AREY, The movements of the visual cells and retinal pigment of the lower vertebrates. *Jour. Comp. Neur.*, **26**, 1916, 121–202. (H.L.)

[5] *Monatsberichte d. K. Akad. d. Wiss. Berlin* 1877, January.

[6] *Sitzungsber. d. K. d. Wiss. Wien* LVI. 1867.

[7] *Arch. f. Anat. u. Physiol.* 1878.

and KÜHNE[1] were all aware that the pigment of the epithelial cells, which consists of fine brown granules and needles, may migrate in between the rods, and that then the individual rods are surrounded by a thick mantle of pigment. What occurs here is not some sort of amoeboid extension and contraction of protoplasmic processes of the pigment-cells, as might be supposed at first thought; for these processes always project between the rods, even when the eye is protected from light. In the dark the pigment simply migrates into the cell-bodies close to the nucleus; but under illumination it migrates away from the light into the processes between the rods. On exposure to light the rods also become somewhat thicker, and consequently the rods and pigment-processes adhere to each other firmly, so that when the retina is removed from the open bulbus, the pigment-layer is torn away with it.

The forward migration of the pigment due to exposure to light takes place more quickly than the return movement in darkness. In sunlight the completed light-position is assumed in about ten minutes, whereas the typical dark-position is not completed for an hour or more. Fifteen minutes is sufficient to complete the migration in an enucleated dark eye in sunlight, while the dark reaction requires a considerably longer time for its completion.[2]

Pigment migration may be observed very distinctly not only in amphibians but in fishes and birds. It is not so distinct in reptiles.[3] Nor can it be said to have been certainly demonstrated in mammals,[4] although likewise in them the retina clings more tenaciously to the choroid when the eye has been illuminated.

The short wave-lengths in the blue part of the spectrum seem to have more effect on pigment migration than the long or red ones. The eye itself being kept dark, it is sometimes sufficient merely to expose the posterior end of the animal in order for the pigment to assume its light-position (ENGELMANN).[5] Various other influences, such as heat, low temperature,[6] irritation of the membrane may also bring about the light-position in the dark, perhaps as a reflex action. But direct action of light on the retina is possible quite independently of the migration

[1] *Untersuchungen aus dem Heidelberger physiol. Institut.*

[2] ¶See L. B. AREY, The function of the efferent fibres of the optic nerve of fishes. *Jour. Comp. Neur.*, **26**, 1916, 213–245. (H. L.)

[3] ¶See S. R. DETWILER, *loc. cit.* Also H. LAURENS and S. R. DETWILER, Studies on the retina. The structure of the retina of Alligator mississippiensis and its photochemical changes. *Jour. Exper. Zool.*, **32**, 1922, 207–234. (H. L.)

[4] ¶See GARTEN, *loc. cit.*, p. 72. (H. L.)

[5] ¶According to L. B. AREY (*loc. cit.*, 1916), this is not the case. See also A. E. FICK, Über die Ursachen der Pigmentwanderung in der Netzhaut. *Viertelj. Naturf. Gesell. in Zürich*, Jahrb. 35, 1890, 83–86. (H. L.)

[6] ¶See L. B. AREY, *loc. cit.*—H. HERZOG, Experimentelle Untersuchungen zur Physiologie der Bewegungsvorgange in der Netzhaut. *Arch. f. Physiol.*, 1905, 413–464. (H. L.)

of pigment in the enucleated eye, as is shown by an experiment described by KÜHNE. He contrived to illuminate the eye of a frog so that, while some parts of the retina were very bright, the other parts got as little light as possible; and he found that the pigment adhered only to the bright parts and not to the dark places.[1]

A change which occurs in the cones of the retina under the action of light and which is easy to demonstrate was discovered by VAN GENDEREN STORT.[2] This consists in a shortening of the inner segment on exposure to light. The effect is very considerable in the case of a frog, the inner segment being shortened more than 50 percent. It is even greater in many fishes. GARTEN observed a shortening of from 50μ to 5μ in the shiner (Abramis brama).[3] Apparently, this reaction does not take place in the eel. In reptiles and birds the amount of shortening is much less.[4] In mammals it is very slight, but GARTEN seems to have shown that it does certainly occur in monkeys. So far as the human eye is concerned, there is no proof of it.

In amphibians and fishes the displacements produced by shortening of the inner segment are so considerable that the cones, whose outer segment in the dark eye is in the outside zone of the layer of rods, are found in the light eye up against the external limiting membrane. However, many of the cones are stationary, for example, in the frog's eye. These stationary cones are generally near the limiting membrane.

Cone contraction takes place more quickly than pigment migration. For example, when the eye of a frog is exposed to bright daylight, the contraction is complete in two minutes. The sensitivity of the reaction is also much greater; the light-position of the cones being brought about by the light of a candle, although the forward migration of pigment is not affected at all, or very slightly anyhow, by such an illumination.[5]

According to HERZOG,[6] cone contraction is induced not only by direct stimulation of light but by all kinds of irritations of the membrane, in a much greater degree than pigment migration. Ignorance of this fact may be responsible for the erroneous idea that displacements of individual parts of the retina are due to the action of light.

[1] ¶R. DITTLER, *loc. cit.* (H. L.)

[2] *Onderzoek. Physiol. Labor. Utrecht* (3), **9**, 145, 1883.

[3] ¶L. B. AREY, *loc. cit.* (H. L.)

[4] ¶C. HESS, Gesichtsinn in WINTERSTEINS *Handb. d. Vergl. Physiol.*, Bd. 4 (1913), 744, 751. — GARTEN, *loc. cit.* — DETWILER, *loc. cit.* — LAURENS and DETWILER, *loc. cit.* (H. L.)

[5] ¶L. B. AREY, A retinal mechanism of efficient vision. *Jour. Comp. Neur.*, **30**, 343–353. (H. L.)

[6] *Arch. f. (Anat. u.) Physiol.* 1905, 413

It is still a moot question as to whether stimulating one eye by light can bring about a similar light-position of the cones of the other eye, when the latter has not been exposed. ENGELMANN[1] believes that it can, but A. E. FICK[2] denies it. It is not easy to avoid sources of error in carrying out the experiment.

Any theoretical explanation of the photo-mechanical movements of the retinal pigment and cones must be advanced with some hesitation. The first thing to bear in mind is that neither of these processes has been absolutely proved so far as the retina of the human eye is concerned; and that even in the case of mammals, where the tests are more easily carried out, usually there are just faint traces of these phenomena, which are much less in evidence than they are in fishes and amphibians. The fact is, there is not a single positive proof that the changes in the mutual arrangement of the parts of the retina, as they are found to occur in microscopical preparations of light-adapted and dark-adapted eyes, really proceed in the same way during life. With very many other contractile tissues and also with individual contractile cells, such as amoeba, it is impossible or extremely accidental to get fixation in the state either of contraction or of expansion as may be desired. In chemical fixation and in the isolation of elements in maceration preparations violent stimulations of the tissues cannot be avoided; and there is always the possibility that with the microscopical preparation light and darkness have merely a quantitative or qualitative effect on the receptivity of the stimulation by the pigment cells and cones. However, supposing (as is plausible) that the displacements in the normal retina *in situ* take place in the same way as they do in microscopical preparations, we must try to connect these phenomena with the change of excitability of the visual apparatus in the transition from light adaptation to dark adaptation. Many different conjectures of this kind have been made. One question is whether the pigment movement has anything to do with the formation of the visual purple. On the other hand, the isolation of the rods by the insertion of pigment between them has been supposed to be a protection against disturbing diffusion of light in the visual epithelium. Detailed discussion of these questions had better be postponed until we come to consider the more modern theories concerning the functions of the rods and cones.

[1] *Arch. f. d. ges. Physiol.* **35**, (1885).

[2] *Vierteljahrsschr. naturforsch. Ges. Zürich* **35**, 1890. **40**, 1894—v. GRAEFES *Arch. f. Ophthalm.* **37**, 1891.

2. *The Bleaching of the Visual Purple*

The purple or rose-red colouration of the outer segments of the rods, which, under some circumstances, makes the entire isolated retina appear coloured, is evanescent under the action of light, or photolabile. This red colouring matter had already been noticed by H. MÜLLER, LEYDIG and MAX SCHULTZE. Its most interesting property, namely, its high sensitivity to light, was discovered by FR. BOLL.[1] More precise data concerning the "visual purple" or "visual red" were obtained by the researches of KÜHNE,[2] KÖNIG,[3] ABELSDORF,[4] TRENDELENBURG,[5] GARTEN,[6] and other workers.[7]

Visual purple has been found in the rods of man and of all vertebrates that have been investigated. Where there are no rods, there is also no visual purple as in the retina of many birds and reptiles, as well as in the rod-free area (*fovea centralis*) of the human retina.[8] In many animals only parts of the retina contain much purple. In the rabbit there is a horizontal streak—the so-called purple ridge—that is particularly deeply coloured. In the rods of the outer margin of the retina, near the ora serrata, no purple is found.

Besides the red rods, isolated green rods have been seen in the case of the frog.

With a view to seeing the purple colour, the animal should be kept in darkness two hours or longer before it is killed, and the retina then removed from the eye. Owing to the sensitiveness of the colouring matter to green and blue light, the preparation of the retina should be made under illumination by red light or the yellow light of a sodium flame. The entire retina of a frog is easily removed by seizing it with a pointed forceps at the entrance of the optic nerve. The retina of an owl, which is extremely rich in visual purple, can be removed by open-

[1] *Berlin. Monatsber.* 12. Nov. 1876; *Acad. dei Lincei* 3. Dec. 1876; *Arch. f. Anat. u. Physiol.* 1887.

[2] *Untersuchungen d. physiol. Instituts Heidelberg.* II. III. IV; also summary in HERMANNS *Handbuch der Physiologie.* II. 1879

[3] A. KÖNIG, *Gesammelte Abhandlungen.* XXIV. Über den menschlichen Sehpurpur und seine Bedeutung für das Sehen (Also in: *Sitzungsber. Akad. Wiss. Berlin.* XXX. 1894).

[4] *Sitzungsber. Akad. Wiss. Berlin.* XXXVIII. 1895; *Zeitschr. f. Psychol. und Physiol. d. Sinnesorgane.* **12,** 1896.

[5] *Zeitschr. f. Physiol. u. Psychol. d. Sinnesorg.* **37,** 1904.

[6] GRÄFE-SÄMISCHS *Handbuch der Augenheilkunde,* 1. Teil, III. Band, XII. Kapitel.

[7] ¶Some recent work in this subject is to be found in the following: V. HENRI, Photochimie de la rétine *Jour. Physiol. et Path. gen.,* **13,** 1911, 841–856. — S. HECHT, Photochemistry of visual purple. I. The kinetics of the decomposition of visual purple by light. *Jour. Gen. Physiol.* **3,** 1920, 1–13; and by same author, The effect of temperature on the bleaching of visual purple by light. *Jour. Gen. Physiol.,* **3,** 1921, 285–290. (H. L.)

[8] ¶EDRIDGE-GREEN and DEVEREUX deny that the visual purple is absent from the fovea of the retina of the monkey's eye (*Trans. Ophth. Soc.,* **22,** 1902, 300), but their observations lack confirmation. See PARSONS *Colour Vision,* 12. (J. P. C. S.)

ing the eye under water (or physiological salt solution) and cutting away the attachment to the optic nerve. The same technique is used with the eyes of mammals. However, in the latter case, after the eye has been sectioned equatorially, it is well to let it stay an hour in 4 percent alum solution to harden the retina and prevent it from tearing. The isolated retina is then laid, rod-side outwards, on a plate of ground glass or, better still, on a little porcelain dish of the same curvature as the eyeball.

In the eye of an animal that has just been killed or paralyzed by curari, a strong contrast image can be produced by pointing the eye, say, at the cross-bar of a window. Under certain circumstances, KÜHNE got in this way sharply delineated bleaching effects or so-called *optograms*, which are miniature reproductions of the objects depicted. The parts that have been exposed to light are bleached, whereas the shaded portions remain distinctly red.

Good optograms are not so easy to obtain in a frog's eye, because the illuminated parts of the retina stick so fast to the pigment epithelium that it cannot be taken out entire and laid on a little porcelain knob. The retina comes away more easily when the frog is first curarized and then made oedematous by being kept in water for several hours. According to S. GARTEN, quinine poisoning is similarly effective.

Visual purple cannot be detected by the ophthalmoscope in the living eye of either man or mammal of any kind, because the transparent retina is seen either on a very darkly pigmented background or on one that shines with a bright colour (the tapetum of predatory animals). However, there are some animals that have a white, or almost white, tapetum; and with these ABELSDORF[1] succeeded in getting ophthalmoscopic proof not only of the unbleached visual purple but also of its bleaching in bright light. The shiner (Abramis brama) among fishes and the crocodile (Alligator lucius) among reptiles are particularly favourable for this observation.

The visual purple is not soluble in water. Alcohol, ether and chloroform, as well as most acids and alkalies, quickly destroy its colour. On the other hand, it is very soluble in solutions of the salts of the bile acids, which dissolve the substance of the rods almost instantaneously.

If unbleached retinas are placed in the dark in a 2 to 5 percent solution of sodium glycocolate, and if the liquid is then filtered and centrifuged, a clear solution will be obtained that shows the purple

[1] *Sitzungsber. Akad. Wiss. Berlin.* XVIII. 1895; *Zeitschr. f. Psychol. u. Physiol. d. Sinnesorg.* **14**, 1897; *Arch. f. (Anat. u.) Physiol.* 1898.

colour very distinctly, when it is evaporated in vacuum over sulphuric acid down to a few drops. This solution is also photo-sensitive. In darkness it keeps its colour; in light it bleaches, in a few minutes in daylight, but more slowly under artificial illumination.

Visual purple both in the fresh retina and in solution has not the same hue in all animals. Thus in the frog and cat it is almost red; in owls and fishes it is a purple containing much violet. In man also, according to KÜHNE, it is more violet than it is in the frog.

The process of bleaching may run through different gamuts of colour. Thus under some circumstances the colour gets more and more whitish until finally the retina is almost devoid of colour. In other instances, the purple changes to red, orange and yellow, in succession, not becoming white for a long time. This second case, when a distinct yellow tint is one of the stages, is characteristic of rapid bleaching under brilliant illumination (sunlight), as GARTEN has shown; whereas simple fading out occurs when the bleaching is pretty gradual.

The visual purple is converted into a very light-stable yellow pigment when it is treated with certain metallic salts, such as zinc chloride or platinum chloride, or with acetic acid. In some circumstances distinct red colouration is left on the retina by action of formaldehyde.

The quantitative results on the absorption of spectrum light by visual purple, as determined by KÜHNE and by KÖNIG and his pupils, and recently by TRENDELENBURG and GARTEN, are of particular interest. Clear solutions were used in making these measurements, and during the determination of the absorption they must be protected from bleaching as much as possible. If the absorption values are represented by a curve plotted with wave-lengths as abscissae, and the amount of absorption of the separate kinds of light as ordinates, a distinct maximum of absorption is found in the green part of the spectrum.

In the fish retina, which is more of a violet purple, the maximum of absorption is more towards the yellow-green, according to the measurements of KOETTGEN and ABELSDORF.[1]

A comparison of the absorption of unbleached solutions with that of solutions which have been more or less bleached shows first of all a diminution of absorption in the green as a regular effect of the action of light. When the conditions were such that the colour of the retina changed to yellow-red, GARTEN[2] found, along with diminution of absorption in the green, increased absorption in the

[1] *Sitzber. Akad. Wiss. Berlin*, XXXVIII. 1895.

[2] v. GRAEFES *Arch. f. Ophthalm.* LXIII. 1906.

violet. The conclusion was that a yellow decomposition product (KÜHNE's "visual yellow") was the result of the bleaching of the purple. It is not unlikely that some such yellow material may also be produced in slight quantities in the live retina. However, according to the evidence of experiments on the retinas of animals and with solutions of visual purple, an assumption of this kind is not justifiable except in the unusual instance where a retina, which has been kept in the dark for a long period so as to develop a large supply of visual purple, is suddenly exposed to a very strong light.

Proceeding on the assumption that in typical total colour-blindness the formation and bleaching of visual purple takes place normally, the writer has carried out experiments on a totally colour-blind girl which, had they succeeded, might have resulted in proving the development of a "visual yellow" intra vitam. The girl was required to equate green and violet with the HELMHOLTZ spectrophotometer, which she did easily of course. Now if under certain conditions her retina contained a more yellowish pigment than under other circumstances, this might be indicated eventually by the fact that the green-violet adjustment which was right under one set of conditions would be wrong under the other conditions. But the writer could get no uniformity in this respect, although comparisons were made: First, between one eye that had been dark-adapted for one hour and the other eye that in the meantime had been exposed to bright daylight (of course, taking into consideration the very much greater sensitivity of the dark eye, for which both halves of the comparison-field were proportionately dimmed); and, second, between one eye, which was in the average state of light adaptation in daylight, and the other eye which had been kept closed for a long time and was then quickly light-adapted under the action of a bright AUER lamp. The object in view was to compare in this way the slow and rapid bleaching of the visual purple. But one circumstance that operated against the successful performance of this experiment is that an eye of a totally colour-blind person that has been really brilliantly illuminated can generally not see anything at first (for example, after being illuminated by an ophthalmoscope, pupil dilated); and therefore the best that can be done under the circumstances is to use a fairly moderate degree of light adaptation.

The more a certain kind of light is absorbed by the visual purple, the greater is its bleaching effect. KÜHNE found that yellow-green light bleaches visual purple most rapidly, while yellow and red act very slowly. Recently W. TRENDELENBURG has made careful experiments with spectral light. He exposed one of two samples of visual purple to the constant illumination of light corresponding to the sodium line ($\lambda = 589\mu\mu$) and the other sample to some other definite kind of light of the same dispersion spectrum, and then, after a certain interval of time, measured the decrease of the absorption by means of the spectrophotometer. The following table gives TRENDELENBURG's "bleaching values" for rabbit visual purple.[1] The value for sodium light is put equal to unity.

[1] ¶H. LAURENS finds maximum of absorption for wave-lengths of equal energy for visual purple of frog at $510\mu\mu$. The method used is described in a paper by H. LAURENS and H. D. HOOKER, Jr., in *Amer. Journ. Physiol.*, **44,** 1917, 504-516. (H. L.)

Wave-length	589	542	530	519	509	491	474	459
Bleaching value	1	3.40	3.62	3.45	3.09	1 69	0.975	0.299

Both in the live eye and also, under some circumstances, in the enucleated eye, there is a *regeneration of visual purple* after bleaching; and to a certain extent even in isolated retinas and solutions. When both eyes of a live frog have been exposed to sunlight for half an hour, and the animal has then been killed and the eyeballs taken out, the retina of the eye that is opened immediately will be found to be without colour; but if the other eye has been kept an hour in the dark in a damp receptacle, the retina will be a purple-red. In the case of the frog, KÜHNE detected the first trace of red after complete bleaching twenty minutes after shutting off the light; whereas in the case of the rabbit there were signs of this colour in about five minutes. The regeneration is by far the best and most complete when the retina is in contact with the pigment epithelium. A retina taken from an eye that is without pigment never regains the perfect red colour.

According to KÜHNE and GARTEN, the most favourable condition for the regeneration of the purple in the isolated retina was when the bleaching had been permitted to proceed as far as the yellow, the retina then being placed in the dark. Apparently therefore, the visual purple is most easily produced anew from the products of its own decomposition before they have lost all colour. When the retina has been bleached completely, regeneration does not pass through all the intermediate stages of yellow, orange and red, but the retina becomes bright lilac, and then pink. In this case, therefore, the process of formation of the purple must be different from that when the purple is recovered from the yellowish product.

Both bleaching and regeneration depend on the temperature.[1] Regeneration, in particular, is much retarded by cold; for example, the retina of a frog at 0° C takes nine hours to regain its purple colour. In warm-blooded animals the regenerative ability is lost a few minutes after death or after circulation ceases. Evidently, the damage is greatest in this case to the pigment epithelium which is so important for regeneration. Whatever our knowledge may be as to the physiology of visual purple in solution and in the isolated retina, it is doubtful how far it can be applied in the case of the eye of a warm-blooded animal with circulation intact.

The fluorescence of the retina[2] when radiated by ultra-violet light is another remarkable property. It is much more pronounced in

[1] ¶See HECHT, *loc. cit.* (H. L.)

[2] HELMHOLTZ, POGG. *Ann.* XCIV (1855); SETSCHENOW, V. GRAEFES *Arch. f. Ophthalm.* V. 1859.

the bleached retina. This is true, as HIMSTEDT and NAGEL found,[1] also with respect to the retina of the pigeon's eye, which certainly contains very little purple. Solutions of purple in bile acid likewise fluoresce. However, bile acid salt solutions are themselves fluorescent, indeed nearly as much so as when they contain unbleached visual purple of a frog's eye. But if a few drops of a solution of sodium glycocolate and of a similar solution containing visual purple are suspended in little platinum dishes and exposed to daylight, what happens is that the solution with the purple in it is subsequently distinctly more fluorescent in the dark than the other solution. Therefore the bleached products of the visual purple are certainly fluorescent, even if there were some doubt as to the visual purple itself.

3. *Electromotive Phenomena in the Eye*

With all vertebrates that have been investigated by electrical methods, there is found to be a difference of potential between the anterior and posterior poles of the eye, both in the living eye as a whole and in the isolated retina as long as it continues to survive; as was observed by E. DU BOIS-REYMOND. If two places in a prepared specimen are placed in contact with suitable electrodes and connected with a sensitive galvanometer, a continuous current will flow through the circuit as long as the preparation stays alive. Most of these experiments have been made on eyes of frogs. This so-called "*Ruhestrom*" ("current of rest") can still be detected for hours after the eye has been taken out of the body of the dead frog. The severed end of the optic nerve is negative with respect to the anterior part of the eye. On the other hand, the optic nerve is positive with respect to the posterior lateral parts of the bulbus.[2]

According to the experiments of KÜHNE[3] and of STEINER,[4] when one electrode is placed on the inner surface of an isolated retina and the other on the outer surface, the rod-layer is found to be electronegative with respect to the layer of nerve fibres.

While a current can be obtained for hours from an enucleated frog's eye, the electromotive force in a fish's eye, and also in the eye of a warm-blooded animal, dies out usually in a few minutes when the blood ceases to circulate.[5]

[1] *Festschr. d. Albert-Ludwigs Universität Freiburg f. Grossherzog Friedrich.* 1902.

[2] ¶See J. H. PARSONS, *An introduction to the study of colour vision.* 1915. p. 15. (J. P. C. S.)

[3] *Untersuchungen über tierische Elektrizität.* II. Abteil. 1. Berlin 1849.

[4] *Untersuch. d. physiol. Inst. Heidelberg.* III and IV.

[5] ¶Some recent literature pertaining to this subject may be noted here as follows:

E. C. DAY, Photoelectric currents in the eye of the fish. *Amer. Jour. Physiol.*, **38, 1915,** 369–397. — C. SHEARD and C. MCPEEK, On the electrical response of the eye to stimulation

The electromotive force varies considerably in different individuals of the same species. For example, HIMSTEDT and NAGEL[1] found variations between 0.0056 and 0.017 volt in frogs' eyes. The values obtained by other investigators are on the average between 7 and 9 millivolts.

But even with the same specimen the current does not stay constant, nor does it diminish uniformly as a rule. For no apparent reason it increases and decreases again. The fluctuation is slow with frogs, taking several minutes; but in birds, especially in pigeons, it is often quick and apparently without any rule, the galvanometer-needle hardly ever being still for a second. When the experiment is long continued, the direction of the current may be reversed, even in the case of the frog.

The above has reference to the so-called "dark current," the eye being supposed to be kept in the dark. If light falls suddenly on an eye of this sort, the current intensity fluctuates, differently, however, under different conditions. This was discovered first by HOLMGREN,[2] and afterwards independently by DEWAR and MCKENDRICK.[3] The following phenomena are most easily demonstrated on the enucleated frog's eye, one electrode being placed on the edge of the cornea and the other on the section of the optic nerve. After an interval of from one to two tenths of a second, the current increases rather quickly until it is between 3 and 10 percent of the strength of the dark current. If the stimulation by light continues longer, in the case of an eye that has been previously made right sensitive by dark adaptation, the current slowly increases still more. With illumination by a bright incandescent lamp, the increase of current can be observed for a minute, and then it begins to fall off, *even if the illumination is continued.* When the illumination is feeble, the current goes on increasing for a longer time. In enucleated eyes taken from light-adapted frogs, the current quickly reaches its maximum strength and then falls off almost just as rapidly again, although as a rule it never quite sinks as low as the dark current.

If the stimulating light is suddenly shut off, a "dark response" occurs after a latent period, as in the light reaction. It is manifested by another quick increase in the flow of current, succeeded by a fairly slow falling off to the strength of the dark current.

by light of various wave-lengths. *Amer. Jour. Physiol.*, **48**, 1919, 45–66. — C. SHEARD, Photoelectric currents in the eye. *Physiol. Review*, **1** (1), 1921, 84–111. —SHEARD used freshly excised eyes from young dogs, never starting an experiment later than two hours after enucleation. (H. L.)

[1] *Ber. naturf. Gesellsch. Freiburg i. Br.* 1901; *Ann. d. Physik.* (4) IV, 1901.

[2] *Upsala Läkares Förhandlingar.* 1866 and 1871.

[3] *Phil. Trans. Roy. Soc. Edinburgh* VII. 1871–72.

The reaction takes place differently in other animals. In reptiles and in diurnal birds (birds of prey, hens), on illumination, there is sometimes a latent period lasting from 1/40 to 1/15 second, succeeded by a strong *negative* variation of the current; or else what happens is, at first a short positive "discharge," succeeded then by the negative variation lasting while the illumination continues. If the eye is not illuminated, the current may drop at once to the strength of the "current of rest," or, before doing this, it will show another negative variation. H. PIPER's work[1] has helped a great deal to clarify this subject. With respect to nocturnal birds of prey, he confirmed the observation of HIMSTEDT and NAGEL, namely, that the only result of illumination in this case is a strong positive variation followed by an equally strong negative variation when the light is shut off. In mammals also the reaction consists chiefly of a positive variation.

Every injury of the eyeball changes its electromotive behaviour and tends to promote negative variations. For example, the retina taken from the eye of a frog responds to the light stimulus at first with negative variation succeeded afterwards by a positive one. The positive dark variation occurs also in injured preparations of this kind. In perfectly fresh eyes of various animals GARTEN found that a very brief negative variation preceded the positive variation of the current as a regular result of stimulation.

The sensitivity of these photo-electric reactions is sometimes very considerable, particularly in such animals as have numerous rods in their retinas and much visual purple. For example, in the eye of a frog the threshold value of the energy that is just sufficient to produce variation of current is perhaps very nearly the same as that which will just elicit the sensation of light in the dark-adapted human eye. Under stimulation by X-rays the eyes of frogs and of several species of owls also give a distinct photo-electric reaction, but a hen's eye does not. The light of a cigar, moonlight, phosphorescent paint, each produce distinct photo-electric variations. Ultra-violet light has the same effect, evidently due to production of fluorescence in the ocular media.

By a careful study of the quantitative relation between the retinal current and the intensity of the stimulating light, DE HAAS[2] showed that the reaction does not obey the WEBER-FECHNER law except for a certain range of rather strong stimuli. For weaker stimuli the relation between current and stimulus is more complicated. The surprisingly long duration of the current variation that is observed after a brief "instantaneous" stimulus of sufficient intensity is curious. The

[1] *Arch. f. (Anat. u.) Physiol.* 1905. Supplement.

[2] *Lichtprikkels en retinastroomen in hun quantitatief verband.* Inaug.-Diss. Leiden 1903.

reaction may last a hundred times as long as the duration of the stimulus.[1]

The effects of light of different wave-lengths have been quantitatively compared and the distribution of the stimulus-value ascertained for the animal eye in the different parts of the spectrum. This is done by exposing the eye for a certain length of time to the various colours of the spectrum in succession and measuring the corresponding deflections of the galvanometer. Provided the periods of the single exposures are not too brief, and a sufficient interval occurs between successive exposures, the magnitude of the total deflection for each exposure is a measure of the specific stimulating effect of the kind of light in question. This was the method of finding the relative stimulus-value that was used by HIMSTEDT and NAGEL[2] on the frog; and by PIPER[3] on a number of warm-blooded animals. The results showed that the photo-electric reaction is distinctly different in different animals. Indeed, different stimulus-values are found in the same animal in the states of light adaptation and dark adaptation. In the case of the dark-adapted eye of a frog, the maximum stimulus-value occurs in the yellow-green for $\lambda = 544 \mu\mu$; whereas in the same eye, light-adapted, the maximum is in the bright yellow at $590 \mu\mu$. Of the various birds examined, the nocturnal birds (different kinds of owls) have a maximum between 535 and $540 \mu\mu$; diurnal birds (mousehawk, hen, pigeon) around $600 \mu\mu$. The maximum for dogs, cats, and rabbits is somewhere near $535 \mu\mu$. This same value was found in the case of the dog, even when the eye had been exposed previously to bright

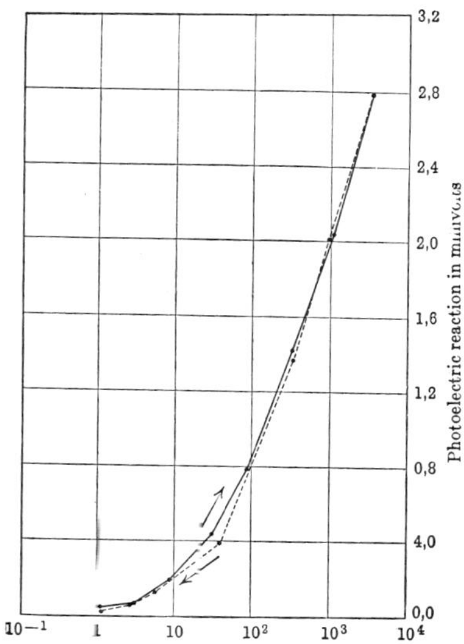

Fig. 7.

Connection between photo-electric response and stimulus (according to DE HAAS).

[1] ¶See paper by E. L. CHAFFEE, W. T. BOVIE and ALICE HAMPSON, The electrical response of the retina under stimulation by light. *Jour. of Opt. Soc. of Amer.*, etc., **7**, 1923, 1–44. — Preliminary report of some of these results in same journal, **6**, 1922, 407 (Paper read at Rochester meeting of Optical Society of America, Oct. 1921)—. (J. P. C. S.)

[2] *Ber. d. Naturf. Ges. Freiburg i. Br.* XI. (1901).

[3] *Arch. f. (Anat. u.) Physiol.* 1905. Supplement.

illumination and was therefore in the state of light adaptation. On the other hand, according to PIPER, the distribution of stimulus-values in the light-adapted eye of a rabbit is similar to that found by HIMSTEDT and NAGEL for the light-adapted eye of a frog, namely, with a maximum in the yellow ($574\mu\mu$).[1]

The relations of these particular facts to the subjective phenomena of colour vision will be discussed more fully hereafter.

The explanation of the photo-electric response is beset with great difficulties. The time relations of the objectively demonstrable electrical phenomena in the retina are so different from the subjective visual sensations that it is difficult to make a comparison between them.

As a matter of fact, however, the duration of the rise ("*Anklingen*") of the visual sensation, as determined by EXNER,[2] and the latent period of the electromotive reaction in the warm-blooded eye, as found by PIPER, GARTEN and others, do not differ very widely; in both cases it is only a few hundredths of a second. But the other parts of the electrical variations have certain characteristics that are at present difficult to reconcile with the subjective behaviour of the visual sensation, particularly when the stimulus is very brief. It should be noted that it is easier to compare the time process of the photo-electric reaction with the visual sensation, the more nearly the retina on which the measurements are made can be kept under normal conditions. Owls' eyes are particularly suitable for carrying out experiments of this kind without disturbing the function of the eye in any marked degree, and the electrical phenomena are simplest in such eyes; namely, positive variation of current under illumination, and negative discharge soon after shutting off the light. It is natural to think of the negative after-image in this connection. However, such comparisons in the present state of our knowledge are rather fruitless, and we must wait until it has been extended by further experiments on the eyes of warm-blooded animals.—N.

[1] ¶In addition to SHEARD's work, previously mentioned, the following are some recent contributions on this subject:

A. BROSSA and A. KOHLRAUSCH, Die Aktionströme der Netzhaut bei Reizung mit homogenen Licht. *Arch. f. Physiol.*, 1913, 449–492. — F. W. FRÖHLICH, Beiträge zur allgemeinen Physiologie der Sinnesorgane. *Zft. f. Sinnesphysiol.*, **48**, 1913, 28–164. — S. GARTEN, Die Produktion von Electrizität. WINTERSTEINS *Handb.* Bd. 3, 2 Hfte; also in GRAEFE-SAEMISCHS *Handb. d. ges. Augenheilk.*, I. Teil, III Bd., XII. Kap., Anhang, p. 213.— A. KOHLRAUSCH and A. BROSSA, Die photoelektrischen Reaktion der Tag- und Nachtvogelnetzhaut auf Licht verschiedener Wellenlänge. *Arch. f. Physiol.*, 1914, 421–431. — A. KOHLRAUSCH, Die Netzhautströme der Wirbeltiere in Abhängigkeit von der Wellenlänge des Lichtes und dem Adaptationszustand des Auges. Ibid., 1918, 195–214. (H. L.)

[2] *Wiener Sitz.-Ber.* **58**, 601.

§19. The Simple Colours

The subject which has now to be considered is the sensations that are excited in the visual mechanism by various kinds of luminous radiation. As has been already stated (§8), there are also other physical distinctions between waves of light of one frequency and those of another frequency, for example, differences of wave-length and refrangibility and differences in the way they are absorbed by coloured media. But the physiological distinction between luminous radiations of different frequencies is manifested generally by the production of sensations of different colours in the eye.[1]

[1] ¶"*Die einfachen Farben,*" which is the title of this section in the original, are the colours of the spectrum corresponding to luminous radiations of a definite period of vibration, "single-frequency light" or homogeneous light. They are sometimes called "pure" colours. No refraction can modify the colour that is associated with homogeneous light.

The physical stimulus of light is one thing; the physiological response or sensation of light is a totally different thing. This is so obvious that it scarcely needs to be stated, and yet many persons fail to make the distinction sometimes and are therefore liable to much confusion. The sensation of light can be aroused by other stimuli besides that of objective light, as has been shown in the previous chapters. On the other hand, only those radiations whose frequencies are comprised within a certain rather narrow range can arouse the sensation of light in the visual organ. When so-called luminous radiations are received on the foveal region of the retina of the human eye, the response, in general, is a very complicated one. It may involve elements of both time and space and give rise to a visual "pattern," and at the same time it may consist of "blends" of simpler sensations. Nearly all sensations of colour are mixed sensations which may be produced in a variety of ways. Exactly the same stimulus may produce a totally different sensation in different parts of the visual organ. If it acts on the extreme peripheral parts of the retina, it will produce only a grey sensation ranging anywhere from white to black depending on circumstances. So far as colour is concerned, vision in this region is similar to the vision of the totally colour-blind eye.

The physicist especially is liable to confusion, because he knows that by mixing coloured lights properly he can get white light; and so he concludes that white light is a mixture either of yellow light and blue light or (what is the same thing) of red and green and blue lights; and, hence is a compound light. But that is no reason to infer that the sensation of white is a complex sensation composed of the sensation of yellow and the sensation of blue. On the contrary, there is every reason to think that the sensation of white or grey is the most fundamental and elementary of all the visual sensations. It is the only sensation of the totally colour-blind eye or of the normal eye in the darkness of night when colour vision is entirely in abeyance.

On the other hand, a given visual sensation may be correlated with a vast number of different combinations of specific light-frequencies; a yellow, which is a perfectly unitary sensation, may be due to a homogeneous (unitary) light or it may be due to any one of countless different mixtures of any red light with any green light. In other words, there is absolutely no one-to-one correspondence between the composition of wave-lengths in a given luminous radiation and the light sensation which will be attached to it (provided every link in the nerve chain from retina to cortex is intact).

In the case of the black sensation the question is more difficult still. The physicist is right in thinking of a black object as sending no stimulus to the retina, but he is wrong when he supposes there is no black sensation correlated with this absence of stimulation. It is absurd to say that black is the absence of visual sensation. Black is a positive sensation just like any other visual sensation and just as real and distinct as the sensation of white or of yellow and blue or of red and green.

All known sources of light emit simultaneously radiations of different frequencies. The best way of separating a homogeneous or *pure kind of light* from such a mixture is to analyze the light by causing it to go through a transparent prism. Thus, for instance, if homogeneous blue light, proceeding from a distant source a (Fig. 8) through a prism P, comes to the observer's eye at O, the rays will be refracted by the prism into a new direction, and the source will appear to be shifted to some point b, in a direction indicated roughly by the direction in which the refracting angle of the prism is pointed. In this instance the colour of the light will naturally be the same as that of the simple light that is radiated by the source, that is, blue. But if the source emits both red and blue light at the same time, the observer will see simultaneously also a red image at r and a blue image at b. And, finally, if the source sends out white light containing radiations of all kinds of refrangibility including red light and blue light, each separate colour will have its own special image, all these images being ranged side by side in regular sequence from red to blue according to the degree of refrangibility. Now if the coloured images that are thus inserted between r and b are very manifold, and if each of them has a certain width, approximately the same as that of the luminous object itself, obviously one image will partly overlap the next. One way of reducing this overlapping and intermingling will be to make the luminous object narrower, and so to diminish the width of each separate image without altering the total width of the spectrum rb. While it is not possible to prevent entirely some overlapping between one image and the next when the source of light sends out radiations of all degrees of refrangibility, the luminous object and its images can be made so narrow that the differences of refrangibility between overlapping colours will be vanishingly small.

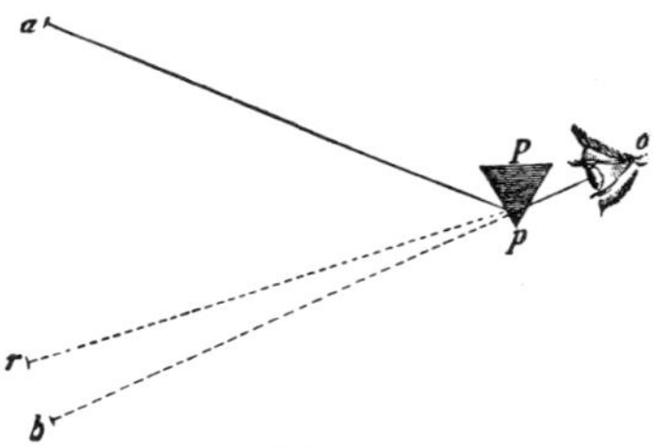

Fig. 8.

According to the above explanation, when the source is a very narrow slit illuminated by composite light, each individual point of the slit contributes a line to the spectrum lengthwise. And so the prismatic spectrum appears in the form of a coloured rectangle, the end nearest

To sum up, the physicist is in the habit of calling the colours of the spectrum "simple" or "primary" colours when all he means is that their physical stimulus is simple. The simple (unitary) sensations, in the order of their development, are white and black, yellow and blue, and red and green; and they do not need for their production "simple" physical lights. (J. P. C. S.)

the source being red and the opposite end violet. In between there is a sequence of other colours, each blending imperceptibly into the next, occurring in the following order, namely, red, orange, yellow, green, blue and violet. The coloured image of a luminous line obtained in this way by a prism is called a *prismatic spectrum*. In the illustration here used the spectrum is a *subjective* one, because the coloured images of the source of light are all virtual images. But a real image can be projected by adjusting a convex lens on the far side of the prism where the observer's eye was at first; so as to converge the rays after they leave the prism into a real image of *rb*, thereby producing an *objective spectrum* in or beyond the second focal plane of the lens. In the original illustration the spectrum projected on the retina of the eye is in a certain sense an objective spectrum. When the emitted light contains luminous radiations of all possible sorts, the spectrum is perfectly continuous. But if the source sends out light of certain definite degrees of refrangibility, the spectrum likewise will be composed of just so many coloured images, one for each particular kind of light; and provided the dimensions of the luminous object and its coloured images are sufficiently small, there will be dark gaps in the spectrum between some of the coloured images. For example, in the foregoing illustration (Fig. 8), when it was supposed that the luminous point *a* sent out simply red and blue light, the red image was formed at *r*, and the blue image at *b*, with an intervening dark space *br*. Of course, the same sort of thing occurs, no matter whether there are two or ten or a hundred or a thousand different kinds of homogeneous light in the light that comes from *a*.

Now the composition of sunlight is of this nature. When the solar spectrum is as perfect as we can get it, it is seen to be subdivided by a great number of dark lines, the so-called FRAUNHOFER lines. Their presence here implies the absence of light of certain refrangibilities in sunlight. The purer the solar spectrum is, the more dark lines are visible in it. FRAUNHOFER and STOKES attached certain letters of the alphabet to the most conspicuous of these lines, because these lines proved to be exceedingly sure and convenient guide-posts for finding the place in the spectrum that corresponded to some perfectly definite kind of radiation. This notation will be used in this treatise whenever the species of colour is to be given exactly. The dark lines of the solar spectrum are exhibited in the figure on Plate I. The relative lengths of corresponding portions of the prismatic spectrum are different for prisms of different substances. They are different also from the corresponding intervals in the diffraction spectrum, in which the distribution of colours is simply a function of the wave-length. Consequently, any representation of the prismatic spectrum is to a certain extent arbitrary. In the illustration in Plate I the intervals are arranged on the principle of the musical scale, because this seemed to be the best method for physiological reasons. Thus, colours whose wave-lengths are in the same ratio as the interval of a semi-tone between two musical notes are always at equal distances apart in the drawing; or to put it mathematically, equal distances in the drawing correspond to equal differences between the logarithms of the wave-lengths. The numerals on the left-hand side indicate the semi-tone intervals, and the letters on the other side are the designations of the more prominent dark lines, according to FRAUNHOFER and STOKES.

As some uncertainty prevails about the denomination of the different colours,[1] the names used in this book will be employed as follows.

Red is the colour at the less refrangible end of the spectrum, which shows no marked difference of hue from the extreme end of the spectrum to about the line *C*. In pigments it is represented by something like vermilion (cinnabar). *Purple-red* is different from red, and in its lighter tints is *pink-red*. As compared with pure red, it is bluish. This hue which in its most saturated stage is what we shall call *purple*, and in its more reddish forms *carmine red*, is not in the spectrum at all, and can only be produced by mixing the extreme colours in the spectrum, red and violet.

From the line *C* to the line *D* the red passes through *orange*, that is, yellow-red with red predominating, into *golden yellow* or red-yellow with yellow predominating. Among metallic dyes, minium and litharge (oxide of lead) are approximately the same as orange and golden yellow, respectively.

There is a rapid transition of colour from the line *D* to the line *b*. First, there is a narrow strip of pure *yellow* which lies about three times as far from *E* as from *D*. Then comes green-yellow and, finally, pure *green* between *E* and *b*. There are two very good representatives of pure yellow and green among pigments used by artists, namely, finely precipitated bright lead chromate known as chrome yellow and arsenite of copper or SCHEELE's green.

[1] ¶"The whole gamut of light-waves is responded to by us subjectively," says CHRISTINE LADD-FRANKLIN (Art. on "Vision" in BALDWIN's *Dictionary of Philosophy and Psychology*), "with only four different sensation qualities—red, yellow, green and blue. These are the sensations which are produced in their purity by about the wave-lengths 576, 505, 470, and a colour a little less yellow than the red end of the spectrum. For all intermediate wave-lengths we have nothing in sensation except combinations of these hues, or *colour-blends*, as reddish-yellow, blue-green, greenish-yellow, etc., but with this very singular peculiarity, that non-adjacent colour-pairs do *not* give colour-blends (red and green reproduce yellow, and blue and yellow give white, or grey); were it not for this latter circumstance, the confusion in the response to ether-radiation distinctions would be far greater than it is now." (J. P. C. S.)

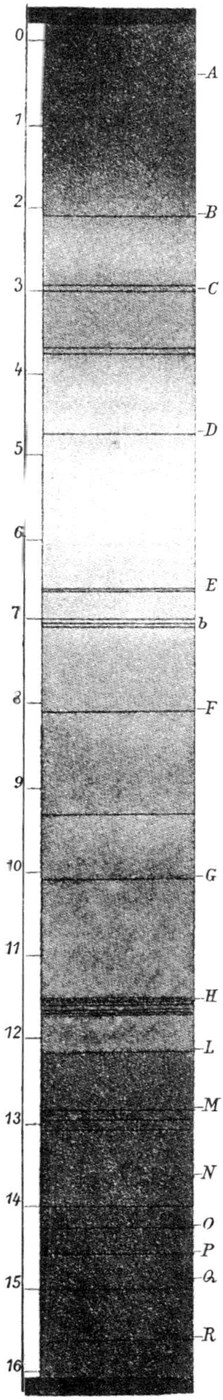

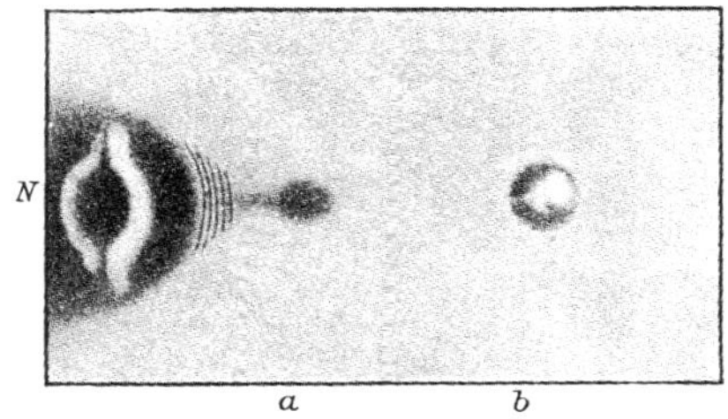

PLATE I.

Between the lines E and F the green becomes blue-green and then blue, and from F to G there are different hues of blue. In the prismatic spectrum of sunlight, as NEWTON observed it, the extent of the blue portion is comparatively great, and so he gave different names to different parts of this region distinguishing them as "blue" and "indigo" in English, and as *thalassinum, cyaneum, caeruleum* and *indicum* in Latin, in the order named; violet, *violaceum*, being last of all.[1] The name *indigo-blue* may be retained for the last two-thirds of the interval extending from F to G; but what is commonly known simply as *blue* is the less refrangible blue in the first part of this interval. It is sometimes described also as *sky-blue*, but that is incorrect. The reason why this blue in a spectrum of the proper brightness resembles sky-blue is simply because of the superior luminosity of the sky. The hue of the sky is really an indigo-blue but this hue as it occurs in the spectrum above mentioned is too dark to match sky-blue. In ordinary language, however, when a thing is said to be blue, it is natural to think of the colour of the blue sky as the principal representative of this hue, and to speak of a less refrangible blue as greenish blue. As it would be unscientific to call this latter hue simply blue as contrasted with indigo-blue, the author proposes to describe the greenish blue part of the spectrum as *cyan-blue*, as suggested by NEWTON's term *cyaneum*. The name *water-blue* might also very well be employed to describe the hue itself, because large masses of very pure water (like the lake of Geneva, glacier ice) do in fact show this colour in their depths. When, for instance, one gazes for a long time into the water of the lake of Geneva on a bright day and then looks up at the sky, the latter appears violet by contrast or even pink-red. But as the colour of a mass of water as it looks ordinarily is very whitish, except possibly in the case of deep crevices of ice, it is preferable to reserve the term water-blue for the lighter shades of *cyan-blue*. The pigments known as Prussian blue (iron ferrocyanide) and ultramarine correspond to cyan-blue and indigo-blue, respectively.

Violet (which is the colour of the flower of that name) is the region in the spectrum from the line G to the line H or L; sometimes called *purple* also. Violet and purple are the hues in the transition from blue

[1] ¶See R. A. HOUSTOUN, NEWTON and the colours of the spectrum. *Science Progress*, 1917. Also, volume on *Light and Colour*, 1923. There is much conjecture about what NEWTON meant exactly by "indigo" which is commonly supposed to be "more akin to green than to violet." The question, as HOUSTOUN propounds it (*Light and Colour*, p. 9), is, whether there is "a colour between blue and violet with as much right to a special name in the spectrum as orange has." He tested it with four of his students who all concurred in discriminating a hue, which they preferred to call "dark blue" and which was more blue than violet. The boundary between it and blue was estimated as falling about at wave-length $465\mu\mu$. (J. P. C. S.)

to red. As above stated, the name purple will be used here simply for the more reddish hues of this gradation that are not present at all in the spectrum.

The last region of all, corresponding to the most refrangible side of the spectrum, is *ultra-violet.*[1] This portion, extending from L to the end of the solar spectrum at R, is invisible unless the brighter parts of the spectrum mentioned above are carefully screened off. The existence of a special kind of radiation here was revealed first by its chemical actions, and consequently these rays were called invisible chemical rays. But, as a matter of fact, they are not invisible, although they certainly do affect the eye comparatively much less than the rays of the luminous middle part of the spectrum between the lines B and H. When these latter rays are completely excluded by suitable apparatus, the ultra-violet rays are visible without difficulty, clear to the end of the solar spectrum. At low intensity their colour is indigo-blue, and with higher intensity bluish grey. The easiest way to demonstrate the existence of these rays is by the phenomenon of *fluorescence.* For example, when a clear solution of sulphate of quinine is illuminated by ultra-violet light, a pale bluish light emanates in every direction from all the places in it where the ultra-violet rays fall, appearing somewhat like a luminous cloud pervading the liquid. Analyzed by a prism, this pale bluish light turns out not to be ultra-violet light at all, but compound whitish light of medium refrangibility. The simplest description of the phenomenon, therefore, is that as long as the ultra-violet rays fall on the quinine solution, it is self-luminous and emits compound whitish light of medium refrangibility. But the eye, being ever so much more sensitive to this kind of radiation than it is to ultra-violet, is entirely unconscious of the ultra-violet light until it falls on some fluorescent substance, and then the light that was previously invisible becomes visible in this material. Besides quinine, other substances that are highly fluorescent are uranium glass (canary glass), aesculin, platinum cyanide of potassium, etc.[2]

The fluorescent substance itself does not appear to be changed in the least, and it can be made to fluoresce over and over again. And

[1] ¶Silver chloride was long known to be sensitive to luminous radiations. J. W. Ritter (1801) found that the greatest effect on this substance was produced beyond the violet end of the visible spectrum. (J. P. C. S.)

[2] ¶Fluorescence is a term derived from fluor spar, which was the first substance that was observed to exhibit this peculiar emission of light. The phenomenon was first investigated by Sir J. Herschel and Sir D. Brewster; and subsequently by Sir G. G. Stokes. In every instance the fluorescent light is found to be light of longer wave-length than that of the incident light that excites the effect; all fluorescent phenomena being cases of the so-called degradation of energy. See R. W. Wood, *Physical Optics*, second edition, 1914, Chapter XX. (J. P. C. S.)

as no heat seems to disappear in the process, the inference from the law of the conservation of energy is that, notwithstanding that the fluorescent light affects the eye more, the actual energy of this radiation is no greater than that of the incident ultra-violet rays. No exact measurements are as yet available as to the ratio between the brightness of the original ultra-violet radiation and that of the same radiation after it has been changed by fluorescence. However, from certain facts to be mentioned later [p. 113] in describing the methods, it may be estimated that the fluorescent light is about 1200 times brighter than the ultra-violet radiation that induces it. Even without making any measurement, it is easy to show that the luminosity of the two kinds of light for the eye is extraordinarily different. This can be done by focusing ultra-violet light that has been completely purified of all more refrangible rays, and letting it fall first on a non-fluorescent screen like white porcelain, and then on quinine. The solar spectrum, at any rate as produced by sunlight that has traversed the atmosphere, does not actually extend beyond the place where the eye, suitably screened from the brighter light, can perceive ultra-violet radiation because even when an objective spectrum is projected by quartz prisms and lenses on a quinine solution or some other fluorescent substance, there is no fluorescence beyond the limit above mentioned. On the other hand, however, STOKES found that the spectrum of the electric arclight, projected on a fluorescent screen by a quartz optical system, extends much farther than the solar spectrum. Thus, in fact, his method is adapted for detecting also still more refrangible light than is contained in sunlight; and hence, it is to be inferred that the spectrum of sunlight that has been filtered out by the atmosphere really ends at the limit indicated by the eye and fluorescent substances.[1] No experiments have as yet been made on the visibility of the most refrangible rays of the electric arclight. The spark in vacuo that is obtained by an induction coil contains, indeed, a relatively large proportion of ultra-violet as compared with the small amount of less refrangible radiation, but the absolute intensity of the light is too slight to be resolved minutely by a prism.[2]

[1] ¶Owing to absorption by ozone in the higher levels of the atmosphere, the ultra-violet region of the solar spectrum ends at about 290$\mu\mu$. (J. P. C. S.)

[2] The arclight may be made richer still in ultra-violet by soaking the carbons in solutions of zinc or cadmium salts. Far better still for obtaining this sort of radiation is the spark of a large induction coil which is discharged between cadmium or magnesium electrodes, especially when means are taken to cut out the ordinary luminous rays. Rays of wave-length 257 cause a still perceptible fluorescence in the eye, and hence produce a sensation of light.—N.

¶The beautiful series of researches, beginning with CORNU's work forty years ago, followed by SCHUMANN's investigations (1890), and continued by LYMAN and by MILLIKAN in very recent years, has resulted in the exploration of the ultra-violet region as far as to

At the other end of the spectrum also, when the brighter light that is ordinarily visible is screened off, it is possible to distinguish parts of the spectrum that usually remain invisible. An adequate screen for this purpose is obtained by interposing a piece of red glass in the path of the rays. Red glass coloured by protoxide of copper transmits much orange light, and hence, if necessary, a piece of blue cobalt glass, which absorbs orange but transmits extreme red light, may be used in combination with the red glass. But as compared with the great extent of the ultra-violet spectrum, there is not much that is gained at the red end by this mode of observation. The strip of red light beyond the line *A* is about as wide as the interval between *A* and *B*. The hue of the red does not change up to the extreme end and is not at all purple.

But, as a matter of fact, the solar spectrum extends on the red side farther than the eye can detect. Hitherto the existence of these infra-red rays has not been made manifest except by their thermal effects, and that is why they are called *dark heat rays*.[1] Glass, water and numerous other substances that are transparent to ordinary light are opaque to infra-red, and so rock-salt prisms and lenses must be used to explore this region of the spectrum. The width of the dark heat spectrum as produced by a prism is certainly limited by reason of the fact that, according to the theory of elastic aether vibrations, the refraction approaches a minimum as the wave-length increases. This minimum cannot be surpassed, and the dispersion of colours terminates at it. In Fig. 9 the wave-lengths are plotted as abscissae from an origin that is just as far to the left of the point *H* as the point *b* is to the right of this point. The capital letters from *B* to *H* correspond to the FRAUNHOFER lines and to their positions in an interference spectrum. The ordinates are the values of the indices of refraction for a flint glass prism used by FRAUNHOFER.

Line	B	C	D	E	F	G	H
Index of refraction	1.6277	1.6297	1.6350	1.6420	1.6483	1.6603	1.6711

The letters $B_{,,}$ $C_{,,}$ etc., on the vertical axis, indicate the positions of the same dark lines in the solar spectrum of this flint glass prism.

wave-lengths of only 20$\mu\mu$, which "is the limit reached today in the study of radiations by optical means." See C. FABRY, Studies in the field of light radiation. *Jour. of the Franklin Inst.*, **192,** (1921), 277-290. For the most recent work in this farthest region, see R. A. MILLIKAN, *Nat. Acad. Sci. Proc.* **7,** 289-294. Oct. 1921. and R. A. MILLIKAN and I. S. BOWEN, Extreme ultra-violet spectra. *Phys. Rev.*, **23,** 1924. 1-34. (In this paper the exploration extends as far as 13.6$\mu\mu$). (J. P. C. S.)

[1] ¶The thermal action of these dark heat rays was detected by Sir WM. HERSCHEL in 1800; who drew the correct conclusion from his experiment at first, but afterwards changed his opinion. See HOUSTOUN, *Light and colour*, 1923, page 37. (J. P. C. S.)

The base-line Hb corresponds to the index 1.6070 which is the minimum value for this particular kind of glass. With increasing wave-length the indices of refraction must approach this minimum value asymptotically.[1] The dotted curve $H_,d$ shows, therefore, the refrangibility of the light as a function of the wave-length, and if it were extended farther, it would approach the base-line Hb asymptotically. Consequently, supposing that the refraction spectrum $H_,B_,$ is extended beyond its red end at $B_,$ so as to include the dark heat rays, its extreme limit must be on the base-line at H;[2] that is, it is about as far from the red end B, visible under ordinary conditions as B, is from the boundary F, between green and blue; which is a distance corresponding roughly to half the length of the ordinary visible spectrum. Another thing that the diagram (Fig. 9) shows clearly is how light at the blue end $F_,G_,H_,$ of the refraction spectrum $B_,H_,$ is spread out, while that at the red end $B_,C_,D_,$ is condensed together, as compared with the interference spectrum BH. Naturally, this condensation of the less refrangible light in the refraction spectrum should be more and more marked, the nearer we come to the limit of the infra-red region. Therefore, towards the blue end where the spectrum is more elongated, the number of visible dark lines becomes greater; and since the same amount of light or heat is spread out here over a larger area, brightness and temperature are less. On the other hand,

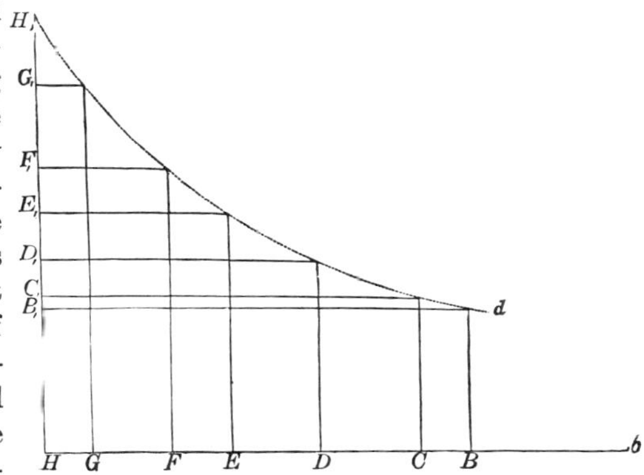

Fig. 9.

[1] The value of this minimum has been taken from BADEN POWELL's calculation (POGGENDORFF XXXVII); as his interpolation formula agrees closely enough with CAUCHY's theoretical formula.

[2] According to a remark of FR. EISENLOHR this limit seems actually to have been reached in MELLONI's experiments. *Kritische Zft. f. Chemie*. Erlangen 1858. S. 229. (In the 2nd edition the following statement is added here: Theoretically, this was what was to be expected. However, LANGLEY *Phil. Mag.*, **21**, 1886, 349, in his observations on infra-red, which go much farther than any previous work of this kind, has not found any such limit.)

¶The longest waves in the solar spectrum that have been ascertained are about 5000$\mu\mu$. But in the spectra of artificial sources of light, it has been possible to explore the infra-red region as far as wave-lengths that are almost 400 times as great as those at the red end of the luminous or visible spectrum, that is, to about 300000$\mu\mu$ (or 0.3mm). This is in the neighbourhood of the upper limit of emission spectra. (J. P. C. S.)

towards the red end there are fewer dark lines, and brightness and temperature are greater, than in the interference spectrum. And although the maximum heat effect in the prismatic spectrum is in the infra-red, it does not follow that these particular dark heat rays are present in sunlight in greater numbers than some of the luminous rays. On the contrary, in the interference spectrum the heat maximum falls on yellow.

On account of the characteristics of the refraction spectrum above mentioned, it is extremely difficult to determine the longest wave-lengths in the infra-red portion of solar radiation. By a method which appears to be fundamentally sound, FIZEAU has measured the lengths of the longer of these waves that are transmitted through flint glass, and found the maximum to be 0.001940 mm. This is more than double the wave-length of the farthest red light in the ordinarily visible region, which according to the author's determinations is 0.00081 mm. Incidentally, these dark heat rays exhibit the phenomena of interference just like the luminous rays, and, consequently, like them, they are due to aether vibrations. They are subject to exactly the same laws of polarisation, which implies that the vibrations are transverse to the direction of propagation. The only physical difference, therefore, between these rays and luminous rays is that the waves are longer and the refrangibility correspondingly less.

A possible explanation of the invisibility of infra-red radiation is either that these rays are absorbed by the ocular media or that the retina is not sensitive to them. MELLONI has demonstrated that the dark heat rays are absorbed to a great extent by water. BRÜCKE and KNOBLAUCH have made experiments on the transparent media of the eye of an ox. The cornea, vitreous humor and crystalline lens were inserted in a convenient tubular mounting, with the vitreous humor in between the cornea and the lens. Sunlight reflected by a heliostat into the dark room was transmitted through this perfectly transparent system and made to fall on a thermopile. The indicated deflections on the amplifying mechanism amounted to between 26° and 30°. Then both sides of the eye were covered with lamp-black over a turpentine flame, which was accomplished successfully, without producing any other change in the cornea and lens, as was ascertained subsequently. Under such circumstances, it was found that no heat at all was radiated through the eye. But lamp-black is transparent to the dark heat rays and opaque to the luminous rays. If, therefore, a part of the radiation that was transmitted through the ocular media had consisted of dark heat rays, some effect from these would have been manifest through the lamp-black. It would hardly be justifiable to say that this experiment proved that the limits of the visible red coincided with the limits of diathermancy of the ocular media, but it certainly does establish the fact that very little, if any, of the infra-red radiation can get to the retina; and this of itself would seem to be a sufficient explanation of the invisibility of this region of the spectrum.

CIMA[1] has made similar experiments, using a LOCATELLI lamp as source of heat, the radiation being transmitted through the ocular media to a thermopile. He found that about 13 percent of the incident heat was transmitted through the crystalline lens, about 9 percent, through the vitreous humor alone, and also about 9 percent through the eye as a whole.[2]

The mere fact that it is possible to see the ultra-violet spectrum with its dark lines shows that this radiation may traverse the ocular media. DONDERS and REES have demonstrated objectively that these rays go through a glass vessel containing vitreous humor of the eye of an ox, and the cornea and crystalline lens also. The ultra-violet light, after having traversed the ocular media, was made manifest by letting it fall on the surface of a solution of sulphate of quinine where blue fluorescence was excited. Similar experiments had been previously made by BRÜCKE, by testing the action of light on guaiacum solution and on photographic paper after it had been transmitted through the media in the eye.

Guaiacum resin, newly evaporated in the dark from the alcoholic solution, appears blue under the action of blue, violet or ultra-violet radiation, but the blue colour disappears when it is illuminated by less refrangible rays. In ordinary daylight the blue effect is predominant. But the colour of this substance under illumination of daylight that has been filtered through the crystalline lens of an ox is simply yellow-green; and a layer of the resin that has already been coloured blue looks

[1] *Sul potere degli umori dell' occhio a trasmettere il calorico raggionante.* Torino 1852.

[2] J. JANSEN (*C. R.*, LI, 128–131; 373–374; *Ann. der Chir.*, (3), XL, 71–93) and R. FRANZ (POGG. *Ann.*, CXV, 26–279) also found that the absorption in the vitreous humor was very similar to that in water and rather more in the cornea and crystalline lens. Similar results were obtained by TH. W. ENGELMANN (*Onderzoek. physiol. Lab. Utrecht.* 3. Reeks, D. VII Bl. 291. 1882).

Concerning absorption of ultra-violet in the eye, it may be added that, acording to the researches of SORET and others, there is little absorption of rays between the FRAUNHOFER lines H and Q, whereas shorter wave-lengths than these are strongly absorbed. See SORET, *C. R.*, **88**, p. 1012; **97**, pp. 314, 572, 642; CHARDONNET, *C. R.*, **96**, p. 509; MASCART, *C. R.*, **96**, p. 571.—N.

¶In connection with this whole subject the following more recent contributions are worth consulting:

W. CROOKES, Preparation of eye-preserving glass for spectacles. *Phil. Trans.*, **213** (1914), 1–25; and VERHOEFF and BELL, The pathological effects of radiant energy on the eye; an experimental investigation. *Proc. Amer. Acad. Arts and Sciences*, Vol. **51**, No. 13, July 1916. (J. P. C. S.)

yellow-green in the same light. This means that the crystalline lens absorbs the bluish rays of daylight more than the others. If ordinary blue and violet light were much absorbed, the lens itself would have to look yellowish; but under normal conditions it is fairly without any colour, and therefore in the light that makes guaiacum look blue it can only be the ultra-violet portion that is absorbed by the lens in any comparatively considerable amount. The results of similar experiments of BRÜCKE's on the cornea and vitreous humor indicate a behaviour of the same kind as that of the lens, only to a much less degree. These conclusions are supported by the fact that the cornea and crystalline lens, as may easily be observed even in the live eye, are themselves fluorescent to a certain extent when violet or ultraviolet light falls on them. Under such circumstances they shine with a pale blue colour like that of the quinine solution. Fluorescent substances, however, always noticeably absorb the rays that make them fluoresce.

Other experiments were made by BRÜCKE with C. KARSTEN's photographic paper. Cornea, vitreous humor, and crystalline lens were mounted for testing in a similar arrangement to that used in the thermo-electric experiments mentioned above. They were traversed by radiations from a spectrum of sunlight produced by a prism, and the sensitive paper was adjusted in the focal plane of the ocular media. After exposure to violet for 90 seconds, a perfectly black point was produced. In the vicinity of the group of lines known as *M* (according to DRAPER) the effect on the paper ceased entirely, so that even after an exposure of ten minutes no action could be detected. However, it should be remarked that even when the rays do not pass through the ocular media, the photographic action on nearly all sensitive preparations falls off rapidly towards the end of the spectrum. Fluorescence, which was not discovered until after BRÜCKE made the experiments here mentioned, is a much more sensitive means of detecting these effects than photographic action, especially in case of the more refrangible rays; and it has enabled us to explore the spectrum much farther than before. In fact, when the eye is properly screened from the light of the brighter portion of the spectrum, direct observation seems to afford more information of the ultra-violet region than is obtained by photographic methods.[1]

Thus, according to BRÜCKE's researches, ultra-violet rays are absorbed to a considerable extent in passing through the ocular media, especially the crystalline lens, as shown particularly by the effect on guaiacum tincture. On the other hand, DONDERS's experiments tend to show that this absorption is not enough to be noticeable in the ordinary comparisons of brightness by the unaided eye. But it has

[1] ¶One difficulty about photographing the ultra-violet spectrum is that glass begins to become very opaque at about wave-length $340\mu\mu$. It can be replaced by quartz, which does very well as far as 185 $\mu\mu$, when it begins to be highly absorbent also. Fluorite enables us to push the limit much farther. Another difficulty is the opacity of the gelatine of the sensitive film which prevents these very short waves from getting to the sensitive salt at all; and, finally, the air itself ceases to be transparent, and it is necessary, therefore, to conduct the experiments in vacuo. (J. P. C. S.)

already been stated that the brightness of ultra-violet light as compared with that of practically the same amount of light emitted by fluorescent quinine solution is about in the ratio of 1:1200. The inference is that absorption in the ocular media cannot be responsible, except to the minutest extent, for the low subjective luminosity of ultra-violet; and that the real explanation of it is probably due to the lack of sensitivity of the retina.

Another thing to be noted is that the colour sensation produced by light of a definite wave-length depends also on its luminosity. Thus, any increase of luminosity tends to make it look more white or pale yellow. This effect is easiest to see with violet; the less blue and the more purple it is, the fainter it gets. On the other hand, with a moderate degree of brightness, such as is obtained by observing the solar spectrum in a telescope, this same colour appears pale grey, with just a faint bluish violet tinge. Another good way to see this, as Moser suggested, is to look at the sun in a half-clouded sky through a piece of fairly dark violet glass. The sun's disc looks just as white through the glass as the brightly illuminated clouds near by appear to the naked eye. So also for low intensity the blue of the spectrum is more like indigo-blue; with higher intensity, sky-blue; and with still greater intensity (provided the eye can stand it without annoyance), pale blue and finally white. This is the explanation of the wrong use of the name sky-blue as applied to the more refrangible and at the same time more luminous cyan-blue of the spectrum. Green becomes yellow-green and then white; and yellow becomes white directly, but the luminosity is dazzling in its brilliancy. The effect is hardest to see in the case of red; and for the highest degrees of brightness, the most the author has been able to do, either by looking at the spectrum or by looking at the sun through a red glass, is to see it change to bright yellow. These tests can all be made equally well with carefully purified simple light or with mixed light of the given colour as it is obtained with coloured glasses.

There is no part of the spectrum where variation of luminosity produces so much change of hue as it does in the violet and ultra-violet regions. The hues of the most refrangible end cannot be very well compared with each other unless the luminosities are approximately equal. When the brightness is dim, the blue tones in the spectrum are nearer indigo, and the violet is more pink, as has been already mentioned. But from about the line L to the end of the spectrum a reversal occurs in the order of the colours, that is, the hue is no longer more like pink, but from here out is again like indigo. On the other hand, with moderate rise of intensity the ultra-violet looks bluish pale grey, paler than equally luminous indigo-blue, and hence it is called sometimes *lavender grey*.

The reversal in the order of colours exhibited by ultra-violet light at low luminosity probably does not depend on the mode of reaction of the nervous mechanism, but seems to be connected with the fluorescence of the retina itself; which, when illuminated by ultra-violet, emits light of lower refrangibility of a greenish white colour. At least, this was the case with the retina of the eye of a cadaver examined by the author,[1] and with the retinas of perfectly fresh eyes from oxen and rabbits that had just been killed, which were examined by SETSCHENOW;[2] the fluorescence being, indeed, very slight, and the colour of the light the same as that mentioned above. The degree of fluorescence was less than that of paper, linen or ivory, but still it seemed to be always sufficient to change the colour of the incident ultra-violet light. The author tried to test it by comparing the radiation from the fluorescent places in this retina with ultra-violet light diffusely reflected from a white porcelain plate. In both cases the light was emitted in all directions in space. The retina and porcelain plate were observed through a weak prism that separated the two kinds of radiation, that is, the changed ultra-violet light from that which was unchanged. Under these circumstances, the light produced by fluorescence in the retina appeared about as bright as the unchanged ultra-violet illumination of the porcelain plate. It can hardly be doubted that the retina is sensitive to light produced in its own substance by fluorescence; and on this assumption, the sensation for ultra-violet radiation must be composed pretty evenly of the sensation directly produced by the ultra-violet light and that excited by the fluorescence. As this latter appears paler and more greenish than ultra-violet light, it would seem that the direct sensation of ultra-violet light on a non-fluorescent retina would have to be more like pure violet. For the lavender grey of the ultra-violet rays can be obtained by a proper mixture of violet and greenish white. The fact that the colour of the fluorescent retina is quite different from lavender grey does not warrant us in supposing that the ultra-violet light does not stimulate the retina at all and that the sensation is due simply to the fluorescent light.

A prismatic spectrum, short enough to be viewed in its entirety all at once, appears to consist of only four coloured sections, namely, red, green, blue and violet, the transition-colours disappearing almost entirely by contrast with these main colours. At best yellow may still be discerned in the green next the red. This separation of colours is enhanced by the fact that three of the more prominent dark lines of the solar spectrum, namely D, F and G, happen to lie about on the

[1] POGGENDORFFS *Ann.* XCIV. 205.

[2] GRAEFES *Archiv für Ophthalmologie*. Bd. V. Abt. 2. S. 205.

boundary lines of the four intervals of the spectrum above mentioned. But even without being able to distinguish these lines, the same separation of colours is manifest. The transition-colours are indeed more easily seen in a longer spectrum, but yet the visual impression of them is always considerably modified by the proximity of such brilliant saturated colours as are seen in the spectrum, which prevents the transition-colours from being seen in their own right. To distinguish exactly the series of pure colours in the spectrum, they must be isolated. A way of doing this is to project a fairly pure spectrum on a screen with a small slit in it, which permits the light of some single region of the spectrum to pass through it and be received on another white screen beyond. By gradually moving the slit from one end of the spectrum to the other, the whole series of hues can be inspected separately one after the other. Then it will be found that there is nowhere any abrupt transition in the series, and that the hues merge continuously each into the next. The richness and intense saturation of the succession of colours and the delicate transition of hues makes this experiment at the same time one of the most splendid spectacles that optics has to show.

Owing to the exceedingly gradual blending of the hues, it is naturally impossible also to assign any definite width to the separate coloured regions of the spectrum. In order to indicate as well as possible the positions and distribution of the colours, the hues corresponding to the FRAUNHOFER lines are given in the following table, together with their wave-lengths in millionths of a millimetre:[1]

Line	Wave-length in $\mu\mu$	Colour
A	760.40	Extreme red
B	686.853	Red
C	656.314	Border of red and orange
D	589.625 589.024	Golden yellow
E	526.990	Green
F	486.164	Cyan-blue
G	430.825	Border of indigo and violet
H	396.879	Limit of violet
L	381.96	Ultra-violet
M	372.62	
N	358.18	
O	344.10	
P	336.00	
Q	328.63	
R	317.98	
U	294.77	

[1] ¶The first determinations of the wave-lengths of light of different refrangibilities or colours were made by YOUNG. The values of these magnitudes as found by him for the two ends of the visible spectrum were 266 and 167 ten-millionths of an inch or 676 and 424 millionths of a millimeter. (J. P. C. S.)

The different sensations of colour in the eye depend on the frequency of the waves of light in the same way as sensations of pitch in the ear depend on the frequency of the waves of sound; and so, many attempts have been made to divide the intervals of colour in the spectrum on the same basis as that of the division of the musical scale, that is, into whole tones and semi-tones. NEWTON tried it first. However, at that time the undulatory theory was still undeveloped and not accepted; and not being aware of the connection between the width of the separate colours in the prismatic spectrum and the nature of the refracting substance, he divided the visible spectrum of a glass prism, that is, approximately the part comprised between the lines B and H, directly into seven intervals, of widths proportional to the intervals in the musical scale,[1] namely, $\frac{9}{8}$, $\frac{16}{15}$, $\frac{10}{9}$, $\frac{9}{8}$, $\frac{10}{9}$, $\frac{16}{15}$, $\frac{9}{8}$; and so he distinguished seven corresponding principal colours; *red, orange, yellow, green, blue, indigo* and *violet.* The reason why two kinds of blue are mentioned here, while golden yellow, yellow-green, and sea-green, which appear to the eye at least just as different from the adjacent principal colours as indigo is from cyan-blue and violet, are omitted, is because of the peculiar variation of the index of refraction mentioned on page 68, which causes the more refrangible colours in a prismatic spectrum to be elongated more than the less refrangible ones. The distribution of colours in the interference spectrum has nothing to do with the character of a refracting medium and depends simply on the wave-length; and here the blue-violet region is much narrower, and if the intervals were determined in the same way, this span would not be resolved into three parts, whereas the red-orange portion would be in about three parts.

In the light of subsequent discoveries and measurements, suppose that the spectrum as we now know it is divided on the same principle as the musical scale using the vibration-numbers of the aether waves, as was done in the case of the solar spectrum exhibited in Plate I;

[1] ¶A clear description of the actual process that NEWTON used in making the division of the spectrum on the basis of the musical scale is to be found in R. A. HOUSTOUN's *Light and Colour*, 1923, pages 12–14. This writer shows that NEWTON divided the spectrum originally into five colours, and then inserted orange and indigo; and, as to the latter, he concludes that the introduction of indigo was due to an "attempt to find a connection between the spectrum and the musical scale," and that although "the attempt failed completely," as NEWTON himself lived to realize, "indigo remains in the list of colours" "as a witness to it." The same writer points out the mystical influence exerted by PYTHAGORAS's discovery (572–492 B. C.) of the laws of harmony as illustrated by the natural modes of division of a vibrating string, which led to the idea that all the laws of nature were harmonies of some kind, as, for example, the so-called "music of the spheres" which cast its spell over a mind as acute as KEPLER's. An additional reason for dividing the spectrum into seven primary colours is to be traced also to the peculiar significance of the number seven as the "perfect number." (J. P. C. S.)

then if the yellow of the spectrum answers to the tenor C in music and the FRAUNHOFER line A corresponds to the G below it, we obtain for the separate half-tones the following scale of colours analogous to the notes of the piano:

$F^\sharp$,	end of Red.	$f^\sharp$,	Violet.
G,	Red.	g,	Ultra-violet.
$G^\sharp$,	Red.	$g^\sharp$,	Ultra-violet.
A,	Red.	a,	Ultra-violet.
$A^\sharp$,	Orange-red.	$a^\sharp$,	Ultra-violet.
B,	Orange.	b,	Ultra-violet.
c,	Yellow.		
$c^\sharp$,	Green.		
d,	Greenish blue.		
$d^\sharp$,	Cyan-blue.		
e,	Indigo blue.		
f,	Violet.		

The hues that comprise octaves are placed side by side. In the figure on Plate I the places corresponding to the tone-intervals are indicated by lines on the left. The end of the infra-red spectrum, according to FIZEAU and FOUCAULT, calculated on the same basis, would be about D, two octaves below cyan-blue; and if CAUCHY's formula for the connection between wave-length and index of refraction can be supposed to be valid so far, the extreme limit of the spectrum of the arclight would be at b', an octave higher than the ultra-violet end of the solar spectrum.

The colour-scale divided in half-tones as above shows that at both ends of the spectrum the colours do not change noticeably for several half-tone intervals, whereas in the middle of the spectrum the numerous transition colours of yellow into green are all comprised in the width of a single half-tone. This implies that in the middle of the spectrum the eye is much keener to distinguish vibration-frequencies than towards the ends of the spectrum; and that the magnitudes of the colour intervals are not at all like the gradations of musical pitch in being dependent on vibration-frequencies.

These physiological studies demand a much more exact differentiation of the homogeneous kinds of light than is usually necessary for purely physical investigations; and hence the theory of refraction by prisms will now be specially considered, to see what are the conditions of obtaining pure spectra by dispersion. Previously, so far as the writer is aware, the theory has been confined to the problem of tracing single rays of light through a system of prisms, without investigating the position and nature of the *images produced by prisms*; and yet in looking through a prism or letting the light that issues from the prism go through lenses and telescopes, the essential thing is to distinguish the prism-images for each kind of homogeneous light. For these images are really to be considered as objects to be imaged by the ocular media and lenses. To supply this lack, we shall proceed to determine the *nature and position of the image in a prism*, although this investigation does not properly belong to physiological optics. However, the results will perhaps be important for everybody who wishes to produce pure prismatic spectra.

In general, a narrow homocentric bundle of incident rays will not be homocentric after emerging from a prism, but will be astigmatic, with two image-points, exactly in the same way as when a homocentric bundle of incident rays is refracted at an ellipsoidal surface or is incident obliquely on a spherical refracting surface.[1] In order to simplify the treatment of the subject, the law of refraction will be used in a form which was given to it by FERMAT soon after its discovery, and which is particularly adapted for investigation of problems in optics where the different portions of the path of a ray are not all in the same plane.

Suppose light traverses a series of refracting media, and consider the path of a single ray. If the length of the path in each medium is multiplied by the index of refraction of that medium, and these products are all added, this sum is what the writer calls the *optical length* of the ray.[2] For example, if r_1, r_2, r_3, etc. denote the lengths of the path of the ray in the first, second, third, etc., medium, respectively, and if n_1, n_2, n_3, etc., are the corresponding indices of refraction, the optical length according to this definition is

$$\Psi = n_1 r_1 + n_2 r_2 + n_3 r_3 + \cdots + n_m r_m .$$

If c_0 denotes the velocity of light in vacuo, and c_1, c_2, c_3, etc. denote the velocities in the different media in succession, then (see §9):

$$n_1 = \frac{c_0}{c_1}, \qquad n_2 = \frac{c_0}{c_2}, \qquad n_3 = \frac{c_0}{c_3} \cdots n_m = \frac{c_0}{c_m},$$

and therefore

$$\Psi = c_0 \left[\frac{r_1}{c_1} + \frac{r_2}{c_2} + \frac{r_3}{c_3} + \cdots + \frac{r_m}{c_m} \right].$$

[1] See end of §14. Vol. I. The theorems that follow are applicable to the monochromatic aberrations of the eye as treated in §14.

[2] ¶It is interesting to note that HELMHOLTZ originated the name for this function that has since been universally adopted. (J. P. C. S.)

Suppose t denotes the time which the light takes to go over the entire path; then

$$t = \frac{r_1}{c_1} + \frac{r_2}{c_2} + \frac{r_3}{c_3} + \cdots + \frac{r_m}{c_m},$$

and therefore

$$\Psi = c_0 t .$$

Accordingly, the optical length is proportional to the time taken by the light to go over the path and is equal to the distance the light would have travelled in vacuo in the same time.

The notion of optical length may be applied also to the case where the ray in the last medium is prolonged backwards beyond the boundary of this medium to a point where a potential image of the luminous point is situated. To find the optical length between the luminous point and its potential image, the same process is employed as before; only the distance from the place where the ray emerges into the last medium to the place where the potential image is must be reckoned as negative. The following theorems will then be perfectly general.

I. *The law of refraction of light is equivalent to the condition that the optical length of the ray from a point on it in the first medium to a corresponding point in the second medium shall have a limiting value, that is, shall be a maximum or minimum.*

The surface of separation of the two media may have any form whatever, provided the curvature is continuous. If the incidence-normal is chosen as the z-axis of a system of rectangular axes, the form of the surface will be given by an equation in which z is a function of x and y; and at the point of incidence

$$x = y = z = 0 , \qquad \frac{dz}{dx} = 0 , \qquad \frac{dz}{dy} = 0 \quad . \quad . \quad . \tag{1}$$

Moreover, let a_1, b_1, c_1 denote the coördinates of a point on the incident ray, and a_2, b_2, c_2 those of a point on the refracted ray. If these points are connected with a point (x, y, z) of the refracting surface, the optical length between them along this route is

$$\Psi = n_1 \sqrt{(a_1 - x)^2 + (b_1 - y)^2 + (c_1 - z)^2} + n_2 \sqrt{(c_2 - x)^2 + (b_2 - y)^2 + (c_2 - z)^2} .$$

In order that this magnitude, which is a function of the independent variables x and y, shall be a maximum or minimum, the first conditions (which here are likewise sufficient) are:

$$\frac{d\Psi}{dx} = 0 , \qquad \frac{d\Psi}{dy} = 0 ,$$

or

$$
\left.\begin{aligned}
0 &= n_1 \frac{x - a_1 + (z - c_1)\dfrac{dz}{dx}}{\sqrt{(a_1 - x)^2 + (b_1 - y)^2 + (c_1 - z)^2}} \\
&\qquad + n_2 \frac{x - a_2 + (z - c_2)\dfrac{dz}{dx}}{\sqrt{(a_2 - x)^2 + (b_2 - y)^2 + (c_2 - z)^2}} \\
0 &= n_1 \frac{y - b_1 + (z - c_1)\dfrac{dz}{dy}}{\sqrt{(a_1 - x)^2 + (b_1 - y)^2 + (c_1 - z)^2}} \\
&\qquad + n_2 \frac{y - b_2 + (z - c_2)\dfrac{dz}{dy}}{\sqrt{(a_2 - x)^2 + (b_2 - y)^2 + (c_2 - z)^2}}
\end{aligned}\right\} \quad (2)
$$

Combining these equâtions with equations (1), we find for the ray that is incident on the surface at the origin of the system of coördinates:

$$
\left.\begin{aligned}
0 &= n_1 \frac{a_1}{\sqrt{a_1^2 + b_1^2 + c_1^2}} + n_2 \frac{a_2}{\sqrt{a_2^2 + b_2^2 + c_2^2}} \\
0 &= n_1 \frac{b_1}{\sqrt{a_1^2 + b_1^2 + c_1^2}} + n_2 \frac{b_2}{\sqrt{a_2^2 + b_2^2 + c_2^2}}
\end{aligned}\right\} \quad (2a)
$$

If the positions of the points (a_1, b_1, c_1) and (a_2, b_2, c_2) are given in terms of polar coördinates, by the ordinary formulae of transformation, namely:

$$
\left.\begin{aligned}
a_1 &= r_1 \sin \alpha_1 \cos \vartheta_1 & a_2 &= r_2 \sin \alpha_2 \cos \vartheta_2 \\
b_1 &= r_1 \sin \alpha_1 \sin \vartheta_1 & b_2 &= r_2 \sin \alpha_2 \sin \vartheta_2 \\
c_1 &= r_1 \cos \alpha_1 & c_2 &= r_2 \cos \alpha_2
\end{aligned}\right\} \quad . \quad . \quad (3)
$$

equatiǫns (2a) become:

$$
\left.\begin{aligned}
n_1 \sin \alpha_1 \cos \vartheta_1 &= - n_2 \sin \alpha_2 \cos \vartheta_2 \\
n_1 \sin \alpha_1 \sin \vartheta_1 &= - n_2 \sin \alpha_2 \sin \vartheta_2
\end{aligned}\right\} \quad . \quad . \quad . \quad . \quad (2b)
$$

Squaring each of these equations, and then adding them, we find:

$$
n_1^2 \sin^2 \alpha_1 = n_2^2 \sin^2 \alpha_2 ,
$$

that is,

$$
n_1 \sin \alpha_1 = \pm n_2 \sin \alpha_2 .
$$

The positive sign is the only one that applies here, because the angle α_1 is between 0° and 90°, whereas, according to our convention, the angle α_2 must be between 90° and 180°. Hence $\sin \alpha_1$ and $\sin \alpha_2$ are both positive; and, since n_1, n_2 are always positive, we have therefore:

$$n_1 \sin \alpha_1 = n_2 \sin \alpha_2 \quad . \quad . \quad . \quad . \quad . \quad . \quad . \quad . \quad . \qquad (4)$$

Combining this equation with equations (2b) we get:

$$\cos \vartheta_1 = -\cos \vartheta_2 \,,$$
$$\sin \vartheta_1 = -\sin \vartheta_2 \,,$$

that is,

$$\vartheta_2 = \vartheta_1 + 180° \quad . \quad . \quad . \quad . \quad . \quad . \quad . \quad . \quad . \qquad (4a)$$

Equations (4) and (4a), which have been derived from the condition that the optical length of the ray is a limiting value, are, however, identical with the two conditions of the law of refraction. Thus, as follows from equations (3), α_1 is the angle of incidence, ϑ_1 is the angle between the plane of incidence and the xz-plane, and ϑ_2 is the angle between the plane of refraction and the xz-plane. Accordingly, the planes of incidence and refraction are inclined to each other at an angle of 180°, that is, they are one and the same plane. Exactly the same mode of proof is used when the ray is reflected from the surface instead of being refracted. All we have to do in this case is to put $n_1 = n_2$, because the ray remains in the same medium, and to suppose that both α_1 and α_2 are comprised between 0° and 90°. Then equations (4) and (4a) become:

$$\sin \alpha_1 = \sin \alpha_2 \quad \text{or} \quad \alpha_1 = \alpha_2 \,,$$
$$\vartheta_2 = \vartheta_1 + 180°,$$

which are the two conditions of the law of reflection.

Having established the above theorem for a single refracting surface, we can readily extend it to any number of such surfaces. *When a ray of light traverses a series of transparent media, separated by refracting surfaces of continuous curvature, its path can be traced from the fact that the optical length along the ray from a point in the first medium to a point in the last medium is a limiting value, that is, is a maximum or a minimum.*

Let x_1, y_1; x_2, y_2; etc. denote the coördinates of the points where the ray meets the various refracting surfaces in succession; x_m, y_m being the coördinates in the case of the mth or last surface. All these systems of coördinates are chosen so that the z-axis coincides with the normal to the surface and the xy-plane is tangent to the surface. Then the first conditions, that the optical length of the ray, which is denoted by Ψ, shall be a limiting value, are:

$$\frac{d\Psi}{dx_1} = 0\,, \qquad \frac{d\Psi}{dy_1} = 0\,,$$

$$\frac{d\Psi}{dx_2} = 0\,, \qquad \frac{d\Psi}{dy_2} = 0\,,$$

etc .

$$\frac{d\Psi}{dx_m} = 0\,, \qquad \frac{d\Psi}{dy_m} = 0\,.$$

By the theorem just proved, the first pair of these equations is equivalent to the condition that the ray shall be refracted at the first surface according to the known law of refraction; and the second pair of equations is equivalent to the same thing for the second surface; and so on for each pair of equations for each surface in turn. And therefore the path of the ray is given by the above theorem in accordance with the law of refraction.

In this case also the investigation of the first derivatives of the optical length is sufficient. Whether the path of the ray is a maximum for all positions of the point of incidence, or a minimum for all positions, or whether it is a maximum for some positions and a minimum for others, can only be ascertained, of course, by investigating the second derivatives; but this is not the question at present. Hence, in the following discussion, we can speak of all values of the optical length of the ray as limiting values, provided the first derivatives of this function satisfy the maximum and minimum conditions; without stopping to consider the sign and magnitude of the second derivative. The influence of the second derivative on the problem here considered will appear presently.

II. *If a bundle of homocentric incident rays is refracted through a series of optical media, separated by surfaces of continuous curvature, the rays that emerge in the last medium will be normal to a surface which is the locus of all points for which the optical lengths along the different rays have the same value.*

The notation will be the same as that used above. The terminus of the ray is in a curved surface for which

$$\Psi = \text{const} \quad . \quad . \quad . \quad . \quad . \quad . \quad . \quad . \quad . \quad . \quad . \qquad (1)$$

The coördinates of the points of this surface will be referred to the systems of axes used for the points of the last refracting surface. Let the coördinates of a point on the surface $\Psi = C$ be denoted by (a, b, c) where c is to be regarded as a function of a and b.

Consider now two adjacent rays of the bundle. The coördinates of the points where the first ray meets each of the surfaces in succession are:

$$x_1\,,\; y_1\,;\; x_2\,,\; y_2\,;\; \text{etc}\,.\,;\; x_m\,,\; y_m\,;\; a\,,\; b\,,\; c\,;$$

and for the second ray the corresponding coördinates will be:

$$\begin{gathered} x_1 + \Delta x_1\,,\quad y_1 + \Delta y_1\,;\quad x_2 + \Delta x_2\,,\; y_2 + \Delta y_2\,;\; \text{etc}\,. \\ x_m + \Delta x_m\,,\; y_m + \Delta y_m\,,\quad a + \Delta a\,,\quad b + \Delta b\,,\; c + \Delta c\,, \end{gathered}$$

The fact that c is a function of a, b is expressed by the equation.

$$\Delta c = \frac{dc}{da}\,\Delta a + \frac{dc}{db}\,\Delta b\,.$$

If Ψ, $\Psi + \Delta\Psi$ denote the optical lengths along the two rays, then for infinitesimal values of the differences:

$$\begin{aligned} \Psi + \Delta\Psi = \Psi &+ \frac{d\Psi}{dx_1}\Delta x_1 + \frac{d\Psi}{dx_2}\Delta x_2 + \cdots + \frac{d\Psi}{dx_m}\Delta x_m + \left(\frac{d\Psi}{da} + \frac{d\Psi}{dc}\cdot\frac{dc}{da}\right)\Delta a \\ &+ \frac{d\Psi}{dy_1}\Delta y_1 + \frac{d\Psi}{dy_2}\Delta y_2 + \cdots + \frac{d\Psi}{dy_m}\Delta y_m + \left(\frac{d\Psi}{db} + \frac{d\Psi}{dc}\cdot\frac{dc}{db}\right)\Delta b\,. \end{aligned}$$

Now as the value of Ψ is constant over the surface whose points are given by the coördinates a, b, c, we must have

$$\Delta\Psi = 0\,.$$

Moreover, by the foregoing theorem,

$$0 = \frac{d\Psi}{dx_1} = \frac{d\Psi}{dy_1} = \frac{d\Psi}{dx_2} = \frac{d\Psi}{dy_2} = \text{etc}\,.\,,$$

hence

$$\left(\frac{d\Psi}{da} + \frac{d\Psi}{dc}\cdot\frac{dc}{da}\right)\Delta a + \left(\frac{d\Psi}{db} + \frac{d\Psi}{dc}\cdot\frac{dc}{db}\right)\Delta b = 0\,.$$

This equation must be satisfied for all values of $\frac{\Delta a}{\Delta b}$; and, consequently, we must have:

$$\left.\begin{aligned} \frac{d\Psi}{da} + \frac{d\Psi}{dc}\cdot\frac{dc}{da} = 0 \\ \frac{d\Psi}{db} + \frac{d\Psi}{dc}\cdot\frac{dc}{db} = 0 \end{aligned}\right\} \quad\ldots\ldots\ldots \tag{2}$$

Let $r_0, r_1, \ldots r_m$ denote the lengths of the portions of the path of the ray in the various media, and let $n_0, n_1, \ldots n_m$ denote the indices of refraction; then

$$\Psi = n_0 r_0 + n_1 r_1 + \cdots + n_m r_m .$$

In this equation r is a function of a, b and c; consequently,

$$\frac{d\Psi}{da} = n_m \frac{dr_m}{da} = n_m \frac{a - x_m}{r_m}$$

$$\frac{d\Psi}{db} = n_m \frac{dr_m}{db} = n_m \frac{b - y_m}{r_m}$$

$$\frac{d\Psi}{dc} = n_m \frac{dr_m}{dc} = n_m \frac{c - z_m}{r_m},$$

and hence equations (2) become:

$$\left.\begin{aligned} (a - x_m) + (c - z_m)\frac{dc}{da} = 0 \\ (b - y_m) + (c - z_m)\frac{dc}{db} = 0 \end{aligned}\right\} \quad \ldots\ldots \quad (2a)$$

The interpretation of these equations is that the straight line drawn from the point (x_m, y_m, z_m) to the point (a, b, c) is the normal to the surface $\Psi = C$ at the latter point.

The easiest way to see this is by recalling that the distance measured along the normal to the surface is itself the longest or shortest distance from a given point to the surface. Now if the distance

$$r_m = \sqrt{(x_m - a)^2 + (y_m - b)^2 + (z_m - c)^2};$$

between (x_m, y_m, z_m) and (a, b, c) is to be a maximum or minimum, then we must have:

$$0 = \frac{dr_m}{da} + \frac{dr_m}{dc}\frac{dc}{da} = \frac{a - x_m}{r_m} + \frac{dc}{da}\cdot\frac{c - z_m}{r_m},$$

$$0 = \frac{dr_m}{db} + \frac{dr_m}{dc}\frac{dc}{da} = \frac{b - y_m}{r_m} + \frac{dc}{db}\cdot\frac{c - z_m}{r_m},$$

These conditions are the same as equations (2a). And so the ray that goes through (a, b, c) is normal to the surface $\Psi = C$ that passes through this same point.

As the light traverses equal optical lengths in equal times, it reaches all points of the surface $\Psi = C$ at the same instant, and hence this surface is a *wave-surface*, that is, it is a surface which contains all those points where the aether-vibrations are precisely in the same phase.[1]

Procedure of an infinitely narrow bundle of rays. Having proved that rays, emanating originally from one point, and undergoing any number of refractions at a system of surfaces of continuous curvature, will all be normal to a certain surface, the so-called *wave-surface,* we have therefore simply to consider a bundle of optical rays as being geometrically the system of normals to a curved surface, and governed therefore by the same laws. Accordingly, if a plane is passed through a certain ray A, it will intersect the wave-surface in a curve whose curvature at the point where the ray meets the surface will generally be different for different azimuths of the plane. According to the theory of curved surfaces, *the sections of greatest and least curvature will be in planes at right angles to each other.* If normals are drawn to the wave-surface that are infinitely near the ray A and that therefore represent adjacent rays, those normals that are in the sections of greatest and least curvatures will intersect the ray A at the centres of the circles of greatest and least curvature, respectively; whereas those normals that do not lie in one or other of these two principal sections will not intersect the ray A at all. Thus, in general, along every ray there are two focal points which are the centres of greatest and least curvature of the wave-surface at the point where the ray crosses it. When the surface has the same curvature at this place in all directions, the two focal points on the ray will coincide; and in this one exceptional case the adjacent rays will all meet the ray A in one point.

In order to formulate these propositions analytically, let us consider a system of rectangular axes whose z-axis coincides with the ray A. The coördinates of any point on the wave-surface will be denoted as follows:

$$x = a \, , \quad y = b \, , \quad z = c \, .$$

The equation of the surface itself will be expressed in such a form that c is given as a function of a and b. For the given set of axes, when

$$a = b = 0 \text{ , then also } \frac{dc}{da} = \frac{dc}{db} = 0 \quad . \quad . \quad . \quad . \quad . \qquad (1)$$

Let x, y, z denote the coördinates of a point on the normal to the wave-surface at the point (a, b, c); then as in proposition II, equation (2a):

[1] ¶This is equivalent to the law of MALUS published in 1808, that *rays of light meet the wave-surface normally;* and, conversely, *The system of surfaces which intersect at right angles rays emanating originally from a point-source is a system of wave-surfaces.* (J. P. C. S.)

$$\left.\begin{aligned}(a-x)+(c-z)\frac{dc}{da}&=0\\(b-y)+(c-z)\frac{dc}{db}&=0\end{aligned}\right\}\quad\ldots\ldots\quad(1\text{a})$$

The magnitudes $a+\Delta a$, $b+\Delta b$ differ infinitesimally from a, b; and if the former are substituted in the above equations in place of the latter, these equations become:

$$(a+\Delta a-x)+\left(c+\frac{dc}{da}\Delta a+\frac{dc}{db}\Delta b-z\right)\frac{dc}{da}$$
$$+(c-z)\left(\frac{d^2c}{da^2}\Delta a+\frac{d^2c}{da\cdot db}\Delta b\right)=0\,,$$
$$(b+\Delta b-y)+\left(c+\frac{dc}{da}\Delta a+\frac{dc}{db}\Delta b-z\right)\frac{dc}{db}$$
$$+(c-z)\left(\frac{d^2c}{da\cdot db}\Delta a+\frac{d^2c}{db^2}\Delta b\right)=0\,.$$

Putting $a=b=0$, and also equating to zero the first derivatives of c, according to equations (1), we shall obtain the following equations of a normal to the wave-surface at a point infinitely near that where the ray A meets this surface:

$$\left.\begin{aligned}\Delta a-x+(c-z)\left(\frac{d^2c}{da^2}\Delta a+\frac{d^2c}{da\cdot db}\Delta b\right)&=0\\\Delta b-y+(c-z)\left(\frac{d^2c}{da\cdot db}\Delta a+\frac{d^2c}{db^2}\Delta b\right)&=0\end{aligned}\right\}\quad\ldots\quad(2)$$

For all points on the ray A, we have $x=y=0$. Therefore, if the ray whose equations are given by (2) does intersect the ray A, x and y must vanish on this other ray also for any value of z. Putting $x=y=0$ in the above equations and eliminating z, we obtain the following equation as the condition of intersection of two adjacent rays:

$$\frac{d^2c}{da\cdot db}\Delta a^2+\left(\frac{d^2c}{db^2}-\frac{d^2c}{da^2}\right)\Delta a\,\Delta b-\frac{d^2c}{da\cdot db}\Delta b^2=0\quad\cdot\quad\cdot\quad(3)$$

If r denotes the distance between the two adjacent points on the surface, and α the angle which it makes with the x-axis (this angle being comprised, therefore, between 0 and π), then

$$\Delta a=r\cos\alpha\,,\qquad\qquad\Delta b=r\sin\alpha\,.$$

For abbreviation, let us write

$$2n = \frac{\dfrac{d^2c}{db^2} - \dfrac{d^2c}{da^2}}{\dfrac{d^2c}{da \cdot db}};$$

then supposing $\frac{d^2c}{da\,.\,db}$ is not equal to zero, equation (3) becomes:

$$\tan^2 \alpha - 2n \tan \alpha = 1 \quad . \; . \; . \; . \; . \; . \; . \quad (3a)$$

and hence

$$\tan \alpha = n \pm \sqrt{1 + n^2} \quad . \; . \; . \; . \; . \; . \; . \quad (3b)$$

The two values of tan α, which are always real, may also be written

$$n + \sqrt{1 + n^2} \quad \text{and} \quad -\frac{1}{n + \sqrt{1 + n^2}}.$$

Accordingly, if α_0 is one of the values of α, the other value is $\alpha_0 + \frac{\pi}{2}$ or $\alpha_0 - \frac{\pi}{2}$. The two angles differ by a right angle. The magnitude r, which denotes the distance of the normals measured along the wave-surface, vanishes from equation (3a). Hence, the ray A is cut by all adjacent rays that lie in planes that make angles α_0 and $\alpha_0 + \frac{\pi}{2}$ with the axis of x.

Thus far the actual positions of the axes of x and y in the plane perpendicular to the ray A have been perfectly arbitrary. Now let us suppose, for simplicity, that the xz-plane and yz plane coincide with the two principal planes of the wave-surface at the origin of coördinates, that is, with the two perpendicular planes in which the adjacent rays lie that intersect the ray A (or z-axis). In this case the two values of tan α will become 0 and ∞, and this means that

$$n = \pm \infty$$

and

$$\frac{d^2c}{da \cdot db} = 0.$$

The condition of intersection of the ray A by the adjacent rays, as expressed by equation (3), reduces, under these circumstances, to

$$\left(\frac{d^2c}{db^2} - \frac{d^2c}{da^2}\right) \Delta a \, \Delta b = 0.$$

Now, as a matter of fact, this latter equation will be always satisfied, provided either $\Delta a = 0$ or $\Delta b = 0$, that is, provided the intersecting normals lie either in the yz-plane or in the xz-plane. And, finally, if at the same time

$$\frac{d^2c}{db^2} - \frac{d^2c}{da^2} = 0 ,$$

the condition that adjacent rays shall intersect the ray A is satisfied for all arbitrary values of Δa and Δb; that is, all the adjacent rays meet the ray A. Still supposing that $\frac{d^2c}{da\,db} = 0$, and then putting either $\Delta a = 0$ or $\Delta b = 0$, we find, as was mentioned above, the distance z of the point where the adjacent rays meet the ray A by putting $x = y = 0$ in equations (3).

For the rays in the xz-plane, $\Delta b = 0$; hence from equations (2) the distance of the point of intersection from the wave-surface is

$$z - c = \frac{1}{\frac{d^2c}{da^2}} .$$

The second of equations (2) becomes $0 = 0$.

For the rays in the yz-plane, $\Delta a = 0$, and

$$z - c = \frac{1}{\frac{d^2c}{db^2}} .$$

Finally, if $\frac{d^2c}{da^2} = \frac{d^2c}{db^2} = \frac{1}{\rho}$, then for all adjacent rays without distinction

$$z - c = \rho .$$

Moreover, in this case the xz-plane and yz-plane are also the principal sections of the surface for which the curvature has its maximum and minimum values; and the values of the corresponding radii of curvature are:

$$\rho_a = \frac{1}{\frac{d^2c}{da^2}}, \qquad \rho_b = \frac{1}{\frac{d^2c}{db^2}},$$

and hence the focal points of the bundle of rays are also at the centres of principal curvature of the wave-surface.

Constitution of an infinitely narrow bundle of rays that meets the wave-surface in a circle. In order to get a clearer notion of the way the rays go in an infinitely narrow bundle of rays, let us consider the constitution of a bundle of rays that cuts out a circle on the wave-surface. Therefore, in equations (2), as before, we put

$$\frac{d^2c}{da \cdot db} = 0 \text{ and } \Delta a = r \cos \alpha \, , \Delta b = r \sin \alpha$$

and obtain:

$$r \cos \alpha - x + (c - z) \frac{d^2c}{da^2} r \cos \alpha = 0 \, ,$$

$$r \sin \alpha - y + (c - z) \frac{d^2c}{db^2} r \sin \alpha = 0 \, .$$

In order to find the curve in which the surface of the bundle cuts a plane perpendicular to its axis, we must put $z =$constant, and eliminate the angle α. By way of abbreviation, let us put

$$p = + r \left[1 + (c - z) \frac{d^2c}{da^2}\right] = + \frac{r}{\rho_a} [\rho_a + c - z] \, ,$$

$$q = + r \left[1 + (c - z) \frac{d^2c}{db^2}\right] = + \frac{r}{\rho_b} [\rho_b + c - z] \, ,$$

so that we may write:

$$\frac{x^2}{p^2} + \frac{y^2}{q^2} = 1 \, .$$

This is the equation of an ellipse, with its axes $2p$, $2q$ parallel to the axes of x and y. The smaller r is, the shorter both axes become; and hence if the bundle of rays meets the first wave-surface not simply in the points in the circumference, but at all the points also in the interior, of a circle, all the rays continue to be comprised inside the space that is formed by the outside rays, so that the form of the bundle is determined by the latter. On the wave-surface itself where the investigation of the bundle begins, we have $c - z = 0$, and therefore the semi-axes $p = q = r$, that is, the cross section at this place is circular. The axis p collapses into a point when

$$z - c = \frac{1}{\frac{d^2c}{da^2}} = \rho_a \, ,$$

that is, when the section of the bundle is made at the focal point of the rays lying in the xz-plane. The other semi-axis at this place is

$$q = \pm \frac{r}{\rho_b} (\rho_a + \rho_b) \, .$$

The section of the bundle, therefore, is a straight line parallel to the y-axis, whose length is twice that of the value of q given above.

On the other hand, the section of the bundle will be a straight line parallel to the x-axis, when

$$z - c = \frac{1}{\frac{d^2c}{db^2}} = \rho_b \,,$$

$$q = 0 \,, \qquad p = \pm \frac{r}{\rho_a} (\rho_a + \rho_b) \,.$$

Finally, there is one other place where the section of the bundle is a circle, namely, the place where

$$p = -q \,,$$

$$1 + \frac{c - z}{\rho_a} = -1 - \frac{c - z}{\rho_b} \,,$$

$$z - c = \frac{2\rho_a\rho_b}{\rho_a + \rho_b} \,.$$

and here

$$p = q = \pm r \cdot \frac{\rho_a - \rho_b}{\rho_a + \rho_b} \,.$$

Between the two circular sections of the bundle one of the linear sections must lie. The longer axes of the elliptical sections that are between the two circular sections are parallel to this linear section; whereas the longer axes of the elliptical sections that are beyond this

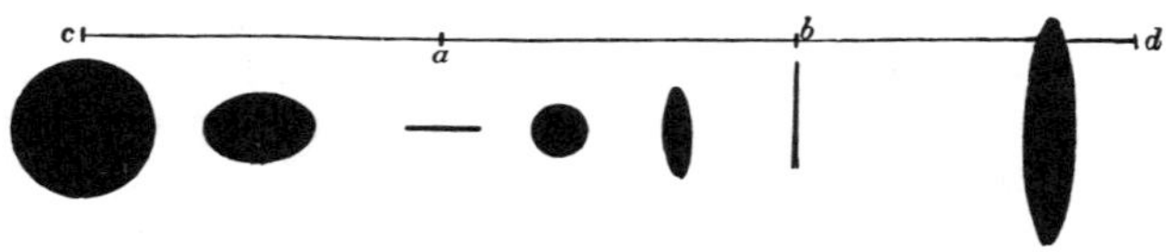

Fig. 10.

interval are perpendicular to this linear section. The horizontal line cd in Fig. 10 is intended to represent the central ray of the bundle. A circular diaphragm is supposed to be at c. The focal points are marked at a and b. Below the line are shown the forms of the cross sections of the bundle corresponding to the points in the ray above.

General analytical condition for the positions of the focal points. Consider a pair of contiguous rays A and B which are supposed to have a common origin. After being refracted at a series of surfaces of continuous curvature, they meet again at a focal point. The optical lengths along these two rays from their starting point to their focal point will be denoted by Ψ and $\Psi + \Delta\Psi$. The different systems of coördinates to which the points on the various refracting surfaces are referred will be chosen again so that the z-axis in each instance is along the incidence-normal belonging to the ray A; the xy-plane being tangent to the refracting surface. The coördinates of the points where the ray B meets the first refracting surface will be denoted by x_1, y_1, z_1; and those of the point where it meets the second refracting surface by x_2, y_2, z_2; and so on, the coördinates for the last surface being x_m, y_m, z_m. However, it will be assumed in this discussion that the optical lengths are expressed as functions of x, y only; that is, that the z's which are themselves functions of x and y have been eliminated. Moreover, as the rays A and B are assumed to be infinitely close to each other, the magnitudes x_1, y_1, etc., to x_m, y_m are regarded as being infinitesimal.

Then by Taylor's theorem

$$\Psi + \Delta\Psi = \Psi + \frac{d\Psi}{dx} x_1 + \frac{d\Psi}{dx_2} x_2 + \cdots + \frac{d\Psi}{dx_m} x_m + \frac{d\Psi}{dy_1} y_1 + \frac{d\Psi}{dy_2} y_2 + \cdots + \frac{d\Psi}{dy_m} y_m \,.$$

Now the optical length along either ray must be a limiting value, according to the first of the theorems proved above, that is, the first derivatives of Ψ and $\Psi + \Delta\Psi$, with respect to each of the coördinates x_1, y_1; x_2, y_2; and so on to x_m, y_m, must be equal to zero. Thus, for the first ray:

$$\frac{d\Psi}{dx_1} = 0\,, \quad \frac{d\Psi}{dx_2} = 0\,, \cdots, \frac{d\Psi}{dx_m} = 0\,,$$

$$\frac{d\Psi}{dy_1} = 0\,, \quad \frac{d\Psi}{dy_2} = 0\,, \cdots, \frac{d\Psi}{dy_m} = 0\,;$$

and, taking account of these relations, we have the following system of equations for the second ray:

$$\left.\begin{array}{l}
\dfrac{d^2\Psi}{dx_1^2} x_1 + \dfrac{d^2\Psi}{dx_1 dy_1} y_1 + \cdots + \dfrac{d^2\Psi}{dx_1 dx_m} x_m + \dfrac{d^2\Psi}{dx_1 dy_m} y_m = 0 \\
\dfrac{d^2\Psi}{dy_1 dx_1} x_1 + \dfrac{d^2\Psi}{dy_1^2} y_1 + \cdots + \dfrac{d^2\Psi}{dy_1 dx_m} x_m + \dfrac{d^2\Psi}{dy_1 dy_m} y_m = 0 \\
\cdots\cdots\cdots\cdots\cdots\cdots\cdots\cdots\cdots \\
\dfrac{d^2\Psi}{dx_m dx_1} x_1 + \dfrac{d^2\Psi}{dx_m dy_1} y_1 + \cdots + \dfrac{d^2\Psi}{dx^2_m} x_m + \dfrac{d^2\Psi}{dx_m dy_m} y_m = 0 \\
\dfrac{d^2\Psi}{dy_m dx_1} x_1 + \dfrac{d^2\Psi}{dy_m dy_1} y_1 + \cdots + \dfrac{d^2\Psi}{dy_m dx_m} x_m + \dfrac{d^2\Psi}{dy^2_m} y_m = 0
\end{array}\right\} \quad (4)$$

Incidentally, the number of terms in these equations will be considerably reduced by the fact that $\frac{d^2\Psi}{dx_f dx_g}$, $\frac{d^2\Psi}{dx_f dy_g}$ and $\frac{d^2\Psi}{dy_f dy_g}$ will vanish whenever the subscripts f and g differ from each other by more than unity.

There are 2 m equations with 2 m unknowns, namely, the x's and y's from x_1, y_1 to x_m, y_m. However, since all these unknown magnitudes cannot be equal to zero (because the ray B is not the same as the ray A) the equations may all be divided by one of these magnitudes, say, x_1, that is not equal to zero, which will have the effect of reducing the total number of unknown quantities to $(2\,m-1)$ ratios all having the common denominator x_1. Thus, when these $(2\,m-1)$ quantities are eliminated from the 2 m equations, one equation will be left that does not any longer contain x_1, y_1, etc., to x_m, y_m, but simply the second partial derivatives of Ψ. This last equation, obtained by putting the determinant of equations (4) equal to zero, is the required equation for the position of the focal point.

This determinant is easily formed by known methods;[1] it consists of a sum of terms the first of which is the product

$$\frac{d^2\Psi}{dx_1 \cdot dx_1} \cdot \frac{d^2\Psi}{dy_1 dy_1} \cdot \frac{d^2\Psi}{dx_2 \cdot dx_2} \quad \cdots \quad \frac{d^2\Psi}{dx_m \cdot dx_m} \cdot \frac{d^2\Psi}{dy_m \cdot dy_m}.$$

The other terms of the series are easily obtained. The denominator of each of the differential coefficients is a product of two factors. What has to be done is to leave the first factor just as it is, while the other factor is varied in every possible way, merely changing the sign of the term whenever two of these factors are interchanged with each other.

Thus, in the language of the calculus of variations, the position of a ray between its two terminal points is found by putting the first variation of its optical length equal to zero. The terminal points will be conjugate foci, provided the second variation of the optical length vanishes also. In the latter case the optical length is not necessarily a maximum or minimum.

Refraction in a Prism

Let the position of the luminous point be given by the coördinates a, b, c, referred to a system of axes in which the c-axis coincides with the refracting edge of the prism and the bc-plane is the same as that of the first face of the prism, the positive direction of the a-axis being outside the prism. Suppose that the ray is incident on the first face

[1] See Jacobi in Crelles *Journ. für Math.* XXII.

at the point whose coördinates are $a=0$, $b=y$, $c=z$. Similarly, let α, β, γ denote the rectangular coördinates of a point on the emergent ray, referred to another system of axes whose γ-axis again coincides with the edge of the prism and whose $\beta\gamma$-plane is the same as that of the second face of the prism, the positive direction of the α-axis being likewise outside the prism. The γ's are all measured from the same point in the edge as the c's, that is, the ab-plane of the first system is identical with the $\alpha\beta$-plane of the second system. Let the coördinates of the point where the ray emerges at the second face be denoted by $\alpha=0$, $\beta=v$, $\gamma=\zeta$. The refracting angle of the prism will be denoted by φ and the relative index of refraction of the two media by n. The lengths of the paths of the three portions of the ray, before entering the prism, inside it, and after leaving it, will be denoted by r_0, r_1 and r_2, respectively; and the total optical length will be denoted by Ψ. Then

$$\left.\begin{aligned} r_0 &= \sqrt{a^2 + (b-y)^2 + (c-z)^2} \\ r_1 &= \sqrt{y^2 - 2yv\cos\varphi + v^2 + (z-\zeta)^2} \\ r_2 &= \sqrt{\alpha^2 + (\beta - v)^2 + (\gamma-\zeta)^2} \\ \Psi &= r_0 + nr_1 + r_2 \end{aligned}\right\} \quad . \quad . \quad . \quad . \quad (5)$$

The coöordinates of the second system in terms of those of the first system are given by the formulæ:

$$\left.\begin{aligned} \alpha &= -a\cos\varphi - b\sin\varphi \\ \beta &= -a\sin\varphi + b\cos\varphi \\ \gamma &= c \end{aligned}\right\} \quad . \quad . \quad . \quad . \quad . \quad . \quad . \quad (5a)$$

If the ray is refracted according to the law of refraction, then, by theorem I above, the following conditions must be satisfied:

$$\left.\begin{aligned} 0 &= \frac{d\Psi}{dy} = \frac{y-b}{r_0} + n\,\frac{y - v\cos\varphi}{r_1} \\ 0 &= \frac{d\Psi}{dv} = \frac{v-\beta}{r_2} + n\,\frac{v - y\cos\varphi}{r_1} \\ 0 &= \frac{d\Psi}{dz} = \frac{z-c}{r_0} + n\,\frac{z-\zeta}{r_1} \\ 0 &= \frac{d\Psi}{d\zeta} = \frac{\zeta-\gamma}{r_2} + n\,\frac{\zeta - z}{r_1} \end{aligned}\right\} \quad . \quad . \quad . \quad . \quad (6)$$

Here let us introduce the following abbreviations:

$$\left.\begin{aligned} \frac{b-y}{nr_0} &= \frac{y - v\cos\varphi}{r_1} = \cos m \\ \frac{\beta - v}{nr_2} &= \frac{v - y\cos\varphi}{r_1} = \cos\mu \\ \frac{c-z}{nr_0} &= \frac{\zeta-\gamma}{nr_2} = \frac{z-\zeta}{r_1} = \cos\nu \end{aligned}\right\} \quad . \quad . \quad . \quad . \quad (6a)$$

where

$$\sin^2\varphi \sin^2\nu = \cos^2 m + 2\cos m \cos\mu \cos\varphi + \cos^2\mu \quad . \quad . \quad (6b)$$

If the second derivatives of Ψ are developed in terms of this notation, the system of equations (4), which give the positions of the focal points and the relations between the infinitely small differences Δy, Δz, Δv, $\Delta\zeta$ of the coördinates y, z, v and ζ for a pair of adjacent rays intersecting each other at conjugate foci, becomes:

$$\left.\begin{aligned} &\left[\frac{1}{r_0}(1 - n^2\cos^2 m) + \frac{n}{r_1}\sin^2 m\right]\Delta y - \left(\frac{n^2}{r_0} + \frac{n}{r_1}\right)\cos m \cos\nu\, \Delta z \\ &\quad - \frac{n}{r_1}(\cos\varphi + \cos m\cos\mu)\,\Delta v + \frac{n}{r_1}\cos m\cos\nu\,\Delta\zeta = 0\,; \end{aligned}\right\} \quad (7a)$$

$$\left.\begin{aligned} &-\left(\frac{n^2}{r_0} + \frac{n}{r_1}\right)\cos m\cos\nu\,\Delta y + \left[\frac{1}{r_0}(1 - n^2\cos^2\nu) + \frac{n}{r_1}\sin^2\nu\right]\Delta z \\ &\quad - \frac{n}{r_1}\cos\mu\cos\nu\,\Delta v - \frac{n}{r_1}\sin^2\nu\,\Delta\zeta = 0\,; \end{aligned}\right\} \quad (7b)$$

$$\left.\begin{aligned} &-\frac{n}{r_1}(\cos\varphi + \cos m\cos\mu)\,\Delta y - \frac{n}{r_1}\cos\mu\cos\nu\,\Delta z \\ &\quad + \left[\frac{1}{r_2}(1 - n^2\cos^2\mu) + \frac{n}{r_1}\sin^2\mu\right]\Delta v + \left(\frac{n^2}{r_2} + \frac{n}{r_1}\right)\cos\mu\cos\nu\,\Delta\zeta = 0\,; \end{aligned}\right\} \quad (7c)$$

$$\left.\begin{aligned} &\frac{n}{r_1}\cos m\cos\nu\,\Delta y - \frac{n}{r_1}\sin^2\nu\,\Delta z \\ &\quad + \left(\frac{n^2}{r_2} + \frac{n}{r_1}\right)\cos\mu\cos\nu\,\Delta v + \left[\frac{1}{r_2}(1 - n^2\cos^2\nu) + \frac{n}{r_1}\sin^2\nu\right]\Delta\zeta = 0\,. \end{aligned}\right\} \quad (7d)$$

Generally, the length of the path of the ray inside the prism (r_1) may be neglected in comparison with the lengths r_0 and r_2 outside the prism. If the four equations above are each multiplied by r_1, and then all terms neglected that contain $\frac{r_1}{r_0}$ or $\frac{r_1}{r_2}$ as a factor, as being infinitely small, the four equations will reduce to three as follows:

$$\left.\begin{aligned} \sin^2 m\,\Delta y - (\cos\varphi + \cos m \cos\mu)\,\Delta v - \cos m \cos\nu\,(\Delta z - \Delta\zeta) &= 0\,,\\ -\cos m \cos\nu\,\Delta y - \cos\mu\cos\nu\,\Delta v + \sin^2\nu\,(\Delta z - \Delta\zeta) &= 0\,,\\ -(\cos\varphi + \cos m\cos\mu)\,\Delta y + \sin^2\mu\,\Delta v - \cos\mu\cos\nu\,(\Delta z - \Delta\zeta) &= 0\,. \end{aligned}\right\} \quad (8)$$

However, one of these three equations can be deduced from the other two; and hence, after eliminating $(\Delta z - \Delta\zeta)$, we get:

$$(\cos\mu + \cos m \cos\varphi)\,\Delta y = (\cos m + \cos\mu\cos\varphi)\,\Delta v$$

or

$$\frac{\Delta y}{y} = \frac{\Delta v}{v} \quad (8a)$$

or after elimination of Δv:

$$(\Delta z - \Delta\zeta)\,(\cos m + \cos\mu\cos\varphi) = \cos\nu\sin^2\varphi\;\Delta y$$

or

$$\frac{\Delta z - \Delta\zeta}{z - \zeta} = \frac{\Delta y}{y} = \frac{\Delta v}{v}\,. \quad \cdot\ \cdot\ (8b)$$

These two equations are simply the conditions that the two rays may be considered as being sensibly parallel during their short routes through the prism; as must obviously be so, provided the point where they meet is infinitely remote as compared with the length of the path inside the prism.

The next step is to express two of the unknown magnitudes Δv and $\Delta\zeta$ in terms of the other two Δy and Δz. This involves obtaining by elimination from equations (7) two new equations which do not contain the small magnitude r_1, and from which the ratios $\frac{\Delta z}{\Delta y}$ and $\frac{r_2}{r_0}$ may be found.

One equation of this kind is obtained by adding (7b) and (7d):

$$\left.\begin{aligned} -\frac{n^2}{r_0}\cos m\cos\nu\,\Delta y + \frac{1}{r_0}(1 - n^2\cos^2\nu)\,\Delta z + \frac{n^2}{r_2}\cos\mu\cos\nu\,\Delta v &\\ + \frac{1}{r_0}(1 - n^2\cos^2\nu)\,\Delta\zeta = 0. & \end{aligned}\right\} \cdot\ \cdot\ (8c)$$

In order to obtain the second equation, multiply equation (7a) by

$$y = \frac{r_1}{\sin^2\varphi}(\cos m + \cos\mu\cos\varphi)\,,$$

equation (7c) by

$$v = \frac{r_1}{\sin^2\varphi}(\cos\mu + \cos m\cos\varphi)\,,$$

and equation (7b) by

$$z - \zeta = r_1 \cos \nu \,;$$

and add the three equations thus obtained. All the terms that are multiplied by $\frac{1}{r_1}$ will vanish, and we shall get:

$$\left.\begin{aligned} &\frac{y}{r_0}\left\{(1 - n^2\cos^2 m)\,\Delta y - n^2 \cos m\ \cos \nu\,\Delta z\right\} \\ &\qquad + \frac{z-\zeta}{r_0}\left\{-n^2\cos m\cos\nu\,\Delta y + (1 - n^2\cos^2\nu)\,\Delta z\right\} \\ &\qquad + \frac{v}{r_2}\left\{(1 - n^2\cos^2\mu)\,\Delta v + n^2\cos\mu\cos\nu\,\Delta\zeta\right\} = 0 \end{aligned}\right\} \quad . \; . \quad (8d)$$

If the values of Δv and $\Delta\zeta$ in terms of Δy and Δz as obtained from equations (8a) and (8b) are substituted in equations (8c) and (8d), two equations will be obtained containing the unknown quantities $\frac{\Delta z}{\Delta y}$ and $\frac{r_2}{r_0}$. When one of them is eliminated, the other is given by a quadratic equation which has two roots. Thus, for any arbitrary combination of values of the angles m, μ, ν, we get at least one definite numerical value of the ratio $\frac{r_2}{r_0}$. Consequently, for a given direction of the bundle of rays r_2 is proportional to r_0, supposing that the latter varies. If r_0 is infinite, so also is r_2. It is not worth while actually to give the elimination equations here. We shall merely investigate certain special cases that are of interest to us.

First, let us inquire in what cases *a homocentric bundle of incident rays will issue from the prism as a homocentric bundle of emergent rays.*[1] If all the rays emanating from the luminous point are to intersect each other, the conditions (8c) and (8d) must be satisfied, no matter what values we take for Δy and Δz. Each of these magnitudes, therefore, may be put equal to zero, and thus the following conditions are obtained.

1. If we put $\Delta y = 0$ in equations (8c), which, according to equations (8a) and (8b), means also that $\Delta v = 0$ and $\Delta\zeta = \Delta z$, then

$$\left(\frac{1}{r_0} + \frac{1}{r_2}\right)(1 - n^2\cos^2\nu) = 0 \quad . \; . \; . \; . \; . \; . \; . \quad (9a)$$

[1] ¶This whole subject has been beautifully treated synthetically by L. Burmester, Homocentrische Brechung des Lichtes durch das Prisma. *Zft. f. Math. u. Phys.*, XL (1895), 65–90. See also: J. P. C. Southall, *The principles and methods of geometrical optics*, 1910, pp. 97–105. (J. P. C. S.)

Since by equation (6a) $n \cos \nu = \dfrac{c - z}{r_0}$, the second factor of the above equation cannot be equal to zero unless $r_0 = c - z$, that is, unless the ray of light grazes the first face of the prism and therefore does not enter it. Consequently, the first factor of equation (9a) must be zero, that is,

$$r_2 = -r_0.$$

2. If we put $\Delta z = 0$ in equation (8d) and $r_2 = -r_0$, then

$$0 = (1 + n^2 \sin^2 \nu + n^2 \cos^2 \nu)(\cos^2 m - \cos^2 \mu) .$$

The first factor is equal to $(1+n^2)$, which is never zero; consequently,

$$\cos m = \pm \cos \mu \quad . \quad . \quad . \quad . \quad . \quad . \quad . \quad . \quad . \quad (9b)$$

3. If we put $\Delta z = 0$ in equation (8c) or $\Delta y = 0$ in equation (8d), and $r_2 = -r_0$, then, taking account of equation (6b), we get:

$$(1 - n^2) \cos v \sin^2 \varphi = 0 .$$

Since φ is the refracting angle of the prism, $\sin \varphi$ cannot be zero; and hence

$$\left.\begin{aligned} \cos v &= 0 , \\ c = z &= \zeta = \gamma . \end{aligned}\right\} \quad . \quad . \quad . \quad . \quad . \quad . \quad . \quad (9c)$$

Accordingly, the ray lies wholly in a principal section of the prism, that is, in a plane perpendicular to its refracting edge. Under these circumstances, let us write equation (9b) in conformity with (6a) in the form:

$$y - v \cos \varphi = \pm (v - y \cos \varphi) ,$$
$$y(1 \pm \cos \varphi) = \pm v(1 \pm \cos \varphi) ,$$

that is,

$$y = v \quad . \quad . \quad . \quad . \quad . \quad . \quad . \quad . \quad . \quad . \quad (9d)$$

Let ϵ, ϵ_1 denote the angles of incidence and refraction at the first face of the prism, and η_1, η the angles of incidence and refraction at the second face (the symbols with the subscript referring to the two angles inside the prism); then

$$\cos \epsilon_1 = \frac{v \sin \varphi}{r_1} , \qquad \cos \eta_1 = \frac{y \sin \varphi}{r_1} ,$$

and hence under the given conditions:

$$\cos \epsilon_1 = \cos \eta_1 ,$$

and therefore also

$$\sin \epsilon = n \sin \epsilon_1 = n \sin \eta_1 = \sin \eta .$$

In other words, the angle of emergence at the second face is equal to the angle of incidence at the first face.

If the bundle of emergent rays is to be homocentric, the path of the chief ray through the prism must be in the direction of minimum deviation.

When the coördinates a, b, c, x and y as referred to the first system of axes are transformed by equations (5a) into those of the second system, the cosines of the angles which the incident ray makes with the axes of α, β and γ in the second system are found to be as follows:

$$-\frac{a\cos\varphi+(b-y)\sin\varphi}{r_0},\quad \frac{(b-y)\cos\varphi-a\sin\varphi}{r_0},\quad \frac{c-z}{r_0};$$

and the cosines of the angles made with these same axes by the emergent ray are:

$$\frac{\alpha}{r_2},\quad \frac{\beta-v}{r_2},\quad \frac{\gamma-\zeta}{r_2}.$$

Let ω denote the angle between the directions of the incident and emergent rays; then

$$\left.\begin{aligned}\cos\omega = -\frac{[a\cos\varphi+(b-y)\sin\varphi]}{r_0}\frac{\alpha}{r_2}&+\frac{[(b-y)\cos\varphi-a\sin\varphi]}{r_0}\frac{(\beta-v)}{r_2}\\ &+\frac{(c-z)}{r_0}\frac{(\gamma-\zeta)}{r_2}\end{aligned}\right\}\quad(10)$$

The variables a, b, c, α, β and γ may be eliminated from formula (10) by means of equations (5) and (6). In the first place

$$\left.\begin{aligned}\frac{a}{r_0} &= \sqrt{1-n^2\frac{(y-v\cos\varphi)^2+(z-\zeta)^2}{r_1^2}} = \sqrt{\frac{n^2v^2\sin^2\varphi}{r_1^2}-(n^2-1)}\\ \frac{\alpha}{r_2} &= \sqrt{1-n^2\frac{(v-y\cos\varphi)^2+(z-\zeta)^2}{r_1^2}} = \sqrt{\frac{n^2y^2\sin^2\varphi}{r_1^2}-(n^2-1)}\end{aligned}\right\}\quad(10a)$$

If one of the two radicals here should be imaginary, the ray will be totally reflected at the corresponding face of the prism. Equations (6) give at once convenient expressions for the quotients $\frac{b-y}{r_0}$, $\frac{c-z}{r_0}$, $\frac{\beta-v}{r_2}$, $\frac{\gamma-\zeta}{r_2}$. When these values are substituted in formula (10), the cosine of the angle of deviation will be given in terms of y, v, z and ζ. Indeed, it can easily be contrived so that the last two of these magnitudes do not occur at all except as they are involved in r_1. Thus, the following expression is found:

$$\left.\begin{aligned}\cos\omega = &-n^2+n^2\frac{\sin^2\varphi}{r_1{}^2}(y^2-y\,\upsilon\cos\varphi+\upsilon^2)\\ &-n\frac{\sin\varphi}{r_1{}^2}(y-\upsilon\cos\varphi)\sqrt{n^2y^2\sin^2\varphi-(n^2-1)r_1{}^2}\\ &-n\frac{\sin\varphi}{r_1{}^2}(\upsilon-y\cos\varphi)\sqrt{n^2\upsilon^2\sin^2\varphi-(n^2-1)r_1{}^2}\\ &-\frac{\cos\varphi}{r_1{}^2}\sqrt{n^2y^2\sin^2\varphi-(n^2-1)r_1{}^2}\sqrt{n^2\upsilon^2\sin^2\varphi-(n^2-1)r_1{}^2}\end{aligned}\right\}\quad(10b)$$

Supposing that x and y are constant, let us try to find the values of υ and ζ for which the angle ω is a maximum; in which case

$$\frac{d\omega}{d\upsilon}=0 \quad\text{and}\quad \frac{d\omega}{d\zeta}=0\,.$$

As ζ does not occur in the expression for $\cos\omega$ except as it is involved in r_1, the second equation above may also be written:

$$\frac{d\omega}{d\zeta} = -\frac{1}{\sin\omega}\,\frac{d(\cos\omega)}{d(r_1{}^2)}\cdot(\zeta-z)=0\,.$$

This equation is satisfied for all values of υ, provided

$$\zeta - z = 0\,.$$

But this condition would not be sufficient, either if $\sin\omega=0$, that is, if the ray were not deviated at all (as would be the case if the prism were a plate with its two faces parallel to each other), or if the derivative of $\cos\omega$ should become infinite due to the fact that one of the denominators in the expression for this function happened to vanish. It is evident from (10b) that r_1 and the two radicals are the only functions that could occur as denominators in the expression for the derivative. But as long as y and υ are positive, even if their values are infinitely small, as they must be if the ray is to go through the prism, r_1 cannot vanish. Moreover, on account of equations (6a), the expressions under the radicals cannot vanish if the ray is to extend on either side of the prism itself. Accordingly, the condition

$$\frac{d\omega}{d\zeta}=0$$

will be satisfied by putting

$$z=\zeta\,.$$

Consequently, we have also by equations (6)

$$z=c \quad\text{and}\quad \zeta=\gamma\,,$$

which means, as above stated, that the ray lies wholly in the plane of a principal section of the prism.

The other condition which has to be satisfied in order that the angle of deviation shall be a maximum is that

$$\frac{d\omega}{dv} = 0 ,$$

and, with this in mind, let us first simplify the expression for cos ω by imposing the condition $z = \zeta$, that is,

$$r_1^2 = y^2 + v^2 - 2yv\cos\varphi .$$

Introducing here a new variable q in place of v by writing

$$v = qy ,$$

we contrive to make both y and v disappear together from the expression for cos ω as given by equation (10b), so that cos ω becomes then simply a function of q, which may be written therefore

$$\cos\omega = f(q) .$$

But as the expression for cos ω is not altered by interchanging the letters y and v whenever they occur, it follows that for any value of q we may write:

$$\cos\omega = f(q) = f\left(\frac{1}{q}\right)$$

Moreover, if $f'(q)$ denotes the derivative of $f(q)$ with respect to q then

$$\frac{d\cos\omega}{dv} = \frac{1}{y} f'(q) = -\frac{1}{y} f'\left(\frac{1}{q}\right)\frac{1}{q^2} .$$

Now for $v = y$, that is, for $q = 1$,

$$f'(q) = -f'(q) , \qquad \text{or} \qquad f'(q) = 0 ;$$

that is,

$$\frac{d\cos\omega}{dv} = 0 .$$

Accordingly, unless sin $\omega = 0$ at the same time,

$$\frac{d\omega}{dv} = -\frac{d\cos\omega}{dv}\cdot\frac{1}{\sin\omega} = 0 .$$

Hence, if

$$z = \zeta \quad \text{and} \quad y = v ,$$

we have both

$$\frac{d\omega}{d\zeta} = 0 \quad \text{and} \quad \frac{d\omega}{dv} = 0\,;$$

and the angle ω is a limiting value. Investigation of the second derivative shows that in this case ω is a maximum. Accordingly, the angle between the prolongation of the incident ray and the refracted ray (which is the real measure of the deviation) is a minimum.

The maximum value of the angle ω is found by putting $y = v$ and $z = \zeta$ in formula (10b); which gives

$$\omega = \varphi + 2\arccos\left[n \sin\frac{\varphi}{2}\right] \quad \cdots\cdots\cdots \quad (10c)$$

The condition that a narrow homocentric bundle of incident rays shall emerge from a prism as a homocentric bundle is that the chief ray shall go through the prism with minimum deviation, that is, shall lie in a principal section of the prism and shall make equal angles with the two faces of the prism.

Under such circumstances, therefore, the prism produces a potential image of a luminous point, lying on the same side of the prism as the luminous point and at an equal distance from the prism. The image, however, is not where the source is, but is displaced towards the refracting edge of the prism through an angle equal to $(\pi/2-\omega)$

Image of a Luminous Line in a Prism

The requirement for a distinct image of a luminous point is that the bundle of emergent rays that enter the eye shall be homocentric. However, when the source of light is a luminous line, deviations (or aberrations) of the rays in the direction of the image of this line do not tend to impair the exactness of the image. Now this is the case that ordinarily occurs in the spectrum. If the luminous line is parallel to the refracting edge of the prism (z-axis), deviations in the direction of z are not objectionable at all, whereas deviations in a plane passed through the ray at right angles to the z-axis do affect the definition of the image. If there are to be no deviations from homocentricity except in the direction parallel to the z-axis, then in equations (8) we must put $\Delta y = 0$, and therefore also $\Delta v = 0$, $\Delta z = \Delta\zeta$; and thus from equations (8c) and (8d) we obtain:

$$\left(\frac{1}{r_0} + \frac{1}{r_2}\right)(1 - n^2 \cos^2 \nu) = 0\,,$$

that is,

$$r_2 = -\,r_0\,; \quad \cdots\cdots\cdots\cdots \quad (11a)$$

and, secondly:

$$(1 - n^2) \cos \nu \sin^2 \varphi = 0 ,$$

whence we derive as above:

$$\cos \nu = 0$$
$$c - z = z - \zeta = \gamma - \zeta = 0 .$$

If the last condition is satisfied the deviations Δy will be in a plane passed through the ray at right angles to the deviations Δz. Accordingly, the second plane of convergence which has to be regarded as perpendicular to the other is the one in which these deviations occur. The place where the rays intersect that lie in the plane at right angles to the edge of the prism is found by putting $\Delta z = 0$ and $\cos \nu = 0$ in formula (8d); consequently $\Delta \zeta = 0$ also, and

$$\frac{1}{r_0}(1 - n^2 \cos^2 m) y^2 + \frac{1}{r_2}(1 - n^2 \cos^2 \mu) v^2 = 0 .$$

If, as formerly, the angles of incidence and emergence are denoted by ϵ and η and the angles inside the glass by ϵ_1 and η_1, that is, if

$$\cos \epsilon_1 = \frac{v \sin \varphi}{r_1} \qquad \cos \eta_1 = \frac{y \sin \varphi}{r_1}$$

$$\sin \epsilon = n \sin \epsilon_1 = n \frac{y - v \cos \varphi}{r_1} = n \cos m$$

$$\sin \eta = n \cos \mu ,$$

then

$$\frac{r_2}{r_0} = - \frac{\cos^2 \epsilon_1 \cos^2 \eta}{\cos^2 \epsilon \cos^2 \eta_1} \quad . \quad . \quad . \quad . \quad . \quad . \quad . \quad . \quad (11b)$$

or

$$r_2 \frac{\cos^2 \eta_1}{\cos^2 \eta} = - r_0 \frac{\cos^2 \epsilon_1}{\cos^2 \epsilon} ,$$

$$r_2 \left[1 + \frac{n^2 - 1}{\cos^2 \eta} \right] = - r_0 \left[1 + \frac{n^2 - 1}{\cos^2 \epsilon} \right] .$$

In this latter form, it is easy to see that when η decreases and ϵ increases, r_2 increases and r_0 decreases. The point of intersection of the rays is therefore more remote from that face of the prism that corresponds to the smaller of these two angles.

For minimum deviation ($\epsilon = \eta$), $r_2 = -r_0$, so that the point of intersection of the rays in the plane perpendicular to the edge of the prism is at the same distance as the point of intersection of the rays in the plane parallel to the edge.

The image of a luminous line parallel to the edge of the prism is formed at the place where, according to equation (11b), the rays meet that lie in the plane of a principal section of the prism. Hence, *the distance between the prism and the image of a luminous line parallel to the edge of the prism is greater than the distance of the object, provided the angle of incidence at the first face of the prism is greater than it is for minimum deviation; and, conversely, the image is nearer the prism than the object, when the said angle of incidence is less than it is for minimum deviation.*

If, therefore, a luminous line adjusted in this way is viewed through a prism either by the naked eye or with the aid of a telescope, the focusing for minimum deviation is for the same distance as that of the object itself. But if the prism is rotated out of this position around an axis parallel to its edge, the focusing of eye or telescope will have to be changed accordingly. The image will not be at infinity unless the object is infinitely distant also; and then the focusing is the same for all positions of the prism.

If the luminous object is a bright vertical line, emitting a definite kind of homogeneous light, say, red light, its image in a vertical prism (that is, a prism with its edge vertical) will be a vertical line. But if the source emits violet light as well as red, there will be a violet image also consisting of a vertical line, but farther from the object than the red image because the violet rays are more refrangible than the red rays. And, if, finally, the luminous line-source emits light of all kinds of refrangibility, there will be a special image of it for each special kind of light, all these linear images being ranged in order side by side between the red image at one end and the violet at the other, and constituting a spectrum in the form of a rectangle. On the supposition that the luminous source sends out light of every possible degree of refrangibility between certain limits, the spectrum will be continuous. On the other hand if light corresponding to some particular wave-lengths is missing, the corresponding images in the spectrum will be missing also, and there will be dark vertical lines at these places, the so-called FRAUNHOFER lines.

Apparent Width of Image in Prism

It is impossible to have a luminous object corresponding to a geometrical line. A luminous object as actually employed in optical experiments necessarily has some superficial extent. Therefore, the image of a so-called luminous line will likewise have a certain width, which we proceed now to determine.

If ϵ, ϵ_1 denote the angles of incidence and refraction at the first face of the prism, and η_1, η the corresponding angles for the second face of the prism, then

$$\left.\begin{aligned} \sin\epsilon &= n\sin\epsilon_1 \\ \sin\eta &= n\sin\eta_1 \\ \eta_1+\epsilon_1 &= \varphi \end{aligned}\right\} \qquad \qquad (12)$$

where φ denotes the refracting angle of the prism. Suppose the slit is very far away and that the angle it subtends at the prism is denoted by $d\epsilon$; then the angles of incidence of light coming from opposite sides of the slit will be ϵ and $\epsilon+d\epsilon$. The angles ϵ_1, η_1 and η for one side of the slit will become, therefore, $\epsilon_1+d\epsilon_1$, $\eta_1+d\eta_1$ and $\eta+d\eta$ for the other side. Differentiating equations (12), we obtain:

$$\begin{aligned} \cos\epsilon\, d\,\epsilon &= n\cos\epsilon_1\, d\epsilon_1\,, \\ \cos\eta\, d\,\eta &= n\cos\eta_1\, d\eta_1\,, \\ d\eta_1+d\epsilon_1 &= 0\,. \end{aligned}$$

Eliminating $d\epsilon_1$ and $d\eta_1$, we have:

$$-\frac{\cos\epsilon\cdot\cos\eta_1}{\cos\eta\cdot\cos\epsilon_1}\,d\epsilon = d\eta \qquad \qquad (12a)$$

Now $d\eta$, as given by this formula, is the angle subtended by the image of the slit in the prism. Suppose the prism is adjusted in the position of minimum deviation; then

$$\epsilon = \eta\,, \qquad \epsilon_1 = \eta_1\,,$$

and, consequently,

$$-\,d\epsilon = d\eta\,.$$

The greatest value of ϵ is a right angle, which happens when the ray grazes the first face of the prism. In this case the other angles continue to be acute angles, so that their cosines do not vanish, and hence

$$d\eta = 0\,.$$

For this focusing, therefore, the image of the slit is infinitely narrow; but although it may be possible in actual experiments to approximate this limiting position, of course, it cannot be perfectly accomplished. The opposite state of affairs occurs when the prism is adjusted so that the light issues from the second face very nearly at grazing emergence, and therefore $\cos\eta$ is nearly zero. In this case

$$\frac{d\eta}{d\epsilon} = -\infty\,.$$

If the distance of the slit from the prism is r_0, and if the apparent distance of its image from the prism for horizontal rays is r_2, then from (11b)

$$\sqrt{r_0} : \sqrt{r_2} = d\eta : d\epsilon \quad . \; . \; . \; . \; . \; . \; . \; . \; . \quad (12b)$$

Purity of the spectrum. The smaller the difference of refrangibility dn for colours in the same part of the spectrum, the purer the spectrum will be; and therefore the magnitude of dn may be taken as a measure of the impurity.

Suppose we consider the refracted ray that comes from the given place in the spectrum to the nodal point of the eye; so that both its position and the angle η are definite and fixed. On the other hand, the angle ϵ varies for rays coming from different parts of the slit, and the index of refraction varies for different colours. In the equations

$$\begin{aligned} \sin \epsilon &= n \sin \epsilon_1 , \\ \sin \eta &= n \sin \eta_1 , \\ \eta_1 + \epsilon_1 &= \varphi \end{aligned}$$

let us treat φ and η as constant and ϵ, ϵ_1, η_1 and n as variable. By differentiation we get:

$$\begin{aligned} \cos \epsilon \; d \epsilon &= \sin \epsilon_1 \, dn + n \cos \epsilon_1 \, d\epsilon_1 , \\ 0 &= \sin \eta_1 \, dn + n \cos \eta_1 \, d\eta_1 , \\ d\eta_1 + d\epsilon_1 &= 0 . \end{aligned}$$

Eliminating $d\epsilon_1$ and $d\eta_1$, we have:

$$\begin{aligned} \cos \epsilon \cdot \cos \eta_1 \cdot d\epsilon &= (\sin \epsilon_1 \cos \eta_1 + \cos \epsilon_1 \sin \eta_1) \, dn \\ &= \sin \varphi \cdot dn . \end{aligned}$$

Now if $d\epsilon$ is the angle subtended by the width of the slit at the prism, the measure of the impurity of the spectrum is

$$dn = \frac{\cos \epsilon \cdot \cos \eta_1}{\sin \varphi} d\epsilon \quad . \; . \; . \; . \; . \; . \; . \; . \; . \quad (13)$$

As the light tends to graze the first face of the prism, the angle ϵ is more and more nearly a right angle, and therefore $\cos \epsilon$ tends to vanish, so that ultimately $dn = 0$. In this case, therefore, for a slit of given size the spectrum is purest, but at the same time the aperture of the prism also becomes very small for such oblique incidence, and the loss of light by reflection is very large. On the whole, therefore, it is better to try to get a pure spectrum by narrowing the slit (that is, by diminishing $d\epsilon$) which can generally be done without special difficulties.[1]

[1] ¶Concerning purity of spectrum and resolving power of prism or a prism-system, see Lord RAYLEIGH, Investigations in Optics. *Phil. Mag.* (5) VIII, 1879, pp. 261–274, 403–411, 477–486, and (5) IX, 1880, pp. 40–55. See also J. P. C. SOUTHALL, *Principles and methods of geometrical optics*, 1910, pp. 492, foll.; and R. W. WOOD, *Physical Optics*, 2nd ed., 1914, pp. 108, foll. (J. P. C. S.)

So far as the *luminiosity* of the spectrum is concerned, the brightness of the slit for any special colour is to that of its image in the inverse ratio of their widths $d\epsilon$ and $d\eta$, provided we leave out of account losses of light by reflections at the sides of the prism, and provided the aperture of the prism is greater than that of the pupil of the eye or than the object-glass of the telescope, if the spectrum is observed through that instrument. Thus, if the brightnesses of slit and image are denoted by h and h_1, then

$$h\,d\epsilon = h_1\,d\eta$$

and hence by equation (12a):

$$h_1 = h\,\frac{\cos\eta\cos\epsilon_1}{\cos\epsilon\cos\eta_1}\,.$$

Let H denote the luminosity at some place in the spectrum; being equal to the sum of the luminosities h_1 of all the separate homogeneous kinds of light that contribute to the illumination at this place. As a rule, it may be assumed that the brightness of simple colours of nearly the same wave-length λ is the same. If, therefore, $d\lambda$ denotes the difference of wave-length and dn the difference of refrangibility for colours that overlap at a certain place in the spectrum, we may write

$$H = h_1\,d\lambda = h_1\,\frac{d\lambda}{dn}\,dn\,,$$

and hence, when the value of dn as given by equation (13) is substituted in this expression, the following formula is obtained:

$$H = h\,\frac{\cos\eta\cos\epsilon_1}{\sin\varphi}\,d\epsilon\cdot\frac{d\lambda}{dn},$$

where $d\epsilon$ denotes the apparent width of the slit. To get a clear idea of the meaning of this expression for H, suppose that we had an actual geometrical line of light instead of an illuminated slit, and that the problem was to find the angular width $d\eta$ of the interval in an absolutely pure spectrum corresponding to the positions of two coloured images whose difference of refrangibility was dn. The ratio $\frac{d\eta}{d\lambda}$, which will be denoted by l, is found by differentiation in the same way as above; thus

$$\frac{d\eta}{dn} = \frac{d\eta}{d\lambda}\frac{d\lambda}{dn} = l\,\frac{d\lambda}{dn} = \frac{\sin\varphi}{\cos\eta\cos\epsilon_1}\,.$$

In this case, therefore,

$$H = \frac{h \cdot d\epsilon}{l}.$$

Accordingly, apart from losses of light by reflection and absorption, *the luminosity of the spectrum*, regarded as something that is independent of the dispersion of the prism and the geometrical conditions, *is directly proportional to the luminosity of the given colours in the spectrum and to the apparent width of the slit, and inversely proportional to the apparent length of the portion of the spectrum under consideration.*

When the prism is adjusted for minimum deviation, the apparent widths of the slit and image are equal, and $l/d\epsilon$ may be regarded as the measure of the purity of the spectrum. Under these circumstances therefore, *the luminosity of the spectrum, for constant luminosity of the slit, is simply inversely proportional to the purity of the spectrum.* This means, therefore, that to get the greatest purity, the light should be as brilliant as possible.

On the other hand, it would be theoretically possible to obtain rather greater luminosity for the same degree of purity in the spectrum, by increasing the angle of incidence at the first face of the prism and making the slit broader; but in order not to vary the length of the spectrum, the refracting angle of the prism would have to be increased also. However, no practical advantage is gained in this way, because there is always more light lost by reflection, and, besides, if the faces of the prism are not truly plane, slight errors of this kind become more manifest in the image at large angles of incidence.

It has been assumed above that the prism was used without the aid of any other optical system. But the prismatic spectrum like any other optical image may be made the object for inspection through a telescope and may be magnified at pleasure. Of course, this will not affect the purity of the spectrum; and if the aperture of the telescope is large enough to show the object in its natural brightness, and if the aperture of the prism is as great as that of the telescope, the luminosity of the magnified image will also remain unchanged. The above rules also concerning luminosity and purity of the spectrum are not altered, provided $d\epsilon$ is understood to mean the apparent size of the slit, $d\eta$ that of its image, and l the length of the portion of the spectrum under consideration, as these magnitudes appear through the telescope. Incidentally, the condition formulated for the luminosity of the spectrum shows why quite small prisms are sufficient in observations where no telescope is employed, whereas the size of the prism increases with the magnifying power of the instrument, when the apparatus is designed to be used with a telescope.

In focusing a telescope on a spectrum, another point to be noticed is that the coloured bands and dark lines appear distinct when the rays that diverge horizontally are brought to a focus (supposing that slit and prism-edge are vertical, which is always to be understood); whereas the upper and lower boundary lines of the spectrum and any other horizontal lines, which may easily be produced in the spectrum by some irregularities in the edges of the slit or by particles of dust on them, will show up distinct if the vertically diverging rays are the ones that are converged to a focus. It is only when the prism is in the position of minimum deviation that the telescope can be focused for vertical and horizontal lines at the same time. And, in fact, provided the faces of the prism are perfectly plane, the focusing is the same as it would be to see the slit distinctly when the prism was out of the way. But if, starting with the prism in the position of minimum deviation, the edge is rotated more towards the object-glass of the telescope, the latter will have to be focused for a greater distance in order to see the coloured bands and dark lines; whereas when the prism is turned the other way, the telescope must be focused for a nearer object. For horizontal lines the focusing is the same, no matter which way the prism is turned.

A spectrum is obtained by letting the light from an illuminated slit go through a prism. The transmitted light may enter the eye directly, or it may be sent through a telescope into the eye, or it may be focused by a lens to project an objective image of the spectrum.

Any luminous body may be the source of light. The luminosity of the various colours, as is well known, is different for light of different sources, terrestrial as well as celestial; and the spectra are different both in the bright portions and in the dark portions. If it is desired to use the spectrum of sunlight for experimental work, and it is not necessary to see any but the more conspicuous dark lines and merely such colours as are ordinarily visible, sky-light reflected from a mirror or a sheet of paper illuminated by sunlight will be sufficient; except that in the former case the yellow and orange will be a little faint. One advantage of this kind of illumination is that it remains the same for a long time. The more prominent dark lines D, F, and G can be seen simply by looking with the naked eye through a flint glass prism of 50° angle at a slit 1 mm wide and 40 cm away. If the observer is twice as far from the prism as the slit is, he will be able to discern most of the lines which FRAUNHOFER designated by capital letters. He must find out, however, precisely the adjustment of the prism for which the eye can be accommodated for the lines.

If a spectrum of greater purity is necessary, so as to see also the finer dark lines, or if the extreme limits of the spectrum are to be made visible, a mirror must be adjusted to reflect light from the parts of the sky close to the sun or from the sun itself through the slit on to the prism; and, as the sun moves in the sky, the mirror will have to be adjusted again about every three minutes unless it is attached to a heliostat and moved by clock-work so as to keep pace with the sun.

The *slit* itself, which is to be illuminated and which is the peculiar object for the prismatic image, may be easily cut out of a piece of opaque paper, if it is to be used for experiments in which the finer dark lines do not matter, or provided it is placed far enough away from the prism. But if a very fine slit has to be used, the GRAVESANDE slit is the best. Two straight bars *ab*, *ab* (Fig. 11) are screwed in a rectangular plate of brass. Between the ends *aa* a plate *aacc* is fastened with its edge *cc* beveled. Opposite it is the beveled edge *dd* of another plate *ddee* which can slide between the two parallel bars. It is adjusted by a screw *f* with a very fine thread, which goes through the nut *g* fastened to the base. Thus the two knife-edges *cc* and *dd* can be nicely adjusted at a very slight distance apart, always being parallel to each other, if the contrivance is properly constructed. The part of the base-plate where the slit is formed by the knife-edges is cut away to let the light pass.

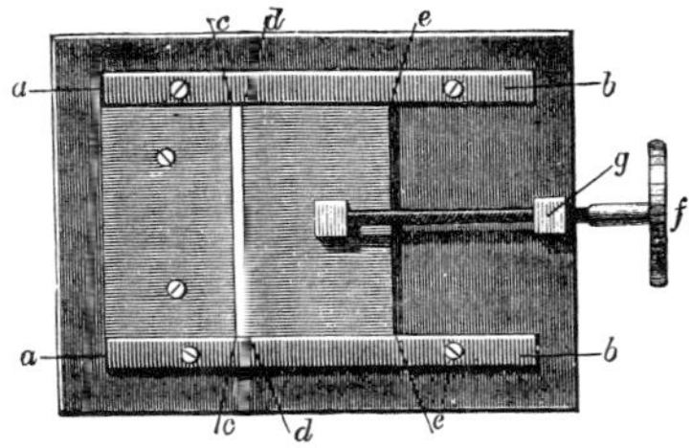

Fig. 11.

The GRAVESANDE slit has to be fastened in the centre of a sufficiently large dark screen and the side towards the observer must be blackened. The screen must be big enough not to allow any luminous object to be visible anywhere in the vicinity of the slit, whose spectrum might extend as far as that of the slit. Unless there is not to be any trace of white light at all, the main thing usually is that the screen where the slit is shall be *uniformly* dark rather than *absolutely* dark. Wherever there is any difference of illumination, even if it be no more than the contrast between velvet black and grey black, the prism shows colours; but when the surface is the same everywhere and evenly illuminated all over, there will be no colours. Many such tests can be carried out perfectly well in a bright room, provided the slit is inserted in a large enough screen that is painted uniformly black.

On the other hand, when it is necessary to get rid of every vestige of white light, as, for example, in the case of experiments that are intended to show that homogeneous light cannot be resolved any further or changed at all, and also in investigations of the limits of the spectrum, the screen where the slit is must be absolutely dark. The easiest way of accomplishing this is to do the work in a dark room specially designed for optical experiments and provided with suitable window-shades that do not allow any light to come through. The brass plate with the slit in it may then be inserted in an opening in the shutter itself. Incidentally, it is often possible to produce the same conditions even in an ordinary living-room by closing the shutters and window-hangings and leaving just a small slit for the light to come through. The slit should be placed in the back of a box painted black on the inside, the open front side of the box being towards the spectator. The sides of the box will keep the lateral light from falling on the back, so that the latter will be very dark. Two strips of black velvet should be glued on the back of the box on either side of the slit, of such dimensions that when the spectrum is supposed to be projected backwards on the velvet it will be wholly on it. In this way the spectrum will be actually seen on a black or non-luminous background. Another useful precaution consists in adjusting dark screens to prevent any possible illumination in different parts of the room from falling on the prism and telescope or on the observer's eye.

But an absolutely dark screen in a dark room is not enough by itself to prevent any visible traces of white light from disturbing the spectrum, if very intense light of several colours is being transmitted through the optical apparatus into the eye. In the theory of the formation of prismatic images, as developed above, it was tacitly assumed that all the light was refracted regularly. But it must not be forgotten that some of the light is also *reflected*; and a small amount of light is always diffused in all directions when light passes through any transparent medium, solid or liquid.

Reflections occur, in the first place, at the base of the prism unless it is painted black with some suitable varnish to prevent this very thing. If it is a ground glass surface, it will usually be illuminated whenever light traverses the prism. In Fig. 12, suppose that *dcba* is the path of a ray through a prism, and that the eye of an observer is at *a*. He will see a reflected image of the base of the prism *fe* in the apparent position $f\epsilon$ and if the base is illuminated, the image will appear bright, and thus diffused white light will be scattered throughout the field of view. But if the base of the prism is also polished, the light will be regularly reflected from it. In a prism whose section is an equilateral triangle, the observer will not only get light along *dcba* but also along the path *dcbgcba* after the light has been reflected three times at *b*, *g* and *c* in succession. This light is not dispersed in colours, but is white. It forms a faint white image of the slit in the field of view, which may be utilized for adjusting the prism in the position of minimum deviation. In a prism of this particular form, the white image of the slit coincides exactly with the colour in the spectrum for which the deviation is minimum. But this faint white image of the slit is usually not a serious matter at all so far as spectrum observations are concerned, because it takes up a comparatively small space in the field and is not so harmful in its effects as the reflected image of the base of the prism when the latter is a ground glass surface. On the other hand, it is possible also for light from surrounding objects to get to the eye through the base of the prism, and this must be carefully avoided. In any case, it is best to paint the whole prism black except the two refracting faces.

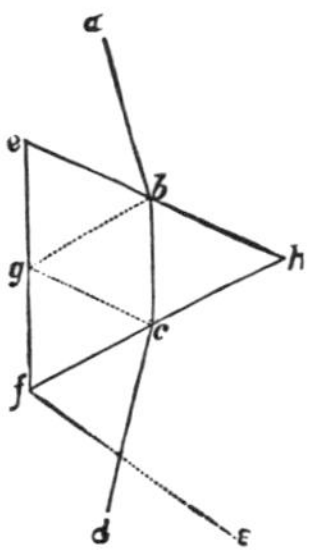

Fig. 12.

When the spectrum is observed with a telescope, the reflections at the two surfaces of each of the lenses have to be taken into account. The effect of these reflections is to produce tiny little regular images of objects in front of the telescope, but most of them are so situated that the observer cannot accommodate for them; so that what they do is to make a faint white illumination in the field. This illumination can be easily noticed by pointing a telescope at a dark black object surrounded by very bright ones. The field will then appear feebly illuminated as contrasted with the black diaphragm of the ocular.

A similar defect, which is much more difficult to obviate, is the diffusion of the light in the body of the glass itself. On close observation any piece of clear glass held against a black background will appear whitish throughout when the sun shines on it, especially if the line of sight is in very nearly the same direction as the transmitted light. The same phenomenon is observed in the cornea and crystalline lens of the human eye (see Vol. I, pages 18, 193). Accordingly, we must take into account that any mass of glass traversed by light scatters some of it, even if it be only a small portion, and that the field of view contains everywhere some diffused illumination of this kind. Similarly, there is always a very small amount of each kind of light that comes into the eye spread over the entire retina. The intensity of this irregularly diffused light is certainly exceedingly low, as compared with that which is regularly refracted or reflected. And yet it may be appreciable when faint portions of the spectrum are under observation. For example, this is the reason why, under ordinary circumstances, the extreme red corresponding to the line A and the ultra-violet region of the spectrum are not perceptible. It is very noticeable when the luminosity of individual parts of the spectrum is considerably lowered by coloured glasses, for then the hues at these places are decidedly altered by the faint light scattered by diffusion throughout the field of view.

In experiments on the faintly luminous parts of the spectrum the only way to surmount these difficulties completely is not to let any light come through the slit and fall on the prism and telescope except light of high intensity of precisely the kind that is to be investigated. All other kinds of light must be excluded as far as possible. In certain special cases this can be done simply by interposing coloured filters between the source and the slit; for example, a piece of red glass in order to see the farthest red in the spectrum. A better and more general method consists in using two slits and two prisms. The image of the second slit is the spectrum. The only kind of light that is allowed to pass through this slit is the particular light that is to be investigated. A diagram of the arrangement is shown in Fig. 13. The incident ray ab

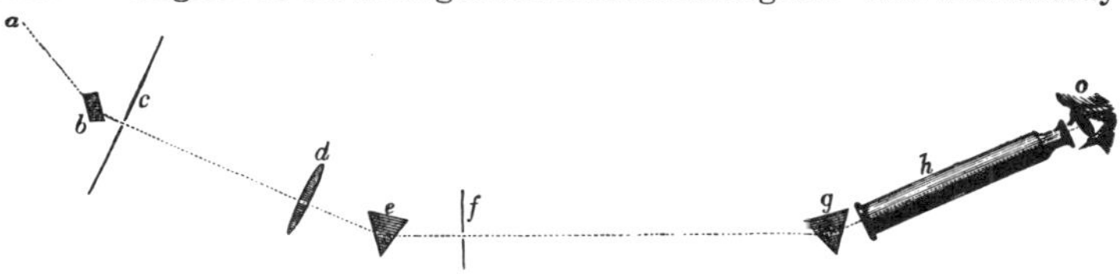

Fig. 13.

falls on the mirror of the heliostat at b and is reflected through a slit in the screen c, which as a rule does not have to be very narrow. The light passes then through the lens d and the prism e, and arrives at the screen f. This screen is adjusted so that the rays that diverge from the slit c are focused on it and form there an image of the slit expanded into a spectrum. This first spectrum does not generally have to be pure. Indeed, if a somewhat extended region of the spectrum is to be studied, say, the ultra-violet portion, this original spectrum should be so impure that all the ultra-violet light falls at one place in it. To regulate this according to circumstances, instead of putting the prism e beyond the lens, it is even better to insert it in the system ahead of the lens. The effect of bringing the screen nearer the prism and moving the lens correspondingly farther from it, will be to make the spectrum shorter and more impure. If the screen and prism are farther apart, the spectrum will be longer and purer. The screen f has a fine slit in it between two GRAVESANDE knife-edges, which is adjusted so that the precise colour of the spectrum that is to be studied is projected on it. Thus, suppose the ultra-violet light is under investigation; then the slit is moved so as to be close to the extreme edge of the visible violet, and then regularly refracted ultra-violet light, as intense as it occurs in sunlight, will pass through the slit, along with it also some little white light diffusely scattered from the glass prism and lens or reflected several times from their surfaces. This latter light is certainly far more feeble than the other, but yet it is intense enough to cover completely the ultra-violet on the screen f. Having passed through the second slit, the light falls now on a second prism g, and then enters the eye directly or after traversing a telescope; unless one prefers to substitute a lens for the telescope and project an objective image of the spectrum on a screen placed in the focal plane of the lens. As some white light after all has come through the slit f, a complete spectrum is produced here also, but it is very faint everywhere except in the ultra-violet region or whatever portion of the spectrum was focused on the second slit. Some light will be irregularly scattered in the second prism g, in the lenses that compose the telescope h, and in the ocular media of the observer's eye o; but all the other light except the ultra-violet is by this time too faint for its small scattered portions to be still perceptible. As a matter of fact, with this arrangement, it is possible to see the spectrum even in a telescope projected on a completely dark black background, so black that it cannot be distinguished from the blackness of the ocular stop, the edge of the latter not being visible except where it crosses the spectrum. Not until this deep blackness of the background has been obtained, can one be certain that it is pure monochromatic light that is seen. Under these circumstances, the ultra-violet light of sunlight also becomes directly visible to the eye; and it is only by such precautions that the unchangeableness of the colour of homogeneous light can be successfully demonstrated when it is filtered through a coloured glass. As long as there is still a small amount of diffused white light mixed in the spectrum, coloured filters, that absorb largely the particular kind of light in question, apparently also change their hue. Blue cobalt glass, for example, absorbs spectrum yellow almost entirely, but allows the blue rays of the diffused white light to pass unaffected, and this latter light, getting mixed with the faint yellow that is not absorbed, produces a white or bluish white compound colour in place of the yellow. However, this compound colour is not due, as BREWSTER supposed, to light of a single degree of refrangibility, because it can be resolved again by another prism into light of different colours and different refrangibilities. On the other hand, if the experiment is performed again with a spectrum that is completely purified of diffused light, the homogeneous yellow is found to remain pure yellow after passing through the blue glass, no matter how faint it is. From these experiments and similar ones we are not justified, therefore, in inferring, as BREWSTER has done, that light of definite refrangibility and wave-length may be still further decomposed into three different kinds of light whose colours are merely mixed in different proportions in different parts of the spectrum, and might be separated from each other by absorption by coloured media. The experiments on which he bases this conclusion depend partly on the circumstance above mentioned, partly too on contrast actions, and lastly on the fact that the hue is a function of the intensity of the light, as has been previously stated.[1]

By the method just described, as represented in the diagram (Fig. 13), the ultra-violet spectrum in its entire length can be made directly visible to the eye, without having to use any fluorescent substance; but for the farthest ultra-violet the prisms and lenses should all be made of rock crystal (quartz), and not of glass, because the latter absorbs to a considerable extent the more extreme ultra-violet radiations of the solar spectrum. But with quartz apparatus, the unusually large number of dark lines that occur in this part of the spectrum can be very distinctly seen. The author supposed the luminosity of the ultra-violet spectrum as seen in the telescope could be increased by inserting in the ocular stop a thin layer of quinine solution between two quartz plates. The spectrum in this case is projected right on the quinine surface causing it to fluoresce. The image as viewed through the ocular is similar to that which is seen without the quinine preparation, except that it does not consist of ultra-violet light but of pale blue light of medium refrangibility. But, contrary to expectation, the luminosity of this image in the actual experiment was no greater than that of the directly viewed ultra-

[1] HELMHOLTZ über D. BREWSTERS neue Analyse des Sonnenspektrums. POGG. *Ann.*, LXXXVI, 501. — BERNARD, *Ann. de Chim.* XXXV, 385–438.

violet image, but almost the same and rather less, and, owing to the thickness of the film of quinine, the dark lines were more indistinct. The reason of this is because, although the cone of light that enters the instrument through the object-glass is a small one, practically all this light comes to the eye and illuminates the retina, provided the film of quinine is not present. But when the ultra-violet light falls on the quinine solution, it is scattered there in all directions in space, and only a very small part of the light that comes from the quinine gets to the observer's eye; and hence, in spite of the greater luminosity of the fluorescent light, the retina itself is not so highly illuminated. This experience is the basis of the estimate given above as to the ratio between the luminosity of the original ultra-violet light and that of the fluorescent light produced when it falls on quinine.

Let a denote the aperture of the object-glass or of the prism in front of it, if it is the latter that determines the cone of rays; and let r denote the distance of the image. With the position of the image as centre, describe a sphere of radius r. Supposing that the ultra-violet light is propagated without any interference, the portion of this spherical surface that is illuminated will be an area not greater than a. But if the image were projected on quinine, the entire spherical surface of area $4\pi r^2$ would be uniformly illuminated. Thus, the light in the first case is more concentrated than in the latter case in the ratio of $\frac{4\pi r^2}{a}$; and if an eye, whose pupil is completely inside the region of radiation for both kinds of light, sees both equally bright, then, for the same manner of distribution, the fluorescent light must be brighter in the ratio above mentioned. In the author's experiment, the value of this fraction, after making the necessary corrections, was 1200. Hence, ultra-violet light received on a quinine screen must appear about 1200 times brighter than it does when it falls on a smooth white surface of porcelain that does not fluoresce.

Fluorescence of highly fluorescing substances may be easily observed and detected in any spectrum. However, when it is a question of perceiving the weakest sort of fluorescence, for example, that of the retina, the apparatus represented in Fig. 12 may be employed with the following modifications. The first spectrum is made very impure by removing the first slit at c altogether and moving the prism e near the screen f. The slit in this screen is opened wide; and adjusted on the edge of the violet. The object-glass is the only part of the telescope that is used here. The substances to be tested are placed at the focus of this lens where the ultra-violet light is most concentrated, and where there is no white light. There is scarcely any material that will not show signs of fluorescence under these circumstances. As the unchanged ultra-violet light may also still be visible in these experiments, the substance under investigation should be examined through a yellow or green glass (uranium glass is best), that absorbs ultra-violet, or through a thin prism that separates ultra-violet from the colours of medium refrangibility. The fluorescence of the cornea and crystalline lens of the eye is easy to demonstrate by focusing ultra-violet light on the eye of a living person. The crystalline lens is so illuminated in this way that its position right behind the iris and its form can be much better distinguished than by illumination with ordinary light (Vol. I, p. 18). Of course, a large amount of bluish white light is uniformly scattered by the fluorescent lens over the entire background of the eye. When, on the other hand, an ultra-violet spectrum is viewed by the eye, it shows up very fine and distinct. It must not be inferred from this that the fluorescence of the lens is what makes ultra-violet light visible to the eye. Fluorescent light would never produce a well defined image on the retina.

The infra-red region is studied in the same way as the ultra-violet.

The methods of measuring wave-lengths of light are described in treatises on physical optics and do not belong here.

Previous to the time of NEWTON, the theory of colour consisted chiefly of vague hypotheses. The intensity of the coloured light that was derived from the total white light being always necessarily less than that of the whole, the old-fashioned idea was that this decrease of intensity was an essential thing about colour, and ARISTOTLE's opinion that colour is a mixture of white and black had many adherents. He himself was undecided whether this mingling of white and black was to be considered as a real blending or more as an atomic superposition or juxtaposition. He supposed that darkness must be due to reflection from bodies, because every time light is reflected, it gets fainter. This was the prevalent view until the beginning of more modern times, as can be seen, for instance, in the doctrines of MAUROLYCUS, JOH. FLEISCHER, DE DOMINIS, FUNK, NUGUET (see GOETHE's history of the theory of colour). And in very recent times GOETHE has tried to uphold it again in his *Farbenlehre*. This theory of colour does not pretend to give an explanation of colour phenomena in the physical sense—considered in that way, GOETHE's propositions would be void of any meaning; but all that he tries to do is to formulate in a general way the conditions of the production of colours. Colours must be manifested distinctly in some original phenomenon ("*Urphänomen*"). He considers the colours of cloudy media as being this original phenomenon. Many media of this sort give a red colour to light that passes through them, whereas the incident light as seen against a dark background looks blue. Thus, although GOETHE follows ARISTOTLE's view in general, and supposes that light is darkened or has to be mixed with darkness to produce colours, he imagines that in the phenomena of cloudy media he has discovered the special kind of darkening that produces, not what we call grey, but colours as usually understood. What change occurs in the light itself under these circumstances, he never does explain. He does, perhaps, say that the cloudy medium gives the light something corporeal, shadowy, which is necessary for the production of colour. But what he means by it, he does not explain more precisely. He cannot possibly suppose that something corporeal escapes from bodies along with the light they emit; and yet as a physical explanation scarcely any other meaning can be attached to it.

Moreover, in GOETHE's way of looking at the matter, all transparent bodies are a little cloudy; and this is true of a prism too. Thus, he assumes that the prism communicates something of its cloudiness to the image which the spectator beholds. Apparently, what he means by this is that the image in a prism is never quite sharp, but is confused and indistinct; for in his theory of colour he classifies such images with the secondary images produced by plane parallel glass plates and crystals of Iceland spar. The image made by a prism certainly is confused in heterogeneous light, but it is perfectly sharp in homogeneous light; but this is something that apparently GOETHE never did see, as he disdained to consider at all the array of arguments that prove this fact. When a bright surface on a dark background is viewed through a prism, his idea is that the image is shifted and clouded by the prism. The anterior edge of the bright area is shifted over the dark background, and being a bright cloud over darkness, looks blue. But the other edge of the bright area, being overlapped by the cloudy image of the black background that succeeds it, is a bright thing behind a dark cloud, and is therefore yellow-red. Why the anterior edge appears in front of the background and the other edge behind it, and not the other way, he does not explain. This presentation of the matter is likewise absolutely meaningless if it is intended to be a physical explanation. For the prism image, as thus seen, is a potential one and is therefore simply the geometrical place where the rays of light that come to the spectator would intersect if they were prolonged backwards; and hence this image cannot have the physical effects of an image seen through a cloudy medium. These representations of GOETHE's are therefore not to be regarded as physical explanations at all, but merely as figurative illustrations of the process. In his scientific work he does not attempt anyhow to go beyond the domain of sensory perception. But every physical explanation must be in terms of the forces that come into play, and these forces, of course, can never be the object of sensory perception, but are simply concepts of the mind.

The experiments which GOETHE uses to support his theory of colour are accurately observed and vividly described. There is no dispute as to their validity. The crucial experiments with as homogeneous light as possible, which are the basis of NEWTON's theory, he seems never to have repeated or to have seen. The reason of his exceedingly violent diatribe against NEWTON[1] was more because the fundamental hypotheses in NEWTON's theory seemed absurd to him, than because he had anything cogent to urge against his experiments or conclusions. But NEWTON's assumption that white light was composed of light of many colours seemed so absurd to GOETHE, because he looked at it from his artistic standpoint which compelled him to seek all beauty and truth in direct terms of sensory perception. The physiology of the sense-perceptions was at that time still undeveloped. The complexity of white light, which NEWTON maintained, was the first decisive empirical step in the direction of recognizing the merely subjective significance of the sense-perceptions. GOETHE's presentiment was, therefore, correct when he violently opposed this first advance that threatened to ruin the "fair glory" (*schönen Schein*") of the sense-perceptions.

The great sensation produced in Germany by GOETHE's *Farbenlehre* was partly due to the fact that most people, not being accustomed to the accuracy of scientific investigations, are naturally more disposed to follow a clear, artistic presentation of the subject than mathematical and physical abstractions. Moreover, HEGEL's natural philosophy used GOETHE's theory of colour for its purposes. Like GOETHE, HEGEL wanted to see in natural phenomena the direct expression of certain ideas or of certain steps of logical thought. This is the explanation of his affinity with GOETHE and of his chief opposition to theoretical physics.

In developing the theory of the rainbow, DESCARTES advanced a new hypothesis, by assuming that the particles of which light is composed not only had a rectilinear motion but also rotated about their axes, and that the resultant colour was due to the velocity of rotation. Moreover, the action of transparent bodies may change the rotation and along with it the colour also. Similar mechanical conceptions were formulated by HOOKE and DE LA HIRE. The latter assumed that the colours were dependent on the force of the impact of the light on the optic nerve.

Finally, NEWTON[2] proved the heterogeneous nature of white light. He isolated homogeneous light from it, and showed that this latter light was coloured. This colour was characteristic homogeneous light, and could not be altered any more by absorption and refraction. He found that light of different colours had different refrangibilities and that the colours of natural bodies were due to peculiarities of absorption and reflection. Incidentally, he explained the colour of the rays of light as being entirely the result of their

[1] ¶In GOETHE's work not only were NEWTON's theories misstated and derided, but NEWTON himself was heaped with abuse, and accused of being no better than a mere charlatan—"this Cossack Hetman," as GOETHE calls him. At the same time GOETHE, in spite of his absurdities, performed a real service, as HELMHOLTZ says. (J. P. C. S.)

[2] ¶NEWTON's original prism experiments on the decomposition of sunlight were carried out at Trinity College in Cambridge in 1666. (J. P. C. S.)

action on the retina. The rays themselves were not red, but they produced the sensation of red. He leaned towards the emission (corpuscular) theory of light; but he advanced no hypotheses as to the physical differences between homogeneous kinds of light of different colours.

About the same time, in 1690, HUYGENS proposed the hypothesis that light consisted in undulations of a delicate elastic medium. EULER showed how NEWTON's discoveries could be explained on this basis, and deduced the result that the simple colours in the spectrum were the effects of light of different frequencies of vibration. As a matter of fact, however, his first assumption was that the red vibrations were the faster ones, but subsequently he discovered the mistake. HARTLEY correctly supported this view in explaining the colours of thin plates. But a crucial test could not be made until the principle of interference had been discovered by YOUNG and FRESNEL; and it was this discovery also that led first to a general acceptance of the undulatory theory.

NEWTON's inference that the colours of the rays depended on the refrangibility, and that light of given refrangibility was homogeneous in all other ways and invariable in colour, was opposed by BREWSTER. He thought he had found that homogeneous light could change colour in traversing a coloured medium, and that in this way it might be possible to get white light from homogeneous light. He was led thus to infer that there were three different kinds of light, corresponding to the three so-called fundamental colours, red, yellow and blue, and that each of these kinds of light gave rays of every degree of refrangibility within the range of the spectrum, in such fashion, however, that red light predominates at the red end, yellow light in the middle, and blue light at the blue end. His idea was that light of given refrangibility would be absorbed by media of various colours to different extents, and that the colours would be separated in this way. BREWSTER's views were opposed by AIRY, DRAPER, MELLONI, HELMHOLTZ and F. BERNARD. Outside of some cases in which, owing to contrast effects with adjacent vivid colours, the hue appeared to be different after the light had been greatly reduced in intensity by being filtered through coloured glasses, and some other cases in which the above mentioned variation of colour was connected with the intensity of light, most of BREWSTER's observations were probably due to the circumstance, to which attention has already been called, that small quantities of white light were diffused over the field of view as a result of repeated reflections at the various surfaces or of scattered reflection inside the prism substance and in the ocular media.

The comparison between the spectrum colours and musical notes was suggested first by NEWTON. But the comparison he made was between the widths of the coloured areas in the spectrum of glass prisms and the musical intervals of the Pythagorean scale. LAMBERT pointed out long ago that this division was largely arbitrary, because there were no fixed limits to the spectrum; and that it does happen to be similar to the musical scale in one respect, inasmuch as the quantitative relations in the distribution of the colours in the spectrum are concerned not so much with the sum of the widths of the coloured bands themselves as with the sum of the ratios of these intervals. DE MAIRAN was of the same opinion. Meantime, however, Father CASTEL tried to make this analogy the basis of a colour-piano, intended to produce pleasing effects by a certain play of colours similar to the effects of music. HARTLEY endeavoured to show that differences of colour were due to vibrations of different amplitudes which enabled him to make a more direct comparison with the vibration-numbers of musical notes. This was what YOUNG had in mind when he said that the entire range of the spectrum as then known was equal to a major sixth, and that red, green and blue correspond about in the ratios of 8 : 7 : 6. More recently, thanks especially to FRAUNHOFER's measurements, the values of the wave-lengths of light of different colours have been ascertained, and, with the aid of these data, DROBISCH has tried again to find a connection between the colour scale and the musical scale. Like NEWTON, he compared the width of the colours with the intervals of the so-called Pythagorean scale: $1 : \frac{9}{8} : \frac{6}{5} : \frac{4}{3} : \frac{3}{2} : \frac{5}{3} : \frac{16}{9} : 2$. But since the width of the ordinary visible spectrum, as measured by FRAUNHOFER, is less than an octave, he raised each of those ratios to a certain power, which had the value $\frac{2}{3}$ at first, afterwards $\frac{6}{7}$. In this way he got the following table, in which the wave-lengths are given in millionths of a millimetre:

Colour	Boundaries	Line
	688.1	
Red		$B = 687.8$
	622.0	$C = 655.6$
Orange		$D = 588.8$
	588.6	
Yellow		
	537.7	
Green		$E = 526.5$
	486.1	
Blue		$F = 485.6$
	446.2	
Indigo		$G = 429.6$
	420.1	
Violet		$H = 396.3$
	379.8	

In this scheme the boundaries between the colours themselves agree fairly well with the natural ones. Possibly, it might be better to use the major third instead of the minor third, that is, to make the whole comparison on the basis of the major scale, as DROBISCH himself suggested. Then the border between orange and yellow instead of being at D in the golden yellow, as in the above arrangement, would fall nearer the pure yellow. Even so, it must not be forgotten that any comparison between sound waves and light waves ceases to have any sense at all as soon as the numerical values of the musical intervals are modified entirely by the process of raising them all to a certain fractional power. Moreover, the spectrum is broken off arbitrarily at both ends, because, as a matter of fact, the faint terminal colours of the spectrum extend much farther on both sides. And, finally NEWTON's division into seven principal colours was perfectly arbitrary from the beginning and deliberately founded on the musical analogies. Golden yellow has just as much right to a place between yellow and orange as indigo has between blue and violet; and the same is true with respect to yellow-green and blue-green. Indeed, there are no real boundaries between the colours of the spectrum. These divisions are more or less capricious and largely the result of a mere love of calling things by names. In the author's opinion, therefore, this comparison between music and colour must be abandoned.

Lastly, quite recently UNGER has endeavoured to establish a theory of aesthetic colour harmony by an analogy between the wave-length ratios and the musical intervals. In his actual statements about harmony of colours there seems to be a good deal of truth, in large measure borrowed correctly from works of art; but the theory itself, the analogy with the musical ratios, is rather far-fetched. On his chromo-harmonic disc he has assembled a lot of hues intended to correspond to the 12 semi-tones of the octave, but for this purpose he has inserted purple reds between violet and red, although the purples do not exist as simple colours. He makes the FRAUNHOFER lines G, H, A fall in these purple hues, whereas G and H are the borders of the violet, and A belongs to pure red. The simple colours that lie beyond violet are, as a matter of fact, blue and not purple-red. The most perfect harmony should correspond to the major chord. This will give on his disc, for example, the composition of red, green and violet that is so common in the work of Italian painters. But if green is taken as major third, the correct major chord would be red, green, and indigo blue. The ancient painters did not have any good red pigment. They used minium (red lead), which is an orange colour, for red, and made the chord consisting of orange, green blue and reddish violet. The effect of the minor chords is milder and sadder, and that of the diminished and augmented triads is piquant and not so pure artistically. The writer's belief is that some other basis must be sought for the correct explanation of the colour effects described by UNGER, instead of forced musical analogies. In fact, the saturated colours do constitute a recurrent series, provided the gap between the ends of the spectrum is filled in by the purple hues; and it seems to be agreeable to the eye when three colours are presented to it that are about equally far apart in the series. The celebrated composition of the Italian painters alluded to above, namely, red, green and violet, which does not correspond to any correct major chord, does indeed correspond to YOUNG's three fundamental colours, and it may be that this is the explanation of the aesthetic effect in this case. Other colours, chosen at correct distances apart, produce a similarly satisfying impression. If two of them come too near together, the impression is less pure. This is probably the significance of UNGER's observations. At any rate, it is clear that in the so-called colour harmony no such absolutely definite relations are to be expected as are characteristic of the musical intervals.

384–322 B. C. ARISTOTELES, *De coloribus.*

1571. JOH. FLEISCHER, *De iridibus doctrina Aristotelis et Vitellionis.* Vitembergae 1571, p.86.

1583. JO. BAPT. PORTA, *De refractione libri novem.* Neapoli 1583. lib. IX.

1590. BERNARDINI TELESII, *Opera.* Venetiis 1590. De Iride et coloribus.

1611. M. ANTONI DE DOMINIS, *De radiis visus et lucis in vitris perspectivis et Iride.* Venetiis 1611.

1613. MAUROLYCUS, *De lumine et umbra.* Lugd. 1613. p. 57.

1637. CARTESIUS, *De meteoris.* Kap. VIII.

1648. JO. MARCUS MARCI, *Thaumantias, liber de arcu coelesti, deque colorum apparentium natura, ortu et causis.* Pragae 1648.

1665. R. HOOKE, *Micrographia.* London 1665. p. 64.

1675. *I. NEWTON in *Phil. Trans.* 1675. (First accounts of his views).—*Optics.* London 1704. (Complete development of his optical discoveries).—*Lectiones opticae.*

1711. DE LA HIRE, *Mém. de l'Acad. des Sc.* 1711. p. 100.

1746. EULER. Nova theoria lucis colorum in den *Opusculis varii argumenti.* Berol. 1746. pp. 169–244.

1752. EULER in *Mem. de l'Acad. de Prusse* 1752. p. 271. Essai d'une explication des couleurs.

Opposed to NEWTON

1727. RIZETTI, *Specimen physico math. de luminis affectionibus.* Ven. 1727.

1737. LEBLOND, *Harmony of colouring.* London.

1740. CASTEL, *L'optique des couleurs.* Paris.

1750. GAUTHIER, *Chroagenesie ou génération des couleurs contre le système de Newton.* Paris. 2 vol. 8.

1752. GAUTIER, *Observations sur l'histoire naturelle, sur la physique et la peinture.* Paris.

1754. COMINALE, *Anti-Newtonianismus.* 4. Napoli.

1780. MARAT, *Découvertes sur la lumière.* Paris. 8.

1784. MARAT, *Notions élémentaires d'optique.* Paris. 8.

1791–92. GOETHE, *Beiträge zur Optik.* Weimar.

1794. WÜNSCH, *Kosmologische Unterhaltungen.*

1810. GOETHE, *Zur Farbenlehre. Entoptische Farben, zur Naturwissenschaft.* 126–190.

Concerning NEWTON and GOETHE

1811. SEEBECK, Von den Farben und dem Verhalten derselben gegeneinander. SCHWEIGGERS *Journal* 1811. p. 1.
MOLLWEIDE, *Demonstratio propositionis quae theoriae colorum Newtoni fundamenti loco est.* Lipsiae 1811.
PFAFF, Über die farbigen Säume der Nebenbilder des Doppelspats mit besonderer Berücksichtigung von GOETHES Erklärung der Farbenentstehung durch Nebenbilder. SCHWEIGGERS *Jahrbücher* VI. 177.
POSELGER in GILBERTS *Annalen* XXXVII. 135.
1817. WERNEBURG, *Merkwürdige Phänomene durch verschiedene Prismen zur richtigen Würdigung* NEWTON*scher und* GOETHE*scher Farbenlehre.* Nürnberg 1817. 4.
1827. BRANDES, Art. Farbe in GEHLERS *neuem physik. Wörterbuch.*
1833. REUTHER, *Über Licht und Farbe.* Kassel 1833.
STEFFENS, *Über die Bedeutung der Farben in der Natur. Schriften alt und neu.*
1835. HELWAG, NEWTONS *Farbenlehre aus ihren richtigen Prinzipien berichtigt.* Lübeck 1835.
MOSER, Über GOETHES Farbenlehre. *Abh. der Königsberger deutschen Gesellsch.*
1853. HELMHOLTZ, Über GOETHES naturwissenschaftliche Arbeiten; in *Kieler Monatsschrift für Wissenschaft und Lit.* 1853. May, S. 383.
1857. GRÄVELL, GOETHE *im Recht gegen* NEWTON. Berlin 1857. — Reviewed by Q. ICILIUS in KEKULES *kritischer Zeitschr. für Chemie, Physik und Math.* Erlangen 1858. 2. und 3. Heft.
1859. GRÄVELL, *Über Licht und Farben.* Berlin. (Rejoinder.)

Individual Theories

1816. READE, *Experimental outlines for a new theory of colours, light and vision.* London 8.
1824. HOPPE, *Versuch einer ganzneuen Theorie der Entstehung sämtlicher Farben.* Breslau. 8.
1828. RÖTTGER, *Erklärung des Lichts und der Dunkelheit.* Halle.
1830. SCHÄFFER, *Versuch einer Beantwortung der von der Akad. zu Petersburg aufgegebenen Preisfrage über das Licht.* Bremen. 8.
WALTER CRUMM, *An experimental inquiry into the number and properties of the primary colours, and the source of colours in the prism.* London, 1830.
1831. D. BREWSTER, Description of a monochromatic lamp with remarks on the absorption of the Prismatic Rays. *Edinb. Trans.* IX. P. II. p. 433.
Idem, On a new analysis of solar light. *Edinb. Trans.* XXII. P. I. 123.—POGG. *Ann.* XXIII. 435.
1834. EXLEY, *Physical optics or the phenomena of optics.* London.

As to BREWSTER's Theory

1847. AIRY in *Phil. Mag.* (3.) XXX. 73.—POGG. *Ann.* LXXI. 393.
BREWSTER, *Reply.* Ibid., XXX. 153.
DRAPER in SILLIMAN *Journ.* IV. 388.—*Phil. Mag.* XXX. 345.
BREWSTER, *Phil. Mag.* XXX. 461.
MELLONI, *Bibl. univ. de Genève.* Aug. 1847.—*Phil. Mag.* XXXII. 262.—POGG. *Ann.* LXXV. 62.
BREWSTER, *Phil. Mag.* XXII. 489.
1852. HELMHOLTZ, Über Herr D. BREWSTERS neue Analyse des Sonnenlichts. POGG. *Ann.* LXXXVI. 501. — *Phil. Mag.* (4) IV.
F. BERNARD, Thèse sur l'absorption de la lumière par les milieux non crystallisés. *Ann. de Chim.* (3.) XXXV. 385–438.
1855. D. BREWSTER, On the triple spectrum, *Athen.* 1855, p. 1156.—*Inst.* 1855, p. 381.—*Report of Brit. Assoc.* 1855. 2. pp. 7–9.

Limits of Sensitivity

1845-46. Brücke in Müllers *Archiv für Anat. und Physiol.* 1845. S. 262. 1846. S. 379.—Pogg. *Ann.* LXV. 593. LXIX. 549.

1852. Cima, *Sul potere degli umori dell' occhio a trasmettere il calorico raggiante.* Torino.

1853. Donders, In Onderzoekingen gedaan in het physiol. Laborat. van de Utrechtsche. Hoogeschool. Jaar VI. p. I Müllers *Archiv* 1853. S. 459.

1854. G. Kessler in Graefe *Archiv für Ophthalmologie* I. Abt. 1. S. 466.

1855. Helmholtz, Über die Empfindlichkeit der menschlichen Netzhaut für die brechbarsten Strahlen des Sonnenlichts. Pogg. *Ann.* XCIV. 205.—*Ann. de Chim.* (3.) XLIV. 74.—*Arch. de sc. phys.* XXIX. 243.

Comparison with Musical Scale

1704. I. Newton, *Optics.* Lib. I. Pars 2. Prop. 3.

1725-35. L. B. Castel, Clavecin oculaire in *Journ. de Trevoux.*

1737. de Mairan in *Mém. de l'Acad. des Sc.* 1737. p. 61.

1772. Lambert, *Farbenpyramide.* Augsburg 1772. §19.

Hartley in Priestley *Geschichte der Optik.* S. 549.

1801. Th. Young in *Phil. Trans.* 1802. p. 38.

1852. Drobisch, *Abhandl. d. sächsischen Gesellsch. der Wiss.* Bd. II. *Sitzungsberichte,* Nov. 1852.—Pogg. *Ann.* LXXXVIII. 519-526.

Unger in Pogg. *Ann.* LXXXVII. 121-128.—*C. R.* XL. 239.

1854. Idem, *Disque chromharmonique.* Göttingen.

1855. Helmholtz in *Sitzber. d. Akad. d. Wiss. zu Berlin* 1855. S. 760.—*Inst.* 1856. p. 222.

J. J. Oppel, Über das optische Analogon der musikalischen Tonarten. *Jahresber. der Frankfurter Vers.* 1854–55. pp. 47–55.

E. Chevreul, Remarques sur les harmonies des couleurs. *C. R.* XL. 239-242.—*Edinb. Journ.* (2.) I. 166–168.

§20. The Compound Colours

We have seen that homogeneous light of different refrangibilities and frequencies produces sensations of different colours in the organ of vision. Now when one and the same place on the retina is stimulated at the same time by light of two or more different vibration-frequencies, new kinds of colour sensations are produced which, generally speaking, are different from those of the simple colours of the spectrum. A peculiarity of these sensations is that it is not possible to recognize the simple colours that are contained in the mixture. The fact is that generally the sensation of any given compound colour may be produced by several different combinations of spectral colours, without its being possible for the most practised eye to tell by itself what simple colours are concealed in the compound light. In this respect, there is a fundamental difference between the eye in its reaction to the aether vibrations and the ear, which responds to sound-waves of different pitch by combining the separate notes, it is true, in a compound sensation of a chord, and yet is able to detect separately each single note in it. Thus, two chords consisting of different musical notes never do appear identical to the ear, as different aggregates of compound colours may be for the eye.

All this applies to the immediate sensory perception, and is not invalidated at all by the fact that an act of judgment sometimes enables us to recognize the composition correctly, at least as to its main features. Anybody who has had some experience with the effect of mixing coloured light may occasionally fancy that he really does see the simple colours that are contained in a compound colour, and may judge whether there is more of one colour in it than of another. But in this case an act of judgment based on experience is confused with an act of sensation. For instance, looking at purple, one may know that it is predominantly red and violet and about in what proportion the two are mixed. But nobody can tell whether there are also subordinate quantities of orange and blue in it. Were it a matter of sensation, and not simply a question of judgment based on experience, the latter could be detected just as well as the former. The sensation of white can be produced by the greatest variety of combinations; but nobody can make out what simple colours are in it, whether two or three or four, and which ones in particular. Green shows how easy it is to be deceived here. Even persons like GOETHE and BREWSTER, misled by the mixture of pigments, have insisted that they could see both yellow and blue in green, although, as a matter of fact, we know now that green cannot be produced at all from these colours, unless modifications of them are employed that already have some green in them.

The most curious illusion is when two simple colours are seen in the same place at the same time, the surface being illuminated simultaneously by two different colours, one of which predominates at certain places and the other at other places; especially if one of them covers the whole background while the other is in the form of a regular figure upon it. When the figures or spots can be made to change their position, the illusion is even more perfect. In a case of this kind we often imagine we see the two colours simultaneously at the same place, one through the other, as it were. The effect is very much as if objects were seen through a coloured veil or mirrored in a coloured surface. We have learned by experience, even under such circumstances, to form a correct judgment of the true colour of the object, and this distinction between the colour of the background and that of the light that is irregularly distributed over it is taken into consideration in all similar cases. To get the sensation of the mixture of colours without any disturbing element, the light must be evenly mixed over the entire field where it is displayed.

Under special circumstances, for example, when two colours that are far apart in the spectrum have sharply defined positions in the field of view, the marginal colours may be recognized as separate by virtue of the chromatic aberration of the eye (see Vol. I, p. 174). This does not amount to any real contradiction of the fact above stated, because in this case the eye itself acts like a prism and causes light of different colours to fall at different parts of the retina.

The following are the methods of mixing light of different colours and of testing the action of compound light in the eye:

1. Two different spectra may be superposed, or different parts of the same spectrum; whereby combinations of pairs of simple colours are obtained.

2. Two coloured surfaces may be adjusted symmetrically with respect to a transparent plate of glass. Rays from one of them traverse the glass obliquely and enter the eye, while rays from the other are reflected into the eye from the side of the plate next the eye. Thus the observer gets light of both colours simultaneously at the same part of the retina of his eye. This method is particularly convenient for mixing the colours of natural bodies.

3. Discs with sectors of different colours can be rotated rapidly on the colour-top. When the rotation is fast enough, the impressions made on the retina by the different colours blend into the sensation of a single compound or mixed colour.

All three methods are equivalent, so far as mixing the colours is concerned. However, the actual process will be described more in detail below. The method of mixing powdered or liquid pigments must not be employed for this purpose, although NEWTON and many other physicists have supposed that it was equivalent to the first method, that is, the method of mixing the colours of the spectrum. For the mixed pigment does not give at all a colour that would be the resultant of mixing the two kinds of lights that are reflected separately from each of the ingredients.[1]

To make this clear, consider first the mixture of coloured liquids. When light passes through them, some of the coloured rays of the white light become so spent after they have gone a short way in the liquid that they vanish entirely, while others can traverse longer paths in the liquid without being appreciably enfeebled. These latter rays predominate in the emergent light, which has therefore the colour of the rays that are least absorbed by the liquid. The absorption of certain colours of the spectrum can be easily demonstrated by sending the light through a prism after it has traversed a coloured liquid or glass. In the spectrum thus obtained a series of colours will be missing; or

[1] ¶"The failure to recognize the fact, first made plain by HELMHOLTZ, that mixtures of pigments are not the same thing as mixture of colour (the former are a phenomenon of subtraction and not of addition) was the cause of much confusion and error"— CHRISTINE LADD-FRANKLIN in article on "Vision" in BALDWIN's *Dictionary of Philosophy and Psychology*. (J. P. C. S.)

these colours will be faint, while the parts of the spectrum corresponding to the colour of the liquid will be bright as usual.

Now suppose two coloured liquids are mixed together without any chemical action taking place, so that each of them has the same power of absorption for light as before. Then no rays will go through the mixture except those which are not absorbed by either of the two ingredients. Ordinarily, the rays which pass will be those that correspond to the portion of the spectrum that is midway between the colours of the two liquids before they were mixed. Most blue substances, like the cupric salts, for example, let the blue rays through unimpaired, and the green and violet also to a less extent, but they absorb most of the red and yellow. Yellow liquids, on the other hand, let almost all the yellow light through without loss, and some red and green also, but intercept most of the blue and violet. When, therefore, a yellow liquid and a blue liquid, like those mentioned, are mixed, on the whole it is green light that is transmitted, because the blue liquid absorbs red and yellow, and the yellow liquid absorbs blue and violet. The effect is similar to that obtained by sending the light in succession through two plates of glass of different colours. In such a case the emergent light is always weaker than it would be if it had passed through two plates of the same colour. But it is obvious that what takes place here is not a summation of the light which is allowed to go through each liquid separately, but, on the contrary, a kind of subtraction, since the yellow liquid takes away from the rays that have traversed the blue liquid those rays which it absorbs itself. As a rule, therefore, mixtures of coloured liquids are much darker, than either one of the liquids by itself.

The behaviour of powdered pigments is quite similar. Each individual particle of the coloured powder must be regarded as a tiny little transparent body which colours the light by absorbing some of it. It is true that the powdered pigment as a whole is exceedingly opaque. But whenever we examine pigments in coherent masses of uniformly thick structure, we find that at any rate in the form of thin sheets they are transparent. Crystallized vermilion, verdigris, lead chromate, blue cobalt glass, etc., which are used as pigments in form of fine powders, are some examples.

Now when light falls on a powder of this kind made up of transparent parts, a certain portion of it will be reflected from the outer layer, but most of it will not be reflected until it has penetrated into the interior to some extent. A single plate of white glass reflects four percent of the light incident normally on it, while two such plates reflects nearly twice as much; and a large pile of plates reflects almost all the light. Consequently, when the glass is pulverized, we must suppose that only four percent of the light that falls perpendicularly on it is reflected from the first layer, and that all the rest is reflected from the interior layers. The same thing must happen when blue light falls on blue glass. Accordingly, only a very small fraction of the light that comes from a coloured powder is reflected from the top layers; by far the greatest part of it being reflected from the deeper layers. In non-metallic reflection, the light that is reflected from the outside surface is white, but that which comes from the interior begins to be coloured by absorption; the farther the light penetrates, the deeper being the colour. This is why a coarser powder is darker in colour than a finer one of the same substance. So far as reflection is concerned, it is simply a question of the number of facets involved, and not a question of the thickness of the particles. The larger the latter, the farther the light has to go inside the material, before meeting the same number of facets as when the particles are smaller. Therefore, in a coarse powder the absorption of light is higher than it is in a fine powder of the same material. The colour of the coarse powder is darker and more saturated. When the powder contains some liquid to make it cohere, the index of refraction of the liquid being nearer that of the particles than that of air, the internal reflection of light is thereby diminished. And so dry pigment powders are usually paler than when they are mixed with water or with oil, which has a still higher index of refraction.

Now if a uniform mixture of two coloured powders merely reflected light from its outer surface, this light would really be the sum of the two kinds of light obtained from each powder separately. But, as a matter of fact, most of the light is reflected back from the interior, and the behaviour is just like that of a mixture of coloured liquids or of a series of coloured glasses. This light has had to pass on its way particles of both sorts, and so it contains merely such rays as were able to get through both elements. Thus, most of the light reflected from a mixture of coloured powders is due, not to an addition of both colours, but to a subtraction in the sense explained above. This is the reason too why mixtures of pigments are much darker than the separate ingredients, especially if the colours of the latter are far apart. Vermilion and ultramarine, for instance, make a dark grey with scarcely a trace of violet, although that is the compound colour of red and blue light. The reason is that one pigment almost completely absorbs the rays that the other lets pass. A convenient way of exhibiting the distinction here is to paint the sectors *a* and *b* (Fig. 14) on the edge of a colour-disc in two different colours, while the central portion *c* is painted with a mixture of the two pigments. Thus, if the

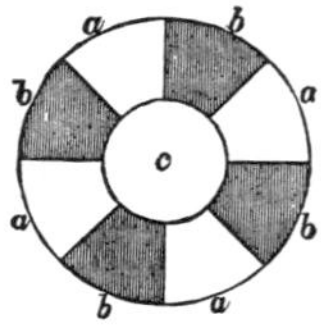

Fig. 14.

sectors are cobalt-blue and chrome-yellow, the appearance is pale grey when the disc is rotated so as to get the impression of both colours at once; but the mixture of the two pigments in the centre is a much darker green.

Evidently, therefore, the result of mixing pigments cannot be used to deduce conclusions as to the effect of combining different kinds of light. The statement that yellow and blue make green is perfectly correct in speaking of the mixture of pigments; but it is not true at all as applied to the mixture of these lights.

"Colour mixture" and "mixed colour" are terms that are employed with reference to mixing pigments; but we shall continue to use them here in speaking of the composition of coloured light, although it is not really correct. However, let us say here once for all, that these terms are not used in this book to mean the mixture of pigments and the result of such a mixture; unless the contrary is expressly stated.[1]

The simultaneous action of different simple colours at the same place on the retina produces a new series of colour sensations that are not excited by the simple colours of the spectrum by themselves. These new sensations are those of *purple* and *white*, together with the transitions of white into the spectrum colours on the one hand and into purple on the other hand.[2]

Purple-red results from mixing the simple colours at the two ends of the spectrum. It is most saturated when violet and red are mixed; and is paler or *pink-red* when blue and orange are used instead of violet and red. Purple-red passes through carmine-red into the red of the spectrum, and is distinctly different from red and violet at the extreme ends of what is ordinarily meant by the visible spectrum. For the eye, however, it is a transition between the two with continuous intermediate gradations. It forms the link that closes the chain of *saturated colours*, that is, the colours that are least like white.

White can be produced by combining different pairs of simple colours. Colours which combined in a definite ratio make white are

[1] ¶Mrs. FRANKLIN insists that the term *mixture* should be used for the physical procedure only. The various different psychological effects of light-ray mixtures may then be referred to as: (1) *colour-blends* when the elements of the mixture are still perceptible in the result, as blue and green in the blue-greens or peacocks; (2) *colour-fusions* (or *colour-extinctions*) when the elements of the colour have disappeared in the process of mixing, as red and green in making yellow, and yellow and blue in making white." — Art. on "Vision" in BALDWIN'S *Dictionary of Philosophy and Psychology*. (J. P. C. S.)

[2] ¶"We have in the purples an example of a sensation which *must* be a blend of several sensations, for there is no physical cause for their production except the combination of the causes of blue and red; we therefore know the character of a colour-blend, and we know that white and yellow are sensations wholly destitute of this character."— CHRISTINE LADD-FRANKLIN in article on "Vision" in BALDWIN'S *Dictionary of Philosophy and Psychology*. (J. P. C. S.)

said to be *complementary*. The complementary colours of the spectrum are:

Red and greenish blue;
Orange and cyan-blue;
Yellow and indigo-blue;
Greenish yellow and violet.

There is no simple colour that is complementary to the green of the spectrum. The complementary colour to green is a compound colour, namely, purple.

In order to discover whether there are any regular connections between the wave-lengths of the simple complementary colours, the writer has measured these magnitudes for various pairs of complementary colours, the values (in millionths of a millimetre[1]) being exhibited in the following table.

Colour	Wave-Length	Complementary Colour	Wave-Length	Ratio of the Wave-Lengths
Red....................	656.2	Green-blue.......	492.1	1.334
Orange..................	607.7	Blue............	489.7	1.240
Golden yellow..........	585.3	Blue............	485.4	1.206
Golden yellow..........	573.9	Blue............	482.1	1.190
Yellow..................	567.1	Indigo-blue......	464.5	1.221
Yellow..................	564.4	Indigo-blue......	461.8	1.222
Green-yellow............	563.6	Violet...........	from 433 on	1.301

On account of the low luminosity in the violet, the extreme rays from wave-length 433 must all be included together.

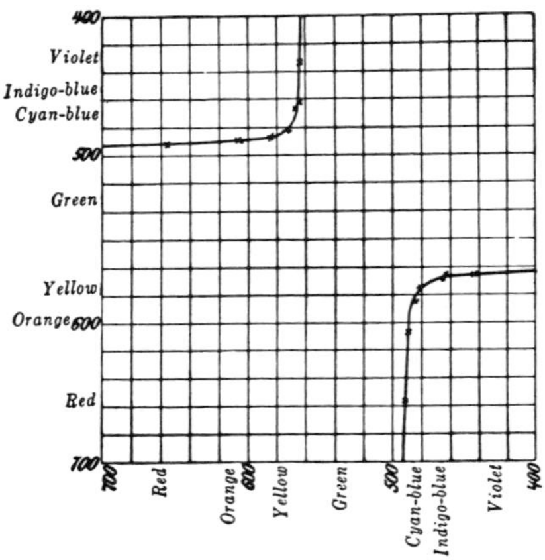

Fig. 15.

The results are plotted in Fig. 15, with reference to a pair of rectangular axes, which indicate the wave-lengths of the colours in the spectrum from 400 to 700. The curve, therefore, shows the wave-length of the complementary colour as a function of that of each of the simple colours. The names of the colours are inserted along the two axes at the proper places. The values actually measured are indicated by the points marked on the curve. This curve, com-

[1] In the first edition the unit used was the Paris inch. In the second edition the numbers were given in terms of the millimetre, and the adjoining figure changed accordingly.—N.

posed of two branches, shows a striking irregularity in the distribution of the complementary colours in the spectrum. Proceeding along the horizontal axis from the violet to the red, we see that the wave-length of the complementary colour changes at first very gradually, as shown by the almost horizontal procedure of the curve. But when we reach the greenish blue colours, the curve suddenly turns downwards almost vertically, and the wave-lengths of the complementary colours change very rapidly. The same thing happens in the yellow; and then in the red the change is again very gradual. This is connected with the fact, mentioned in the preceding chapter, that at the ends of the spectrum the hue changes exceedingly slowly as compared with the wave-length, whereas it changes very fast in the middle of the spectrum. The result is that there is no simple or constant connection between the wave-lengths of the pairs of complementary colours. To use the musical notation, it varies between that of the fourth (1.33) and that of the minor third (1.20).

Incidentally it may be added that the intensities of the light of two complementary simple colours, that are together just equivalent to white, do not by any means appear always equally bright to the eye. The only case where we have to take quantities of the two colours that appear to the eye to be about equal is when we mix cyan-blue and orange. But violet, indigo-blue and red all appear to be darker than the complementary amounts of greenish yellow, yellow and greenish blue, respectively. It will be shown in the next chapter that when proportional amounts of light of two different colours appear to the eye to have the same luminosity, the absolute luminosity is in fact very different; and, consequently, no definite figures are available for the relative luminosity of complementary amounts of two different colours.

The colours of the spectrum, therefore, have different colour-values in mixtures, and are, so to speak, colours of different degrees of saturation. The most saturated is violet, and the order of the others is about as follows:

	Violet	
	Indigo-blue	
Red		Cyan-blue
Orange		Green
	Yellow	

Note by W. NAGEL.

We have now some more recent determinations of pairs of complementary colours by other observers. Their discrepancies with each other and with HELMHOLTZ's data are perhaps not to be attributed to errors of observation so much as to individual peculiarities of the colour system. Some of these measurements were noted by HELMHOLTZ in the second edition. The subjoined table exhibits side by side the data of HELMHOLTZ, v. FREY and v. KRIES (*Arch. f. Anat. u. Physiol.* 1881. 336. The numbers were converted into wave-lengths by A. KÖNIG.), KÖNIG and DIETERICI (WIED. *Ann.* XXXIII. 1887), and ANGIER and TRENDELENBERG (*Zft. f. Psychol. u. Physiol. d. Sinnesorg.* XXXIX. 284. 1905).[1]

Table of the Pairs of Complementary Colours according to several observers

HELMHOLTZ		v. KRIES		v. FREY		KÖNIG		DIETERICI		ANGIER		TRENDELENBURG	
656.2	492.1	656.2	492.4	656.2	485.2	675	496.5	670	494.3	669.3	490.9	669.4	491.2
607.7	489.7	626	492.2	626	484.6	663	495.7	660	494	654.6	489.0	654.9	490.5
585.3	485.4	612.3	489.6	612.3	483.6	650	496.7	650	494.3	641.2	490.2	641.3	490.4
573.9	482.1	599.5	487.8	599.5	481.8	638	495.9	635	494	628.1	487.9	628.4	489.2
567.1	464.5	587.6	484.7	587.6	478.9	615.3	496	626	493.1	616.2	487.4	616.2	487.9
564.4	461.8	579.7	478.7	586.7	478.7	582.6	483.6	610	492.2	604.8	487.0	604.8	487.3
563.6	from 433 on	577	473.9	577.7	473.9	578	476.6	588	485.9	593.8	484.7	593.9	485.7
		575.5	469.3	572.8	469.3	576	467	585.7	485.7	583.3	480.6	583.5	482.8
		572.9	464.8	570.7	464.8	574.5	455	578	476.6	572.9	473.3	572.4	469.1
		571.1	460.4	569	460.4	573	450	575.6	470				
		571	452.1	568.1	452.1			571.5	455				
		570.4	440.4	566.3	440.4			571.3	448				
		570.1	429.5	566.4	429.5			571.4	442				

Lastly, there is still to be considered the effect of mixing colours that are not complementary. Concerning this matter, the following rule may be given: When two simple colours are mixed that are not so far apart in the spectrum as complementary colours, the mixture matches one of the intermediate colours in hue; being more nearly white in proportion as the two components are farther apart, and more saturated the nearer they are together. On the other hand, the mixture of two colours that are farther apart than complementary colours, gives a purple hue or a match with some colour comprised between one of the given colours and that end of the spectrum. In this case the resultant hue is more saturated when the two components are farther apart in the spectrum, and paler when they are nearer together; provided, of course, that the interval between them always exceeds that of a pair of complementary colours.

For instance, when red, whose complementary colour is greenish blue, is mixed with green, the result is a pale yellow, which for different proportions of the two components may pass either through orange into red or through greenish yellow into green. A mixture of orange

[1] ¶The curve shown in Fig. 15 suggests the form of a rectangular hyperbola. From the results found by various observers, as given in the table above, GRÜNBERG has derived the following empirical formula connecting the wave-lengths of the pairs of complementary colours in the spectrum:

$$(\lambda - 559)\ (498 - \lambda') = 424;\ (\lambda > \lambda').$$ (J. P. C. S.)

and greenish yellow may also match pure yellow, but it is more saturated than that produced by red and green. On the other hand, by mixing red and cyan-blue, we get pink (pale purple-red); and by changing the proportions of the mixture, we can make this pink pass into red or through violet and indigo-blue into cyan-blue. But red mixed with indigo-blue or, better still, with violet gives a saturated purple-red.

These results are exhibited in the subjoined table. The pure simple colours are at the tops of the vertical columns and at the left-hand ends of the horizontal rows. The place where a column and row intersect contains the resultant mixed colour; but this latter can be changed through the intermediate colours in the spectrum series into either of the two simple component colours by varying the proportions of the mixture.

	Violet	Indigo-blue	Cyan-blue	Blue-green	Green	Green-yellow	Yellow
Red	Purple	Dark pink	Pale pink	White	Pale yellow	Golden yellow	Orange
Orange	Dark pink	Pale pink	White	Pale yellow	Yellow	Yellow	
Yellow	Pale pink	White	Pale green	Pale green	Green-yellow		
Green-yellow	White	Pale green	Pale green	Green			
Green	Pale blue	Water-blue	Blue-green				
Blue-green	Water-blue	Water-blue					
Cyan-blue	Indigo-blue						

Incidentally, too, it appears that in these mixtures the degree of saturation of the colours of the spectrum is different. Thus red mixed with green of equal brightness gives a reddish orange; and violet mixed with green of equal brightness gives an indigo-blue close to the violet. On the other hand, when equally saturated colours of the same luminosity are mixed, the resultant compound colour is about midway between the two components.

No new colours are obtained by mixing more than two simple or homogeneous colours. The number of different colours is exhausted by mixing pairs of simple colours. Indeed, in these latter mixtures we have seen already that most compound colours can be produced in this way. In general, the result of mixing compound colours is the same as that of mixing spectrum colours that are similar to them; except that the mixture is paler than the spectrum colours to the same degree that the mixed colours themselves are already paler.

Accordingly, with all possible combinations of systems of aether-waves of different frequencies of vibration, there is after all a comparatively small number of different states of stimulation of the organ of vision which can be recognized as different colour sensations. First of these are the series of *saturated* colours, composed of the colours in the spectrum, along with purple which links the ends of the series. Each of these hues again may occur more or less pale in different gradations. The paler it is, the less saturated it appears. The palest degrees of these hues pass at last into pure white. Thus, we have here two kinds of differences between colours, namely, first, differences of *hue*, and, second, differences of *saturation*. Differences of hue are such as are exhibited by the differences of colour in the spectrum.[1] When these colours are supposed to be mixed with more or less white light, we obtain the different degrees of saturation of each one of the hues. Thus, the degree of saturation may be denoted by the proportion between the amounts of the saturated colour and white that go to make up the mixture. Ordinarily, we do not have special names for these pale colours, as, for example, pink in speaking of pale purple, flesh-colour when we mean pale red, sky-blue to denote pale blue; but we describe them by adding some word like bright, pale or white. For instance, "bright blue" is about the same as sky-blue, "pale blue" is a whiter blue still, and "white blue" is a blue that is hardly to be distinguished from white. The use of the word "bright" in connection with these paler tones deserves special comment, because, strictly speaking, it means high luminosity, whereas this mode of speech here does not differentiate between luminosity and paleness (or whitishness). It was mentioned in the preceding chapter that even the saturated colours in the spectrum appear to be whitish when the luminosity is high.

Lastly, in ordinary speech we are wont to describe differences of luminosity as differences of colour; however, only in case colour is considered as a characteristic of bodies. Absence of light is called *darkness*. But when a body does not reflect any light that falls on it, we say it is *black*. On the other hand, a body that scatters all incident light is said to be *white*. A body that reflects an equal share of all the incident light without reflecting all of it is called *grey*; and one which reflects light of one colour more than that of another is said to be a coloured body. Accordingly, in this sense *white, grey* and *black* are colours also. Saturated colours of low luminosity are said to be "dark," as, for example, dark green, dark blue. But when the luminosity is extremely low, the same names are used as for pale colours of low luminosity. Thus, red, yellow and green of low luminosity are called *red-brown, brown* and *olive-green*, respectively; whereas exceedingly pale colours of low luminosity have names such as *reddish grey, yellow-grey, blue-grey*, etc.

[1] ¶Thus change of hue may be said to involve primarily change of wave-length. (J. P. C. S.)

Black is a real sensation, even if it is produced by entire absence of light. The sensation of black is distinctly different from the lack of all sensation. A spot in the field of view which sends no light to the eye looks black; but no light comes to the eye from objects that are behind it, whether they are dark or bright, and yet these objects do not look black,—there is simply no sensation so far as they are concerned. With eyes shut, one is perfectly conscious that the black field of view is limited, and it never occurs to anybody that it extends behind him. It is simply those parts of the field whose light we can perceive when there is any light there, that appear black when they do not emit any light.

The easiest way of recognizing that grey and brown and red-brown are the same as white and yellow and red of low luminosity, respectively, is to analyze the light by a prism. It is more difficult to do this by projecting light of the given colour intensity on a screen, because there is a continual tendency to distinguish between the part of the colour and appearance of the body that depends on the illumination and the part that is characteristic of the surface of the body itself. Therefore, the experiment should be arranged so that the observer is prevented from knowing that there is any special illumination. A grey sheet of paper exposed to sunlight may look brighter than a white sheet in the shade; and yet the former looks grey and the latter white, simply because we know very well that if the white paper were in the sunlight, it would be much brighter than the grey paper which happens to be there at the time. But if a round grey spot is on a sheet of white paper, and if light is concentrated on it with a lens, so that it is highly illuminated while the surrounding white paper is not, the grey spot can be made to look whiter than the white paper. In a case of this kind, where the unconscious influence of experience is excluded, the quality of the sensation appears to be dependent on the luminosity alone.

Similarly, the writer has succeeded in making the homogeneous golden yellow of the spectrum look brown. The method, which will be described presently, consisted in illuminating a little rectangular area with this yellow light on an unilluminated white screen; while at the same time a larger neighbouring portion of the screen was illuminated by brighter white light. Red used in the same way gave red-brown, and green olive-green.

Accordingly, taking the intensity of the light into account also, we find that the quality of every colour impression depends on three variable factors, namely, *luminosity, hue* and *saturation*.[1] There are no other differences of quality in the impression made by light. This result may be expressed as follows:

The colour impression produced by a certain quantity of mixed light x *can always be reproduced by mixing a certain amount of white light* a *with a certain other amount of some saturated colour* b (*spectrum colour or purple*) *of definite shade.*

This statement limits the number of different kinds of colour impressions; for while they may still be infinite in variety, they must be fewer than they would be if every possible combination of different simple colours were responsible for a separate colour impression. The complete determination of the objective nature of a mixture of different kinds of light involves telling how much light it contains for every different value of the wave-length. But the number of different wave-lengths is infinite; and therefore the physical quality of such an admixture of light must be represented as a function of an endless number of unknowns. But *the impression that any admixture of light makes on the eye may always be represented as simply a function of three variables,* capable of being given numerically, namely: 1. The quantity of saturated coloured light; 2. The quantity of white light which mixed with it gives the same colour sensation; and 3. The wave-length of the coloured light. This enables us also to obtain finally a basis for arranging all the colours systematically in order. Thus, suppose at first we leave out of account differences of luminosity; then there will still be two variables left on which the quality of the colour depends, namely, the hue and the proportion between coloured light and white light. All the various colours may be represented, therefore, according to their two dimensions, by points lying in a plane, just like the values of any other function of two variables. The saturated colours constitute a closed series, and hence they must

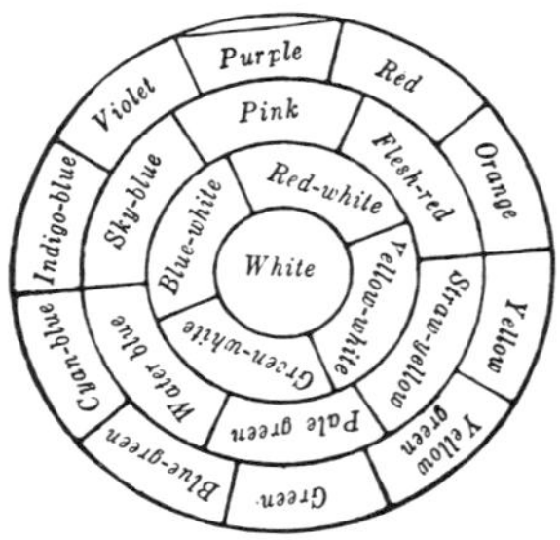

Fig. 16.

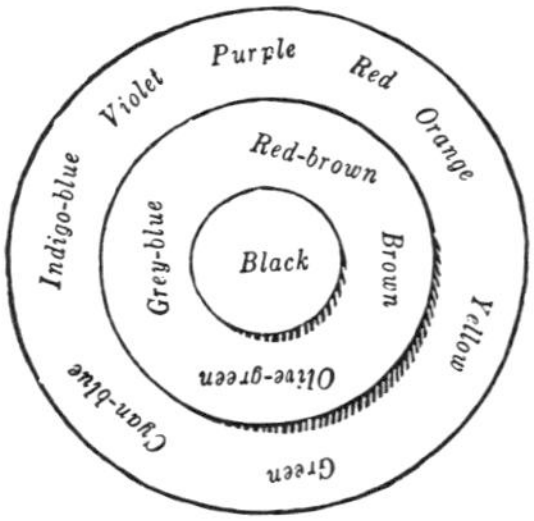

Fig. 17.

[1] ¶This is what painters mean when they say that colour may vary in hue, tint (saturation) and shade (luminosity). The substitution of the word "brilliance" in place of "brightness" or "luminosity" is recommended in the report of the committee on colorimetry of the Optical Society of America published in the *Journal,* VI, 1922, 527–596. (J. P. C. S.)

lie on a closed curve, which NEWTON chose to be a circle with white placed in the centre (Fig. 16). The transitions from white to any saturated colour at a point on the circumference of the circle lie along the radius drawn to this point, so that the paler shades of this hue are nearer the centre, and the more saturated ones nearer the circumference. Thus we get a *colour chart* containing all possible kinds of colours of the same luminosity arranged according to their continuous transitions. When the different degrees of luminosity are also to be taken into account, the third dimension of space has to be utilized, as was done by LAMBERT. The darkest colours for which the number of discriminable shades continually gets less and less may be made to culminate in a point corresponding to black. In this way we get a *colour pyramid* or a colour cone. Three sections of a cone of this sort are shown, one on top of the other, in Fig. 17. The largest, corresponding to the base, ought to show the same distribution of colours as the colour circle in Fig. 16. The middle one, taken from the middle of the cone, contains red-brown, brown, olive-green, grey-blue on the outer edge, with grey in the centre. And, lastly, the smallest, from near the apex of the cone, is black, as shown in the figure.

NEWTON also used the arrangement of the colours in one plane to express the law of colour mixture. He conceived the intensities of the mixed colours as being expressed by weights, which were attached at the places in the chart occupied by the colours concerned. Then the centre of gravity of these weights was to be the place of the compound colour in the chart, and the sum of the weights its intensity. GRASSMANN has developed and formulated the principles that are really implied in NEWTON's method. In addition to the proposition mentioned above, namely:

1. *Any mixed colour, no matter how it is composed, must have the same appearance as the mixture of a certain saturated colour with white*; the following laws are also necessary:

2. *When one of two kinds of light that are to be mixed together changes continuously, the appearance of the mixture changes continuously also*; and

3. *Colours that look alike produce a mixture that looks like them.*

On the basis of these three assumptions, an arrangement of the colours in one plane may be made that enables us to find the mixed colour by a centre-of-gravity method. A colour chart on which the mixed colours are found by the principle of centre-of-gravity constructions, will be called a *geometrical colour chart*. Now as there is no general unobjectionable method of making quantitative comparisons by the eye of the intensities of light of different colours, in constructing a chart of this kind it must be stipulated in advance, that the unit quantity of light for each colour is to be determined by NEWTON's rule of colour mixture itself. Suppose three colours are taken at random, except that no two of them can be combined to make the third. If three arbitrary places are assigned to them on the chart, not all on one straight line, and if the unit intensity for each of these colours is determined in any convenient way, then the position and unit of intensity for every other colour on the chart will be uniquely determined.[1]

Construction of the Colour Chart. Suppose A, B, C are the three colours with which we start; and suppose the units of their intensities have been defined and also their positions a, b, c (Fig. 18) in the colour chart. Let us take the amounts α and β of the colours A and B, and put the mixed colour at the centre of gravity of the weights α and β, that is, at a point d in the straight line ab such that

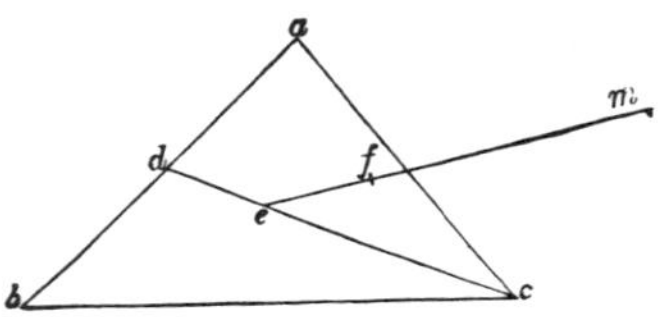

Fig. 18.

$$\alpha . ad = \beta . bd \ .$$

Thus, all mixtures of the colours A and B will lie on the line ab.[2] Suppose now that a quantity γ of some third colour C is to be mixed with the quantities α and β of the colours A and B; all we have to do is to consider that we have a quantity $(\alpha+\beta)$ of the mixed colour corresponding to the point d, and to find the position of the centre of gravity e of the weights $(\alpha+\beta)$ at d and γ at c. This is the place where the compound colour lies, its quantity being

$$\epsilon = \alpha+\beta+\gamma \ .$$

At the same time the unit luminosity of this mixed colour is also determined from the fact that

$$1 = \frac{\epsilon}{\alpha+\beta+\gamma} \ .$$

Evidently, by this method every mixture of the three colours A, B and C must lie *within* the triangle abc; for both its position and its unit luminosity are found as above described.[3]

[1] ¶Thus any colour that differs from one of the colours in the chart in luminosity only is referred to this colour as a unit, and in this way its quantity is estimated. (J. P. C. S.)

[2] ¶Thus the rule for finding the result of mixing two colours in the diagram is to divide the line joining them inversely in the ratio of the quantities of the two components. (J.P.C.S.)

[3] ¶The result obtained in this way would be a triangular portion of NEWTON's diagram. (J. P. C. S.)

On the assumption that the positions and units of measurement have been found for all the results that can be obtained by mixing together the three colours A, B and C, we may then proceed to determine the same functions for all the colours that cannot be produced by combinations of these three. Suppose M is such a colour. It is always possible to choose the amount μ of this colour so small that the result of mixing it with one of the colours of the triangle will be also a colour lying inside the triangle. For example, it may be mixed with an amount ϵ of the colour defined by the point e, where ϵ is measured in terms of the unit found above. If the amount of M is supposed to be infinitesimal at first and then to be increased gradually up to the value μ, the mixture will begin by having the same colour as E itself and will continually change into the adjacent colours, according to the fundamental axioms assumed above. When the amount of M finally gets to be equal to μ, let us suppose that the result is a certain colour defined by a point f that is still inside the triangle. In the first place, the amount of this colour F, according to the rule, must be

$$\varphi=\epsilon+\mu\ .$$

The quantity μ is dependent therefore on the units of measurement that have been previously determined. Moreover, the position of the point f must be the same as that of the centre of gravity of μ at m and ϵ at e; which means that the point m must lie in the prolongation of the line ef, so that

$$\frac{mf}{ef}=\frac{\epsilon}{\mu}\ .$$

This proportion, therefore, settles also the place and amount of the colour M; and the same method is pursued for finding all the other colours that cannot be produced by mixing A, B and C.

Proof of the above construction. What has to be demonstrated now is that in a chart thus constructed, for which also the metrical units of the amounts of the different colours are determined in the manner above described, the colour obtained by mixing any two colours is at the centre of gravity of the component colours, and its luminous intensity, as measured by the stipulated units, is equal to the sum of the amounts of the parts of the mixture.

Let m_1, m_2, etc. denote the masses of a set of particles whose positions are given by the coördinates x_1, y_1; x_2, y_2; etc., respectively. The coördinates X, Y of the centre of gravity are given by the equations:

$$X(m_1+m_2+m_3+\text{etc .})=m_1x_1+m_2x_2+m_3x_3+\text{etc .}\ ;$$
$$Y(m_1+m_2+m_3+\text{etc .})=m_1y_1+m_2y_2+m_3y_3+\text{etc .}$$

The coördinates of a point designated by a letter n will be denoted by x_n, y_n.

A. Two colours E_0 and E_1 are to be mixed, each of which can be obtained by a mixture of three originally chosen colours A, B and C. Let ϵ_0 and ϵ_1 be the amounts of the colours E_0 and E_1 obtained by mixing the amounts α_0, β_0, γ_0 and α_1, β_1, γ_1 of the colours A, B and C, respectively. Then, if x_0, y_0 denote the coördinates of the point where E_0 is in the colour chart, and x_1, y_1 those of the point where E_1 is, according to the above rule:

$$\begin{aligned} x_0(\alpha_0+\beta_0+\gamma_0) &= \alpha_0 x_a+\beta_0 x_b+\gamma_0 x_c \\ x_1(\alpha_1+\beta_1+\gamma_1) &= \alpha_1 x_a+\beta_1 x_b+\gamma_1 x_c \\ y_0(\alpha_0+\beta_0+\gamma_0) &= \alpha_0 y_a+\beta_0 y_b+\gamma_0 y_c \\ y_1(\alpha_1+\beta_1+\gamma_1) &= \alpha_1 y_a+\beta_1 y_b+\gamma_1 y_c \\ \epsilon_0 &= \alpha_0+\beta_0+\gamma_0 \\ \epsilon_1 &= \alpha_1+\beta_1+\gamma_1 \,. \end{aligned}$$

Now by the axiom that two colours that look alike give a mixture of the same colour, the mixture of ϵ_0 and ϵ_1 is the same as that of $\alpha_0\beta_0\gamma_0$ and $\alpha_1\beta_1\gamma_1$; and in the construction of the colour chart, the coördinates X, Y of the point that belongs to this latter mixture are given by the equations:

$$\begin{aligned} X(\alpha_0+\beta_0+\gamma_0+\alpha_1+\beta_1+\gamma_1) &= (\alpha_0+\alpha_1)x_a+(\beta_0+\beta_1)x_b+(\gamma_0+\gamma_1)x_c \,, \\ Y(\alpha_0+\beta_0+\gamma_0+\alpha_1+\beta_1+\gamma_1) &= (\alpha_0+\alpha_1)y_a+(\beta_0+\beta_1)y_b+(\gamma_0+\gamma_1)y_c \,. \end{aligned}$$

If the six equations above are used to eliminate x_a, x_b, x_c and y_a, y_b, y_c from these two last equations, then

$$\begin{aligned} X(\epsilon_0+\epsilon_1) &= \epsilon_0 x_0+\epsilon_1 x_1 \,, \\ Y(\epsilon_0+\epsilon_1) &= \epsilon_0 y_0+\epsilon_1 y_1 \,. \end{aligned}$$

Thus, the coördinates X, Y of the colour obtained by mixing E_0, E_1 are the same as those of the centre of gravity of ϵ_0 and ϵ_1.

The total amount of light (q) obtained by mixing E_0, E_1 in the quantities ϵ_0, ϵ_1, respectively, must also be equal to that obtained by mixing the amount $\alpha_0+\beta_0+\gamma_0$ and the amount $\alpha_1+\beta_1+\gamma_1$; that is,

$$q=\alpha_0+\beta_0+\gamma_0+\alpha_1+\beta_1+\gamma_1=\epsilon_0+\epsilon_1 \,.$$

Thus the rule given for constructing all colours obtained by mixing A, B and C in the same way as they are located on the colour chart is proved to be correct.

B. *Two colours M_0 and M_1 are to be mixed, neither of which can be obtained by a mixture of A, B and C.* Let x_0, y_0 denote the coördinates of the position of the colour M_0, and let μ_0 denote the amount of this colour; similarly, let x_1, y_1 denote the coördinates of the position of the colour M_1, and let μ_1 denote the amount of it. Suppose the position of M_0 in the colour chart has been found from the fact that when the quantity μ_0 is mixed with the quantity ϵ_0 of a colour E corresponding to the point e, the result is a quantity φ of a colour F corresponding to the point f; then

$$\begin{aligned}\epsilon_0+\mu_0&=\varphi\\ \varphi x_f&=\epsilon_0 x_e+\mu_0 x_0\\ \varphi y_f&=\epsilon_0 y_e+\mu_0 y_0\,.\end{aligned}$$

Likewise, suppose the position of M_1 has been found from the fact that when the quantity μ_1 of this colour is mixed with the quantity ϵ_1 of the same colour E as above, the result is a quantity ψ of a colour G corresponding to the point g; then

$$\begin{aligned}\epsilon_1+\mu_1&=\psi\\ \psi x_g&=\epsilon_1 x_e+\mu_1 x_1\\ \psi y_g&=\epsilon_1 y_e+\mu_1 y_1\,.\end{aligned}$$

To get the place of the result of mixing μ_0 and μ_1 in the same way, they must be mixed with the quantity $\epsilon_0+\epsilon_1$ of the colour E. But this is equivalent to mixing the quantities φ and ψ of the colours F and G. Let the coördinates of the position of this mixed colour be denoted by ξ, v, as given by the equations

$$\begin{aligned}(\varphi+\psi)\xi&=\varphi x_f+\psi x_g\\ (\varphi+\psi)v&=\varphi y_f+\psi y_g\,.\end{aligned}$$

The coördinates X, Y of the position of the colour obtained by mixing μ_0 and μ_1 will be given then by the following equations, in which η denotes the as yet undetermined amount of this colour:

$$\begin{aligned}(\varphi+\psi)\xi&=(\epsilon_0+\epsilon_1)x_e+\eta X\\ (\varphi+\psi)v&=(\epsilon_0+\epsilon_1)y_e+\eta Y\\ \varphi+\psi&=\epsilon_0+\epsilon_1+\eta\,.\end{aligned}$$

Eliminating φ, ψ, x_e, and y_e by the aid of the previous equations, we obtain:

$$\begin{aligned}\mu_0 x_0+\mu_1 x_1&=\eta X\\ \mu_0 y_0+\mu_1 y_1&=\eta Y\\ \mu_0+\mu_1&=\eta\,,\end{aligned}$$

Consequently, the resultant colour obtained by mixing μ_0 and μ_1 is found to be actually, as desired, at the centre of gravity of these two masses, and its amount is equal to the sum of those of the two components.

C. *Two colours are to be mixed, one of which can be obtained by mixing A, B, and C, while the other cannot be so obtained.* The procedure is similar to that in the preceding case. Let μ_0 denote the amount of the colour that cannot be obtained by mixing A, B and C; and suppose that the coördinates x_0, y_0 of the position of this colour have been found from the fact that when it is mixed with an amount ϵ_0 of a colour E

corresponding to the point e, the result is a quantity φ of a colour F corresponding to the point f; then

$$\mu_0 x_0 + \epsilon_0 x_e = \varphi x_f$$
$$\mu_0 y_0 + \epsilon_0 y_e = \varphi y_f$$
$$\mu_0 + \epsilon_0 = \varphi \ .$$

In order to find the point corresponding to the colour obtained by mixing μ_0 with an amount η of the colour that can be produced by mixing A, B and C, we mix η with ϵ_0, and then proceed by the rule given above. But since η is composed of μ_0 and μ_1, another way would be to mix μ_0 and ϵ_0 first, which, according to the assumption, will give the amount φ of the colour F, and then mix φ with μ_1. The centre of gravity of the two is the position of the colour resulting from the mixture of η and ϵ_0, its coördinates ξ and υ being given by the following equations:

$$(\varphi + \mu_1)\xi = \varphi x_f + \mu x_g$$
$$(\varphi + \mu_1)v = \varphi y_f + \mu y_g \ .$$

The coördinates X, Y of η, according to the above rule, are to be found by the equations:

$$(\varphi + \mu_1)\xi = \eta X + \epsilon_0 x_e$$
$$(\varphi + \mu_1)v = \eta Y + \epsilon_0 y_e$$
$$\varphi + \mu_1 = \eta + \epsilon_0 \ .$$

Thus we derive finally:

$$\eta X = \mu_0 x_0 + \mu_1 x_g$$
$$\eta Y = \mu_0 y_0 + \mu_1 y_g$$
$$\eta = \mu_0 + \mu_1 \ ,$$

which was to be proved.

Hitherto, the places of colours that cannot be obtained by mixing A, B and C have been always found by mixing them with a single colour E; but the last proposition shows that if any other colour G is used, the place found for a given colour of the kind mentioned will turn out to be the same as before.

It is not possible to tell in advance the form of the curve on which the simple colours will lie in a construction of this kind. Indeed, these curves can be very manifold, depending on the choice of the three colours to begin with and on the arbitrary units by which they are measured. One unit has to be arbitrary always, and likewise the positions of the points where two of the chosen colours are placed. When two points on the curve are specified, the form of the curve will depend simply on the other four selections. Accordingly there are four more conditions which may in general be satisfied by a corresponding choice of the four arbitrary factors that are still left. Thus, for instance, it might be desirable that the distances of five certain simple colours from white should all be equal on the chart. Then the curve containing the simple colours would hardly be appreciably different from NEWTON's colour circle as shown in Fig. 16; except that the chord which is drawn in that diagram between the extreme red and violet would have to be the continuation of the curve instead of the arc, because purple, which can be obtained only by mixing red and violet, would have to lie on the straight line joining those two colours. Besides, from the methods of the construction, each pair of complementary colours must be at opposite ends of a diameter of the circle, because the mixed colour white must always be on the lines connecting the two colours that produce it. This condition is also satisfied in Fig. 16.

As to the units of luminosity of light of different colours, in those cases where the boundary of the colour chart has been designed in the form of a circle, complementary amounts of the complementary colours, that is, amounts which produce white when they are mixed, must be considered as being equal, because by hypothesis their mixed colour, white, is midway between them. Moreover, such amounts of other non-complementary colours would be considered as being equal which when combined with a sufficient quantity of their complementary colour in each case will produce equal quantities of white. From what has been stated about the different saturation of the colours of the spectrum, it follows that the quantities which are here considered as equal do not by any means look equally bright to the eye. However, in the next chapter it will be seen that very different results are obtained in estimating brightness by the eye at different absolute intensities of illumination; and that, on the contrary, a determination of the unit of measurement of different colours by the results of mixtures continues to be valid in a certain sense at least for all degrees of luminosity.

On the other hand, supposing quantities of light of different colours are to be considered as being equal, when for a certain absolute intensity of light they look equally bright to the eye, we shall get an entirely different form for the curve on which the simple colours lie, like that shown in Fig. 19. Saturated violet and red must be farther from white than their less saturated complementary colours, because, as the eye estimates it, it takes less violet than yellow-green when we mix these two complementary colours together to get white. Hence, if white is to be in the position of their centre of gravity, the

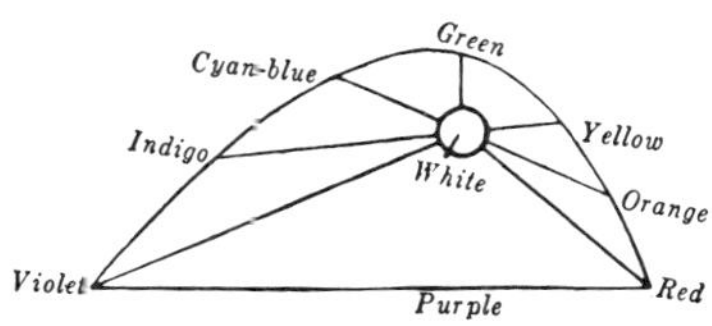

Fig. 19.

smaller amount of violet must have a longer lever-arm than the larger amount of yellow-green. Incidentally, here also, as before, the spectrum-colours themselves would be ranged along a curve and the purples on a chord; complementary colours lying at the opposite ends of chords that all pass through the point on the chart that corresponds to white. In these respects the chart is similar to the colour circle in Fig 16.

NEWTON's device of exhibiting the laws of colour mixture by the method used for constructing centres of gravity was intended originally simply as a kind of mathematical picture for expressing graphically a large mass of facts; the justification for it consisting in the fact that the results as found by this process were qualitively in accordance with experimental realities, even if they had not been tested quantitatively. However, quite recently the method has been still further supported by careful quantitative measurements made by MAXWELL. He used a colour top, for which two sets of circular sectors were made, the radii of one set being longer than those of the other set. These sectors were cut out of highly coloured papers (vermilion, bright chrome yellow, Paris green or emerald-green, ultramarine, white, and black). They were placed on the surface of the colour top in such a way that the angles of the sectors at the centre of the circle could be adjusted at pleasure; and also so that one combination of colours could be produced out towards the edge of the disc and at the same time another combination in towards the centre. The experiment generally consisted in varying the angular widths of the sectors until the outer and inner colours appeared exactly alike when the colour top was spinning rapidly; and then the angles of the exposed portions of the sectors were measured. Endless varieties of combinations of colour can be made in this way, and the laws of colour mixture may be tested by them. In accordance with the ideas developed above, the plan of this method may be easily explained as follows. Suppose a colour chart is constructed with three fundamental colours, say, red, green and blue, the same colours as used on the colour top; the brightness or luminosity of each of them being put equal to unity according to some arbitrary system of measurement. Hence, in any experimental combination of these three colours, the luminosities must be equal to the ratios between the arcs of the coloured sectors and the circumference of the circle. The first thing we can do is to make a grey by a combination of the three fundamental colours, and to match it by a combination of black and white. This experiment will enable us to locate the position of white in the colour chart and to find also unit luminosity for white.

The next experiment might be to make two grey-yellow matches, by using red and green for one mixture and yellow, white and black for the other. By the aid of the rules given above, this determination will enable us to locate where yellow belongs on the chart and to find its unit luminosity. This preliminary work being out of the way, we can proceed then, either by construction with the chart or by calculation, to show how any mixture of three of the five colours, red, yellow, green, blue and white, can be matched by mixing any other three of them; and these predictions can then be verified experimentally, every such test tending to confirm the correctness of the centre-of-gravity method of finding the result of mixing colours. Incidentally, the colour top is nicely adapted also for numerical evaluations of the colours of natural bodies.

Every difference of impression made by light, as we have seen, may be regarded as a function of three independent variables; and the three variables which have been chosen thus far were (1) the luminosity, (2) the hue, and (3) the saturation, or (1) the quantity of white, (2) the quantity of some colour of the spectrum, and (3) the wave-length of this colour. However, instead of these variables, three others may also be employed; and in fact this is what it amounts to, when all colours are regarded as being mixtures of variable amounts of *three so-called fundamental colours,* which are generally taken to be *red, yellow* and *blue.* To conceive this theory objectively, and to assert that there are simple colours in the spectrum which can be combined to produce a visual impression that will be the same as that produced by any other simple or compound light, would not be correct. There are no such three simple colours that can be combined to match the other colours of the spectrum even fairly well, because the colours of the spectrum invariably appear to be more saturated than the composite colours. Least suited for this purpose are red, yellow and blue; for if we take for blue a colour like the hue of the sky, and not a more greenish blue, it will be impossible to get green at all by mixing these colours. By taking a greenish yellow and a greenish blue, the best we can get is a very pale green. These three colours would not have been selected, had it not been that most persons, relying on the mixture of pigments, made the mistake of thinking that a mixture of yellow and blue light gives green It would be rather better to take *violet, green* and *red* for fundamental colours. Blue can be obtained by mixing violet and green, but it is not the saturated blue of the spectrum; and a dead yellow can be made with green and red, which is not at all like the brilliant yellow in the spectrum.

If we think of the colours as plotted on a colour-chart by the method sketched above, it is evident from the rules given for the construction that all colours that are to be made by mixing three colours must be contained within the triangle whose vertices are the places in the chart where the three fundamental colours are. Thus, in the adjoining colour circle (Fig. 20), where the positions of the colours are indicated by the initial letters of their names (I = indigo-blue, C = cyan-blue, Y = yellow, G = green, etc.), all the colours that can be made by mixing red, cyan-blue and yellow are comprised within the triangle RCY. Thus, as we see, two large pieces of the circle are missing, and all that could be obtained would be a very pale violet and a very pale green. But if, instead of cyan-blue, the colour of the blue sky, indigo-blue, were taken, green would be missing entirely. The triangle VRG comprises the colours obtained by mixing violet, red and green, and a larger number of the existing colours would indeed be represented. But, as the diagram shows, large portions of the circle are still missing, as must always be the case according to the results of experiments on the mixture of the colours of the spectrum. The conclusion is that the boundary of the colour chart must be a curved line which differs considerably from the perimeter of the triangle.

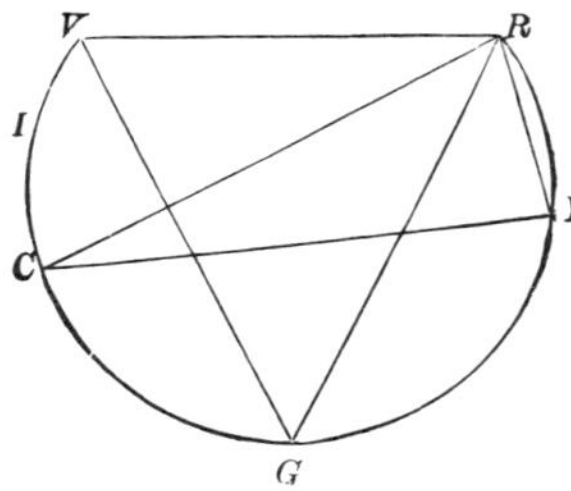

Fig. 20.

BREWSTER, endeavouring to defend the objective nature of three fundamental colours, maintained that for every wave-length there were three different kinds of light, red, yellow and blue, mixed merely in different proportions so as to give the different colours of the spectrum. Thus, the colours of the spectrum were considered as being compound colours consisting of three kinds of light of different quality; although the degree of refrangibility of the rays was the same for each individual simple colour. BREWSTER's idea was that light of all three fundamental colours could be proved to exist in the different simple colours by the absorption of light by coloured media. His entire theory is based on this conception, which was shown in the preceding chapter to be erroneous.

Apart from BREWSTER's hypothesis, the notion of three fundamental colours as having any objective significance has no meaning anyhow. For as long as it is simply a question of physical relations, and the human eye is left out of the game, the properties of the compound light are dependent only on the relative amounts of light of all the separate wave-lengths it contains. When we speak of reducing the colours to three fundamental colours, this must be understood in a subjective sense and as being an attempt to trace the *colour sensations* to three *fundamental sensations*. This was the way that YOUNG regarded the problem;[1] and, in fact, his theory affords an exceedingly simple and clear explanation of all the phenomena of the physiological colour theory. He supposes that:

1. The eye is provided with three distinct sets of nervous fibres. Stimulation of the first excites the sensation of red, stimulation of the second the sensation of green, and stimulation of the third the sensation of violet.

2. Objective homogeneous light excites these three kinds of fibres in various degrees, depending on its wave-length. The red-sensitive fibres are stimulated most by light of longest wave-length, and the violet-sensitive fibres by light of shortest wave-length. But this does not mean that each colour of the spectrum does not stimulate all three kinds of fibres, some feebly and others strongly; on the contrary, in order to explain a series of phenomena, it is necessary to assume that that is exactly what does happen. Suppose that the colours of the spectrum are plotted horizontally in Fig. 21 in their natural sequence, from red to violet, the three curves may be taken to indicate something like the degree of excitation of the three kinds of fibres, No. 1 for the red-sensitive fibres, No. 2 for the green-sensitive fibres, and No. 3 for the violet-sensitive fibres.

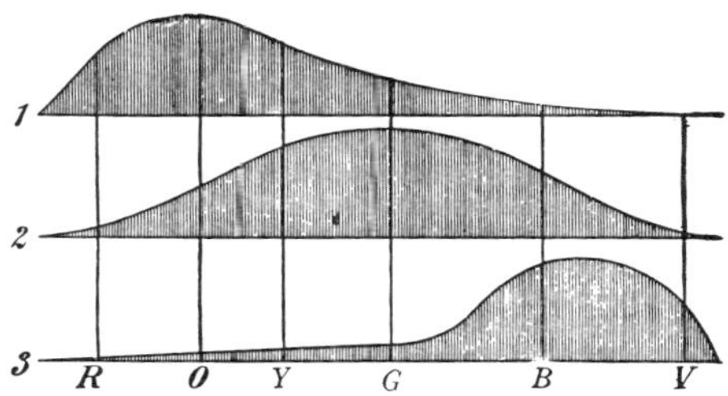

Fig. 21.

[1] ¶MAXWELL in his lecture "On colour vision" at the Royal Institution (see *Scientific Papers of* JAMES CLERK MAXWELL, II. pp. 266–279), speaking of YOUNG's theory, says:

"We may state it thus:—We are capable of feeling three different colour-sensations. Light of different kinds excites these sensations in different proportions, and it is by the different combinations of these three primary sensations that all the varieties of visible colour are produced. In this statement there is one word on which we must fix our attention. That word is, Sensation. It seems almost a truism to say that colour is a sensation; and yet YOUNG, by honestly recognizing this elementary Truth established the first consistent theory of colour. So far as I know, THOMAS YOUNG was the first who, starting from the well-known fact that there are three primary colours, sought for the explanation of this fact, not in the nature of light, but in the constitution of man. Even of those who have written on colour since the time of YOUNG, some have supposed that they ought to study the properties of pigments, and others that they ought to analyse the rays of light. They have sought for a knowledge of colour by examining something in external nature—something out of themselves." (J.P.C.S.)

Pure *red* light stimulates the red-sensitive fibres strongly and the two other kinds of fibres feebly; giving the sensation red.

Pure *yellow* light stimulates the red-sensitive and green-sensitive fibres moderately and the violet-sensitive fibres feebly; giving the sensation yellow.

Pure *green* light stimulates the green-sensitive fibres strongly, and the two other kinds much more feebly; giving the sensation green.

Pure *blue* light stimulates the green-sensitive and violet-sensitive fibres moderately, and the red-sensitive fibres feebly; giving the sensation blue.

Pure *violet* light stimulates the violet-sensitive fibres strongly, and the other fibres feebly; giving the sensation violet.

When all the fibres are stimulated about equally, the sensation is that of *white* or pale hues.

It might be natural to suppose that on this hypothesis the number of nervous fibres and nerve-endings would have to be trebled, as compared with the number ordinarily assumed when each single fibre is made to conduct all possible colour stimulations. However, in the writer's opinion there is nothing in YOUNG's hypothesis that is opposed to the anatomical facts in this respect; because we are entirely ignorant as to the number of conducting fibres, and there are also quantities of other microscopical elements (cells, nuclei, rods) to which hitherto no specific functions could be ascribed. But this is not the essential thing in YOUNG's hypothesis. That appears to the writer to consist rather in the idea of the colour sensations being composed of three processes in the nervous substance that are perfectly independent of one another. This independence is manifested not merely in the phenomena which are being considered at present but also in those of fatigue of the nervous mechanism of vision. It would not be absolutely necessary to assume different nervous fibres for these different sensations. So far as mere explanation is concerned, the same advantages that are afforded by YOUNG's hypothesis could be gained by supposing that within each individual fibre there might occur three activities all different from and independent of one another. But the form of this hypothesis as originally proposed by YOUNG is clearer in both conception and expression than it would be if it were modified as suggested, and hence it will be retained in its original concrete form, for the sake of exposition if for nothing else. Nowhere in the physical (electrical) phenomena of nervous stimulation either in the sensory or motor nerves can there be detected any such differentiation of activity as must exist if each fibre of the optic nerve has to transmit all the colour sensations. By YOUNG's hypothesis it is possible even in this connection to transfer directly to the optic nerve the simple conceptions as to the mechanism of the stimulation and its conduction which we were led to form at first by studying the phenomena in the motor nerves. This would not be the case on the assumption that each fibre of the optic nerve has to sustain three different kinds of states of stimulation which do not mutually interfere with one another. YOUNG's hypothesis is only a more special application of the law of specific sense energies. Just as tactile sensation and visual sensation in the eye are demonstrably affairs of different nervous fibres, the same thing is assumed here too with respect to the various sensations of the fundamental colours.

The choice of the three fundamental colours is somewhat arbitrary. Any three colours which can be mixed to get white might be chosen. YOUNG may have been guided by the consideration that the terminal colours of the spectrum seem to have special claims by virtue of their positions. If they were not chosen, one of the fundamental colours would have to have a purplish hue, and the curve corresponding to it in Fig. 21 would have two maxima, one in the red and one in the violet. This would be a more complicated assumption, but not an impossible one. So far as the writer can see, the only other way of determining one of the fundamental colours would be by investigating the colour-blind. To what extent such investigation confirms YOUNG's hypothesis for red at least, will be shown later.

That each of the three chosen fundamental colours of the spectrum stimulates not only the nervous fibres that are designated by the same name as the colour in question but the other fibres also in a less degree, has been already proved by the results of colour mixture, certainly in the case of green. For if we think of all the colour sensations that are composed of the three fundamental colours as being plotted on a plane chart according to NEWTON's system, it follows from what has been stated above that the colour area must be enclosed in a triangle. This triangle must include within it the colour area shown in Fig. 22 which comprises all colours that are miscible from the colours of the spectrum. It would be possible to do this by shifting the sensation of pure green towards A, as is done in Fig. 22, on the assumption that spectrum red and violet, R and V, are pure fundamental colours. In this case the colour triangle that contains within it all possible colour sensations would be AVR. This

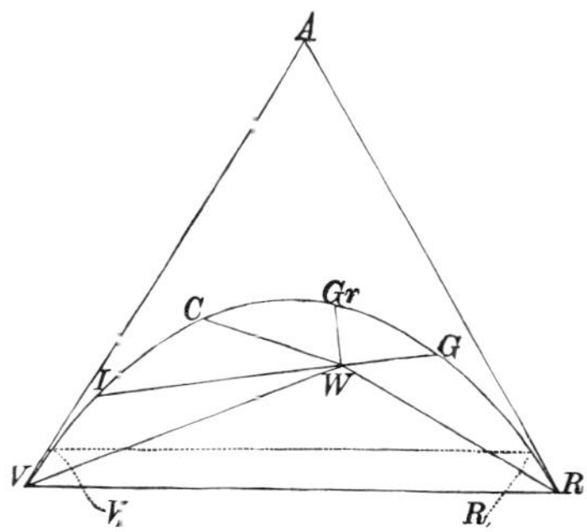

Fig. 22.

assumption, as stated, would satisfy the actual facts of colour mixture. On the other hand, however, certain other facts to be mentioned presently, in connection with colour blindness and the change of hue due to increase of intensity of the light and the phenomena of after-images, render it necessary to assume that neither spectrum red nor violet corresponds to a simple sensation of one fundamental colour, but to a slightly mixed sensation. Accordingly, the positions of spectrum red and violet in the colour triangle Fig. 22 would have to be displaced about to $R_,$ and $V_,$ and the closed curve $ICYR_,V_,$ would then embrace all possible colours of objective light.

Thence it follows that there must be a series of colour sensations still more saturated than those which are evoked under ordinary circumstances by objective light, even by that of the spectrum. In Fig. 22 the colours aroused in the normal eye by external light are comprised within the area bounded by the curve and the straight line $V_,R_,$. The rest of the triangle corresponds to colour sensations that cannot be excited directly by external light. Since these latter sensations are all farther separated from white than the colours of the spectrum, they must be even more saturated than those colours themselves, which are the most saturated objective colours of which we have any knowledge. And, as a matter of fact, when we come to the theory of after-images, produced by fatiguing the eye by the complementary colours, we shall see how to produce colour sensations beside which the colours of the spectrum look pale.

The fact above mentioned, that the different colours of the spectrum do not appear to be all saturated to the same degree, is easily explained by this theory.

The eyes of some individuals are not able to distinguish as many colours as those of ordinary persons. The visual perceptions in cases of *colour blindness* (*achromatopia, achrupsia*) are of particular interest for the theory of colour sensations. A. Seebeck has demonstrated that there are two classes of colour-blind people. Individuals belonging to each group confuse the same set of colours, and differ from each other merely in the degree of their difficulty. On the other hand, individuals in one class recognize most of the mistakes made by those of the other class.

The most numerous cases, especially in England, appear to belong to Seebeck's second division. Their trouble is often called *Daltonism* (or *anerythropsia*, by Goethe), after the celebrated chemist J. Dalton, who himself belonged to this group and was the first to investigate this condition carefully.[1] As some English scientists have protested against this mode of perpetuating the name of their renowned countryman by one of his defects, let us call this condition *red blindness*.[2] Individuals in whom it is completely developed see only two colours in the spectrum, which they usually describe as blue and yellow. They include in the latter all of the red, orange, yellow and green. They call the green-blue hues grey, and the remainder blue. They do not see the extreme red at all when it is faint, but they may do so when it is intense. Thus, they usually put the red end of the spectrum at a place where normal eyes still see distinctly a faint red. In pigments they confuse *red* (that is, vermilion and reddish orange) with *brown* and *green*; whereas to the normal eye in general the confused red hues are much brighter than the brown and green. They cannot distinguish between *golden yellow* and *yellow* or between *pink-red* and *blue*. On the other hand, all mixtures of different colours that appear alike to the normal eye appear alike also to the red-blind. With regard to DALTON's case, Sir J. HERSCHEL[3] has already advanced the opinion, that all colours discriminated by him might be considered as being composed of two fundamental colours instead of three.[4] This view has been recently confirmed by MAXWELL by his method of measuring colours with the mixtures on the colour top. In the case of a healthy eye, as has been shown, a colour match may be formed between any given colour and three suitably chosen fundamental colours, plus white and black. In case of red-blind persons, as the writer himself has verified, only two colours are needed besides white and black (for instance, yellow and blue) to make a colour match on the colour top with any other colour.

In the author's experiments with Mr. M., who was a student in the polytechnic institute, accustomed to physical investigations and fairly sensitive to such differences of colour as he could still recognize at all, chrome yellow and ultramarine were used as principal colours. A mixture on the colour top of 35° of yellow and 325° of black, which was olive-green to an average person, seemed to him identical with a *red* about like that of sealing wax. The experiments indicated that a

[1] According to more modern determinations, SEEBECK's second form of colour blindness, called "green blindness" by HELMHOLTZ, is more frequent, and the only reason why it is more often unnoticed than "red blindness" is because its symptoms are rather less striking. As to new suggestions for designating the forms of colour blindness, see the section on this subject in the appendices at the end of this volume.—N.

[2] ¶Red-blind individuals, as HELMHOLTZ calls them, are termed *protanopes* by v. KRIES, and green-blind *deuteranopes*. (J. P. C. S.)

[3] In a letter quoted in G. WILSON, *On colour Blindness*. Edinburgh 1855. p. 60.

[4] ¶These types of colour blindness have colour systems that are functions of two variables, whereas normal colour vision, as has been stated, is a function of three variables. Thus normal individuals are said to be *trichomats* as distinguished from these abnormal *dichromats*. (J. P. C. S.)

mixture of 327° of yellow and 33° of blue, which looks grey-yellow to the normal eye, was to him the same as *green* of hue about corresponding to the line E in the spectrum. And 165° of yellow mixed with 195° of blue, which ordinarily gives a faint reddish grey, was the same to him as *grey*. As all other hues could be mixed from red, yellow, green, and blue, the result was that, so far as Mr. M. was concerned, they could all be obtained by mixing yellow and blue.

From GRASSMANN's laws of colour mixture, as applied to an eye that confuses red with green, it follows directly that the hues which it does differentiate can all be obtained by mixing two other colours, say, yellow and blue. For if red and green appear to be the same, necessarily all mixtures of these two colours will appear to be the same. Moreover, since colours that look alike produce a mixture that looks like them, every mixture of a given amount of yellow with a given amount of any one of the colours made by mixing red and green, which has the same appearance to a colour-blind person, will give a resultant colour that looks the same to him. But for the healthy eye one of the colours obtained by mixing red and green can be made also by mixing yellow and blue; and hence for the colour-blind eye this colour can be substituted for all combinations of red and green. Consequently, all mixtures of yellow, red and green may be produced also by mixing yellow and blue, so far as the colour-blind eye is concerned; and the same thing may be proved likewise for all mixtures of blue, red and green. And, lastly, since all hues for the healthy eye can be obtained by mixing red, yellow, green and blue, all hues for the colour-blind eye can be obtained by mixing yellow and blue.

If the colours are plotted on a plane chart by the method of constructing the centres of gravity, all such colours as appear to a colour-blind person to be the same at suitable luminosity will be ranged along a straight line, since a mixture of two colours must be on the straight line joining these two points, and the mixture must appear to have the same hue as its components, if the latter look alike. Moreover, it may be proved that all these straight lines intersect in one point (which may be at infinity, in which case they will all be parallel), and that the colour corresponding to this point must be invisible to the colour-blind eye.

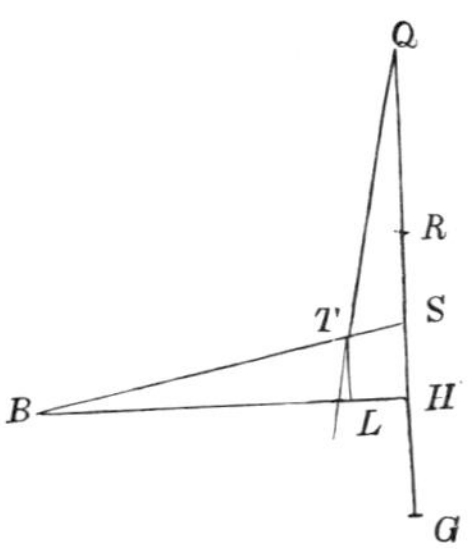

Fig. 23.

To the colour-blind person the quantity r of the colour at R in Fig. 23 appears the same as the quantity g of the colour at G. Now

$$r = nr + (1-n)r.$$

The quantity ng of the colour G looks just like the quantity nr of the colour R. Thus, supposing that n is a proper fraction, the quantity r of the colour R appears the same as the mixture of the quantity $(1-n)r$ of the colour R with the quantity ng of the colour G. In the colour chart this mixed colour will be at the point S in the line RG such that

$$RS : SG = ng : (1-n)r \quad . \quad . \quad . \quad . \quad . \quad . \quad . \quad . \quad (1)$$

and the quantity of this mixed colour will be

$$s = ng + (1-n)r.$$

So far as the colour-blind eye is concerned, the appearance of this colour will not depend on the value of n.

Suppose now that the quantity b of the colour B is mixed with the quantity s of the colour S; the result will be a mixed colour whose appearance to the colour-blind eye is independent of the variable magnitude n. Let T be the place of this mixed colour in the chart and t its quantity; then

$$t = b + s = b + ng + (1-n)r$$
$$TS : BT = b : s = b : [ng + (1-n)r] \quad . \quad . \quad . \quad . \quad . \quad . \quad (1a)$$

From B let fall the perpendicular BH on RG and from T the perpendicular TL on BH; and put

$$LH = x \qquad BH = h$$
$$TL = y \qquad RG = a$$
$$RG = c$$

Then by (1a):

$$\frac{x}{h} = \frac{LH}{BH} = \frac{TS}{BS} = \frac{b}{b + ng + (1-n)r} \quad . \quad . \quad . \quad . \quad . \quad . \quad (1b)$$

$$\frac{y}{h-x} = \frac{TL}{BL} = \frac{SH}{BH} = \frac{SG - a}{h}.$$

Since from equation (1)

$$SG = c \,.\, \frac{(1-n)r}{ng + (1-n)r},$$

therefore

$$\frac{y}{h-x} = \frac{(c-a)\,(1-n)r - ang}{h[ng + (1-n)r]} \quad . \quad . \quad . \quad . \quad . \quad . \quad (1c)$$

Eliminating the variable n from (1b) and (1c), we obtain an equation connecting the rectangular coördinates of the point T, as follows:

$$0 = ybh(g-r) - x[crg + br(c-a) + abg] + bh[(c-a)r + ag] \quad . \quad . \quad . \quad (1d)$$

As this is a linear equation between x and y, the locus of the point T is a straight line, and all the mixed colours that lie on this line appear to be the same to the colour-blind eye. Suppose TQ is this straight line meeting the straight line RG in the point designated by Q; then the value of y for $x=0$ will be $QH=y_0$ as given by the equation

$$y_0 = \frac{(c-a)r+ag}{r-g} \quad . \quad . \quad . \quad . \quad . \quad . \quad . \quad . \quad . \quad (1e)$$

This value of y_0 is independent of the amount b of the colour that is to be mixed with S; and every straight line that is the locus of points corresponding to colours that all look alike, obtained by mixing the colours R, G and B, will pass through this point Q; and in case $r=g$, that is, when y_0 becomes infinite, the point Q will be the infinitely distant point of the straight line RG, and the system of lines TQ will be a pencil of parallel lines.

The distance of Q from the point R is

$$y_0 - c + a = \frac{cg}{r-g} = QR \quad . \quad . \quad . \quad . \quad . \quad . \quad . \quad (1f)$$

When an amount q of the colour Q is mixed with the amount g of the colour G so as to make the colour R, then we must have

$$\frac{QR}{RG} = \frac{g}{q}$$

and therefore by equation (1f), since $RG=c$:

$$\frac{g}{r-g} = \frac{g}{q}$$

$$q = r-g \; .$$

The amount of the mixed colour R in this case is:

$$r=g+q \; .$$

But by hypothesis r looks the same to the colour-blind eye as g; and since, in general, the amount $q=r-g$ is different from zero, the conclusion is that *the colour-blind eye cannot be sensitive to the colour Q at all.*

The point of intersection of the straight lines that are the loci of points corresponding to colour-mixtures that look alike falls therefore at the place of the colour which is missing in the colour sensations of the colour-blind eye.

On YOUNG's hypothesis this colour that is not visible to the colour-blind person is necessarily one of the fundamental colours; for if there were sensation for all the fundamental colours, no other colour sensation composed simply of these fundamental ones could be lacking.

Now when we try to discover those colours that look like white (or grey), they will be found to be those which for the normal eye are colours of the hue of the missing fundamental colour or of its complementary colour, mixed with white in different proportions. For all these colours that look like white must lie on a straight line. But every straight line drawn through the point on the chart that corresponds to white contains on its two opposite sides colours of the same hue in different degrees of saturation. But the colours on one half of the line are complementary to those on the other half. Every line of this sort containing colours that all look alike must, however, as just proved, pass through the point where the missing fundamental colour is, and, consequently, must contain on one of its two halves colours of the same hue as the fundamental colour. In the experiments which the writer conducted with Mr. M. it was found that the same appearance as pure grey was produced by a red which corresponded very nearly to the extreme red of the spectrum in hue (38° of ultramarine and 322° of vermilion), perhaps leaning a little to the purple side, and by a corresponding complementary blue-green (59° ultramarine and 301° emerald-green). MAXWELL has obtained similar results, namely, 6 percent ultramarine and 94 percent vermilion for the red, and 40 percent ultramarine and 60 percent emerald-green for the green. And, besides, as for normal eyes the red appeared much darker than the grey and green, with equal luminosity, there can be no more doubt that it is red, and not green, that is the missing colour. On YOUNG's hypothesis, therefore, red blindness would be explained as a paralysis of the red-sensitive nerves.

If a red not far from the extreme red of the spectrum is really one of the fundamental colours, the two others cannot be very far anyhow from the green and violet as chosen by YOUNG.

The result of this would be that people who are red-blind are not sensitive except to green and violet and blue, which is a mixture of the first two. The *red* of the spectrum which seems to stimulate the green-sensitive nerves just a little and the violet-sensitive nerves almost not at all, according to this, would have to appear to red-blind persons as a *saturated green of low luminosity*, containing appreciable amounts of the other colours mixed with it. Red of low luminosity which is still adequate to excite the red-sensitive nerves of the normal eye is, on the contrary, no longer adequate to excite the green-sensitive nerves, and therefore this sort of light appears black to red-blind individuals.

The *yellow* of the spectrum will appear as *brilliant saturated green*, and doubtless it is just because it does give the saturated and more luminous shade of this colour that the red-blind select the name of this colour and describe all these peculiar hues of green as being yellow.

Green as compared with yellow begins to show an admixture of the other fundamental colours, being therefore indeed a more luminous but yet a pale shade of green, like that produced by red and yellow. According to SEEBECK's observations, the most luminous part of the spectrum for red-blind individuals is, not in the yellow as for normal vision, but in the green-blue.[1] As a matter of fact, on the assumption that green stimulates the green-sensitive nerves most, as must be the case, the maximum of the total stimulation for red-blind persons will be rather towards the side of the blue, because here the stimulation of the violet-sensitive nerves increases. What a red-blind individual means by white is naturally a mixture of his two fundamental colours in some definite proportion, which looks green-blue to us; and therefore he regards the transitional shades in the spectrum from green to blue as being grey.

Farther on in the spectrum the second fundamental colour, which a red-blind person calls blue, begins to predominate, because although indigo-blue is still rather pale to him, yet by its luminosity it appears to his eye to be a more striking representative of this colour than violet. Such an one can distinguish the difference of appearance between blue and violet. The subject H. who was examined by SEEBECK knew where the boundary came, but explained that he preferred to call violet *dark blue*. Incidentally, the blue hues must look to the red-blind pretty much as they do to normal persons, because with the latter also there is not much admixture here with red.

All these colours of the spectrum must appear to the red-blind to have certain differences, even if they are less marked. Evidently therefore by paying more attention to them and by practice, they may even learn to call very saturated colours by their right names. But for paler colours the distinctions above mentioned must be too much for them, as they cannot get rid of the confusion.

With respect now to the other group of colour-blind persons comprised in SEEBECK's first division, there are not yet sufficient observations to enable us to define their condition perfectly. According to SEEBECK's data, the difference between them and the red-blind is that they have no difficulty in detecting the transition between violet and red, which to all red-blind persons appears blue. On the other hand, they are confused between green, yellow, blue and red. Both classes confound the same hue with green, but individuals of SEEBECK's first class choose a more yellow-green than red-blind persons do. They

[1] In the yellow-green is more correct.—N.

are not insensitive to the farthest red, and the brightest part of the spectrum for them is in the yellow.[1] They also discriminate only two hues in the spectrum, which they call, probably quite correctly, blue and red. Accordingly, it may be conjectured that their difficulty is due to insensitivity of the green-sensitive nerves, but further investigations on this point are desirable.

Besides total insensitivity, of course, also there may occur all kinds of degrees of lowered sensitivity of the nerves of one sort or the other, with a resultant inability of discriminating colours to a greater or less extent. Cases have also been reported by WILSON and TYNDALL where the trouble was not congenital, but appeared suddenly as a result of serious injuries to the head or eye-strain.

So far as examination of colour-blind persons is concerned, naturally extremely little information can be obtained by asking them how they call this or that colour; for these persons are obliged to use the system of names to describe their sensations which has been devised for the sensations of the normal eye, and which therefore is not adapted for their case. It is not only not adapted because it contains the names of too many hues, but because in the series of colours in the spectrum the differences we speak of are differences of hue, but to colour-blind persons these are merely differences of saturation or of luminosity. Whether what they call yellow and blue corresponds to our yellow and blue, is more than doubtful. Hence, their replies to questions about colours are usually hesitating and perplexed, and seem to us muddled and contradictory.

SEEBECK's method of giving colour-blind persons a selection of coloured papers or worsteds with instructions to arrange them according to their similarity is much better, though still far from satisfactory. But the number of colour tests would have to be enormous in order to include the hues that are characteristic of the difficulty, in their precise admixture with white and of the right luminosity, so that it would be possible to formulate the complete equation for the colour-blind eye. But as long as it is merely a question of similarity, it will be hard to tell whether the difference is one of hue, saturation or luminosity. It is simply by accident, therefore, that a few definite results can be obtained.

On the other hand, the colour top as designed by MAXWELL enables us to obtain quickly the requisite data with great accuracy, because it is easy to make a set of colours by mixture that to the colour-blind eye appear to be *absolutely alike*. Here the chief thing denoting the fundamental character of the trouble is to ascertain

[1] A little towards the orange, about at wave-length $600\mu\mu$.—N.

what two colours are confused with pure grey as obtained on the colour top by mixing white and black. One of them, which in this case appears comparatively much darker to the colour-blind eye than it does to the normal eye, is the missing fundamental colour. At the same time it will be easy too to discover whether there is still some trace of sensitivity for the missing fundamental colour, or not.

To test the theories here propounded, it will be necessary besides to determine whether every given colour, and especially the principal colours of the spectrum, can be compounded for the colour-blind by mixing two suitably selected colours.

G. WILSON has directed particular attention to the danger for navigation and railways that might be caused by not being able to detect coloured signals on account of colour blindness. He found on the average one colour-blind person in every 17.7 individuals.

Lastly, one other thing must be mentioned: colours cannot be distinguished by the eye unless they occupy a certain extent of field, and unless a certain amount of coloured light comes to the eye. The farther the coloured field lies towards the borders of the visual field and of the retina, the larger it has to be for its colour to be recognized. If the coloured field is too small, it will look grey or black on a brighter background, and grey or white on a darker background. Yet it is possible also to recognize the colours of infinitely small fields, in case the amount of light emitted is finite. For example, the colours of the fixed stars may be distinguished. According to AUBERT's experiments,[1] a blue square and a red square, each one millimetre along the side, looked black on a white background at distances of 10 and 20 feet, respectively. A yellow and a green square fused completely into the white background at a distance of 12 feet. On the other hand, when the background was black, the green and yellow squares looked like grey points 16 feet away, and the red square looked the same way 12 feet away. Blue looked blue when it was seen at all.

According to the same observer, the colour of coloured squares disappears at a distance of 200 mm on the average, for the following angular distances from the visual axis:

	Red				Blue			
Side of square	1.	2.	4.	8.	1.	2.	4.	8.
White ground	16°	19°	26°	37°	15°	22°	36°	49°
Black ground	30	32	42	53	36	48	54	72
Average	23	26	34	45	26	35	45	61

[1] *Archiv für Ophthalmologie.* Bd. III. Abt. II. S. 60.

	Yellow				Green			
Side of square	1.	2.	4.	8.	1.	2.	4.	8.
White ground	21°	31°	44°		20°	36°	44°	50°
Black ground	30	32	49	47°	24	27	35	45
Average	26	32	42		22	32	40	47

Before the colours quite disappear, they undergo a change of hue similar to that produced by increasing their intensity. Thus, red and green become distinctly yellow, blue seems to pass directly into grey-white, and in the purple mixtures of blue and red, blue predominates on the edges of the field of view. PURKINJE had found that purple looked blue on the farthest edges of the field, and became violet as it was moved more into the field until it finally took its own colour. To the author pink likewise looks bluish or violet-white on the borders of the field. The last mentioned effect is most noticeable with mixtures of pairs of colours. For instance, if by the method to be described farther on a small coloured field is illuminated by simple red and green-blue so that it looks white in direct (foveal) vision, seen indirectly even a slight distance from the point of fixation, it looks green-blue. From these experiments it appears that the peripheral parts of the retina are more sensitive to blue and green than to red. To a certain extent, it approaches there the state of red blindness.

Here also may be mentioned OPPEL's experiment,[1] in which he found that an orange-yellow spot on a blue ground viewed from a distance looks brighter than the background, whereas viewed close by, where the blue extended more towards the borders of the visual field, it looks darker.

Besides YOUNG's colour theory, reference should be made to certain theories of colour mixture based directly on the undulatory theory of light. Attempts in this direction have been made by CHALLIS and GRAILICH. The latter in particular has tried to develop this in a very laborious work. He investigates the compound vibratory movements of the aether particles resulting from two trains of waves of different wave-lengths, and computes the intervals of time during which the particles are on one side or the other of their positions of equilibrium. These intervals are, in general, different in the case of a compound vibration, whereas for a simple colour they are equal. Now GRAILICH assumes that every displacement of the aether particles on one side of their positions of equilibrium produces the same colour impression as would be produced by that simple colour for which the displacement

[1] *Jahresbericht des Frankfurter Vereins.* 1823–1854. S. 44–49.

from the position of equilibrium lasts just as long. Thus, on his assumption, the compound wave motion stimulates different colour impressions in the eye in rapid succession which are combined into a single sensation; and, as a rule, this sensation is paler in colour in proportion as the number of different successive sensations is greater. The impression of white itself should be composed of the rapidly alternating impressions of the middle hues of the spectrum from yellowish green to orange. In the case of the compound waves there are vibration-frequencies also that are outside the range of the visible spectrum; and so GRAILICH assumes that they produce the impression of purple.

GRAILICH's calculations are based on the intensity-ratios of the colours of the spectrum of the flint glass prism as measured by FRAUNHOGER; and if GRAILICH's last two assumptions are admitted, his results are in agreement with the writer's experiments with the v-shaped slit on mixing the colours of the spectrum. However, it should be added that in these experiments the luminosity of the colours of the spectrum was by no means maintained unchanged, but that in most cases what was attempted was to produce those mixed colours that are equally far apart from their two primaries.

Now in those cases where the amplitudes of the two colours are different, the result cannot be determined in advance by a general theory; and all that can be done is to calculate it for certain numerical examples, as GRAILICH did. In each special case then the calculation gives a series of different colour impressions which have to succeed one another; and by following GRAILICH's principles, the nature of the total impression may be estimated by them, but only in a pretty vague way. Bad for this theory, however, in the writer's opinion, is the fact, that when the two trains of waves are supposed to have equal amplitudes, in which case the mathematical theory may be actually worked out, the agreement with experiences is very far wrong, as GRAILICH himself admitted. Let $\lambda_{,}$ denote the wave-length of one train of waves and $\lambda_{,,}$ that of the other; and let x denote the distance of any point measured along the ray; then the displacement s of the aether particle from its position of equilibrium at any given instant will be:

$$s = A\sin\left(\frac{2\pi}{\lambda_{,}}x + c_{,}\right) + A\sin\left(\frac{2\pi}{\lambda_{,,}}x + c_{,,}\right)$$

$$= 2A\cos\left[\pi x\left(\frac{1}{\lambda_{,}} - \frac{1}{\lambda_{,,}}\right) + \frac{c_{,} - c_{,,}}{2}\right]\sin\left[\pi x\left(\frac{1}{\lambda_{,}} + \frac{1}{\lambda_{,,}}\right) + \frac{c_{,} - c_{,,}}{2}\right]$$

This expression may be abbreviated by making the following substitutions:

$$\frac{2}{l_{,}} = \frac{1}{\lambda_{,}} - \frac{1}{\lambda_{,,}} \qquad 2\gamma_{,} = c_{,} - c_{,,}$$

$$\frac{2}{l_{,,}} = \frac{1}{\lambda_{,}} + \frac{1}{\lambda_{,,}} \qquad 2\gamma_{,,} = c_{,} + c_{,,}$$

Thus we obtain:

$$s = 2A\cos\left(\frac{2\pi x}{l_{,}} + \gamma_{,}\right) \sin\left(\frac{2\pi x}{l_{,,}} + \gamma_{,,}\right).$$

The distances of the points for which $s = 0$ are easy to find in this case. Thus the points for which the factor $\sin\left(\frac{2\pi x}{l_{,,}} + \gamma_{,,}\right)$ vanishes are at intervals apart equal to $\frac{1}{2}l_{,,}$; and those for which the cosine function vanishes are at much greater intervals apart, namely, $l_{,,}$. These latter points may be in between the others or they may coincide with them. In the latter instance, particularly, we should get by GRAILICH's theory genuinely equal wave-lengths in the compound motion which would all produce the same colour impression; and even if the zero-points of the two factors did not fall together, the less frequent ones of the cosine factor could not essentially disturb the impression due to the shorter waves of the sine factor. But by GRAILICH's own calculation the result of this is that violet and red would have to give green, although as a matter of fact they make purple-red. And on the whole the results for small differences of wave-length are in accordance with experience, but for large differences the discrepancies are considerable, since the value of $l_{,,}$ must always come between $\lambda_{,}$ and $\lambda_{,,}$ and must correspond to one of the middle hues of the spectrum. In the writer's opinion, therefore, the assumptions in GRAILICH's theory will have to be considerably modified yet before it can be made to agree with the facts of experience, supposing the attempt is made to explain things in this fashion.

The easiest way to mix simple colours of the prismatic spectrum, and at the same time to get all combinations of pairs of these colours, is to use a v-shaped slit like that shown in Fig. 24, which is inserted in a dark screen with its two arms *ab* and *bc* inclined at 45° on both sides of the vertical. This slit placed in front of a bright background is viewed through a prism with its refracting edge vertical. Thus, two spectra will be obtained, partly superposed, as represented in Fig. 25 where $\alpha\beta\beta_{,}\alpha_{,}$ is the spectrum from the slit *ab* and $\gamma\beta\beta_{,}\gamma_{,}$ is that from *bc*. The bands of colour in the first are parallel to *ab* and $\alpha\beta$, and in the second parallel to *bc* and $\beta\gamma$, as shown by the dotted lines. In the central triangular field $\beta\delta\beta_{,}$, common to both spectra, all the coloured bands of one spectrum intersect all the coloured bands of the other spectrum, and thus at these places all mixtures are produced of all the pairs of simple colours. If the widths of the slits cannot be altered, the relative amounts of the mixture can be varied by inclining the prism and thereby causing the

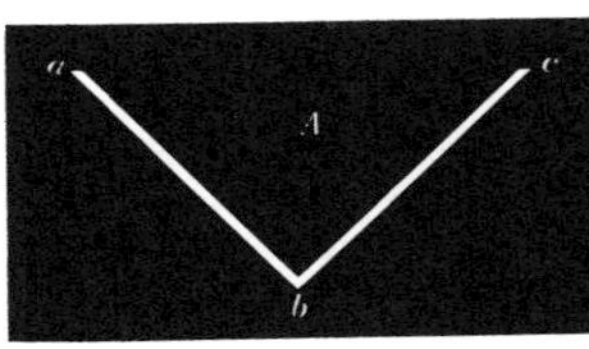

Fig. 24.

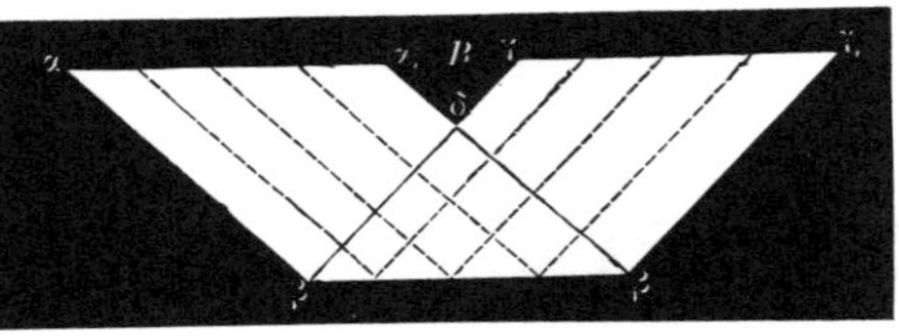

Fig. 25.

spectra to have the forms shown in Fig. 26. The spectrum $\beta\gamma\beta_{,}\gamma_{,}$ in which the same amount of light is distributed over a smaller area becomes brighter, while the other spectrum whose area is increased gets dimmer.

Most of the results mentioned above can be found by this method. But it is difficult to form accurate judgments of the mixtures, particularly the paler ones, first, because the single colours take up too much room, even when a telescope is used, and, second, because a number of other brilliant colours are near by in the field of view, and the contrast effects alter the appearance of the less saturated colours.

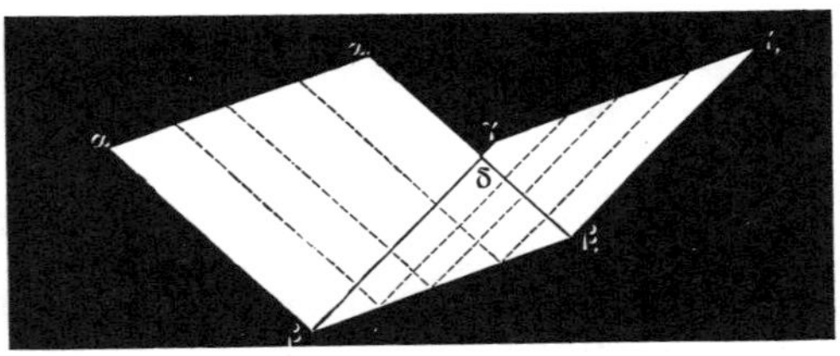

Fig. 26.

These disadvantages are avoided by another method, which requires a more complicated apparatus, as shown in plan in Fig. 27. Sunlight is reflected by a heliostat through a vertical slit into a dark room. The beam of light is made to pass through a prism P and an achromatic lens $L_{,}$. In the focal plane of this lens there is a screen $S_{,}$, on the front side of which an objective spectrum is projected. Between lens and screen there is a rectangular diaphragm D. There are two vertical slits in the screen $\gamma_{,}$ and $\gamma_{,,}$ which let through two coloured bands of

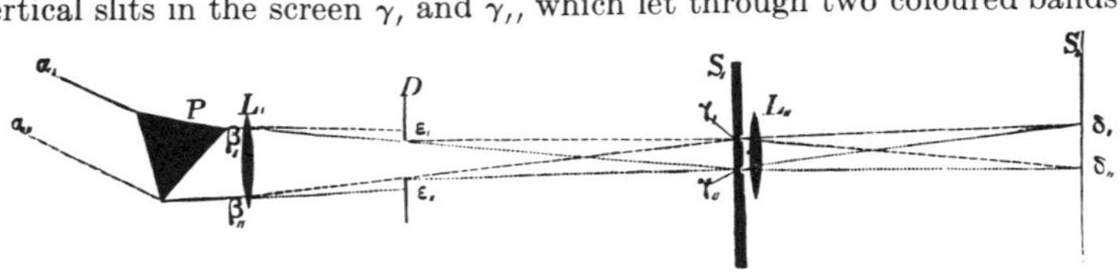

Fig. 27.

the light that is converged here in the spectrum; while all the other coloured light is retained on the screen. Beyond the screen there is another achromatic lens $L_{,,}$ of shorter focus, which projects on a second screen $S_{,,}$ an image $\delta_{,}\delta_{,,}$ of the diaphragm D. The width of the incident beam of white light is $a_{,}a_{,,}$. Beyond the lens $L_{,}$ the outside rays of the two bundles of coloured rays whose focal points are at the two slits $\gamma_{,}$ and $\gamma_{,,}$ are shown by dashed lines for the more refrangible light, and by dotted lines for the less refrangible. The opening in the diaphragm D must be made so narrow that it is completely filled by rays of both bundles, so that from every point in the aperture rays of the given colour go to every point in the slits $\gamma_{,}$ and $\gamma_{,,}$. If the anterior side of the diaphragm is painted white, the beam of light will be projected on it as a white spot with coloured edges (blue at $\epsilon_{,}$, and red at $\epsilon_{,,}$.) To fulfil the required condition, the opening in the diaphragm must be wholly in the white centre of the illuminated place. Under these circumstances, the opening is, so to speak, the luminous object which sends two kinds of light through the slits in the screen S, to the lens $L_{,,}$. Light of both kinds will thus be uniformly distributed over the image $\delta_{,}\delta_{,,}$ of the diaphragm projected by the lens on the farther screen; and hence this area will be coloured by a mixture of the two kinds of light. By screening off one of the slits, the image can be made to show up in one of the two simple colours that are to be mixed.

To enable the investigator to vary the hue and intensity of the mixture very gradually, just as he chooses, a special construction of the screen S, will be needed, as depicted in detail in Fig. 28. The screen itself is a rectangular brass plate *AABB* supported at *C* by a cylindrical rod, which is fitted in an upright split tube *D*. This tube is mounted on a baseboard or tripod with three leveling screws. The screen can be raised or lowered by its holder *C* and clamped at any convenient height by means of the collar and thumb-screw at *E*. The brass plate *AABB* carries two bed-plates *aa* and $\alpha\alpha$ that can be displaced on it in a diagonal direction between the guides *bb*, $\beta\beta$, *c* and *c*, by means of the thumb-screws *d* and δ, which work in nuts *e* and ϵ that are fastened to the plate *AABB*. The ends of these screws turn in the blocks *g* and γ which are attached to the sliders *aa* and $\alpha\alpha$. The slider *aa* in its turn carries another slider *f* that can be adjusted between a pair of parallel guides by means of the screw *m*; and likewise the slider $\alpha\alpha$ carries another slider φ which is adjusted in the same fashion by means of the screw μ. Between the parallel edges of the plates *f* and φ that are nearest to each other there are also the two small triangular pieces *l* and λ, the former being fastened to *aa* and the latter to $\alpha\alpha$. The adjacent parallel knife-edges of *f* and *l* and of φ and λ form two pairs of GRAVESANDE slits (cf. Fig. 11).

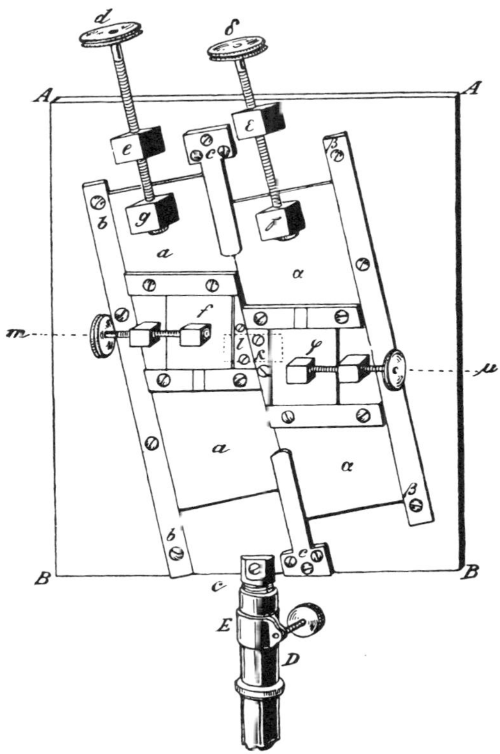

Fig. 28.

A corresponding portion of the large plate *AABB* behind the slits is removed in order to let the light pass through them and go on its way. The front sides of f, l, φ and λ are tarnished with a rough surface of silver to receive the projection of the spectrum. The position of the spectrum is indicated by the little dotted rectangle.

By shifting the plates *aa* and $\alpha\alpha$ by means of the screws d and δ, the slits can be transferred to other places in the spectrum to let light of other hues go through them. The widths of the slits themselves can be adjusted by the screws m and μ so as to regulate the amounts of the transmitted portions of light.

The main thing is that the focus of rays of a given colour, after they have traversed the lens $L_{,,}$ shall be exactly in the plane of the screen $S_{,}$; otherwise, the colour-field on the screen $S_{,,}$ will show different hues from right to left. The slits must be parallel to the dark lines in the spectrum, which can be accomplished by means of the leveling screws in the base-board of the screen. Moreover, both lens and prism must be carefully cleansed of any spots that might make patches in the colour-field. NEWTON's rings may easily be formed between the two parts of the achromatic double lens $L_{,}$ and be projected on the colour-field; but this can be prevented by using Canada balsam between the two components of the lens. However, the farther the diaphragm D is away from the lens $L_{,,}$ the more confused will be the image of spots and imperfections in the glasses, and the less troublesome they will be. Therefore, the disposition of the apparatus as sketched here is better than that which the writer has described elsewhere.

In this method the extent of the field is larger than it was in the first one, and all the other colours that might give trouble by contrast effects are removed. However, in numerous instances there are still a number of handicaps that make it hard to form a calm, sure judgment of the mixed colour. In the first place, the colour dispersion of the eye itself is much more noticeable for combinations of just two simple colours that are far apart than it is with white light (see Vol. I, p. 175). And so the margin of the colour-field may easily show one of the two simple colours while the other one predominates in the centre. Then with some white mixtures, particularly in the case of white made from red and green-blue, the eye is extraordinarily sensitive to the smallest admixtures of one of the original colours; so that the slightest irregularities in the apparatus, or possibly after-images that may be in the eye, especially with higher luminosity, can be very disconcerting. In this connection, also, differences of impression between the central and peripheral parts of the retina are very pronounced. Comparatively speaking, it is easiest to make white with yellow and indigo, harder to make it with yellow-green and violet or with golden yellow and water-blue, and hardest of all to make it with red and green-blue.

The way the writer determined the wave-lengths of the pairs of complementary simple colours was by removing the lens $L_{,,}$ and the screen $S_{,,}$ and using a telescope to see the slit in the screen $S_{,}$ at some distance away. A glass plate with fine, equidistant vertical lines ruled on it was inserted in front of the object-glass. Then diffraction-spectra of the slits will be seen; their apparent distances from the corresponding slits being proportional to the wave-lengths. Therefore, all that is necessary is to measure in the same way the distances of the diffraction-spectra for one of the dark lines in the spectrum, for which the wave-length has been found by FRAUNHOFER, in order to calculate immediately the wave-lengths of the two mixed colours concerned.

The simplest way of mixing the coloured light of pigments and other natural bodies is as follows. At some distance (1 foot) above a black table-top, adjust a small plane-parallel plate of glass a in a vertical plane which meets the top of the table in a point designated by *d* in Fig. 29. An observer looking obliquely downwards through the glass plate will see a portion of the table on the far side of *d* by means of light that is transmitted through the plate to his eye; and at the same time he will see another portion of the table on the other side of *d* by means of light that is reflected from the plate into his eye. The two fields are thus apparently coincident. Suppose now that two coloured wafers are placed at *b* and *c* at equal distances on opposite sides of *d*; then the reflected image of *c* will coincide with *b*. The coloured light from *c* is reflected from the anterior side of the glass plate along exactly the same path to the eye that is pursued by the coloured light from *b*, so that the two kinds of light are mixed in the eye at *o*; and therefore the common image of *b* and *c* which the eye gets must be seen in the mixed colour. The relative intensity can be regulated by shifting the two wafers; the closer they are to *d*, the more intense will be the reflected light that comes from *c* and the less intense will be the transmitted light that comes from *b*.

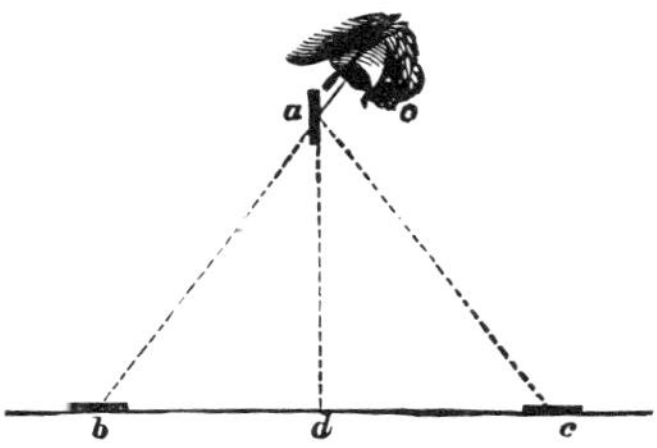

Fig. 29.

In this way too light may be used for mixing that has been filtered through coloured glasses or liquids. Apertures can be made in the table-top *bc* to let the light pass for this purpose. Thus, also, the light of the blue sky reflected by a mirror can be mixed with that of chrome-yellow and proved to give a reddish white like that obtained by mixing ultramarine and chrome-yellow; showing that sky-blue, therefore, is a pale indigo-blue and is not the same as the less refrangible so-called cyan-blue in the spectrum.

This last method is preferable in one respect to the method of mixing colours on the colour top, because the pale mixtures we get in this way are not grey but white. The colour top will be described more in detail in §22. Among other methods of mixing coloured light, an experiment of VOLKMANN's, in which he looked at coloured surfaces through coloured fabrics held close to his eye, deserves to be mentioned. However, it is hard to get a very uniform mixture of the two colours in this way, and the transparency of the fibres themselves is a difficulty, because to some extent it is like looking at a coloured surface through a coloured glass. CZERMAK adapted SCHEINER's experiment by looking through two stenopaic openings in a screen which were covered by glasses of different colours. As long as the objects were seen single, they appeared also in the mixed colour. HOLTZMANN caused light that was diffusely reflected from two pieces of coloured paper to fall on a sheet of white paper. CHALLIS mentions some experiments, which incidentally had already been tried by MILE; in which papers with bands of different colours painted on them were observed from such a distance that the bands could no longer be separately recognized. Lastly, DOVE has described methods of mixing colours produced by interference and absorption. He used silvered mirrors made of coloured glasses. The anterior surface of a mirror of this kind gives polarised white light, while the posterior surface gives unpolarised light coloured by absorption. When the light thus mixed is passed through a mica plate and a NICOL prism, the latter kind remains unchanged. But the polarised white light is so coloured by interference in the crystal between the ordinary ray and the extraordinary ray that it corresponds to one of the shades of colour in NEWTON's coloured rings. The two kinds of light are mixed when they enter the observer's eye.

The theory of colour-mixture originated from the experience of painters in mixing pigments. PLINY tells us that the ancient Greek painters knew how to prepare all colours with four pigments, although in his day, when there were many more materials available, it was not possible to get as many different effects as formerly. And even in the celebrated fresco, The marriage of the Aldobrandini, dating from Roman times, the profusion of pigments is very small, as DAVY's chemical investigations showed. (See GILBERT's *Ann.* LII. 1.) Besides black and white, which, however, are not colours in the peculiar sense, LEONARDO DA VINCI names four simple colours, yellow, green, blue and red; but in another place he speaks also of orange (*lionato*) and violet (*morello, cioè pavonazzo*) as being needed in painting. It is curious that he invariably reckons green as being a simple colour, knowing that it could be mixed; because by his own definition simple colours were such as could not be mixed. Could he have observed that unmixed green is much more vivid than mixed? Before NEWTON's time the three fundamental colours were commonly assumed to be red, yellow and blue. This is mentioned by WALLER, in an attempt at classifying colours and pigments, as being at that time a generally accepted scientific fact. The fact that three fundamental colours were found to be sufficient implies already the discernment of the truth that the character of a colour is a function of only three variables. Experiences with mixed pigments exercised the most decisive influence on the choice of the fundamental colours, which not until long afterwards WÜNSCH and YOUNG tried to change. It was supposed that green could be made from yellow and blue. This is true with respect to pigments, but not for coloured light.

NEWTON was the first to mix the light of different colours in the prismatic spectrum; but in conjunction with this method he used mixtures of coloured powders to get the law of colour-mixture, and did not attach much importance to the differences between the two processes, although he seems to have perceived them to some extent. He did not have the experimental facilities for tracing the facts more accurately. He states that only a pale green can be obtained from *subflavum* and *cyaneum* (that is, from greenish yellow and cyan-blue). NEWTON was the first too to formulate a more exact expression of the law of colour mixture, as it was he who traced it back to the graphical method and centroid-constructions that have been explained above. His law tallied with the available experimental data, and he did not try any more exact test. His colour circle was a development of the system of three objective fundamental colours; but as to the insufficiency of the latter system he has not anywhere expressed any opinion.

On the other hand, most of the later physicists, in their efforts to arrange the system of colours, returned to the notion of three fundamental colours; for example, LE BLOND (1735), DU FAY (1737), TOBIAS MAYER (1758), J. H. LAMBERT (1772), D. R. HAY and J. D. FORBES. Usually, their colour-systems were carried out practically by mixing given pigments in given proportions by weight. MAYER used vermilion, royal yellow (lead chromate) and mountain-blue (cobalt glass); and LAMBERT used carmine, gamboge, and Prussian blue (iron ferri-ferrocyanide). The latter also determined the saturation-ratios of these pigments, by finding the weights in which they had to be mixed in pairs to produce a mixed colour which was equally far from the colours of its two constituents. He found it was necessary to take 1 part of carmine, 3 parts of Prussian blue and 10 parts of gamboge. He chose these weights then as units for making the mixtures. Incidentally, mixtures of pigments so far apart from one another always turn out to be rather unattractive and grey.

Later observations, which, in circumstances where mixing of coloured light was to be expected, gave results that differed from the previous rule, were made by PLATEAU in 1829 with the colour top and by VOLKMANN in

1838 on blurred images; without leading, however, to a closer investigation of the discrepancy. The author's own experiments on mixing the colours of the spectrum resulted in his finding out that a mixture of light and a mixture of pigments were two different things; and he tried to find the reason for it. In these experiments the colours of the spectrum were mixed by means of the v-shaped slit, and white was not obtained from any other pair of colours except yellow and indigo-blue. This contradicted NEWTON's law of mixtures and led GRASSMANN to make a more minute examination of the principles of this law. The writer's improved method of investigating the mixtures of the colours of the spectrum removed the apparent contradiction of NEWTON's rule so far as it related to the centre-of-gravity construction; but on the other hand, in spite of GRASSMANN's arguments, the circular form of the colour-field was left as a question that was still debatable. Finally, the principles of NEWTON's law of mixture were experimentally confirmed by MAXWELL in 1857.

YOUNG's theory of the colour sensations, like so much else that this marvellous investigator achieved in advance of his time, remained unnoticed, until the author himself and MAXWELL again directed attention to it. It is sufficient to assume that the optic nerve is capable of sensations of different kinds, without trying to find out why the system of these visual sensations is just what it is.

1519. LEONARDO DA VINCI, *Trattato della pittura.* Paris 1651.

1686. R. WALLER, A catalogue of simple and mixte colours. *Phil. Trans.* 1686.

1704 *NEWTON, *Optice.* Lib. 1. P. II. Prop. IV–VI.

1735. LE BLOND. *Il Coloritto.* London.

1737. DU FAY, *Mém. de l'Acad. roy. de Paris* 1737.

1758. T. MAYER in *Göttinger gelehrte Anz.* 1758. St. 147.

1772. J. H. LAMBERT, *Beschreibung einer Farbenpyramide.* Berlin.

1792. WÜNSCH, Versuche und Beobachtungen über die Farben des Lichts. GILBERTS *Ann.* XXXIV. 10.

1807. TH. YOUNG, *Lectures on natural philosophy.* London.

1829. PLATEAU, *Dissert. sur quelques propriétés des impressions produites par la lumière sur l'organe de la vue.* Lüttich.

1836. CHALLIS in POGGENDORFFS *Ann.* XXXVII. 528.

1838. VOLKMANN in J. MÜLLERS *Archiv für Anat. und Physiol.* 1838. S. 373.

1839. MILE, Ibid., 1839. S. 64

D. R. HAY, *Nomenclature of colours.*

1843. J. MÜLLER, Zusammensetzung des weissen Lichts aus den verschiedenen Farben. POGGENDORFFS *Ann.* LVIII. 358, 518.

1847. DOVE, Über die Methoden aus Komplementärfarben Weiss darzustellen, und über die Erscheinungen, welche polarisiertes Licht zeigt, dessen Polarisationsebene gedreht wird. *Berliner Monatsber.* 1846. S. 70; POGGENDORFFS *Ann.* LXXI. 97; *Phil. Mag.* XXX. 465; *Inst.* Nr. 712. p. 176; *Arch. d. sc. ph. et nat.* V. 276.

CHEVREUL, Exposé d'un moyen de définir et nommer les couleurs d'après une méthode rationelle et expérimentale. *Quesneville revue scient.* XXIX. 382.—*C. R.* XXXII. 693.—*Inst.* No. 906. p. 155.—*Dingl. polyt. J.* CXXI. 337.—*Athen.* 1851. p. 272.

1848. BRÜCKE, Über das Wesen der braunen Farbe. POGGENDORFFS *Ann.* LXXIV. 461.—*Phil. Mag.* XXXIII. 281.—*Inst.* No. 785. p. 21.

HARLESS, *Physiologische Beobachtung und Experiment.* Nürnberg 1848. S. 45. (One colour seen through the other.)

CHR. DOPPLER. *Versuch einer systematischen Klassifikation der Farben.* Prag 1848, from *Abhandl. der böhm. Ges.* V. 401.

1849. J. D. FORBES, Hints towards a classification of colours. *Phil. Mag.* XXXIV. 161.

1852. *H. Helmholtz, Über die Theorie der zusammengesetzten Farben. Müllers *Archiv für Anat. und Physiol.* 1852. S. 461–482.—Poggendorffs *Ann.* LXXXVII. 45-66.—*Phil. Mag.* (4) IV. 519–534.—*Cosmos.* II. 112–120.—*Ann. de chim.* (3) XXXVI. 500–508.—Fechner, *Zentralblatt* 1853. pp. 3–9.

1853. L. Foucault, Sur la récomposition des couleurs du spectre en teintes plates. *Cosmos.* II.232.—Poggendorffs *Ann.* LXXXVIII. 385–387.

*H. Grassmann, Zur Theorie der Farbenmischung. Poggendorffs *Ann.* LXXXIX. 69–84.—*Phil. Mag.* (4) VII. 254–264.

Holtzmann, Apparat zur Darstellung von Farbenmischungen. *Tagblatt der deutschen Naturforscherversammlung* 1853.

J. Plateau, Reklamation. Poggendorffs *Ann.* LXXXVIII. 172–173—*Cosmos.* II. 241.

Fechner, *Zentralblatt.* 1853. p. 365.

H. Helmholtz, On the mixture of homogeneous colours. *Athen.* 1853. pp. 1197,1198—*Cosmos.* III.573–575.—*Rep. of Brit. Assoc.* 1853. 2. p. 5.—Poggendorffs *Ann.* XCIV. 1–28.—*Ann. de chim.* (3) XLIV. 70–74.—*Arch. d. sc. phys.* XXIX. 242.

1854. J. Grailich, Beitrag zur Theorie der gemischten Farben. *Wiener Ber.* XII. 783–847. —XIII. 201–284.

J. Czermak, Physiologische Studien. *Wien. Ber.* XII. 322. §6 and XVII. 565.

1855. *J. C. Maxwell, Experiments on colour, perceived by the eye, with remarks on colour blindness. *Edinb. Trans.* XXI. 275–297.—*Edinb. Journ.* (2) I. 359–260.—*Proc. of Edinb. Soc.* III. 299–301.—*Phil. Mag.* (4) XIV. 40.

G. Wilson, Observations on Mr. Maxwell's paper. *Edinb. J.* (2) I. 361.

J. D. Forbes, Observations on Mr. Maxwell's paper. *Edinb. J.* (2) I. 362.

1856. J. C. Maxwell, On the theory of compound colours with reference to mixtures of blue and yellow light. *Athen.* 1856. p. 1093.—*Edinb. J.* (2) IV. 335–337.—*Inst.* 1856. p. 444.—*Rep. of British Assoc.* 1856. 2. p. 12-13.

Challis, On theory of the composition of colours on the hypothesis of undulations. *Phil. Mag.* (4) XII. 329–338 and 521.

G. G. Stokes, Remarks on Challis's paper. *Phil. Mag.* (4) XII. 421.

1857. Dove, Eine Methode Interferenz- und Absorptionsfarben zu mischen. *Berl. Monatsber.* March 11, 1857.—Poggendorffs *Ann.* CII.

Colour Blindness

Huddart, *Phil. Trans.* LXVII. I. 14.

Collardo, *Journ. de Physique.* XII. 86.

Whisson, *Phil. Trans.* LXVIII. II. 611. *Journ. de Phys.* XII.

Giros v. Gentilly, Theorie der Farben (published in English and French under the pseudonym G. Palmer). Lichtenberg *Mag.* I. 2. 57.

Harvey, *Edinb. Phil. Trans.* X. 253. *Edinb. J. of Sc.* VII. 85.

J. Butters, *Edinb. Phil. Journ.* XI. 135.—*Archiv für Physiol.* v. Meckel. V. 260.

Nicholl, *Medico chir. Trans.* VII. 477. IX. 359.—*Ann. of Phil.* N. S. III. 128.

v. Goethe, *Zur Naturwiss. und Morphologie.* 1. Heft. 297. *Zur Farbenlehre.* 1. §103.

Meckel, *Archiv für Physiol.* I. 188.—*Ann. of philos.* 1822. Feb. p. 128.

Wardrop, *Essays on the morbid anatomy of the human eye.* London 1818. II. 196.

Meckel, *Archiv für Physiol.* V.

Brewster, *Edinb. Journ. of Sc.* VII. 86. XIX. 153.—*Edinb. phil. J.* VI.—Poggendorffs *Ann.* XXIII. 441.

J. Herschel, Art. Light in *Encyclop. metrop.* p. 434. §507.

Colquhoun from *Glasgow Med. Journ.* in Frorieps *Notizen.* XXIV. 305.

J. Dalton, *Memoirs of Lit. and Phil. Soc. of Manchester.* V.—*Edinb. Journ. of Sc.* IX. 97.

Sommer in Graefe und Walther *Journal für Chirurgie.* V.

Salzburger mediz. chirurg. Zeitung. IV.

Gall, *Anat. et Physiologie du système nerveux.* IV. 98.

REZIER, *Observ. sur la physique.* XIII.
BREWSTER, *Briefe über natürl. Magie.* German translation, S. 44.
HELLING, *Prakt. Handbuch der Augenkrankheiten.* I. S. 1.
1837. *A. SEEBECK, Über den bei manchen Personen vorkommenden Mangel an Farbensinn. POGGENDORFFS *Ann.* XLII. 177–233.
1849. WARTMANN, *Bull. de Brux.* XVI. 1. 137—*Inst.* XVII. No. 799. p. 131.
D'HOMBRE FIRMAS. *C. R.* XXIX. 175.XXX. 60 376.—*Inst.* No. 815. p. 259
1852. SCHNETZLER, *Arch. d. sciences phys.* XXI. 251–252.
BURCKHARDT, *Verh. der naturf. Gesellsch. in Basel.* X. 90–93.
1854. WILSON, *Proc. of Edinb. Soc.* III. 226–227.
EICHMANN in FECHNER *Zentralblatt* 1854. pp. 294–295.—*Med. Z. S. des Ver. f. Heilkunde in Preussen.* 1853. 224.
1853–55. G. WILSON in *Monthly J. of med. science.* Nov. 1853 to Dec. 1854. *Edinb. Journ.* (2) IV. 322–327.
Idem, *Researches on Colour-Blindness.* Edinb. 1855. Here also MAXWELL, *On the Theory of Colours in relation to Colour-Blindness.* p. 153.
1856. W. POLE, *Proc. of Roy. Soc.* VIII. 172–177.—*Phil. Mag.* (4) XIII. 282–286.
J. TYNDALL, *Phil. Mag.* (4) XI. 329–333.—SILLIMAN *J.* (2) XXII. 143–146.—*Arch. d. sc. phys.* XXXIII. 221–225.

Supplement by HELMHOLTZ, *from the first edition*

J. C. MAXWELL carried out an important series of experiments on mixture of spectral colours, for the purpose of determining the hues of the three fundamental colours and the form of the three intensity-curves, Fig. 21, which in YOUNG's theory express the intensity of the separate fundamental colours for each place in the spectrum. In his method white light was admitted in a dark box through three slits whose width and position could be altered. The light then traversed a couple of prisms and was focused on a screen by a lens, where consequently three partially overlapping spectra were produced. A slit in this screen permitted a mixed colour to proceed through it to the observer's eye. Looking through the slit, the observer saw the surface of the lens uniformly covered by the given mixed colour. The same white light was admitted through another section of the box, but in this case it did not pass through a prism. A mirror of black glass was adjusted so as to reflect this white light also to the observer's eye, so that he would see it as a white field right by the lens. His problem was to vary the adjustments and widths of the three slits which were responsible for the light resolved by the prisms until the mixture of the three prismatic colours gave a perfect match with the white light reflected by the mirror.

Subsequently, MAXWELL gave the apparatus a more convenient form by causing the light after it had passed through the prisms to be reflected back through them by a concave mirror. Thus the whole contrivance can be made shorter, and the observer can sit so close to the slit that he can adjust it himself, which is a great advantage.

The fundamental colours selected by MAXWELL were: (1) a *red* between the FRAUNHOFER lines C and D, twice as far from D as from C; according to our previous colour notations, this would be a scarlet-red tending to orange; (2) a *green* near the line E; and (3) a *blue* between F and G, twice as far from G as F, about where cyan-blue changes to indigo-blue.

In one set of observations, from time to time white was made again with these three colours, the required widths of the slits being noted, so as to verify in this way the unchanged mixture of the normal white light. The amounts of light necessary for this were measured by the widths of the slits. Between these tests white was made by mixing each pair of fundamental colours with some arbitrary third colour, the place of this third colour in the spectrum, along with the widths of the three slits, being noted on a scale placed near them.

If the white remained sufficiently constant, a set of colour-equations was obtained in this way, by which, the positions of the chosen three fundamental colours on a colour-chart being arbitrarily taken, the corresponding places of the observed colours of the spectrum could be found. Thus, by actual experiments, the form of the curve in Fig. 22 was obtained, which, as there represented, is simply a rough estimate. The curves traced by MAXWELL and his assistant enclose the perimeter of the triangle ARV very much more nearly than is the case in Fig. 22, two portions of the curve being almost straight. Its most marked bendings appear, therefore, to be nearest the vertices of the complete colour-triangle about corresponding with the aforesaid three fundamental colours. But the results obtained by one of the observers made the blue fall more towards the end of the spectrum, whereas those of the other observer made the red fall more towards the end. But it was precisely in case of the faint extreme colours of the spectrum that observation was difficult. Another difference between MAXWELL's projections and Fig. 22 is that the colour curve with its two ends in the red and violet appears to lie along the third side of the triangle.

On page 129 the writer stated as the result of actual experiments that the mixture of two colours of the spectrum is invariably rather paler than the simple colour in the spectrum that comes nearest to it in hue; but MAXWELL's result does not entirely agree with this. If the above were correct, it would mean that the colour curve cannot have straight portions anywhere; for colours lying on a straight line can be obtained by mutual mixtures with one another. The explanation of the contradiction between MAXWELL's results and those of the writer may be that the relative change of hue is necessarily greatest right at the borders of the colour triangle; hence, although the convexity of the sides of the triangle (which is not definitely shown to exist in Maxwell's more indirect method of investigation) may be very slight, so that there is very little difference between the chord and the arc, still the colours that lie along the chord may be appreciably different in appearance from those on the arc.

Moreover, from the results of his experiments Maxwell calculated the amounts of the three fundamental colours chosen by him that were present in the individual colours of the spectrum; and by these data he constructed curves corresponding to the schematic curves shown in Fig. 21. His curves have rather sharper peaks than those, and the red curve ascends again at the violet end of the spectrum, and the blue curve ascends a little at the red end.

Experiments like Maxwell's should be made to see whether it is possible actually to make perfectly pure spectrum yellow from yellowish green and golden yellow, perfectly pure spectrum violet from the extreme red and indigo-blue, etc.; so as to determine still more directly from these results the contour of the spectrum colour chart. Incidentally, the mixtures of the spectrum colours that gave white were not precisely the same as found by Maxwell and his assistant. Each insisted that what was white for the other was not exactly white for him. Also, the curve of brightness, as obtained by Maxwell himself, showed more of a drop in the region of the F line than the other observer found. Maxwell thinks it probable that this was due to different degrees of pigmentation of the yellow spot, as the yellow pigment (see §25) seems to absorb especially the light of the line F. Hence, white mixtures that contain this particular kind of blue, do not look white any longer even in indirect vision; as the writer himself had likewise previously noticed. (See p. 160.)

Thus, for different individuals the spectral lights have to traverse layers of yellow substance of different degrees of intensity in order to get to the central parts of the retina, and hence the effect must be different in different cases. This explains why the colour triangles for two different persons show variations in the distribution of colours such as would occur when the units of luminosity of the three fundamental colours were varied. These units, by the way, are determined arbitrarily. Thus in Maxwell's eye the red effect was comparatively greater and the blue comparatively weaker than in the other observer's eye.

Schelske finds also that the colour sensations produced by steady *currents of electricity* may be compounded with objective colours, and that the results are similar. The ascending current admixes a bluish violet light with the externally visible colours, whereas the descending current removes some of this colour from them. In fact, colour matches can be made with two colour discs, whose images are formed on two halves of the retina that are traversed by a current in opposite directions.

The *red blindness* in the periphery of the visual field, mentioned on page 154, has been studied more closely by SCHELSKE. He made colour matches for the peripheral parts of the retina between a yellow-blue mixture and red or grey or green. The colours of the spectrum in indirect vision were such that the region of the F line looked almost white, and the violet dark blue, while the intermediate portion appeared blue. On the other side of the F line, the appearance was green, until for the farthest red it was very faint and colourless or greyish.

Numerous observers have confirmed the fact that all colours that can be discriminated by a colour-blind eye may be compounded out of two fundamental colours. However, these experiments have not as yet established more definitely which one of the fundamental colours is missing, because experiments with coloured discs at different times and with different individuals give quite variable results. Sometimes change of the external illumination or light reflected from coloured objects or from the walls of the room may have much to do with the matter, as not only MAXWELL but E. ROSE also has noted. The pigmentation of the yellow spot caused too just the same sorts of differences in colour-blind persons as were found by MAXWELL in healthy eyes. This absorption in the yellow pigment does not alter simply the luminosity of the colours on the painted colour disc, but it alters their combinations also. Thus, assuming that the positions of the two real fundamental colours and of black are fixed in the colour triangle, the colours on the disc will have different positions depending on the intensity of the pigmentation of the eye. But if definite positions in the triangle are assigned to three of the pigment colours considered as being fundamental colours, then, conversely, the real fundamental colours and black will be found to have different positions for different individuals. Variations of this kind in the position of black have been observed by E. ROSE in colour-blind persons even when the tests were all made at the same time and under the same external conditions in every respect. He inferred therefore that YOUNG's theory could not be correct. But the supposed contradictions admit of simple explanation on the basis of the given conditions. ROSE himself states that constant matches could not be obtained unless the colour-blind persons always fixed the same place on the colour disc; the colour match being different for many of them every time the fixation-point was changed. This shows the difference in the colour sensation produced by the pigmentation of different parts of the same retina.

Cases of incomplete colour blindness also occur, as described by Mr. GLADSTONE and as found also by Mr. HIRSCHMANN in the author's laboratory in the case of a student. In this instance the admixture of quite large amounts of red in a colour was not noticed; although beyond a certain limit it was perceived. Leaving this out of account and considering such an eye as being totally red-blind, we cannot expect that its colour matches will agree exactly with the theoretical requirements.

E. Rose's experiments made by daylight located the black point near that of scarlet-red more towards its bluer side; which agrees with Maxwell's result and that of the writer's. But in most of Rose's observations with artificial illumination he used a kerosene oil lamp, which is rather unfortunate perhaps, because this light is comparatively poor in blue rays, especially when the flame varies owing to variations of the supply of air. Now as colour-blind folks lack the red sensation, and as the amount of blue in the flame itself is very slight and variable, this being true also with respect to the more refrangible blue which is particularly liable to absorption in the pigment of the yellow spot, the result is that with this illumination green must be by far the most predominant colour for red-blind persons; and hence it is not surprising that under these circumstances the colour matches made by different colour-blind observers even in the same evening were not very concordant. In every case the position of black in the colour triangle was located between blue and red, but, being dependent on the feebleness of the blue, it was nearer the position of blue than it was with daylight illumination.

E. Rose's experiments, therefore, are quite inadequate to shake the validity of Young's theory.

Some of the methods which this same observer used for investigating colour blindness deserve to be mentioned. In the first place, interference spectra were produced by ruling fine parallel lines on a plate of glass through which the patient looked at an illuminated slit. The series of spectra on each side of the slit that are obtained in this way are quite familiar. But the first of these spectra is the only one that is entirely isolated from the others, and the red of the second begins to overlap the violet of the third. The red end of the spectrum is shortened for colour-blind persons, and so they see the second spectrum too as separate from the third. However, naturally much will depend here on the intensity of illumination of the slit. But for a preliminary idea of the peculiarity of an eye under examination this procedure appears to be quite useful.

In the second place, instead of using the colour top, which takes much time and patience to adjust properly, E. Rose hit on the happy idea of employing the colours of quartz plates with polarised light. The instrument, called a colorimeter, consists of the following parts arranged in order in a tube: A NICOL prism A, a rectangular slit B, a double refracting prism C, a quartz plate D 5 mm thick, and a second NICOL's prism E; then the eye of the observer. The latter sees two images of the slit B, projected close together by the double refracting prism C. In consequence of the rotation of the plane of polarisation by the quartz plate, the two images are exactly complementary in colour, and their colours may be altered by turning the NICOL prism A. Without changing the composition of the colours, their luminosity may be varied by turning the other NICOL prism E. This prism is needed to make both images equally bright. With a quartz plate of the given thickness, a normal eye cannot get any colour match, but a red-blind eye may do so. The colours that are found to match in this case are red and blue-green; but even here different red-blind persons make somewhat different matches. By using thicker plates of quartz or a set of several plates all rotating the plane of polarisation the same way, and adding also a plate of variable thickness composed of two prisms, as employed in SOLEIL's saccharimeter, the apparatus can be used also for making colour matches with the normal eye, as one white can be made from red, green and violet and another white from yellow and blue. But even with this arrangement, as was to be expected from MAXWELL's experiments, a difference was found to exist between Dr. HIRSCHMANN's eye and the writer's eye, neither of which was colour-blind.

Incidentally, by using santonin a healthy eye may be made temporarily blind to violet. From 10 to 20 grains[1] of *sodium santoninate* is about right to get quick action that does not persist too long. The change begins in 10 or 15 minutes and lasts several hours. It is accompanied by nausea, great lassitude and visual hallucinations, so that there are attendant hardships in this experiment. Animals are killed by larger doses. Persons under the influence of santonin see bright objects as green-yellow, whereas dark surfaces appear to be covered with violet. The violet end of the spectrum disappears. Their colour system is dichromatic or at least approximately so. Experiments made with the quartz plate indicated that for moderate intensity of illumination colour matches could be made by such persons with only two fundamental colours, but not when the intensity of illumination was higher. But the matches that were obtained did not last long, and the condition varied continuously in a quite noticeable manner. Yellow and violet compound colours were declared to be alike. The disc of the optic nerve, as viewed by the ophthalmoscope, was not

[1] A grain = 0.06 g.—N.

coloured yellow, and hence there was at least no appreciable yellow stain present in the ocular media. On the other hand, the blood-vessels of the retina were much gorged.

Interpreting these phenomena on the assumptions of YOUNG's colour theory, we should infer that the inherent sensibility of the violet-sensitive nerve-fibres has not been lost, but that the terminal organs (cones of the retina) have become insensitive to the action of violet light. Thus violet and blue light ceased to affect the eye, notwithstanding all darker objects appeared to be violet; which was evidently due to internal causes of stimulation. It reminds one of the green appearance of all dark surfaces when a red glass is held close up to the eyes. It is hard to tell whether there is just the ordinary degree of internal retinal stimulation in the santonin poisoning or a higher degree of it. Indeed, it is a question whether it is not simply here a matter of stimulation of the violet-sensitive fibres by the santonin, which lowers the sensitivity of the eye to objective violet light by fatiguing it, and thus produces an incomplete violet blindness.

The change in the objective colours may be considered on the whole as violet blindness. Whether the fluctuations of judgment as observed by E. ROSE both with colour discs and with polarisation-colours of quartz are due to the variable injection of blood in the retinal vessels, which might act to some extent like a coloured absorbing medium, cannot yet be decided from the experiments.

Here also, just as in the case of persons who are naturally colourblind, it is possible to suppose that the power of the nerve-fibres to function has not been abolished, but that the form of the intensity curves (Fig. 21) for the three kinds of light-sensitive elements has been changed; in which case there might be then a much greater variability in the appearance of objective colours to the eye. In favour of this idea, it may be mentioned that in santonin poisoning, as was noticed several times by E. ROSE, red and yellow light were seen, but taken for violet, as if the cones of the violet-sensitive fibres had become similar to those of the red-sensitive fibres in their reaction to light. On the other hand, according to HIRSCHMANN's observations, it appears sufficient to explain this phenomenon by the diffusion of subjective violet light over the whole field of view, which is one of the effects of santonin.

1858. DE MARTINI, Effets produits sur la vision par la santonine. *C. R.* XLVII, 259–260.
— A. v. BAUMGARTNER, Ein Fall ungleichzeitiger Wiederkehr für verschiedene Farben. *Wiener Ber.* XXIX. 257–258.
— G. WILSON, A note on the statistics of colour blindness. *Year book of facts.* 1858. pp. 138–139.

1859. J. F. Herschel, Remarks on colour blindness. *Proc. of R. Soc.* X. 72–84.—*Phil. Mag.* (4) XIX. 148–158.
— W. Pole, On colour blindness. *Phil. Trans.* CXLIX. 323–339.—*Ann. de chimie.* (3) LXIII. 243–256.
— T. L. Phipson, Action de la santonine sur la vue. *C. R.* XLVIII. 593–594.
— LeFèvre, Action de la santonine. Ibid. 448.
— E. Rose, Über die Wirkung der wesentlichen Bestandteile der Wurmblüten. Virchows *Arch.* XVI. 233–253.
1860. J. J. Oppel, Einige Beobachtungen und Versuche über partielle Farbenblindheit. *Jahresber. d. Frankfurter Vereins.* 1859–1860. S. 70–114.
— Gladstone, On his own perception of colour. *Athen.* 1860. II. 24.—*Rep. of Brit. Assoc.* 1860. (2) pp. 12–13.
— E. Rose, Über die Farbenblindheit durch Genuss der Santonsäure. Virchows *Archiv.* XIX. 522–536. XX. 245–290.
— A. de Martini, Sur la coloration de la vue et de l'urine produite par la santonine. *C. R.* L. 544–545.—*Inst.* 1860. pp. 108–109.
— Guépin, Note sur l'action de la santonine sur la vue et son action thérapeutique. *C. R.* LI. 794–795.
— J. C. Maxwell, On the theory of compound colours and the relations of the colours in the spectrum. *Proc. Roy. Soc.* X. 404–409; 484–486.—*Phil. Trans* CL. 57–84.—*Phil. Mag.* (4) XXI. 141–146.—*Cimento* XII. 33–37.—*Rep. of Brit. Assoc.* 1860. (2) p. 16.
1861. J. J. Oppel, Nachträgliche Bemerkungen zu dem vorjährigen Aufsatze über Farbenblindheit. *Jahresber. d. Frankf. Vereins.* 1860-1861. S. 42-47.
— J. Z. Laurence, Some observations on the sensibility of the eye to colour. *Phil. Mag.* (4) XXII 220–226.
— E. Rose, Über stehende Farbentäuschungen. *Archiv. für Ophthalm.* VII. (2) 72–108.
1862. J. J. Oppel, Zur Veranschaulichung der Achromatopsie für nicht damit Behaftete. *Jahresber. d. Frankf. Vereins.* 1861–1862. S. 48–55.
1863. R. Schelske, Über Farbenempfindungen. *Archiv für Ophthalm.* IX. 3. S. 39–62.
— E. Rose, Über die Halluzinationen im Santonrausch. Virchows *Archiv.* XXVIII.
1864. Aubert, *Physiologie der Netzhaut.* Breslau. S. 154–186.
1865. Schelske, Über Rotblindheit infolge pathologischen Prozesses. *Archiv. für Ophthalm.* XI (1) 171–178.
— C. Bohn, Über das Farbensehen und die Theorie der Mischfarben. Poggendorffs *Ann.* CXXV. 87–118. (Attempt at a theory similar to that of Grailich.)
1866. E. Brücke, *Die Physiologie der Farben für die Zwecke der Kunstgewerbe.* Leipzig.

§21. On the Intensity of the Light Sensation

The intensity of the objective light itself is measured by the kinetic energy of the aether vibrations; which for monochromatic simple polarised light is proportional to the square of the maximum velocity of the particles. When the light comes from different sources or is polarised in different planes, the total intensity is equal to the sum of the separate intensities.

We shall inquire, first, how the intensity of the light sensation is affected by changes in intensity of the objective light, without any change of colour. These relations may be studied with white light. The behaviour of pure coloured light in this respect is in no wise different.

The first thing is to show that the smallest perceptible gradations of the light sense do not correspond to equal changes of objective luminosity. Let the luminosity of a white surface illuminated by a dim light be denoted by h. Now interpose an opaque body that casts a shadow on the surface so that the parts of it in the shadow do not get any light from the source. Then add a second light which by itself would produce the luminosity H; which may be varied by varying the distance between the surface and the second light. Then the objective luminosity in the shadow is H, and outside the shadow it is $(H+h)$.

Now if the luminosity H is very low, the eye will detect the shadow, that is, it will distinguish between the luminosity H and the luminosity $(H+h)$. But no matter how great h may be, it appears that there is always a higher luminosity H for which the shadow is invisible; for which, therefore, the difference h in the objective luminosity no longer produces any perceptible increase of sensation.

A light equivalent to moonlight makes a perceptible shadow on white paper; but if a good lamp is brought near the paper, the shadow disappears. Again, the shadow made by the lamplight disappears in sunlight. Indeed, the luminosity of the surface of the flame of a bright oil-burner with a circular wick can hardly be distinguished by the eye from double the luminosity. A flame of this sort is transparent enough to show this, as may be easily seen by looking at its faint image reflected in a plate of glass, and then shoving a second flame behind the first. The outlines of the latter can be quite distinctly recognized. But if both flames are viewed by the naked eye, the farther one will not be seen through the other, at least not through its brightest part; unless, perhaps, the intensity of the sensation gets blunted by looking at the lights too long. It is just as hard to tell by the naked eye that the edge of the flame as seen lengthwise through the glowing film of gas is very much brighter than the middle where the film is least deep; but it is easy to see this by looking at the image of the flame reflected in a plate of glass. The same explanation applies also to the disappearance of the stars in the daytime, and to the disappearance of images in a glass plate when it reflects other light, etc.

So far the difference of luminosity has been kept constant while the absolute value of the total luminosity has been varied; but we may also let the difference increase in the same proportion as the luminosity. Suppose a drawing is made on a transparent plate of glass in very dilute black india ink; or suppose the plate is covered with a thin film of lampblack and lines drawn on it; or, best of all, suppose we have a photograph made on transparent glass, with soft shadows in some places and deep ones in others. If a diagram of this sort is held against a bright background of steadily increasing luminosity, the soft shadows will be found to be invisible when the luminosity is low; and then as the luminosity gradually increases, they will begin to be seen and continue at about the same degree of distinctness for quite a time, until finally they begin to disappear again. The deeper the shadow in the picture, the less is the luminosity that is needed to make it show up, and the higher it must be to make it disappear. Now the difference between the luminosity of the bright portions and that of the shaded portions is a definite fraction of the total luminosity. If the luminosity of the bright portions is denoted by H, the luminosity of the shaded portions may be put equal to $(1-\alpha)H$, where for a given place on the drawing α denotes a constant proper fraction. Thus the difference of luminosity between the part of the drawing under consideration and the bright background, which is equal to αH, increases or decreases along with the luminosity H. Therefore, notwithstanding that with increasing luminosity the differences of absolute luminosity between the various shaded parts of the drawing become greater, there are no longer perceptible differences of sensation corresponding to the variations of the light itself. Hence there must be certain medium degrees of intensity of illumination for which the eye is most sensitive to a small percentage variation of the total luminosity. These are the degrees of illumination ordinarily used in reading, writing and working that are convenient and agreeable to the eye; ranging, therefore, from about that luminosity at which reading is feasible without difficulty to that of a white surface illuminated by direct sunlight. Within these limits of luminosity where the sensitivity for fractional differences reaches its maximum, the degree of sensitivity is also nearly constant; as is characteristic in general of the gradual mode of variation of a continuous function in the neighbourhood of its maximum value. This is apparent even in ordinary observation of such objects as paintings and drawings, where there are numerous gradations of shadows. They are about as easy to discern by candlelight as in broad daylight; so that usually nothing new in the way of objects and nuances is revealed in the picture by bright illumination that was not visible before when the illumination was dim. Thus, FECHNER remarked that when bright objects like the sky and sunlit clouds are viewed through dark grey glasses, none of the gradations of shadow disappear that were visible before, nor do new ones make their appearance. More accurate photometrical measurements confirm these observations. In general, such measurements indicate that for very different degrees of luminosity the difference of luminosity that can just be made out is nearly the same fraction of the total luminosity. The way that BOUGUER and FECHNER tried to find out the amount of this difference was by illuminating a white screen by two equal candles, so that when a rod was placed between the screen and the candles there were two shadows of it on the screen. Then one of the lights was moved away from the screen until the shadow it made ceased to be visible. Suppose the distance of the nearer candle from the screen is denoted by a, and that of the farther candle by b; then the intensities of illumination of the screen due to the candles are about in the ratio $a^2 : b^2$. BOUGUER found that one of the lights had to be about 8 times farther than the other for the shadow to vanish. FECHNER, aided by VOLKMANN and other observers, found that the distance of one of the candles was about 10 times farther than that of the other. Thus, BOUGUER could just discern a difference amounting to 1/64 of the total luminosity, whereas FECHNER's associates discerned a difference of 1/100. ARAGO noticed that when movement was also involved even finer differences could be recognized; and under the most favourable conditions he obtained the value 1/131 for this fraction. MASSON's experiments were made with rotating white discs with small black sectors; and he found that when the vision was poor, sometimes a difference of not more than 1/50 was the best that could be recognized, but that when the vision was good, the fraction might occasionally be even less than 1/120. Another thing he discovered was that the limit of sensitivity for instantaneous illumination by electric sparks was fairly independent of the luminosity. Indeed, with this illumination, if it were sufficient, the black and white sectors were visible for an instant. Suppose the rotating disc is steadily illuminated by a lamp of brightness L, and then also by an electric spark of brightness l; then for an instant on the white sectors the luminosity is $(L+l)$, whereas on the black sectors it is only L; so that the sectors can just be recognized provided $(L+l)$ can be distinguished from L. When the distances of the two sources of light from the disc were altered, L and l had to be altered in the same proportion in order to keep at the limit of the sensitivity of the eye. Consequently, the same law holds for the ability of perceiving instantaneous difference of light as in the case of constant light.

The fact that within a wide range of luminosity the smallest perceptible differences of the light sensation correspond to (nearly) constant fractions of the luminosity, was used by FECHNER in formulating a more general, so-called *psycho-physical* law, which is found to be true also in other regions of the sensations. Thus, for example, differences of pitch seem to us to be equal when the difference between the vibration-numbers is a certain fraction of the vibration-number itself. According to E. H. WEBER's investigations, the case is similar also with our ability to recognize differences between weights and linear magnitudes.[1] Now as pitch is measured by the logarithm of the

[1] ¶"The next great step in the advance of our knowledge of the relationships subsisting between physical stimuli and sensations was made by E. H. WEBER, the founder of modern psycho-physical methods. In general terms it may be stated that a stimulus must attain

vibration-number, it is natural to measure the intensity of sensation in the same way; since here also, just as in the other case, differences dE of intensity of sensation E that can be perceived equally distinctly may be considered as being equal in amount. Accordingly, therefore, within a wide range of luminosity H, we have approximately

$$dE = A\frac{dH}{H} ,$$

where A denotes a constant. By integration, we obtain:

$$E = A \log H + C ,$$

where C is the constant of integration.[1] If e denotes the intensity of the sensation corresponding to a certain luminosity h, then

$$E - e = A \log \frac{H}{h} .$$

This mode of estimating brightness by the eye has been shown by FECHNER to have had a definite influence in the determination of the magnitudes of the stars. The orders of magnitude of the stars were determined at first by the impressions they made on the human eye, without photometrical measurements of the objective quantities of light emitted by them. Such measurements were not made until recent times, but now we are in a position to compare the actual brightness of a star with its supposed order of magnitude. With the aid of the photometrical determinations of J. HERSCHEL and STEINHEIL, FECHNER made a comparison of this sort; the result being that the order of magnitude (G) for HERSCHEL's measurements is given by the formula

$$G = 1 - 2{\cdot}8540 \log H ,$$

---

a certain intensity in order to excite a sensation and that stimuli of greater intensity excite stronger sensations. There is therefore a quantitative relation between the stimulus and the sensation. The minimum effectual intensity of stimulus is called the *general threshold* or *the general liminal value.* A higher value may arouse a sensation differing in quantity from the other; this value is called the *specific threshold* or the *specific liminal* value. Thus a coloured light of low intensity may excite a colourless sensation; when of higher intensity it may excite a sensation of colour." J. H. PARSONS, *An introduction to the study of colour vision.* 1915. page 19. (J. P. C. S.)

[1] ¶"FECHNER's law states that the *sensation varies as the logarithm of the stimulus;* i. e. the sensation changes in arithmetical proportion as the stimulus increases in geometrical proportion." It is a "questionable assumption that it is permissible to integrate small *finite* quantities," such as are involved here. PARSONS (*loc. cit.*, p. 20), commenting on this law, says: "The bases are insecure on mathematical as well as on physiological grounds. So far as the latter are concerned we have no unit of sensation, and the variations, though quantitative, are only relative. The chief difficulty, however, is to be found in the ever-changing condition of the sensory apparatus. The deductions are not without value, for *some* quantitative relationship certainly exists, even if it be not so simple as FECHNER's law implies." (J. P. C. S.)

and for STEINHEIL's measurements by the formula

$$G = 2.3114 - 2.3168 \log H .$$

Both formulae are in agreement with those given above, since the orders of stellar magnitudes are higher for the fainter stars. The formulae also agree well enough with the results of observation. FECHNER found too that STRUVE's measurements harmonized sufficiently well with his law. Incidentally, the same law was found by BABINET,[1] who, according to observations of JOHNSON and POGSON, gives 2.5 for the value of the coefficient of log H in FECHNER's formula.

FECHNER thinks that the effect of disturbing circumstances accounts for the fact that the law proposed here for the intensity of the sensation is not obeyed either at very low or at very high luminosities. For example, at very low luminosities the influence of the intrinsic light of the eye must make itself felt. Together with the stimulation due to external light, there is always besides a stimulation due to internal causes, the amount of which may be considered as being equivalent to the stimulation by a light of luminosity H_0. It would be more accurate, therefore, to write the formula for the least perceptible degrees of the intensity of the sensation as follows:

$$dE = A \frac{dH}{H + H_0}$$

or

$$dH = \frac{1}{A}(H + H_0)dE .$$

This implies that the increment of luminosity must be rather more in order for it to be perceptible than it would be if H_0 were neglected; and particularly when the values of H are small, the difference will become considerable. FECHNER devised a method of comparing the intensity of the intrinsic light H_0 with that of the objective light, which is based on this formula; but this method unquestionably takes for granted that at the so-called threshold of the light sensation there is nothing else to invalidate the law above mentioned but this intrinsic light of the eye. Suppose an eye that is able to detect a change of luminosity of one percent views a surface, part of which gets no external light at all, while another part is illuminated, and has a luminosity denoted by h; then, taking into account the intrinsic light, the apparent luminosity of the unilluminated portion is H_0, and that of the illuminated portion is (H_0+h). Now if h is the least perceptible luminosity, then, by FECHNER's method of reasoning, we must have $h = \frac{1}{100}H_0$; and thus the brightness

[1] *Comptes rendus*. 1857. p. 358.

H_0 of the intrinsic light would be measured in terms of an objective light. VOLKMANN made some experiments and found that the intensity of the intrinsic light H_0 was equal to the luminosity of a black velvet surface illuminated by a tallow candle at a distance of 9 feet.

The discrepancy between the law and the facts for the upper limits of luminosity may be attributed to the fact that the visual organ begins to be impaired, as FECHNER supposes. The internal changes in the nerves that have to communicate the impression of the stimulus to the brain cannot exceed a certain definite limit without destroying the organ; and hence every action of the stimulus has an upper limit set for it, to which must necessarily correspond also a maximum intensity of sensation.

But, moreover, whatever may be the circumstances that tend to upset the validity of FECHNER'S law at the upper and lower limits of luminosity, the same conditions make their influence felt in accurate measurements even with medium degrees of luminosity; although, of course, that is no reason why the law should not still be regarded as being a first approximation to the truth. Unquestionably, most paintings, drawings and photographs of ordinary objects can be seen equally well under very different degrees of illumination. And yet in some photographs the writer has discovered gradations of shade which do not come out perfectly clear except for a very definite intensity of light. This is particularly noticeable in pictures of landscapes in which far distant chains of mountains are represented as half floating in cloud. But, so far as the writer is concerned, the most striking instance of this peculiarity was in the case of some stereoscopic views of Alpine scenery photographed on glass, which showed parts of glaciers or peaks covered with snow. By lamplight or moderately bright daylight such surfaces of snow look like uniformly white areas; but when they are turned towards the bright sky delicate shades appear, indicating a sort of moulding of the white fields of snow; and then they disappear again with still brighter light. Of course, delicate shades of this kind are found in photographs simply by accident; and in paintings and drawings they are unexpected. But the rotating disc affords an easy way of producing very delicate shades, of any desired luminosity as compared with the white background. MASSON also has already made use of them for photometrical experiments. These shades are easy to get by making a

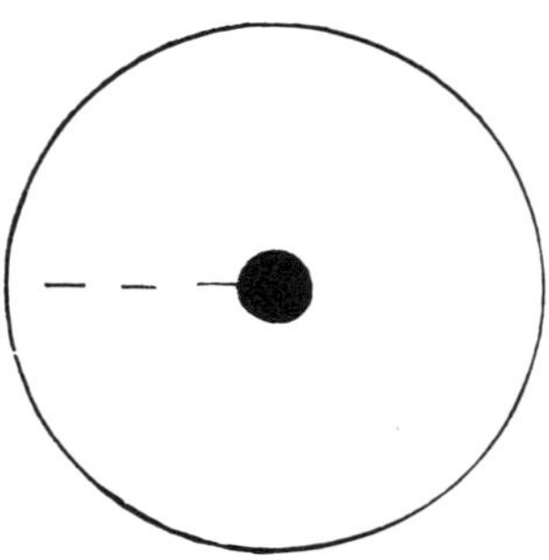

Fig. 30.

pattern on the disc like that shown in Fig. 30. A broken line is made with a pen along one or two radii, the parts being all of the same thickness. When the disc is set spinning, these black marks make grey circles on the disc. Suppose d denotes the width of the mark, and r denotes the distance of a point on it from the centre; then if the luminosity of the disc itself is taken as unity, the luminosity h of the grey ring produced by rotating the disc will be

$$h = 1 - \frac{d}{2r\pi}.$$

Thus the greater r is, the less the grey rings will differ in luminosity from that of the disc. The inner rings are darker and the outer ones brighter, and so we get a series of very delicate gradations. In the experiment all that has to be done is to see how far the edges of the grey rings can still be discerned. They can be seen better by looking here and there at different places on a circle than by looking steadily at one place. When the gaze is fixed on one spot, the fainter circles disappear quickly, even if they have been seen before. Usually, however, the differences are not all equally apparent at first sight, and it is necessary, to gaze at the disc a long time at first. Incidentally, the disc must be made to turn fast enough for the grey rings to be continuous in appearance, and not to flicker. In the latter case even the fainter rings can be discerned, because, every time a black mark goes by, the impression it makes lasts too short a time for the darkening to be appreciated. On bright summer days near a window, it was possible for the writer by shifting his gaze still to see one edge sharply where the difference of luminosity was $\frac{1}{133}$. Another edge could be vaguely seen where the difference was $\frac{1}{150}$, and for a single instant one for which the difference was $\frac{1}{167}$. Up to $\frac{1}{150}$ with the disc illuminated by direct sunlight, the perceptions were rather more troublesome and fatiguing. In the interior of the room at the same time, the writer could not detect edges for which the difference was less than $\frac{1}{117}$, although occasionally and uncertainly he could go as far as $\frac{1}{133}$.

Thus here, too, a certain narrowly limited range of luminosity is indicated as being required in order to get the greatest sensitiveness of perception. Accordingly, in the equation

$$dE = A\frac{dH}{H}$$

the coefficient A should not be considered as absolutely invariable, not even within the ordinary range of illumination; but it must rather be regarded as dependent on H, although for medium degrees of illumination it is nearly constant. Likewise the formula derived from it by integration

$$E = A \log H + C$$

will be only approximately correct for medium values of the luminosity. In comparing the intensity of sensation for different colours, it will be still more apparent that a formula of this kind cannot be sufficient.

Even when the intrinsic light of the eye is taken into account and the equations are written as follows:

$$dE = A\frac{dH}{H + H_0}$$

$$E = A \log(H + H_0) + C ,$$

the formula does not quite fit the facts, because it implies that the sensitivity would have to increase as long as the luminosity increased; whereas the facts stated above indicate that the intensity of the sensation gets to be a maximum value which is not exceeded even when H continues still to increase. Therefore for this value, $\frac{dE}{dH}$ must vanish. Hence, also in the last differential equation we should have to consider A as still being a function of H, which for moderate values of H is approximately constant, but which vanishes when H gets to be infinite. The simplest function of this sort would be

$$A = \frac{a}{b + H} ,$$

where b must be supposed to be very great. If, therefore, we put

$$dE = \frac{a dH}{(b + H)(H_0 + H)} ,$$

then

$$E = \frac{a}{b - H_0} \log \frac{H_0 + H}{b + H} + C .$$

The phenomena might, perhaps, be completely expressed by some such formula as this. The magnitude denoted here by C would be the maximum intensity of sensation corresponding to an infinitely great value of H, and the maximum of sensitivity would occur for $H = \sqrt{b H_0}$.

The connection between the intensity of the sensation and the luminosity which has thus been shown to exist explains a fact which the writer has often noticed, namely, that on dark nights bright objects are much brighter as compared with their environment than they are in daytime, so that sometimes it is hard not to believe they are self-luminous. When the luminosity is very low, the intensity of the sensation is considered as being proportional to the luminosity; whereas with greater illumination the sensation for brighter objects is relatively weaker. Now since we are accustomed to compare the brightness of familiar objects when they are highly illuminated, under feeble illumination bright objects are relatively too bright, and dark ones relatively too dark. Painters also make use of this circumstance in moonlight scenes, to produce the impression of faint illumination. They bring out the light places much more brilliantly than when they are representing daylight.

Let us turn now to the comparison of the intensities of light of different colours. If the intensity of objective monochromatic light of different kinds is supposed to be measured by the kinetic energy of the aether-vibrations, then, according to the general law of the conservation of energy, it must be proportional to the amount of heat developed by absorption of the light in question. Hitherto this has been the only available physical method of comparing the intensities of aether-waves of different frequency. When the luminous power of aether-waves of different frequency is estimated by the eye, the result, as has been explained in §19, is that the intensity of the light sensation is by no means proportional to the kinetic energy of these aether-vibrations as measured by the development of heat. When a spectrum is projected by a prism of rock salt, which of all substances is most uniformly transparent to different kinds of rays, the maximum thermal effect, as found by MELLONI, was in the infra-red where the eye is no longer sensitive to light at all; and the thermal effect in the spectrum rises steadily from the violet to the red, whereas the maximum luminous effect is in the yellow. Likewise, attention has already been called to the fact that the luminous power of ultra-violet light is extraordinarily augmented by transforming it by fluorescence into light of medium refrangibility; although it is not to be supposed that the kinetic energy is increased by this process. Thus, *the intensity of the light sensation depends not only on the kinetic energy of the aether-vibrations, but also on their vibration-frequency.* The consequence is that all comparisons of the intensities of compound light of different kinds that are made entirely by the eye have no objective value apart from the nature of the eye.

We have seen that for a given kind of light the sensation does not increase in the same ratio as the objective intensity of the light, but that the intensity of the sensation is a more complicated function of the intensity of the light. But on comparing light of different colours, it appears that *the intensity of the sensation for different kinds of light is a different function of the intensity of the light.* Purkinje[1] has already noticed that blue was visible for the weakest light and red only at higher illumination. Subsequently, Dove called attention to the fact that in comparing the luminosities of surfaces painted in various colours for different degrees of illumination, sometimes one was brighter and sometimes another. As a rule, the less refrangible red and yellow colours predominate when the illumination is high, and the more refrangible blue and violet colours when it is low. If pieces of red and blue paper appear equally bright by day, then with the approach of night the blue gets bright while the red often looks perfectly black. So also in picture galleries towards evening (supposing the sky is cloudy and twilight disappearing) the red colours are the first to fade away, and the blue ones continue visible longest. And on the darkest nights, when all other colours are gone, the blue of the sky can still be seen.[2] If the apparatus described in the preceding section, as sketched in Fig. 27, is used for mixing the colours of the spectrum, and if a vertical rod is held in front of the field illuminated by the two colours, it casts two differently coloured shadows. Since the two coloured lights fall on the illuminated field from different directions, proceeding from the two slits in the last screen (S', Fig. 27), they cast their shadows at different places. Thus, if, say, violet and yellow were mixed, there would be one shadow which would not be illuminated by violet but would be illuminated by yellow, so that it would look yellow; and there would be another shadow which would not be illuminated by yellow but would be illuminated by blue, so that it would look blue; whereas the rest of the field would be white or whitish. Now if the slit in the screen which lets the violet through is made wider, the violet becomes stronger, and the violet shadow therefore is more luminous. By regulating the two slits properly, the violet shadow can be made to look just as bright as the yellow one. Now if the single slit in the first screen, through which the light reflected from the heliostat passes to the prism, is opened more or closed more, the total amount of light that goes through the apparatus will be increased or diminished, all the different colours in the same proportion; and hence the amounts of light in the yellow and violet

[1] *Zur Physiologie der Sinne.* II. 109. With respect to this, see the notes in the Appendix I. I. B., 1–5.—N.

[2] ¶The explanation of these phenomena is to be found in the difference between "cone vision" and "rod vision." In darkness the colour sensation disappears entirely and all that remains is a pure luminosity sensation. As evening descends all colours begin to fade and disappear and in moonlight even bright red tile-roofs cannot be distinguished as to colour. (J.P.C.S.)

shadows will be changed in equal fashion also. The result is that even a slight increase of light will make the yellow come out brighter, and a slight decrease will make the violet look fainter. The difference here will be much less when the two colours are both in the less refrangible half of the spectrum, and much more when both colours are from the more refrangible half; it will be greatest of all when they are taken from the two ends of the spectrum.

In Fig. 31 the abscissae measured along the line *ad* are supposed to be proportional to the objective intensity of the light, whereas the ordinates represent the intensity of the light sensation. Let the curve *aebg* indicate the intensity of the sensation for yellow light; and suppose that the units for yellow and violet light are so chosen that for the quantity of light *ac* the intensity of the sensation is the same for both kinds of light.

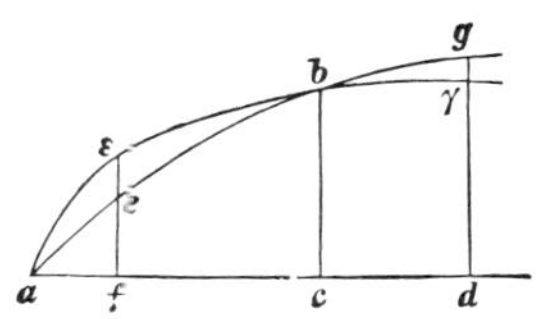

Fig. 31.

Then from the facts given above the curve representing the intensity of sensation for the violet light must have the position $a\epsilon b\gamma$ as compared with the other curve. If the two quantities of light are diminished in the ratio $af:ac$, the intensity of sensation for yellow light as given by the ordinate fe will be less than that for violet light as represented by the ordinate $f\epsilon$. Conversely, if the quantities of the two kinds of light are increased to the amount ad, the intensity of the sensation for yellow as given by the ordinate dg will be greater than that for violet as given by the ordinate $d\gamma$.

Consequently, it is not possible to devise units for measuring light of different colours, such that, for equal amounts of two kinds of light as measured in terms of them, the intensities of the sensations produced in the eye will be also always equal. The fact is that the mathematical functions that exhibit the connection between the intensity of the sensation and the objective intensity of the light are of different degrees for light of different colours.

Suppose white has been obtained by combining two complementary colours. If then the intensities of the two coloured lights are increased or decreased in the same proportion so that the ratio of the mixture remains the same, the mixed colour remains, too, unchanged white; in spite of the fact that under such circumstances the intensities of the sensation for the two simple colours may be materially altered. For example, if, with the apparatus described above, violet and green-yellow are mixed to give white, the amount of green-yellow light may be reduced by narrowing the slit until it appears of the same luminosity as the violet. Since the amount of transmitted light is proportional to the width of the slit, by measuring the latter the ratio by which the quantity of light has been diminished can also be obtained. Thus, the writer has found that violet, which mixed with a certain amount of green-yellow gives white, looks as bright as one-tenth of the green-yellow when the intensity of illumination is increased; whereas it looks as bright as one-fifth of the green-yellow when the intensity of illumination is diminished; although in both cases the ratio of the objective amounts of light is the same. When indigo-blue and yellow were mixed, the blue appeared to be one-fourth as bright as the yellow for higher intensity of illumination, and one-third as bright for lower intensity. The differences were too small to be measured for the less refrangible complementary colours. Thus, when whites of different luminosity are obtained by mixture, the amounts of the complementary colours are in a constant ratio to each other in objective intensity, but in a very variable ratio to each other in subjective luminosity. Consequently, when units of measurement for light of different colours are obtained by mixing the colours, as was explained in the preceding chapter, these units will have little or no connection with the intensity of the light.

Apparently, the reason why mixed colours look about the same for different intensities of light, whereas the comparative intensity of the action on the organ of vision is entirely different, is because sunlight, which is regarded as normal white by day, must itself change its colour for different intensities of light in the same manner as other white or pale mixtures of colours with which it is compared. A colour mixture, which looks just like sunlight toned down to the same degree of luminosity, is white to us. Thus in case of the colour mixture in question, although the impression of blue is stronger in dim light than it is in bright light, still it does not appear to be a bluish white, because when sunlight is toned down to just the same extent, the impression of blue must preponderate about as much. However, the fact that the impression of blue really does prevail in dim sunlight, and the impression of yellow in bright sunlight, may be easily verified with a little pains. In paintings the effect of brilliant sunshine is always conveyed by predominating yellow hues, and the effect of moonlight or starlight by blue hues. The painter tries to imitate nature, but he does not have at his command all the gradations of intensity of light; and so by copying the variegated hues he endeavours to supplement the impression of intensity of light. A similar effect is the impression of brilliant sunshine illumination produced by looking at a landscape on a cloudy day through a piece of yellow glass; whereas a sunlit scene viewed through a piece of blue glass seems to have what is sometimes called a cold illumination.

It has already been stated that the impression made by the pure colours varies also in the same way, so that with increasing intensity of light they look as if they were mixed with yellow. Red and green pass right over into yellow, but blue becomes pale just as it would do if it were mixed with yellow.

The consequence is that with very great intensity of light the discrimination of hues is more imperfect than it is with moderate brightness. But likewise this discrimination is also imperfect with very low intensity of light; which agrees with the fact that it is more imperfect also in the case of colours that occupy a very small portion of the visual field than it is when the colour fields are more extensive. Thus, if the retinal image of a coloured field is smaller than the sensitive elements of the retina, the element in question is no longer stimulated to full intensity; the stimulation being less in proportion as the part of the element which is occupied by the image of the coloured area is smaller.

These variations of the colour sensation with the intensity of the light are accounted for in YOUNG's theory by the assumption that there are three kinds of nerves in the retina, red-sensitive, green-sensitive and violet-sensitive; provided we suppose, as we have done, that each kind of nerve is stimulated by light of any kind, even by homogeneous light, but in very different degrees, and that in each of the three sets of nerves the intensity of sensation is a different function of the intensity of the light. Thus, in the violet-sensitive nerves with rising intensity of the light it increases more rapidly at first, and then more slowly, than it does in the green-sensitive nerves; and the same way in the latter as compared with the red-sensitive nerves.

If the violet light of the spectrum stimulates the violet-sensitive nerves highly, the green-sensitive nerves feebly, and the red-sensitive nerves more feebly still, the sensation of violet will predominate in dim light; but in bright light where the violet sensation approaches its maximum, the green sensation may succeed in becoming more appreciable by comparison with the other, and subsequently even the red sensation. Thus, initially the sensation of violet light must pass through mixed green into blue, and ultimately through mixed green and red into white.

Suppose, moreover, that the green light of the spectrum stimulates the green-sensitive nerves highly and the red-sensitive and violet-sensitive nerves moderately. Then the sensation of green must pass first into yellow, because the sensation of red grows more rapidly with the intensity of light than that of violet; and ultimately when all three sensations approach their maxima, the green sensation becomes white. Moreover, as to the red rays, we have supposed that they stimulate the red-sensitive nerves highly, the green-sensitive feebly, and the violet-sensitive more feebly still or not at all. This would explain how the sensation of bright red light passes into that of yellow.

Thus, discrimination of hue would depend on the fact that the relative amount of light that stimulates each of these sets of nerves is perceived by comparing the intensities of their sensations. Now we have seen that the relative intensity of two quantities of light can be judged best in a certain medium illumination. Hence, also, the discrimination of hues must be most accurate with medium illumination. The application of this consideration to very luminous colours will be obvious already from what has been stated. If with mixed colours all three sets of nerves are near the maximum degrees of stimulation, necessarily, each colour will have to become more and more nearly white. On the contrary, supposing that the violet-sensitive nerves were stimulated to the faintest perceptible extent, we could not possibly tell whether it was accompanied by a somewhat slighter degree of stimulation of the other two sets of nerves, that is, whether the colour of the light was pure violet or indigo-blue or purple or bluish white. Thus here, too, when the light is quite dim, discrimination of hues will be imperfect.

Another series of facts, heretofore classified as phenomena of *irradiation*, can be explained by the fact that the intensity of the light sensation is not proportional to the objective intensity of the light. What is common to them all is that highly illuminated areas appear to be larger than they really are, whereas the adjoining dark areas appear to be correspondingly smaller.

The phenomena themselves are very varied depending on the form of the patterns observed. Generally, they are easiest to see and most pronounced, when the eye is not exactly accommodated for the observed object. It makes no difference whether the accommodation is too much or too little, or whether the eye is provided with a glass lens, convex or concave, which is not suited for the particular distance of the object. But even when the eye is accommodated exactly, irradiation will be manifest to some extent. In fact, it can be distinctly noticed even then, provided the objects are very bright and particularly small. Evidently, the reason why the effect is more marked in the case of small objects is because the size of a small object is relatively more enlarged by the small blur circles than that of a bigger object. The diameters of such tiny blur circles as are present when the eye is well accommodated are practically negligible in comparison with the dimensions of large objects.

1. *Bright areas appear magnified.* The dimensions of narrow apertures and slits illuminated by light from behind are never estimated correctly. They invariably look wider than they really are, even with the most perfect accommodation. Similarly, too, the fixed stars seem to be small bright surfaces, even when we look at them through a concave glass in order to be able to accommodate exactly. In a grating of fine dark bars with intervals exactly as wide as the bars themselves (ordinary wire grating for interference experiments), held in front of a bright background, the intervals appear to be wider than the bars. When, in addition, the accommodation is not perfect, the phenomena are much more striking and are visible even with larger objects. Fig. 32 shows a black square on a white background alongside of a white square on a black background. With good illumination and insufficient accommodation, the white square appears larger, although they are both equal.

Fig. 32.

Fig. 33.

2. *Adjacent bright areas tend to flow over into each other.* A fine wire held between the eye and the sun or a bright flame disappears, because the two bright surfaces adjacent to it in the field of view encroach on it from both sides and fuse together. With patterns composed of white and black squares like a chess-board, as shown in Fig. 33, the white squares fuse by irradiation at adjacent corners and separate the black ones. PLATEAU used squares of this sort also to measure the spread of irradiation. The white fields were cut out of a dark screen and illuminated from behind. One of the two black squares could be shifted horizontally by a screw, and was adjusted so that the two middle vertical lines appeared to the spectator to coincide in one line. For measurements at longer distances the black fields were made of little boards, but for shorter distances they were made of little steel plates. The error made in adjusting the square was called the spread of the irradiation.

3. *Straight lines become broken.* If the edge of a ruler is held between the eye and a bright flame or the sun, it seems to have a break in it where the bright object protrudes above it, as represented in

Fig. 34. With reference to this particular effect, it may be noted here at the same time, that when the bright body is a lamp flame with cylindrical wick, the indentation at the edge of the flame, where, as above stated, the absolute brightness is higher, appears to be deeper than in the middle of the flame, in spite of the fact that the eye is not consciously aware of the difference of brightness in the two parts of the flame.

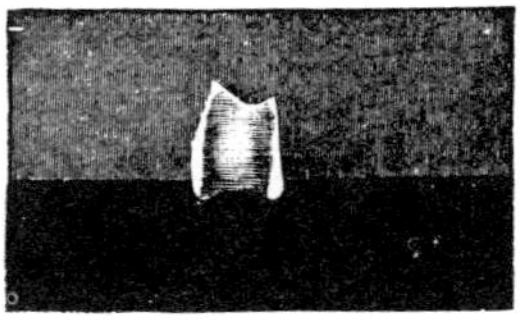

Fig. 34.

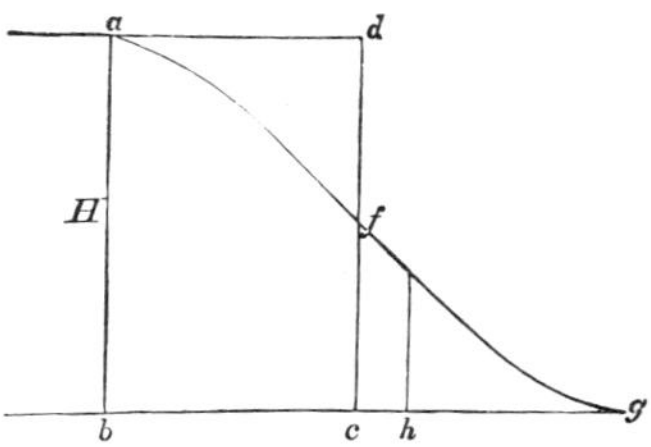

Fig. 35.

The fundamental thing about all these appearances is that the edges of bright areas in the field of view are, as it were, shifted and tend to encroach on the adjacent darker areas. This encroachment is more and more noticeable in proportion as the accommodation is more inexact, in which case the blur circles projected into the eye from each point of the bright area get bigger and bigger. However, even when accommodation is most exact, blur-circles are not entirely lacking, because we know they must be present on account of colour dispersion and the so-called monochromatic aberrations of the eye, which were discussed in § 14. Now the effect of these blur-circles is to make the light spread beyond the geometrical edge of the retinal image of a bright area; but the darkness also infringes over the edge in the sense that the light begins to fade within the contour where it should still have its full strength. In Fig. 35 suppose that c is a point on the edge of a bright area, and that bg is a straight line drawn perpendicular to the edge. At right angles to this line let ordinates be erected that are proportional to the objective brightness at the corresponding places along bg. If the image of the area were perfectly exact, the broken line $adcg$ would represent the magnitude of the luminosity. Thus from b to the edge of the surface at c it would be uniform and equal to H, and from c to g the luminosity would be zero. But if, through lack of accommodation, the image were blurred, then, as was explained in connection with Fig. 71 in Vol. I, the luminosity falls off as shown by the curve afg. Thus, from c to g bright encroaches on dark, and from b to c dark encroaches on bright; naturally, whatever light spreads beyond the edge being borrowed from the bright portion inside. Accordingly, as long as it is simply a question of objective brightness, the bright areas cannot appear magnified by the blur circles. On the contrary, the area of full brightness has been reduced by the blurring, although the total area that gets light in some fashion has been augmented. But now taking into consideration the fact that the light sensation is practically, if not actually, the same for the higher degrees of objective luminosity, the effect is that the falling off of the illumination inside the surface is less noticeable than the illumination of places beyond the edge that were previously dark. Thus, so far as sensation is concerned, the spread of brightness, and not that of darkness, must be the result. The phenomenon will be most striking when the surface is bright enough for the light sensation inside the blur-circle to reach its maximum there. For instance, if this were the case for the point h in Fig. 35, the apparent brightness at h would no longer be distinguished from the full brightness in the interior of the area; and hence the full brightness of the area would appear to extend to h; and even beyond h it would fall off very gradually until it vanished entirely at g. This makes it plain too why the effect of irradiation is easier to get when the brightness is great, and when, therefore, the place where the maximum light sensation is reached lies nearer to g. This is the explanation also of why the irradiation continues to increase when the brightness of the background is enhanced, even if the sensation of this brightness cannot keep pace with it. With increase of objective luminosity all the ordinates of the curve ag will be elevated in the same proportion as H is, which means that the ordinate corresponding to the maximum of the sensation of sufficient brightness will lie farther out nearer to g. Quantitative experiments on the influence of brightness were carried out by PLATEAU; the result being that the amount of irradiation was found not to increase in proportion to the brightness, but less than this. As the brightness is increased, the irradiation approaches a maximum asymptotically, as follows likewise from the explanation given here.

Another result of this theory is the explanation of the fact that the irradiation spreads more when the blur-circles are larger.

With most persons the blur-circles of points that are too far away are wider vertically than they are horizontally. Hence, a bright square on a dark background which is a little too far for the accommodation seems to be higher than it is broad; and a black square on a bright background just the reverse. Even with exact accommodation, most people see the vertical elongation of a white square, because it seems that in this case they accommodate for the vertical lines. On the other hand, white rectangles, whose horizontal dimension is somewhat longer than the vertical, look like squares. A. FICK[1] found in his

[1] HENLE und PFEUFFER, *Zeitschrift für rationelle Medizin*. Neue Folge. II. S. 83.

experiments that, for a practised eye that was not near-sighted, a rectangle, 450 cm away, whose horizontal side was 22 mm and vertical side 20 mm, appeared to be a square; and that another one with a horizontal side of 21 mm and a vertical side of 20 mm was taken for a vertically elongated rectangle. With other eyes that see a distant point of light as a star with three rays, there are also manifested in the other cases of irradiation three main directions in which the effect is greatest; as described by JOSLIN.[1]

In the preceding discussion, the term irradiation has been used merely with respect to those cases where there is no consciousness of blurring as such, but where the area of full illumination is apparently enlarged. However, very recently this term has come to be applied to the formation of blur-circles generally, even where they are recognized as being fainter parts of the image. Perhaps, however, a special new name is not needed for these cases. Incidentally, new boundary-lines may also be produced by the blur-circles, causing the object to appear changed in size, otherwise the intensity of the light itself having no special influence. As a case in point, VOLKMANN[2] found that very fine black threads on a white background, and also white threads on a black background, were regarded as being thicker than they really were; whereas with the kind of irradiation considered thus far it is only the brighter area that is magnified. VOLKMANN's threads were 0.0445 mm thick, and the eye was one-third of a metre away. Consequently, they should have appeared to the eye much smaller than the smallest perceptible width. By means of a micrometer-screw, the threads could gradually be brought closer together, and the problem consisted in adjusting them until the interval was just the same as the width of the threads. But every individual made the interval too wide, even when it was bright and the threads were dark. Hence, VOLKMANN also used these measurements to determine the width of the little blur-images with good accommodation. He himself found the interval to be on the average equal to 0.207 mm, whereas the thickness of the thread, which should have been the same as the width of the interval, actually was only 0.0445 mm. From this he calculated the width of the blur-image on the retina as being 0.0035 mm. For other persons, in case of a bright background, this latter quantity varied between 0.0006 and 0.0025. These dimensions are smaller than the least perceptible width (0.0044 mm) and than the diameters of the cones in the yellow spot (0.0045 to 0.0054). Possibly, therefore, the latter persons may have determined the width of the black image. Doubt-

[1] POGGENDORFFS *Ann.* LI. Ergänzbd. S. 107.

[2] *Berichte der sächsischen Ges. d. Wiss.* 1857. S. 129–148.

less, in the case of such a subtle exercise it is not surprising that such wide discrepancies should occur in the results.

But even black stripes of discernible width, viewed with such insufficient accommodation that the blur-circles are much wider than the stripes, will look wider than they are. The writer is disposed to think that this is due to the distribution of the light in the blur-circle. In Fig. 36 let *ab* represent the section of a sheet of paper on which a black line is drawn as indicated in the diagram by the point *c*. Owing to faulty accommodation, suppose there are blur-circles of radius *fc*. Then according to the principles developed in § 13, leaving out of account disturbances due to the asymmetry of the crystalline lens, the curve representing the light-intensity at the various points on the line *ab* as it is reproduced in the retinal image will be shown by the line $\alpha\varphi\gamma\delta\beta$. The light-intensity at φ and δ undergoes here a sudden drop, and hence these places appear as border-lines. If the line *c* where a white line on a black background, $\alpha\beta$ would have to be taken as abscissa-axis, and the negative ordinates of the curve $\varphi\gamma\delta$ would show the light-intensity. Then too there will be a sudden falling off of light-intensity at *f* and *d*. Incidentally, the rotating disc will prove that the lines that appear as border-lines are those for which the derivative of the light-intensity becomes infinite. When a white disc with a round circular spot on it like that represented in Fig. 37 is made to rotate the black spot looks like a grey circle whose light-intensity would be expressed by a curve quite similar to $\alpha\varphi\gamma\delta\beta$ in Fig. 36; as follows from the laws to be developed in the next chapter. The grey ring in this case appears to be perfectly sharply defined on both sides. The unequal degrees of brightness in its interior are scarcely noticed, and it appears rather to be coloured almost uniformly grey. Incidentally, in the blurred images of small black lines usually there is an admixture of double images due to the asymmetry of the crystalline lens (see Vol. I, Fig. 73), whereby the distribution of light in the blurred image is indeed changed, but yet in every case the width of the image continues to be greater.

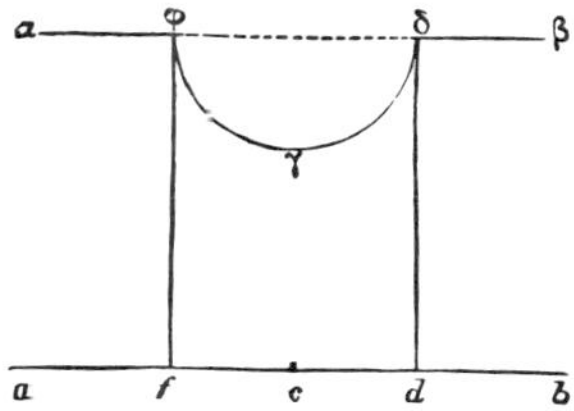

Fig. 36.

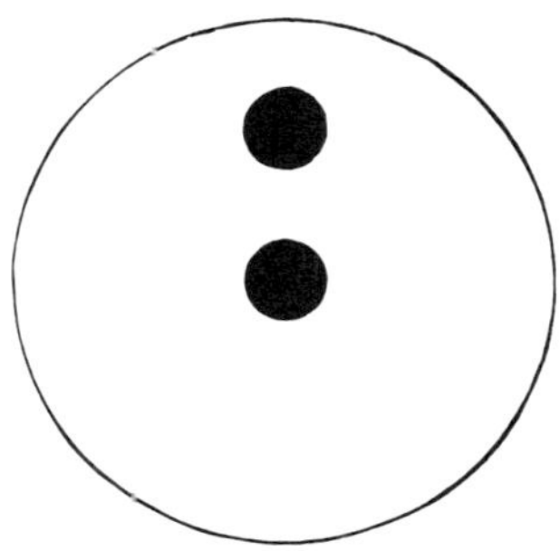

Fig. 37.

As soon as the black line ceases to be very small as compared with the width of the blurred image, the brightness also falls off gradually at its edge, as in Fig. 35, and then the edges appear a vague grey, and the middle black. Then the existence of blur-circles is recognized and the illusion vanishes. The distinction is strikingly manifest in one of VOLKMANN's experiments. When the diagram in Fig. 38 is viewed at a distance for which the accommodation is considerably out, it will be found that the middle white stripe, which is the same width everywhere, seems to have the shape of a club, broadening out at the lower end between the wide black areas, and on the other hand closing in at the upper end between the narrow black strips, thus making, so to speak, the handle of the club. The white portion between the wide black surfaces spreads out in the ordinary way of irradiation. The narrow black pieces, on the other hand, change into wider grey ones and thus encroach on the extent of the white in between them. Similar phenomena have been described by PLATEAU, but he argued that the irradiation of two adjacent white edges was mutually restrictive.

Fig. 38.

These last mentioned phenomena of the spreading out of dark lines are therefore simple cases of blurred images, not dependent on the degree of illumination and on the laws of the intensity of the light sensation. The writer would prefer, therefore, not to include them under irradiation and to reserve this term for those cases where the effect is dependent on the degree of illumination.

Very many physicists and physiologists have adopted another explanation of the phenomena of irradiation, which was specially advocated and elaborated by PLATEAU. The idea is that in the retina a fibre that has been stimulated is in the position to induce the state of stimulation in adjacent fibres also, so that these latter likewise arouse the light sensation without being acted on by any objective light. This would be a case of so-called *synaesthesia*. Sympathetic sensations of the same sort occur in other sensory nerves. For instance, many people feel tickling in the nose when vivid light falls on the eye, or a cold shiver down the back on hearing shrill or squeaky tones. In these and other cases the transference of the stimulation from the primarily excited nervous fibres to the others cannot happen except within the central organ, because the optic nerve has no anatomical communication with the sensory nerves of the nose (*nervus trigeminus*) nor the auditory nerve with the cutaneous nerves of the trunk except through the central organ. Incidentally, cases of synaesthesia of this kind invariably occur in fairly isolated instances only, and the explanation of it as given above cannot be considered as firmly established, because possibly also reflex discharges in the secretory glands of the nose or the vascular muscles of the blood-vessels in the skin might evoke similar sensations directly. In the great majority of cases it is evident from common experience that the stimulation of a sensory fibre is not transferred to other fibres, because individual impressions that are communicated to the organs of sense can be recognized as isolated sensations. If a place on the skin is pricked and the corresponding nervous fibre thereby stimulated, diffused sensations of pain would be aroused at many places in the skin, on the supposition that there was a regular and constant transference to other nervous fibres; and we would not know how to distinguish the first place that was stimulated from the secondary stimulations. But as a rule the sensation due to stimulation of a single place on the skin is felt just simply at the place where the stimulation acted, and nowhere else; and so there is no synaesthesia. But if the local pain is very acute and lasts a long time, undoubtedly there may be pains also in the adjacent parts which are commonly supposed to be sympathetic; but they may indeed be due also to the spread of the mischief that causes the pain or to inflammation. PLATEAU also recalls the fact that when the image of a black spot on a sheet of white paper falls on the place where the optic nerve comes into the eye, the only sensation in the corresponding place in the visual field is that of white; and he assumes that here also the stimulation spreads over the optic disc. But we shall see later that this phenomenon is of quite a different kind. Thus, if irradiation in the eye is explained as being synaesthesia, this view will have to be supported merely by analogies in other parts of the nervous system that are themselves still doubtful. On the other hand, the phenomena of irradiation in the eye are all such that objective light also falls or may fall on the parts of the retina where the synaesthesia is presumed to take place. The amount of irradiation is proportional throughout to the size of the blur-circle, and the whole effect in all its details can be deduced from other well established principles of explanation. In a case like this the author believes it is unjustifiable to employ new modes of explanation that are not themselves securely established.

Here some account must be given of the methods of photometry[1], in so far as the physiological properties of the eye enter into the question. In this survey all methods in which comparisons of brightness are made, not by the eye, but by means of photochemical effects or absorption of heat, will be omitted. It is worth saying that the eye can be employed very well to make

[1] With regard to this subject, see Appendix II at the end of this volume.—N.

a comparison between two quantities of light of the same quality, for example two quantities of white light or two quantities of light of the same simple colour. For when two quantities of light of the same quality under the same circumstances produce equal effects in the eye, the inference may be drawn that the objective intensities are likewise equal. In such cases the eye may be used as a convenient and sensitive reagent, with the special characteristics of which we do not need to be concerned, and hence the results obtained are objectively valid. Strictly speaking, therefore, this part of photometry is not in the domain of physiological optics according to the limitations of this science as prescribed in Vol. I, p. 47. The subject will be treated here merely in so far as the physiological idiosyncrasies of the eye have any influence on the sensitivity of the photometrical measurements.

On the other hand, as has been clearly enough brought out by the facts cited above, we must keep steadily in mind that any comparison of light of different colours as made by the eye has merely a physiological value, and tells us nothing about the objective strengths of the lights that are compared; so that all photometrical measurements of this kind remain entirely within the field of physiological optics.

Generally speaking, the procedure in photometry is as follows. Suppose it is required to determine the ratio of two luminosities A and B, where B, say, is greater than A. The intensity of B is then lowered, by any process that enables us to determine in what ratio it is diminished, until B looks just as bright as A. Suppose the reduced luminosity of B is nB, where n must be a proper fraction of known size; then

$$A = nB,$$

and thus the ratio between A and B is found. The various methods of photometry differ from one another, in the first place, by employing different means for reducing the brighter light in a known ratio. So far as this point is concerned, the method to be selected will necessarily always depend chiefly on the nature of the problem. However, they also differ in the ways and means of presenting the two luminosities to the observer's eye for comparison. With respect to this matter, it should be stated that the eye discriminates best between the luminosities of the two surfaces when they are directly juxtaposed, so that there is nothing to indicate the border between them save the difference of brightness. Moreover, the sensitiveness seems to be more increased still by not having a simple straight line to separate the two luminous areas, but when one of them forms a complicated design in the other (rings, letters, etc.), with manifold alternations of bright and dark. Lastly, the two areas to be compared must have also a certain spatial extent, not too small. Of course, those methods are very much more disadvantageous in which the intensity of a light is measured by reducing its effect on the eye by some means until it vanishes. For, evidently, the limits of sensitivity of the eye are not so definite and so constant that measurements can be made to depend on it. In different circumstances (intensity of illumination, motion, etc.) the same eye will perceive a difference of light-intensity of 1/60, and then again of 1/120. If, therefore, the sensitiveness of the eye was used as a gauge, quantities of light might be put equal to each other when one of them was twice as great as the other or perhaps more still.

Bouguer[1] had two white surfaces illuminated by the lights to be compared, and placed himself so that he saw them both in perspective near each other. Then he altered the distance of one of the surfaces from the light until

[1] *Essai d'Optique* 1729 in 12 mo. — *Traité d'Optique sur la gradation de la lumière.* Paris 1760. Latin translation, Wien 1762.

the illumination was the same. In LAMBERT's famous *Photometria*[1] the first complete system of theoretical photometry was expounded with marvelous acumen and resourcefulness. Along with various other methods adapted to special purposes, the particular process he used was to illuminate a white surface by two lights which cast two shadows of an opaque rod on it. Then the distance of one of the lights was varied until the two shadows were equally bright. RUMFORD[2] also used the same method, and the necessary apparatus for the purpose is known as RUMFORD's photometer. To enable the observer to have a more convenient position, POTTER[3] used two transparent surfaces instead of the two opaque white ones; and RITCHIE[4] added besides two mirrors inclined at 45°, which threw the light on the white surfaces and allowed the sources of light to be placed opposite each other. Sir. J. HERSCHEL[5] insisted on fulfilling the condition of close contact between the two surfaces that have to be compared in RITCHIE's photometer, which meant increased accuracy. Incidentally, in these cases there are two disturbing factors in the use of the law of the inverse square of the distance for measuring the illumination. In the first place, when this rule is employed, the extent of the source of light is supposed to be infinitely small as compared with its distance from the illuminated surface; and this is not the case when the intensities of light are great, and the light must be very close. In the second place, especially when the light is far away, there must not be in the back of the room any appreciably luminous objects, and this condition will always be hard to satisfy when the experiments are conducted in a room. PERNOT[6] modified POTTER's method by illuminating the two transparent illuminated surfaces from the opposite side by still a third light, which was gradually brought closer. If the two areas are equal, they must disappear at the same time. In BUNSEN's photometer a piece of paper which is partly soaked in kerosene is exposed to light on both sides. When the light on one side is faint, the transparent spot appears dark, and when the light is too strong, the spot appears bright.

Absorption of light was used by DE MAISTRE[7] for reducing the intensity. He combined two equal prisms, one of blue glass and the other of white glass, in such a way that the two external surfaces were parallel. The light traversed them without being deviated but it was absorbed differently in different parts of the double prism. Similarly, QUETELET[8] used two blue glass prisms which could be shifted with respect to each other so as to make a plane parallel plate of variable thickness. But the colour of the transmitted light is changed by the blue glass plate employed in this arrangement, and it has already been stated that it is not possible to make an accurate measurement when lights of different colours are compared. More questionable still are two other instruments in which the measurement does not consist in a comparison between two different lights, but absolute intensities of light are to be determined from the fact that they disappear entirely with definite amount of absorption. This method was proposed by LAMPADIUS.[9] He looked at the bright object through a number of thin horn-plates and added to them until

[1] *Photometria sive de mensura et gradibus luminis, colorum et umbrae.* Augustae Vindelicorum 1760.

[2] *Philos. Transact.* LXXXIV. p. 67.

[3] *Edinb. Journal of Science,* New Ser. III, 284.

[4] *Annals of Philosophy.* Ser. III. Vol. I. 174.

[5] *On light.* p. 29.

[6] DINGLERS *polyt. Journ.* CXIX. 155.—*Moniteur industr.* 1850. No. 1509.

[7] *Bibl. univ. de Genève.* LI. 323.—POGGENDORFFS *Ann.* XXIX. 187.

[8] *Bibl. univ. de Genève.* LII. 212.—*POGGENDORFFS *Ann.* XXIX. 187–189.

[9] GEHLERS *Wörterbuch.* 2. Auflage. VII. 482.

the object just vanished. Instead of horn-plates, de Limency and Secretan[1] used paper discs. The other instrument is the *lamprotometer*, proposed by an unknown person[2], for measuring the brightness of daylight. The method consisted in finding the strength of a litmus solution required to cause a platinum wire illuminated by daylight to disappear when it was viewed through a glass cell filled with this substance. The limit of sensitivity of the eye is, however, too indefinite for measurements of this kind, and the errors might be three or more times as great as the magnitudes concerned. A photometer designed by Albert[3] and another one by Pitter[4] depend on the same principle.

On the other hand, there were two other ways along which the more perfected methods now in use were gradually developed. One of these ways was intended for measuring the brightness of stars. By inserting a diaphragm in front of the telescope, Sir. J. Herschel reduced the aperture of the instrument and thus diminished the amount of light coming from the brighter star at which the telescope was pointed. A. v. Humboldt's astrometer is based on the same principle. This is an ordinary mirror sextant. The telescope in the instrument is pointed towards a mirror, one half of which is silvered and the other not, one star being seen directly through the unsilvered portion and the other star by means of the silvered portion and another mirror. By shifting the telescope at right angles to the dividing line between the two halves of the mirror, more rays will be received from one star and fewer from the other, and thus the images of two stars or the two images of one star can be made equal or unequal at will, and the intensities of the light compared in the two cases. The advantage of Humboldt's method is that the two stars to be compared are seen close together in the field of the same telescope. But the comparison of such small intense point-sources of light is more difficult than the comparison of bright surfaces. This fault is remedied in Steinheil's objective photometer.[5] This is a telescope with its objective divided in half. In front of each half there is a reflecting right-angle glass prism. The instrument is so adjusted that the observer sees one of the stars to be measured through one half of the objective, and the other star through the other half. Then the two halves of the objective are each shifted a little so that the images of the two stars are blurred and no longer distinct; the intensity of illumination being diminished in proportion as the areas covered by the images are increased by shifting the two halves of the objective. Each half is provided with a rectangular diaphragm which may be exchanged for another one of a different size. When the adjustment is right, the two images of the stars appear juxtaposed as two large rectangles of nearly the same size and of equal brightness, so that the conditions are the most favourable for detecting any small differences of brightness. This was the first instrument that enabled us to make accurate measurements of the light of the fixed stars and planets. Schwerd[6], on the other hand, used diffraction effects of small circular diaphragms to produce bright surfaces.

But for researches in physics, where we are concerned with ascertaining how much of the light has been lost on the way by refractions, reflections and other adventures, a good way of reducing the more intense light is by re-

[1] *Cosmos.* VIII. 174.—*Polyt. Zentralblatt* 1856. 570—Dinglers *polyt. Journ.* CXLI. 73.

[2] Poggendorffs *Ann.* XXIX. 490.

[3] Dinglers *polyt. Journ.* C. 20 and CI. 342.

[4] *Mechanics Magazine.* XLVI. 291.

[5] Poggendorffs *Ann.* XXXIV. 646. — *Denkschriften der Münchener Akad. Math.-phys. Klasse.* Bd. II. 1836. — Johnson's method in *Cosmos.* III. 301–305 is similar.

[6] *Bericht über die Naturforscherversammlung* 1858.

fraction and reflection with unsilvered plates of glass. BREWSTER[1] and QUETELET[2] employed multiple, nearly perpendicular reflections, in order to compare a strong light with a weak one. Thus, for example, sunlight is extinguished by 28 or 29 such reflections. DUWE[3] employed in the same way reflections from plates of black glass such as are used in polarisation apparatus. POTTER[4] utilized the different degrees of reflection for different angles of incidence. His source of light was a white screen in the form of a semi-cylinder. It must be supposed to be uniformly illuminated, but this is difficult to contrive. The cleverest application of this principle was made by ARAGO in his photometer; which has been used for making very accurate measurements of light-intensity.[5] The source of light in this instrument is a plane, vertical, transparent paper screen, placed at a window, and necessarily uniformly illuminated all over; although, incidentally, this can be regulated by the instrument itself. A plane parallel plate of glass is mounted vertically at right angles to the screen. Underneath the middle of this plate there is a rod, and around this rod as axis a tube can be turned in a horizontal plane. The tube is pointed horizontally at the centre of the plate, and the observer looking through it sees partly through the plate a portion of the paper screen, and partly reflected from the plate another portion of the screen. To right and left of the glass plate, and between it and the screen horizontal black bars are mounted at somewhat different levels. These are seen close together, partly through the plate and partly by reflection in it. Where the reflected image of a black bar is, the observer gets simply light from the screen that is transmitted to him through the plate; but where he sees a black bar through the plate, the light he gets from the screen is light that has been reflected from the plate. The tube is now adjusted so that the two black bands appear equally bright, and the angle between the axis of the tube and the plate is measured by a suitable scale on the instrument. The incident or reflected light may now be subjected to all sorts of other actions, the result being that generally another angle will be found for which the two images appear to be of equal brightness. In order to calculate from this angle how much the intensity of the light has been reduced, the instrument has to be calibrated by preliminary experiments, and the ratio between the amounts of reflected and transmitted light determined empirically for different angles of incidence. For this purpose, ARAGO proposed a special method depending on the fact that the two beams of light in a double refracting crystal are equal in intensity, each being half as strong as that of the undivided beam. Thus, by halving or quartering one of the two beams of light by double refraction, the positions were found for which the transmitted light was one-fourth and one-half that of the reflected light, and twice as much and four times as much. Then by interpolation the required values of the ratio can be found also for all the intermediate angles.

ARAGO had also suggested another method for reducing the intensity of light depending on polarisation in double refracting crystals. If completely polarised light of intensity I falls on a crystal of this sort, and if ϕ denotes the angle between the plane of polarisation and the corresponding principal section of the crystal, the intensities of the two emergent beams will be $I\cos^2\phi$ and $I\sin^2\phi$. If this angle can be measured, the relative intensity of the

[1] *Edinburgh Transactions*. 1815.

[2] *Bibl. univ. de Genève*. LII. 212.—POGGENDORFFS *Ann* XXIX. 187–189.

[3] POGGENDORFFS *Ann*. XXIX. 190 Anm.

[4] *Edinburgh Journal of Science*. New Ser. IV. 50 and 320. — POGGENDORFFS *Ann*. XXIX. 487.

[5] *Oeuvres* de FR. ARAGO. X. pp. 184–221.

beam of refracted light can be directly determined by it. One of the two refracted beams is eliminated entirely by a NICOL prism, the other one being all that is left. This is the principle of F. BERNARD's photometer.[1] The two beams to be compared with each other are parallel. Each of them passes through a separate pair of NICOL prisms that can be rotated. Then they are totally reflected by a right-angle glass prism so as to enter the observer's eye as two parallel beams close together. By properly adjusting the angle between the principal sections of the pair of NICOL prisms traversed by the more intense beam, the observer endeavours to make the intensities the same. When the beams to be compared originate from the same source of light, one pair of NICOL prisms can be dispensed with, and instead of them a double refracting prism can be employed, which splits the light from the source into two equal halves polarised in different planes. BEER's photometer[2] is very similar in principle. The two beams arrive at the instrument horizontally from the right and left; and each of them traverses a NICOL's prism. A double mirror, made of steel, with two reflecting surfaces inclined at 45° to the horizon, makes the beams vertical; and now they both traverse a third NICOL prism and enter the observer's eye. What he sees is a circular field, the right and left halves of which correspond to the two reflecting surfaces of the double mirror. By turning the NICOL prism the brightness of the two fields can be matched. ZÖLLNER's photometer[3] is similar to this also.

BABINET[4] employed a means of comparing the intensities of two beams of polarised light which greatly facilitated the operation. His photometer was primarily intended for comparing the brightness of gas flames. A tube branches out in two arms, one being the prolongation of the tube itself and the other making an angle of 70° with it. They are both closed by ground glass plates. At the junction of the two arms a set of glass plates is placed in the tube along the bisector of the angle. If sources of light are adjusted in front of the ends of the two branches, the light from one source enters the common tube after having traversed the pile of plates; and is polarised by refraction in a plane at right angles to the plane of incidence. The light from the other source, being reflected at the pile of plates, is polarised in the plane of incidence when it enters the common tube. At the end of this tube there is a SOLEIL polariscope. As long as the intensities of the two amounts of light polarised at right angles to each other are unequal, four complementary semi-circles will be visible. The colours disappear when the two quantities of light are equalised by varying the distance of the flames. In this instrument, therefore, the comparison of the light-intensities for the eye is reduced to comparing the colours of adjacent surfaces.

WILD's photometer[5], based on an idea of NEUMANN's, is similar in principle; but, by changing the physiological part of the apparatus, the highest degree of sensitivity appears to have been attained in this instrument. The two beams of light to be measured are parallel to each other when they come to the instrument. They are finally united, one of them being reflected at the polarising angle, first from a glass plate A, and then from a pile of glass plates B parallel to A, being thus completely polarised; and the other traversing the pile of glass plates. But before this second beam arrives at the pile B at the polarising angle, it has already traversed a similar pile C. This latter pile

[1] *Annales de Chemie.* (3) XXXV. 385–438.—*Cosmos.* II. 496–497 and 636–639—*C. R.* XXXVI. 728–731.

[2] POGGENDORFFS *Ann.* LXXXVI. 78–88.

[3] *Photometrische Untersuchungen.* Dissertat. Basel 1859.

[4] *C. R.* XXXVII. 774.

[5] POGGENDORFFS *Ann.* XCIX. 235.

can be turned around an axis so that the light can traverse it at various angles, which can be measured accurately; the result being that the quantity of transmitted light and its polarisation ratio can be altered. Incidentally, the pile of plates C is so adjusted that the polarisation which it produces in the beam of light is opposite to that which would be produced by the pile B. If the second beam is allowed to traverse C perpendicularly, it falls on B without being polarised, and here it is oppositely polarised to the first or reflected beam. The two beams are here united and proceed on their way together. By inclining C more and more, the quantity of polarised light in the second beam will be more and more diminished, in a ratio that can be calculated from the measurement of the angle of incidence. Thus, the light in the first beam is completely polarised, and mixed with it is the light of the second beam composed of variable quantities of oppositely polarised light and natural light. Finally, this mixture of light traverses a plate of Iceland spar cut perpendicular to the axis and a tourmalin plate. If the amounts of polarised light in the two beams are equal, the familiar cross with rings characteristic of Iceland spar will not be seen at all; but if the amounts of polarised light are not equal, this cross will be visible. The sensitivity of the eye was found to be extraordinarily high, so far as recognizing the polarisation pattern of the crystal was concerned, and the difference of intensities for repeated settings did not amount to more than half of one per cent. A higher accuracy still has been attained by WILD[1] in his new photometer. In this instrument the piles of glass plates are replaced by double refracting crystals, and the polariscope consists of two crossed plates of rock crystal cut at an angle of 45° to the axis. The rays that are transmitted are made parallel by means of lenses. Plates of this sort show a system of rectilinear fringes. When the instrument is properly adjusted, it is simply a diagonal portion of this system that is extinguished, the colours on the two sides being complementary. It is possible to focus the cross-wires very sharply on the middle of the extinguished fringes. The error for a single setting, according to WILD's data, does not amount to more than between one and two thousandths of the total intensity of the light.

TALBOT[2] used a rotating disc with opaque and transparent sectors for reducing the intensity of the light. This same means was also utilized by BABINET and SECCHI[3] for measuring stellar brightnesses.

In order to relieve the physiological part of the photometrical process, POUILLET[4] suggested using daguerreotypes made on polished silver plates. In order to get a positive view of such an image, it must be illuminated from the side, but the observer must be placed so as to see some dark body reflected in the plate, but none of the incident light. If, on the contrary, he sees a very bright body reflected in the plate, the image looks like a negative, the parts that should be bright being dark, and *vice versa*. In between, however, there is a certain brightness of the surface for which the image disappears entirely, and with the slightest change of brightness one way or the other, the image comes out positive or negative as the case may be.

SCHAFHÄUTL[5] has employed a physiological principle of photometry entirely different from any used before; but as yet he has not proved that the

[1] *Mitt. der bernischen naturf. Ges.* 1859. No. 427–429.

[2] POGGENDORFFS *Ann.* XXXV.[1]457, 464.—*Phil. Mag.* Nov. 1834. p. 327. With reference to this: PLATEAU in *Bullet. de l'Acad. de Bruxelles*, 1835. p. 52.

[3] *Arch. d. sc. phys. de Genève.* XX. 121–122.—Memorie dell' osservatorio di roma. *Cosmos.* I. 43.

[4] *C. R.* XXXV. 373–379.—POGGENDORFFS *Ann.* LXXXVII. 490–498.—*Inst.* 1852. p. 301.—*Cosmos.* I. 546–549.

[5] Abbildung und Beschreibung des Universal-Vibrations-Photometer. *Münchner Abhandl.* VII. 465–497.

principle is correct. He states that the time that can elapse between two similar light-impressions without the eye noticing the interruption is proportional to the square-root of the intensity of the light. His apparatus consists of a strip of steel clamped at its lower end so that it stands vertically when in equilibrium. On its upper end it carries a rectangular screen made of thin blackened copper-foil with a rectangular hole in the centre of it. Through a horizontal tube with peepholes at both ends the observer sights the screen on the top of the strip of steel. Beyond it the source of light is adjusted so that its light can only get to the observer's eye when the aperture in the screen is directly in line with the axis of the tube. The strip is set to vibrating and shortened until there is no flicker in the light, and it seems to shine steadily. The intensity of the light is supposed to be (inversely?) proportional to the square of the vibration time of the spring, that is, to the fourth power of its length. Even admitting that the intensity does vary in this manner, the latter part of the above statement would not apply to the case of a vibrating spring loaded at the top.

Lastly, it remains to mention the method used by FRAUNHOFER[1] for comparing with each other the intensities of the light of different colours of the spectrum made by a glass prism. The spectrum was observed as usual by a telescope, a prism P being placed in front of the objective A (Fig. 39). The ocular is at B. A small steel mirror s, inclined at 45° to the axis of the telescope, is mounted in the ocular tube. A sharp edge of this mirror lies in the focal plane of the ocular lens and intersects the axis of the telescope. In the half of the ocular opening that is not screened by the mirror a part of the prismatic spectrum may be seen. The mirror, however, reflects the light that comes from a small lamp flame L, which is in a tube inserted in the ocular tube at right angles to it. This lateral tube is open above and below, and the flame L can be shifted in position. A little diaphragm b in front of the flame limits the portion of it that can be seen. The observer sees this light merely as a large blurred circle whose intensity of illumination is inversely proportional to the square of the distance sb. The lamp is now moved until the illuminations of the two semi-circular fields as seen through the ocular appear to be equal, that is, until the border between them is most indistinct. The numerical results of FRAUNHOFER's measurements of the brightness of the different portions of the spectrum agreed very poorly, mainly, no doubt, because he was ignorant of the effect of the absolute intensity on the relative luminosity of the colours.

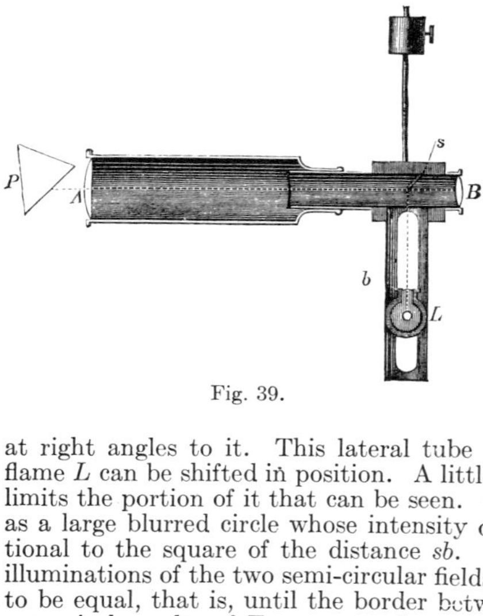

Fig. 39.

The first measurements on the sensitivity of the eye to differences of light were made by BOUGUER; the result being, according to him, that the least perceptible difference is a nearly constant fraction of the total intensity. Subsequently, the same result was obtained by STEINHEIL, MASSON, ARAGO and VOLKMANN by photometrical measurements; and the law was thoroughly investigated by FECHNER.

[1] GILBERTS *Ann.* 1817. Bd. 56. S. 297.

Observations concerning the different relative luminosities of colours were made to a certain extent by PURKINJE, and afterwards more thoroughly by DOVE, and by HELMHOLTZ with respect to the colours of the spectrum.

None of the matters treated in this section has been the subject of so much investigation and controversy as the phenomenon of irradiation. Naturally, the fact that bright objects appear to be magnified under certain circumstances, was bound to attract attention very soon. PLATEAU quotes a letter from EPICURUS to PYTHOCLES in which it is mentioned that a far away light looks smaller by day than it does by night, and that therefore perhaps the stars might look too large also; and then he quotes the beginning of the third satire of PERSIUS:

Iam clarum mane fenestras
Intrat et angustas extendit lumine rimas.

Later it was the astronomers especially who investigated the phenomena of irradiation, because it was a serious difficulty in observations of the sizes of the celestial bodies. KEPLER[1] attributed it mainly to faulty accommodation, and unquestionably he put his finger on the essential thing in most of the phenomena of this kind. GALLIEO[2] likewise studied it more in detail. He stated that the effect was more pronounced when there was more contrast between the bright object and the dark background; that, while bright objects were invariably magnified, dark objects on a bright background (for example, Mercury and Venus in front of the Sun) were diminished in size; and that the magnification of very small objects was the greatest of all. At first, like GASSENDI[3], he believed that it was necessary to suppose that luminous objects illuminated the surrounding air, but afterwards he endeavoured, more correctly, to find the explanation in irregular refractions in the eye. GASSENDI subsequently changed his opinion also and supposed that the reason why the stars look larger at night is because the pupil is dilated. In his own case, the apparent diameter of the moon varied between 33′ and 38′, depending on the brightness of the background. The reduction in the size of small objects on a bright background was discussed especially by SHICKARD,[4] who suggested at the same time that at the edge of dark objects light spread over into the region where the shadow was. This explanation was like that offered later by LE GENTIL,[5] who tried to show that irradiation was due to diffraction. On the other hand, HORROCKES[6] tried to show that the origin of the phenomenon of irradiation was in the eye in the same sense as GALILEO maintained. DESCARTES argued that the pupil was contracted in looking at a bright object, the eye becoming like a near-sighted eye, and that, therefore, the judgment of the distance and size of such objects was altered; but that, besides this, the disturbances of the retinal elements, especially if they were very considerable, could be transmitted to adjacent elements, and that consequently the image perceived would look bigger. Thus, DESCARTES was the originator of the theory of irradiation based on the transference of the nervous stimulation. Later on, when astronomers began to use high-power telescopes of good construction, it was discovered that irradiation was hardly noticeable any more in the case of the bigger stars; and its existence, while

[1] *Paralipomena.* p. 217, 220, 285.

[2] *Opere di Galilei.* T. II. p. 18; 255–257, 396, 467–469. *Systema cosmicum.* Lyon 1641. Dial. III, p. 248.

[3] *Opera omnia.* Florenz 1727. T. III. p. 385, 567, 583–585. T. I. p. 499–508.

[4] *Pars responsi ad epistolas* P. GASSENDI *de Mercurio sub sole viso.* Tubingae 1623.

[5] *Mem. de l'Acad. d. Sc. de Paris.* 1784. p. 469. (Communicated in 1743).

[6] *Venus in sole visa.* Cap. XVI. Inserted after HEVELIUS *Mercurius in sole visus.*

recognized by some astronomers,[1] began to be doubted and denied.[2] In the case of astronomical observations, the effects of the chromatic and spherical aberrations in the telescope were generally involved with those of the imperfections of the eye; and, necessarily, opinions on this subject were apt to be different, depending on the quality of the telescope in each case. In particular, in the observation of the transit of Mercury across the Sun in 1832, BESSEL showed that, with the best telescopes, irradiation is no longer appreciable in the measurements.

While astronomers were engaged chiefly in discussing whether there was any such thing as irradiation, without inquiring into the causes of it, other natural philosophers began to take up the latter question. At first J. MÜLLER[3] considered irradiation, as we have done above, as due to a spreading out of objective light. Later on, like most other physiologists of that time, when the theory of synaesthesia was being developed, he also was influenced by PLATEAU's very complete work on irradiation,[4] and attributed this effect to a transference of the stimulation from one element of the retina to the other. The phenomena that PLATEAU described as irradiation are such as an eye that is a little near-sighted ought to see in looking at more distant objects, that is, they are generally phenomena of imperfect accommodation. But PLATEAU rejected this explanation, because he had also observed the slight irradiation exhibited by very bright objects at the distance of distinct vision, and because he was ignorant then of the other reasons for the spreading out of light in the eye, which in this case are the effective ones. Moreover, he relied on the fact that, according to his experiments, the apparent spread of the irradiation was always the same for objects at different distances; but in his measurements the distances did not exceed more than 60 cm, that is, they were distances within which the errors of accommodation are not appreciable any longer. It is curious that his experiments with lenses, that produced the correct distance of distinct vision and thus abolished irradiation, did not lead him to the correct explanation. Likewise, his statement, that two adjacent irradiations mutually enfeeble each other, is hard to reconcile with any theory of synaesthesia. For if the parts of the retina lying along the image of the black band are stimulated from both sides, they must be stimulated more than if a bright field invaded them from one side only. PLATEAU must have made the above statement in order to explain why a fine black line in a bright field cannot usually be seen when the line is narrower than the irradiation-fringe. But on the supposition that irradiation is due to blurred images, the explanation is simple.

PLATEAU's work was reviewed and criticized by FECHNER; and, subsequently, a more thorough criticism was made by H. WELCKER.[5] The latter went back to KEPLER's explanation, which, as a matter of fact, does include by far the greatest number of cases of irradiation. The only thing to be added to WELCKER's work is that even at the distance of most distinct

[1] HASSENFRATZ, *Cours de physique céleste*. 1810. p. 23 — J. HERSCHEL, *On light*. T. I. §697. — QUETELET, *Positions de Physique*. 1829. T. III. p. 81. — BRANDES in GEHLERS *physikal. Wörterbuch*. Revised ed. V. 796.—ROBISON, *Mem. of the Roy. Astron. Soc. of London*. V. p. 1.

[2] BIOT, *Traité élémentaire d'astronomie physique. edit.* 2me. p. 534, 536. — DELAMBRE. *Astronomie théorique et pratique*. T. II. chap. 29. §12. — BESSEL, *Astronom. Nachrichten*, 1832. No. 228.

[3] *Zur vergleichenden Physiologie des Gesichtssinnes*. 1826. S. 400.

[4] *Mém. de l'Acad. de Bruxelles*. T. VI.—POGGENDORFFS *Ann.* Ergänzungsband. I. S. 79, 193, 405.

[5] *Über die Irradiation und einige andere Erscheinungen des Sehens*. Giessen 1852.

vision irradiation is exhibited by very minute bright objects, due to the other kinds of aberration of light in the eye. Other investigators agreed with WELCKER's conclusions and explained irradiation by means of various kinds of scattering of light in the eye. FLIEDNER,[1] H. MEYER (of Leipzig),[2] and CRAMER, in particular, directed attention to the monochromatic aberrations, and FICK to the chromatic aberrations in the eye. But all the previous objective explanations of irradiation failed to show why the spread of brightness is perceived on a dark background without perceiving at the same time a reduction of brightness at the edge of the bright surface. The author ventures to think the reason for this has been demonstrated in the treatment given above.

Measurement of the sensitivity

1760. BOUGUER, *Traité d'Optique sur la gradation de la lumière, publ. par Lacaille.* Paris 81.

1837. STEINHEIL, *Abhandl. der math.-phys. Klasse der bayr. Akademie.* 1837. S. 14.

1845. MASSON, *Ann. de chim. et de phys.* XIV. 150.

1858. ARAGO, *Oeuvres complètes.* X. 255.

1858. *G. TH. FECHNER, Über ein wichtiges psychophysisches Gesetz. Leipzig. From the *Abhandl. der sächs. Gesellschaft der Wissenschaft. Math.-phys. Klasse.* IV. 457.—Supplemented in *Berichte des sächsischen Gesellschaft* 1859. S. 58.

Comparison of Luminosities of Different Colours

1814. J. FRAUNHOFER in *Denkschr. der bayr. Akad.* V. 211.

1825. PURKINJE, *Zur physiologie der Sinne.* II. 109.

1852. *DOVE, Über den Einfluss der Heiligkeit einer weissen Beleuchtung auf die relative Intensität verschiedener Farben. *Berl. Monatsber.* 1852. S. 69–78.—POGGENDORFFS *Ann.* LXXXV. 397–408.—*Inst.* 1852. p. 193.—*Phil. Mag.* (4) IV. 246–249.—*Arch. d. sc. phys.* XXI. 215–219.—*Cosmos.* I. 208–211.
POUILLET, *C. R.* XXXV. 373–379.—POGGENDORFFS *Ann.* LXXXVII. 490–498.—*Inst.* 1852. p. 301.—*Cosmos.* I. 546–549.

1855. H. HELMHOLTZ in POGGENDORFFS *Ann.* XCIV. 18–21.

Irradiation

1604. KEPLER *ad Vitellionem Paralipomena.* Frankfurt 1604. p. 217.

1619. GALILEI, Discorso delle comete di Mario Guiducci. *Opere.* II. 256, also *Op.* II. 18, 396, 467–469. *Systema cosmicum.* Lyon 1641. *Dial.* III. p. 248.

1632. SCHICKARD, *Pars responsi ad epistolas* P. GASSENDI *de Mercurio sub sole viso.* Tubingae 1632. (The planet is reduced in size by irradiation.)

1637. DESCARTES, *Dioptrique.* Leyde 1637.—*Discours.* VI. pp. 67 and 68.

1642. GASSENDI, Epistola III de proportione, qua gravia decidentia accelerantur. *Opera omnia.* III. 585.

1738. JURIN. *On distinct and indistinct vision.* §53, in SMITH's *Optics.*

1743. LE GENTIL, *Mém. de l'Acad. des sc. Paris.* 1784. p. 469.

1810. HASSENFRATZ, *Cours de physique céleste.* 1810. p. 23.

1811. BIOT, *Traité élémentaire d'astronomie physique.* édit. 2me. pp. 534, 536.

1814. DELAMBRE, *Astronomie théorique et pratique.* T. II. Chap. 26. §197. T. III. Chap. 29. §12.

1826. J. MÜLLER, *Zur vergleichenden Physiologie des Gesichtssinns.* S. 400.

---

[1] POGGENDORFFS *Ann.* LXXXV. 348.

[2] POGGENDORFFS *Ann.* LXXXIX. 540.

1828. Brandes in Gehlers *Neuem physik. Wörterbuch.* V. 796.
J. Herschel, *On light.* I. §697.
1829. Quetelet, *Position de physique.* III. 81.
1832. Bessel in *Astronom. Nachrichten.* 1832. No. 228.
1838. *Plateau, Mémoire sur l'irradiation. *Nouv. mém. de l'Acad. de Bruxelles.* T. XI.—Poggendorffs *Ann.* Ergänzungsbd. I. S. 79, 193, 405.
1840. Fechner, Von den sogenannten Irradiation. Poggendorffs *Ann.* L. 195.
1849. Baden Powell, Sur l'irradiation. *Inst.* 1849. No. 818. p. 288.—*Memoirs of the London astron. Society.* XVIII. p. 69.—*Inst.* No. 840. p. 47.—*Report of British Assoc.* 1849. 2. p. 21.
1850. Haidinger, Das Interferenzschachbrettmuster. *Wiener Ber.* VII. 389.—Poggendorffs *Ann.* LXXXV. 350.—*Cosmos* I. 252, 454. (Case of irradiation, mixed with monochromatic aberrations.)
1851. Dove, Über die Ursache des Glanzes und der Irradiation. Poggendorffs *Ann.* LXXXVIII. 169.—*Berl. Monatsber.* 1851. p. 252.—*Phil. Mag.* (4) IV. 241.—*Arch. d. sc. phys. et nat.* XXI. 209.—*Inst.* No. 991. p. 421.
1852. *H. Welcker, *Über Irradiation und einige andere Erscheinungen des Sehens.* Giessen 1852.
Fliedner, Beobachtungen über Zerstreuungsbilder im Auge, sowie über die Theorie des Sehens. Poggendorffs *Ann.* LXXXV. 348.
Trouessart, Note concernant ses recherches sur la théorie de la vision. *C. R.* XXXV. 134–136.—*Arch. d. sc. phys.* XX. 305–306.
L. L. Vallée, Mémoire XIII. De la vision considérée dans les influences en quelque sorte moléculaires, exercées dans les réfractions, et du phenoméne de l'irradiation. *C. R.* XXXV. 679–681.
1853. H. Meyer, Über die sphärische Abweichung des menschlichen Auges. Poggendorffs *Ann.* LXXXIX. 540–568.—Fechner, *Zentralblatt.* 1853. p. 864.
1854. F. Burckhardt, Zur Irradiation. *Verh. der naturforsch. Gesellschaft zu Basel.* I. 154–157.
1854. J. J. Oppel, Über den Einfluss der Beleuchtung auf die relative Lichtstärke verschiedener Farben. *Jahresber. des Frankf. Vereins.* 1853–54. S. 44–49.
1855. A. Cramer, Beitrag zur Erklärung der sogenannten Irradiationserscheinungen. *Prager Vierteljahrsschrift* 1855. IV. 50–70.
1856. A. Fick, Einige Versuche über die chromatische Aberration des menschlichen Auges. *Archiv für Ophthalmol.* II. 2. p. 70–76.
1857. A. W. Volkmann, Über Irradiation. *Bericht der sächs. Gesellschaft* 1857. p. 129.
1858. A. C. Twining, *The relation of illumination to magnifying power, when visibility is maintained.*
1861. H. Aubert, Beiträge zur Physiologie der Netzhaut. *Abhand. der schlesischen Gesellsch.* 1861. S. 49–103.
Volkmann, Über den Einfluss der Extension eines Lichtreizes auf dessen Erkennbarkeit. *Göttinger Nachrichten.* 1861. S. 170–176.
Idem, Über die Irradiation, welche auch bei vollständiger Akkommodation des Auges statt hat. *Münchener Ber.* 1861. (2) 75–78.
1862. Aubert, Über subjektive Lichterscheinungen. Poggendorffs *Ann.* CXVII. 638–641
1863. v. Wittich, Über die geringsten Ausdehnungen, welche man farbigen Objekten geben kann, um sie noch in ihrer spezifischen Farbe wahrzunehmen. *Königsberger Mediz. Jahrbücher.* IV. S. 23–25.
Volkmann, *Physiologische Untersuchungen im Gebiete der Optik.* Heft 1. Leipzig 1863.
1864. G. Th. Fechner, *Über die Frage des psychophysischen Grundgesetzes mit Rücksicht auf* Auberts *Versuche.* Leipziger *Ber.* 1864.
Aubert, *Physiologie der Netzhaut.* Breslau. S. 23–153.

§22. Duration of the Sensation of Light[1]

When a motor nerve is excited by a brief electric shock, a short time (about a sixtieth of a second) elapses before the effect of the stimulus on the muscle disappears again. The change produced in the organic structures as a result of the stimulation lasts much longer, therefore, than the electric discharge which causes the reaction. The same thing takes place in the eye. We cannot as yet demonstrate, it is true, that the sensation arises after the light has begun to act, but we can easily show that it continues after the light has quite ceased to act.[2]

The more intense the light, and the less fatigued the eye, the longer the after-effect will persist. After glancing towards the sun or at a bright flame for an instant, and then suddenly closing the eyes and covering them with the hand, or looking at an absolutely dark background, one will continue to see for a short time a bright appearance in the dark field, of the same form as the object itself was, which gradually fades away, changing its colour at the same time. The after-images of very bright objects are easiest to see because they last longest. Incidentally, even with less brilliant objects after-images such as those described can be obtained provided the eye has been rested sufficiently beforehand in the dark, and the object is then observed for a moment. The after-image of a bright body on a dark ground has the colour of the object at first, and often shows very accurately still the details of the object in correct form and shading. For example, when a person takes a last look at a lamp just in the act of extinguishing it and before being left completely in the dark, he

[1] See Appendix B. 7 by W. Nagel at the end of this volume.—N.

¶See also Appendix II. 4 by v. Kries. (M. D.)

[2] ¶That there is a demonstrable latent period following the instant of stimulation, before the visual sensation develops, is brought out by Helmholtz a little later in the discussion of rotating discs. Moreover, that the instant of stimulation of the retina is followed by a measurable latent period has been known since the work of Dewar and McKendrick (*Trans. Roy. Soc. Edinb.* 1873. XXVII. 141). They made the first approximate measurements using eyes of birds, mammals, fishes, frogs, lobsters. Einthoven and Jolly (*Quart. Jour. Exp. Physiol.* 1908. I. 373) state that for the frog's eye the latent period, as indicated by the electrical variation following stimulation, varies from 0.01 to 2.0 sec, depending on the intensity of the stimulus. They say, also, that these values agree well with the latent period of the perception of light by the human eye. Temperature is also a factor, as shown by Hecht (*Journ. Gen. Physiol.* 1919. I. 657–667.) What the latent period of the cerebral centres involved may be, we have no way of knowing exactly.

For recent work on the latent period of the organ of vision consult Einthoven and Jolly (loc. cit.); Brücke (*Arch. f. d. ges. Physiol.* 1906. CXIV. 569); Fröhlich (*Zft. f. Sinnesphysiol.* 1913. XLVIII. 28); Hecht (*loc. cit.*); Chaffee, Bovie and Hampson (*Journ. Op. Soc. Amer.*, 1923, VII. 1); Bills (*Psy. Monog.* 1920. XXVIII, No. 127, 101) (a lengthy paper on the 'lag' or inertia, in the rise of a visual sensation to its maximum intensity). (M. D.)

continues to see the bright image of the flame afterwards in the dark, surrounded by the rather fainter glow of the globe and the other portions. When the eye moves, the after-image moves in the same way, always occupying the place in the field of view that corresponds to the part of the retina on which the light fell originally. In order for the after-image to be sharply delineated, a single point on the object should be steadily focused. If the eye has wavered, the image will be faded, or perhaps there may be two or three images of the object partly overlapping. If the after-image is quite sharply delineated, details in it can be noticed under proper conditions that had not attracted attention in gazing at the object itself and had therefore been overlooked.

After-images of bright objects, in which the light portions appear light and the dark parts dark, are called *positive* after-images. As they gradually begin to vanish, they are usually mingled with other images, in which the light parts of the object look dark, and the dark light, so-called *negative* after-images. Apparently, these latter effects are due mainly to the fact that the sensitivity of the retina for light has been altered also by its previous stimulation. The two kinds of phenomena cannot be kept strictly separate from each other in a description of them. Accordingly, the more detailed description of both positive and negative after-images will be reserved for the next chapter; while this chapter will be devoted to describing the effects of quickly recurrent light stimuli, in which the persistence of the light impression appears pure and simple, without being affected essentially by changed sensitivity of the retina.

The leading fact here is that *intermittent light stimuli of a uniform kind, occurring with sufficient rapidity, produce the same effect on the eye as continuous illumination.* For this purpose, all that is necessary is that the repetition of the impression shall be fast enough for the after-effect of one impression not to have died down perceptibly before the next one comes.

The easiest way to show this is with revolving discs. When a black disc with a bright white spot on it is rotated fast enough, a grey ring appears instead of the revolving spot. This ring looks everywhere perfectly uniform, and there is no longer any movement to be seen. As the eye looks steadily at some one place of the apparently stationary ring, the places of the retina on which the image of the ring is formed receive in swift repetition the image of the white spot that traverses the circle. They get, therefore, a light impression which appears continuous on account of the rapidity with which it is repeated. Of course, it is not as strong as it would be if continuous white light fell on the retina; and so the ring looks grey instead of white. On the other hand, if the eye itself is moved, carrying the point of fixation around with the white spot, the latter may become visible, and thus the apparent continuity of the grey ring will be destroyed. Obviously, if the point of fixation of the eye moved so as to keep pace with the white spot and remained fastened on it, the image of the spot would stay right on the fovea, and on all the other parts of the retina there would be simply the image of the black disc. Under these circumstances the eye realizes the existence of a white spot instead of the grey ring. This is the case even when the motions of the fixation point and the white spot are not absolutely concordant, provided, however, the relative motion with respect to each other is comparatively slight.[1]

If there is another white spot at the same distance from the centre of the disc as the first, it likewise will be drawn out in a bright ring, which will coincide with that of the first spot. The impressions of the two points on the retina are summated. It is the same way when there are a number of white spots all on the same ring. Suppose, therefore, that on a disc of this sort concentric rings are drawn around the centre of the disc where the axis of rotation is. Then when the disc is rotated, all the places of one circular ring taken separately produce the impression of a uniformly illuminated ring, and all these circular images of the separate points fall on the same part of the retina and unite there into a total image. The following law can now be stated in regard to this phenomenon: *The appearance of each circle whose centre is in the axis of the rotating disc is the same as if the entire light due to all the points in it by themselves were uniformly distributed over the whole circumference.* This law appears to be just as valid for monochromatic light as for polychromatic. With reference to the activity of the retina itself, it may be stated as follows: *When a certain place on the retina is stimulated always in the same way by regular periodic impulses of light, then, provided each recurrent stimulus is sufficiently short-lived, the result is a continuous impression, equivalent to what would be produced if the light acting during each period were uniformly distributed over the entire time.*

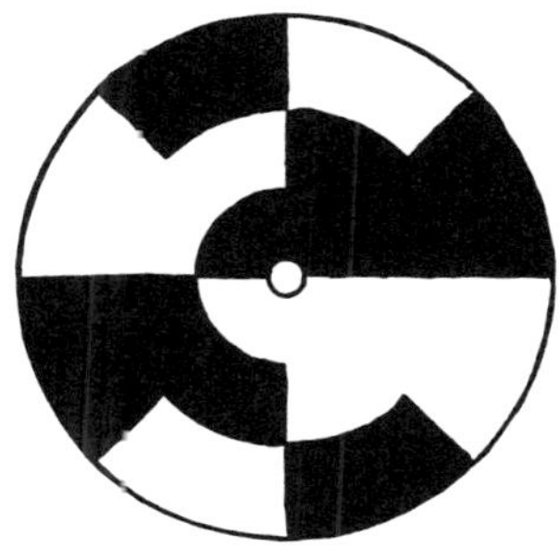

Fig. 40.

The truth of this law can be tested by making discs like that represented in Fig. 40. One half of the central ring is white and the other half black; two opposite quadrants (that is, again half) of the middle ring are white; and in the outside ring four octants of white alternate with four octants of black. When the disc is set in rotation, it

[1] See DOVE in POGGENDORFFS *Ann.* LXXI. 112. — STEVELLY in SILLIMAN'S *J.* (2) X 401. — MONTIGNY, *Bull. de Bruxelles.* XVIII. 2. p. 4. *Institut.* 1847. No. 928. p. 332.

looks a uniform grey all over. All that is necessary is to turn it fast enough for the inside ring to make a continuous impression. As far as that goes, the white can be distributed over the circumference in any way we like, provided that in each ring the sum of the lengths of the white arcs is equal to the sum of the lengths of the black portions. The same grey colour will always be obtained. Different colours can be used instead of black and white; and then the same mixed colour will be obtained in all the rings, provided the sum of the arcs of one colour in each ring is equal to the sum of the arcs of the other colour.

By such devices the law can easily be tested in a great variety of ways. Of course, the comparison must always be made between one intermittent light and another; and obviously the conditions must be such that the quality of the two alternating impressions is the same in the various comparisons.

In order to test the validity of the law also for comparisons between intermittent and continuous light, the author has used the disc illustrated in Fig. 40, on which the white and black arcs are equal. When it is in rotation, the grey has half the luminosity of the white. On the other hand, a grey of this sort can be obtained by laying a white strip of paper on a black ground and looking at it through a double refracting prism. In this case there will be two images of the strip, each half as bright as the strip itself. A larger grey surface of this sort is produced by placing alternate stripes of white and black of equal width side by side, and then holding the prism at such a distance that the double images of the white stripes come exactly over the black ones. Then the whole surface looks grey and half as bright as the white stripes. Now this grey is exactly the same as that produced by rotating the disc in Fig. 40. Of course, for this comparison the same black and white must be used on the disc as that with which the parallel stripes were made, and the two surfaces must be illuminated exactly alike. The rotating disc must also be viewed through the double refracting prism, but without separating its two images, in order that the light from the disc shall also be subjected to just the same reflection and absorption in the prism as that of the white stripes. The way PLATEAU demonstrated this law was as follows. He adjusted two rotating discs, one with white and black sectors and the other pure white all over, at different distances from a light until they looked equally bright. If n denotes the number of white sectors, and if w denotes the angular width of each of them, the width of all taken together will be nw. Let H denote the luminosity of the white at unit distance from the illuminating light, and suppose that the light which it gives out is distributed uniformly over the whole disc; then the luminosity will be diminished in the ratio of the area of the white sectors to that of the whole disc; and hence it will be $\frac{nw}{360} H$. Let r and R denote the distances of the sectored and pure white discs, respectively, from the source of light, when they both look equally bright; then

$$\frac{nw}{360} \frac{H}{r^2} = \frac{H}{R^2} \quad \text{or} \quad \frac{r^2}{R^2} = \frac{nw}{360}.$$

PLATEAU's measurements agree very well also with this law.

The author has used another method too as follows. In the case of a disc covered with narrow black and white sectors, an apparently uniform distribution of the light from the white sectors can be produced over the entire disc by viewing it through a convex lens that prevents accommodation. If the pupil of the eye is at the second focal point of the lens, so that the image of the disc in the lens falls on the surface of the pupil and more than covers it, the light from the bright sectors will appear to be spread out uniformly over the whole field seen through the lens. On the other hand, when the lens is nearer the disc, the separate white and black sectors will be seen more or less distinctly as long as the disc is stationary. If it is in motion, the luminosity is just as great, whether the lens is nearer the eye or nearer the disc; and hence it is evident that the eye is affected by the intermittent light just as much as it would be by an equal amount of steady light.

The truth of the above proposition as far as coloured light is concerned is shown by DOVE's experiments on the phenomena of rotatory polarisation. When double refracting crystals are inserted between two NICOL prisms, well-known colour effects are obtained in many instances for certain positions of the latter. To some extent the colours are distributed uniformly over the entire field, but they may also form coloured patterns. In all these phenomena, however, the colour of each point of the pattern, as can be demonstrated by the theory of the polarisation of light, will become precisely complementary to what it was at first, if one of the NICOL prisms is turned through a right angle. This is experimentally confirmed by turning one of the prisms quickly, in which case what the eye sees is white. If a coloured glass is also inserted, then for two positions of the prism 90° apart the colours are such that together they must give the colour of the glass; as is actually found to be the case when the prism is turned quickly.

Incidentally, the law is verified for intermittent coloured light also by the agreement between the results of colour mixing on colour tops with those obtained by direct composition of coloured light, as by the agreement between the results of colour mixing with colour tops and those obtained by direct composition of coloured light, as was stated in §20 in the theory of colour mixing. Usually, in order to see the whole disc uniformly covered with the mixed colour, it is divided into sectors of different colours, each of which however must be of the same colour all over. Then the entire disc will appear in the mixed colour when it is rotated. But in this case the luminosity of the mixed colour, according to the above law, is always the average luminosity of the colours separately. Now with the same illumination all pigments look more dark than white, because they reflect only certain colours that constitute a part of the total white light. Hence also the mixed colour is invariably less bright than white; and so, if it is not much saturated, it looks grey.

If a coloured star on a ground of another colour (Fig. 41) is mounted on a colour top and rotated, the colour of the star will be seen in the centre and that of the background out on the edge, while in between there will be all continuous transitions from one colour through the series of mixed colours into the other. In fact, by proper choice of the contours of the sectors of the colour top, the luminosity or the colour mixture can be made to vary in any way we please from the centre out to the edge. This was done, for example, in Fig. 37, so as to get a certain distribution of the half-tone.

Fig. 41.

The individual points of a rotating disc describe circles. Of course, the same continuity of effect is obtained by making a bright spot traverse any other closed curve. Consider for example, a taut metal wire painted black, except at one point which shines out by contrast when it is properly illuminated. When the wire is set to vibrating, the path of this point will show up as a continuous line of light, often very complicated in form. When the luminous point describes thus a path that does not exactly close in on itself but in each successive revolution comes very near to its previous course, the eye sees a line of light which gradually changes its form and position. This illustration enables us to study the motion of a stretched string; and there are many other useful applications of the same principle in physics.

If while the luminosity of the moving point is the same at every place in its path, its speed varies, the brightest places in the line of light will be at the points where the motion is slowest. The bright point loiters at these places, as it were, comparatively longer, and hence its light has a longer time to act on the corresponding places of the retina than on the places where the motion is faster. An illuminated vibrating string, for instance, looks brightest farthest from the position of equilibrium, where for a moment its speed is zero.

Here also belong the characteristic effects of intermittent illumination, which are produced best by the regular periodic sparks of an electro-magnetic induction coil with a rotating armature or with NEEF's vibratory interrupter. Every single spark of this apparatus lasts for a very short interval and seems instantaneous as compared with the movements of material things; and yet the light from these sparks is strong enough during this extraordinarily brief moment to make a perceptible impression on the retina. Illuminated by a single electric spark, all moving objects appear instantaneously at rest. Of course, the eye can perceive them only as they were at the moment when they were illuminated. As to their position before and after this moment, it has nothing to go by. Hence, if the duration of the illumination is so short that during this time no perceptible displacement of the body could occur, its outlines will look just as sharply defined as if it were absolutely at rest.

If a series of electric sparks succeed each other at very short intervals, stationary objects illuminated in this way look just as they do by steady illumination in continuous light, whereas moving objects appear manifold. That is, each single spark reveals the moving object in the position in which it is at the given moment; and as all these impressions last a little time, they are all present simultaneously and cause the moving object to appear multiple. The quicker the movement, the farther apart the images of the body will be, because the distance it goes during each interruption of the light gets greater.

Multiple images are obtained in the same way by moving the eye instead of the object. If there is a steady luminous point in the field of view, and the eye is moved, the image of the point glides over to another part of the retina. During the movement it traverses in succession all continuously adjacent points of a line connecting its first position with its last. All these points are stimulated, and thus for an instant the same sensation must be produced on the retina as would be aroused in the immobile eye by a line of light. Ordinarily, we are not cognizant of this sensation. It must accompany every movement of the eye when there are bright objects in the field. But we notice it when the continuity of the line is interrupted in some unusual way by intermittent light. If we employ as a bright object the region of the induction apparatus where the sparks occur, the bright point will appear multiple when the eye is moved. If we think of the line described by the image of the place of the spark as being traced on the retina, then only single places on this line will be stimulated by the intermittent sparks. Corresponding images will be projected in the field of view.

When a moving object illuminated by intermittent light describes a closed path and happens to be at exactly the same place at each flash, it appears single and stationary. For example, the vibrating spring or the rotating armature of the familiar induction coil appears to be standing still in the light of its own sparks. The same thing happens when any other body of periodically changing form is illuminated by intermittent light, provided the flashes always occur during the same phases of the change; as, for example, when a jet of water in the act of breaking up in drops is illuminated so that at the instant of the flash a new drop arrives regularly at exactly the same place; and hence the spectator sees the jet resolved into motionless drops. This happens when the period of the flash coincides exactly with that of the formation of the drop, or is a multiple of it. If these periods do not correspond exactly, or if one is not a multiple of the other, for example, if the period of the flash is a little longer, there appears to be a slow movement of the drops, similar to the actual movement except that it is very much retarded. In this case the phases of the drop formation revealed by successive flashes will not be exactly the same, but each phase of the phenomenon will be a little later one than the preceding. On the other hand, if the frequency of the flashes is greater than that of the formation of the drops, the phenomenon is reversed, and the spectator seems to see the drops ascending and vanishing in the jet. By such means periodic phenomena of this sort that are too rapid to be discerned directly can be exhibited in their single stages and analyzed. Various artificial contrivances of the same kind will be considered presently when we come to describe the apparatus.

The duration of the impression on the eye can be most easily determined with colour tops, whose periods of revolution can be varied and measured. In this method all that can be measured accurately is the rate of revolution needed to make the appearance of the top perfectly uniform all over. Thus it is found that the speed of revolution has to be increased for higher luminosity. Under these circumstances there seem to be differences also between different colours. PLATEAU made a disc with twelve white or coloured sectors and twelve black ones of equal width; and rotated it in ordinary daylight. Thus the time required for a black sector to pass was one twenty-fourth of the period of revolution. When the impression was uniform in appearance, this period was:

	PLATEAU	EMSMANN[1]
For white.	0.191 sec.	0.25 sec.
For yellow.	0.199 sec.	0.27 sec.
For red.	0.232 sec.	0.24 sec.
For blue.	0.295 sec.	0.22 to 0.29 sec.

Not much value can be attached to the comparison of the different colours here, because there was no way of giving them all exactly the same luminosity; and this latter factor has a very great influence on the duration of the after-effect. This is easily realized by rotating a colour top a few feet away from a lamp, at a rate that just gives a continuous impression, and then bringing the lamp nearer; at once the revolving surface begins to flicker again. In direct sunlight still greater rates of revolution are required. Incidentally, PLATEAU's numerical results are strikingly large. Using a disc with black and white sectors all of the same width and illuminated by the strongest lamplight, the author finds that the transit of black does not take more than about one forty-eighth of a second for all flicker to cease; and only one-twentieth of a second by the dim light of the full moon. PLATEAU noticed, by the way, that when the ratio between the widths of the white and black sectors is changed without altering the number of sectors, the period of revolution needed to get a uniform impression is the same. This can be shown very easily by a disc like that in Fig. 42, where the black sectors are wider in towards the centre and the white out towards the edge. With increasing speed of revolution the flicker ceases all over almost at the same time. The sensation is stronger when the white sectors are wider, and hence it diminishes more rapidly when the stimulus ceases; and so the interruption, that is, the width of the black sector, must be less than it is when the white sectors are narrower. And so perhaps it is better to make the measurements by investigating the duration of an entire period of the change of illumination, that is, the

Fig. 42.

[1] POGGENDORFFS *Ann.* XCI. 611. (As to importance of state of adaptation, see NAGEL's Appendix I. B. 6, at end of volume.—N.)

sum of the times of transit of a white and black sector. Thus in the author's experiments this time was a twenty-fourth of a second with strongest lamplight, and a tenth of a second in dim light. LISSAJOUS observed the path of a very bright luminous point attached to the prong of a vibrating tuning fork and found an even shorter time for the higher degree of illumination, namely, that it took a thirtieth of a second for the entire curve to look continuous.

Accordingly, if a rotating disc is to give an entirely uniform impression, it must make from 24 to 30 revolutions a second. But the same result can be obtained with slower speeds by repeating the pattern regularly at equal angular intervals. Thus, for instance, in case of the disc in Fig. 42, it takes only six revolutions per second to fuse the black and white of the eight sectors of the outside ring into a uniform grey; while the middle ring takes twelve, and the inside ring takes twenty-four revolutions per second. The more the luminosity is diminished, the harder it is to determine how long the impression persists before it completely disappears. Obviously, from what has been stated, this time depends also on the luminosity. The after-image of the sun may last even several minutes. Thus, while the effect of bright light decreases most rapidly at first, yet on the whole it persists longest; just as a hot body cools off more rapidly in cold surroundings, the hotter it is, although it takes a longer time to cool down to the temperature of the surroundings. With his colour tops PLATEAU also made measurements of this effect by finding the time of transit of a black sector when the colour of the bright sectors was so distributed over the black that the black no longer appeared pure anywhere. The results were as follows:

For white.....0.35 sec; for yellow....0.35 sec
For red.......0.34 sec; for blue......0.32 sec

The duration of the after-effect is found to be different also for the different colours with respect to the changes of colour that occur in the after-image of a white light on a dark ground before it disappears entirely. However, these phenomena are related in many ways to those that will be described in the next chapter, and therefore it will be better to postpone their description until then.

From the facts described here it follows that light that has fallen on the retina leaves behind it a primary effect on the nervous mechanism of vision, which is not transformed into sensation until some moments later. The magnitude of the primary change which is left behind by an instantaneous impression of light depends simply on the amount of light which has impinged on the part of the retina in question; and hence it is just the same whether very intense light has acted for a short time or weaker light for a longer time, provided that in any case the time the light acts is not less than the fifteenth of a second. Thus the instantaneous primary action of very intense light does not prove to be relatively weaker than that of moderate light, as seems to be the case with prolonged sensation of light of different intensities.

There is no conflict here, as might be supposed at first. For the lack of proportionality was found to be between the objective intensity of the light and the resultant sensation. But here we have simply to do with the instantaneous primary action that is not transformed into sensation until afterwards; and there is nothing to prevent us from supposing that the magnitude of the instantaneous primary action in the nervous substance follows another law from that of the secondary effect or the sensation. Probably the best way to explain the connection is by comparison with the magnetic needle of a galvanometer which is deflected by a current that is frequently interrupted. In this case also the deflection depends simply on the total quantity of electricity that traverses the coil in the unit of time, but it is not proportional to it. But here too there is an effect proportional to the quantity of each single impulse of electricity; and that is the slowness with which the magnet turns and is forced by the earth's magnetism to wait until the next impulse arrives if the deflection is to be kept constant. The magnet seems to be deflected to a stationary position, when its fluctuations due to the single impulses are too small to be perceived. And an intermittent light produces a continuous sensation when the fluctuations in the intensity of the sensation are less than the smallest perceptible degrees of sensation.

The simplest contrivance of rotating disc, mentioned first by MUSSCHENBROEK,[1] is the top. In most of his experiments the author uses a simple top turned out of brass; shown in elevation in Fig. 43 in one third actual size. It is set spinning by hand; and so it can be easily started at any time without preparation, and its velocity can be regulated at will, although the greatest velocity that can be imparted by hand is not more than about six revolutions per second; which is enough to keep it going three or four minutes. Owing to the slow speed, the disc has to be divided in four or six sectors, with the same distribution of colour, light and shade in each of them to get a perfectly uniform impression of light. If the number of identical repetitions of the patterns is less than this, the result is, with strong light at least, an

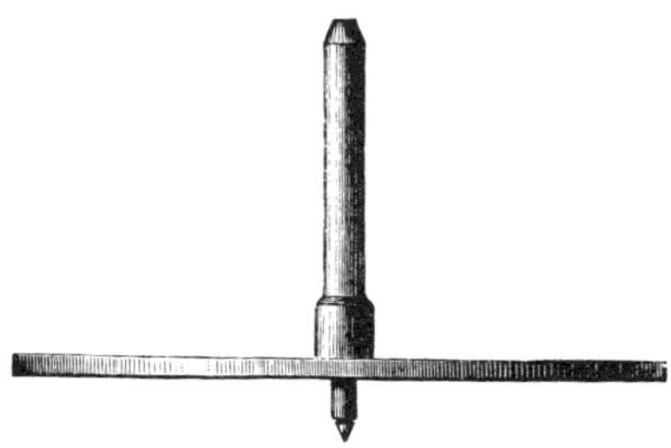

Fig. 43.

[1] *Introductio.* §1820.

effect of more or less play of colours. The patterns can easily be mounted on the top while it is spinning, and they can also be readily altered by mounting a disc with sectors cut out of it on top of a complete disc. The upper one can be adjusted on the lower one by touching it with the finger or by blowing on it with the lips. Thus a great many variations can be made while the top is in action. Suppose, for instance, the disc has blue and red sectors of equal width; and another one is placed on top of it with black sectors alternating with gaps all of the same width. When the top is set spinning, it will all look blue, provided the black sectors of the upper disc coincide with the red ones of the lower disc; but it will look red, if the black sectors of the upper disc cover the blue ones of the lower disc. In intermediate positions different mixtures of red and blue will be obtained, and so during the motion one colour can be made to change gradually into the other by varying the position of the upper disc by touching it with the finger or blowing on it. If the borders of the various sectors are not straight lines but curved or broken, very manifold and variegated systems of rings can be easily produced.

Fig. 44.

In order to make the top spin faster, a cord can be wound around the stem and pulled; the simplest contrivance for this purpose being that shown in Fig. 44; where *c* is a hollow wooden cylinder held by a handle *d*. It has two holes *b* and *c* on opposite sides of its surface, and half way between them a piece is cut out. The stem of the top goes through the two holes. The end of a strong cord is inserted in a hole in the stem, and the top is turned by hand until the cord is wound up, which makes the stem so thick at this place that it cannot slip out of the cylinder *c*. Now hold the top by the cylinder a little above a table, and pull the cord quickly, which sets the top to spinning fast; and as soon as the cord is unwound, the top drops down on the table and spins for a long time. The construction of the top is shown in Fig. 45. The discs can be clamped on it by means of the handle, as has to be done in MAXWELL's experiments for verifying NEWTON's law of colour mixing. For this purpose what is needed is a set of round discs of different sizes made of stiff paper, each of which contains a central opening and a radial slit, as shown in Fig. 46. Each disc is made in one uniform colour. By using two or more of them together and adjusting them with respect to each other, sectors of the separate discs will be exposed on each side, whose width can be varied at pleasure, and thus the proportions in which the colours are mixed can be altered as desired.

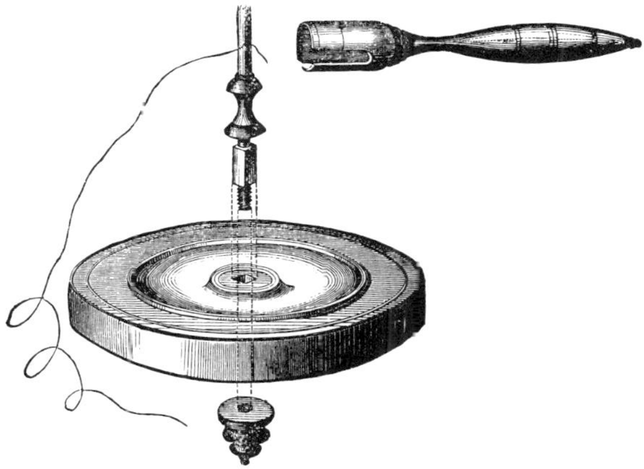

Fig. 45.

BUSOLT's colour top (Fig. 47) is the most perfect construction, when it is necessary to have it spin very fast. The disc, made of an alloy of zinc and lead, is a decimetre in diameter and weighs five pounds. The brass axle ends beneath in a nicely rounded point of unhardened steel. The cylindrical part of the axle is made rough so that the cord will not slip on it. To set the top in motion, the cord is wound round the axle, and the latter is inserted in the notches in the iron arms *dd*, and a plate placed underneath it. The operator seizes the cord with his right hand and pulls it hard, at the same time holding the lever *e* with the

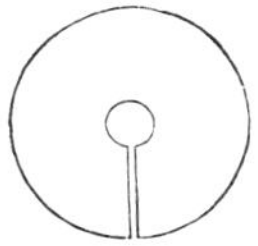

Fig. 46.

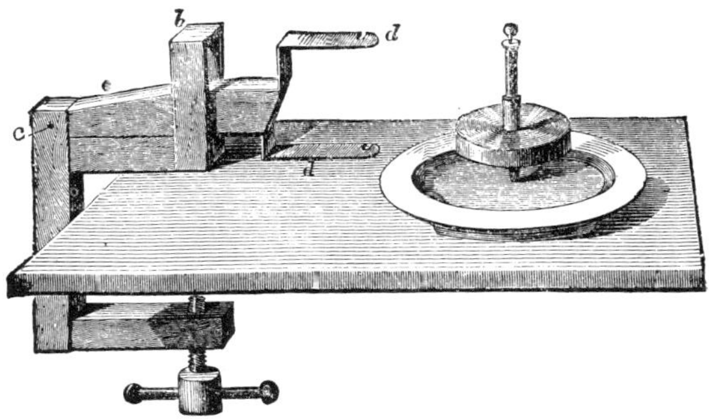

Fig. 47.

other hand. Before starting, the top must be as near the edge of the plate as possible; and the length of the cord should be half a foot shorter than the length of the operator's arm, and it should have a handle at its end. The plate with the top spinning in it is pulled out from under the arms of the lever *e*, whereupon the latter rises, being pivoted at *c*. By pulling the cord quickly, the top can be made to execute as many as 60 revolutions a second, and it will continue in motion for 45 minutes.

Various other kinds of revolving discs have been used also, with their axles in cone bearings, which are operated by clock-work or by an endless chain or by a draw-cord like that used with a top. Generally, there is an inconvenience about this kind of mechanism, because the discs cannot be changed without stopping the apparatus and taking them from their support. On the other hand, there is the advantage of being able to rotate the disc in a vertical plane, so that the phenomena can be exhibited in a large auditorium; and with tops this is not so easy to do. MONTIGNY also used a rotating prism for colour mixing. The spectrum was made to traverse a white screen on which it was projected.

The *thaumatrope* is a little rectangular card, which can be made to revolve around an axis midway between it longer sides. A bird, say, is painted on one side, and a cage on the other. When it is rapidly revolved, the bird is seen sitting in the cage. The device, now used as a toy, was invented by Dr. PARIS.[1] (But see p. 5 of HOPWOOD and FOSTER's *Living Pictures* (1915)).

There are various more complicated devices for observing a rotating disc through slits which are in rotation at the same time. Here may be mentioned STAMPFER's *stroboscopic discs*, which were independently invented by PLATEAU about the same time, and for which the name *phenakistoscope* was used.[2]

Stroboscopic discs are paper discs from six to ten inches in diameter (Fig. 48), on which from eight to twelve figures are arranged in a circle at equal distances apart, representing successive instants of some periodic motion. One such disc is laid concentrically on another disc that is dark and a little bigger. This second disc has as many apertures ranged along the circumference as there are figures on the smaller disc. The two discs are fastened together by a nut that screws on the front end of a small iron axle, which is attached to the upper end of a suitable handle. The instrument is held in front of a mirror with the figures towards it, and the eye is adjusted to see the image of the figures in the mirror through one of the holes on the edge of the larger disc. The discs are then set in rotation, and the

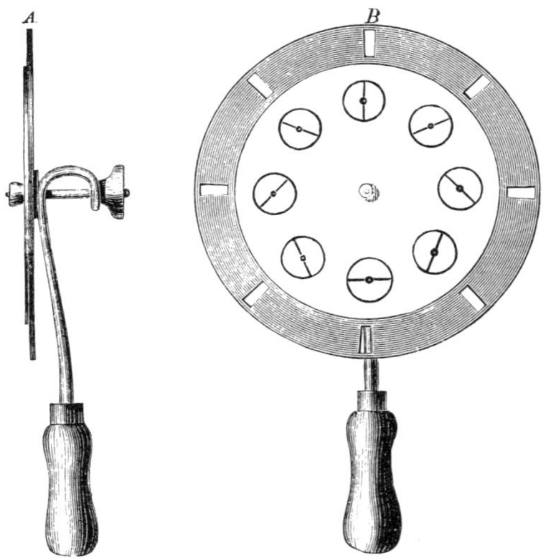

Fig. 48.

[1] *Edinb. Journal of Science.* VII. 87.—POGGENDORFFS *Ann.* X. 480.

[2] As early as November 1830, PLATEAU sent a model of it to FARADAY by QUETELET. STAMPFER completed his first one in December 1832. PLATEAU described his invention in a communication dated January 20, 1833, in the *Correspondance math. et physique de l'observat. de Bruxelles*. VII, 365; STAMPFER in a special paper, *Die stroboskopischen Scheiben oder optischen Zauberscheiben, deren Theorie und wissenschaftliche Anwendung*, the preface of which is dated July 1833.

figures seen in the mirror appear to execute the movement, which they are intended to represent, although each of them stays at the same place on the disc.

If the apertures are designated by numbers, so that the eye looks first through number one, then, as the disc turns, through number two, etc.; and if also the diagrams along the radii drawn to these apertures are designated by the same numbers; then when the spectator looks through aperture No. 1, he will see in the mirror the image of diagram No. 1 opposite the image of his eye. As the disc turns, aperture No. 1 goes past his eye, and for the time being the image in the mirror is completely hidden by the dark disc, until aperture No. 2 arrives in front of his eye, and then he sees an image again. But now diagram No. 2 is there instead of diagram No. 1, that is, on the radius drawn to the eye. Then there is darkness again until aperture No. 3 comes in front of the eye, and diagram No. 3 appears at the same place where the other two diagrams were just before. If all the diagrams were exactly alike, the spectator would get a series of separate but equal visual impressions; and if they were repeated fast enough, they would fuse into a prolonged sensation of an object at rest. But if each diagram is a little different from the preceding one, the separate impressions will be fused also in the image of an object; which, however, seems to be continually changing in the way that the images succeed one another.

When the number of diagrams is not the same as the number of apertures, the diagrams appear to be moving forwards or backwards. Suppose there are n apertures and m diagrams, the numbers n and m, however, not being very different; and that at first one of the diagrams is located on the radius directed towards the observer's eye opposite one of the apertures. When the disc is turned through the arc $\frac{2\pi}{n}$, the next aperture arrives in front of the eye. But the distance of the second diagram from the radius above mentioned is equal to an arc $\left(\frac{2\pi}{n}-\frac{2\pi}{m}\right)$. Now if this arc is small enough so that the second diagram is nearer the place, where the first diagram was first seen, than any other diagram now visible, this second digram now in sight will be identified with the first one that was seen before; and the impression is that the first diagram has been seen to traverse the corresponding arc. Usually, m is taken equal to $(n+1)$ or $(n-1)$. In the former case, the diagrams advance with the motion of the disc, and in the latter case the movement is in the opposite direction.

The narrower the apertures in the larger disc, the less of the images will be seen; but the fainter they will be too. UCHATIUS[1] constructed an apparatus for projecting the images on a screen. A useful application of the stroboscopic disc was made by J. MÜLLER[2] for exhibiting the phenomena of wave motion.

The *daedalion* of W. G. HORNER is a similar affair, except that the holes are on the surface of a hollow cylinder, and the figures partly on the inner surface (preferably transparent), and partly on the base.

In the devices described so far the figures and apertures revolve with the same angular velocity. We obtain another set of phenomena when the angular velocities are different.

One of the simplest contrivances of this kind is the top (Fig. 44) made by J. B. DANCER of Manchester, when a second disc is attached to the upper part of the axle, as in Fig. 49. This disc has holes in it of different shapes; and a piece of string is tied to the edge, as shown in the cut. Owing to the friction

[1] *Sitzungsberichte der k. k. Akad. zu Wien.* X. 432.

[2] POGGENDORFFS *Ann.* LXVII. 271.

on the axle, the upper disc turns with the top, but not so fast on account of the considerable resistance of the air to the thread flying around. If the lower disc contains several sectors of different colours, the patterns cut out of the upper disc appear multiplied and executed in the different colours of the lower disc; a much variegated picture, which appears to move sometimes continuously, sometimes in a jerky fashion.

Consider a single aperture of the upper disc, and reckon its angular distance from the position it has at the beginning of the time of observation. An eye placed in the prolongation of the vertical axis of the top will see one of the colours of the lower disc through the aperture, and this place is the index-point for the measurement of the angle. Suppose the upper and lower discs make m and n revolutions per second, respectively, both in the same direction; then the angular displacement of any point of the upper disc in the time t will be $2\pi mt$, and the corresponding angle for a point on the lower disc will be $2\pi nt$. Accordingly, at the end of this time t a point on the lower disc will be in advance of a point on the upper disc, which was vertically above it at the beginning of this time, by an angle equal to $2\pi(n-m)t$. Hence, the angular distance of the portion of the lower disc that is visible through the given aperture in the upper disc at the time t will be $2\pi(m-n)t$ from the place that was seen at first; the angles being reckoned positive in the sense of the rotation, and negative in the opposite sense. Therefore, at the end of the time $t=\dfrac{1}{n-m}$ all the colours of the lower disc will have appeared once in the hole in the upper disc, and then the same series will begin again and be repeated. However, during this time the hole itself has turned through the angle $2\pi mt=2\pi\dfrac{m}{n-m}$, and the series of colours that succeeded one another in the aperture must appear distributed over this angle, and in fact in the reverse order from what they are on the disc, if $n>m$, as is here supposed. The same series of colours occurs again during each sucessive interval of time in which the aperture turns through an angle equal to $2\pi\dfrac{m}{n-m}$. Now if

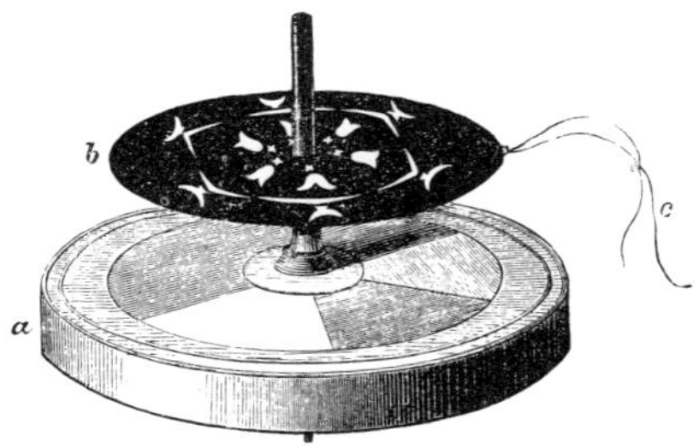

Fig. 49.

$$\frac{m}{n-m}=\frac{1}{p}, \quad \text{that is, if} \quad n=(p+1)m,$$

and if p is a whole number, then in one complete revolution of the upper disc the series of colours in the aperture will have been repeated exactly p times, and with each successive revolution will appear again at exactly the same place as during the first one. What we see on the upper disc in this case is a stationary coloured ring with p repetitions of the colours of the lower disc. If p is not a whole number exactly, the positions of the colours in the second revolution will not be precisely the same as those in the first revolution, and the coloured ring will appear to be gradually turning.

If

$$\frac{m}{n-m}=\frac{2}{2p+1}, \quad \text{that is, if } n=\left(p+\frac{3}{2}\right)m,$$

and if p is a whole number, the colours will have new positions during the second revolution, but during the third the same positions as in the first, and in the fourth the same as in the second. Thus a stationary colour effect may be obtained, provided the top spins fast enough for the impression on the eye to outlast the time it takes the hole to make two revolutions. In this case there are $(2p+1)$ repetitions of the same sequence of colours; but this sequence is not now the same as the sequence of colours on the lower disc, but consists of mixtures of pairs of colours that are opposite each other on the lower disc. For instance, if $p=1$, that is, if $\frac{m}{n-m}=\frac{2}{3}$, the initial colour will appear again for the following angles:

0°, 240°, 480° (or 120°), 720° (or 0°), 960° (or 240°), etc.,

that is, always for 0°, 120° and 240°. On the other hand, the colour on the lower disc on the other half of this same diameter will appear as being midway between the above angles, that is, at 120°, 360° (or 0°), 600° (or 240°), etc.; and hence at the same three places as the other colour so as to be mixed with it.

And, in general, when the fraction $\frac{m}{n-m}$ reduced to its lowest terms is equal to $\frac{q}{p}$, and the impression on the eye outlasts q revolutions of the lower disc, there will be p repetitions of a sequence of colours caused by mixing pairs of colours on the lower disc that are at the distance q apart. But if the impression on the eye does not persist so long, the colours appear to dance to and fro.

By varying the number, shape and size of the holes in the upper disc, of course very variegated kaleidoscopic patterns may be obtained in this way. In the case of this particular contrivance, these pictures are more variegated still, and the patterns are sometimes very delicate, on account of peculiar oscillations that take place in the upper disc. As soon as the upper disc is dropped in its place, the top begins to hum loudly; and if the lower disc is pure white, the pattern on the upper disc does not change into a system of concentric rings, as it would have to do if the upper disc revolved uniformly, but what is seen is a great number of repetitions of the form of the hole. From this it may be inferred that the rate of revolution of the upper disc is retarded and accelerated in regular alternation. These oscillations must be due to the friction of the upper disc against the axle. Moreover, there is another system of oscillations in which the centre of the upper disc moves to and fro horizontally; as is shown by certain peculiarities of the pattern as it appears against a white background.

These phenomena are exhibited in a more orderly way by PLATEAU's *anorthoscope*. Two small pulleys of different diameters, whose axles lie one directly behind the other in the same straight line, are driven by two endless cords, both of which run around the periphery of a larger disc. The latter is turned by a crank. A transparent disc with a distorted diagram on it is fastened to one pulley, and on the other there is a black disc with one or more slits. When the discs are revolved, the drawing appears in its correct form.

Letting m and n denote the number of revolutions per second made by the screen and the pattern, respectively, we saw that all points of the pattern that are at the same distance from the centre will appear in turn on an arc traversed by a point in the slit in the screen whose angular measure is $2\pi\frac{m}{n-m}$. But in the distorted drawing on the transparent disc these points are made to take in the entire periphery. Suppose, therefore, that the points of the original object and its distorted drawing are given by polar coördinates, and that ρ denotes the radius vector drawn from the centre as pole, and ω denotes the angle between it and a fixed radius; and let ρ_0 and ω_0 be the coördinates for the correct figure and ρ_1 and ω_1 for the distorted figure; then

$$\rho_0 = \rho_1$$
$$\omega_0 : \omega_1 = m:(m-n).$$

By means of these equations the distorted figure can be constructed by varying the angles ω in the given ratio. In order that the same figures shall be visible again with every revolution of the disc, the arc $2\pi\frac{m}{m-n}$ must be an aliquot part of the circumference, as before; that is, $\frac{m}{m-n}$ must be a positive or negative whole number.

If both discs turn the same way, that is, if m and n are positive and $n>m$, then ω_0 and ω_1 have opposite signs, and so they must be taken in opposite directions. Now $\frac{m-n}{m}=1-\frac{n}{m}$ will be a negative whole number, provided $\frac{n}{m}=p$ is a whole number; that is, if the transparent disc makes p complete revolutions, while the dark disc makes one. The image is repeated $(p-1)$ times on the circumference of the disc. In this case the dark disc can have p equidistant radial slits.

When the two discs turn in opposite directions, that is, $m=-\mu$, then

$$\omega_0 : \omega_1 = \mu:(n+\mu).$$

The two angles, therefore, are to be taken on the same side. If $\frac{n}{\mu}=p$, and if p is a whole number, the number of images will be $(p+1)$, and the dark disc may again have p slits.

Finally, if the discs turn the same way, that is, if m and n are positive, but $m>n$, then ω_0 and ω_1 again have the same sign, but whereas before ω_1 was equal to or greater than ω_0, now it is less. In the other cases the distorted figure could occupy the whole circumference, and then each separate correct figure took up an aliquot part of the circumference. But in this case evidently the greatest value of ω_0 is 2π, and consequently the greatest value of the other angle is $\omega_1=\left(1-\frac{n}{m}\right)2\pi$. Hence, the distorted figure may be reproduced on the transparent disc several times, and indeed it will be an advantage, because more light will be obtained. In order therefore for this same appearance to recur always, the maximum value of ω_1 must be, as above given, an aliquot part of the circumference, that is, $\frac{m}{m-n}$ must be a whole number, p; and hence

$$\frac{n}{m}=\frac{p-1}{p}.$$

In this case the number of possible repetitions of the distorted figure will be p, and the correct figure will be single. The number of slits may be made equal to $(p-1)$.

However, in this case also a single slit can be used, and by slight variations in the repetitions of the distorted figure to make it represent successive stages of a movement, a correct image can be obtained that appears to execute this motion.

If the required relations between the frequencies m and n are to be exactly maintained, the axles must be driven by gear wheels. With pulleys the ratio between the diameters and the special peculiarities of the cord can never be so nicely adjusted together as to prevent the gradual occurrence of small deviations from the required relations, and then the restored images on the disc will gradually revolve around their central point PLATEAU, by the way, utilized this unavoidable inaccuracy in the procedure of the cord to produce a very gradual change of colour. He made two pulleys as nearly alike as possible; a transparent disc with coloured sectors of equal width was fastened to one of them, and a black disc with one or two equal sectors cut out of it was fastened to the other. If in the beginning the opening stands right in front of one of the coloured sectors of the rear disc, then on rotation the entire field will appear in this colour. However, little by little the discs will be displaced relatively to each other; at first a little bit of another sector of the coloured disc will be exposed, and by and by more and more, the colour of this sector continually being added to the mixture, while that of the first disappears to the same extent. Thus the colour can be made to change by imperceptible degrees.

Here also should be mentioned certain curves that are seen when two sets of straight and curved rods are moved one behind the other. The first case of this kind to attract attention consisted of certain figures that appear on a carriage wheel when it goes past a row of palings.[1] The simplest illustration is the phenomenon observed by FARADAY. He made two equal toothed wheels revolve rapidly in opposite directions one behind the other, their axes being in the same straight line. Owing to the rapidity of the motion, it was not possible to see the separate teeth of either wheel, but when he observed them so that one row of teeth could be seen through the other, he beheld a stationary wheel with double as many teeth. If we think of the teeth as being bright on a dark ground, then a certain amount of light will be distributed apparently uniformly over the dark ground by the rapidly moving bright teeth of each wheel, and the amount of light due to the two rows of teeth will be doubled at those places where first a tooth of one wheel and then a tooth of the other wheel goes by. But where a tooth of the front wheel hides one in the other wheel, the light from the latter will be removed for a moment, because it cannot reach the observer's eye, and a place like this appears, therefore, only half as highly illuminated as the adjacent places where the two sets of teeth send their light to the eye unmolested. Thus, in the bright effect produced by the two sets of teeth those places look darker where a pair of teeth are superposed during the motion of the wheels. If ω is the angular interval between one tooth and the next, then, supposing the two sets of teeth are superposed to begin with, another superposition will occur when one wheel has moved through an angle $\frac{\omega}{2}$ in one direction, and the other has moved through an equal angle in the opposite direction. Hence, the angular distance apart of the dark bands will be only $\frac{\omega}{2}$, and there will be twice as many of them as there are teeth. One of the wheels can be dispensed with, as was pointed out by BILLET SÉLIS, provided a concave mirror is adjusted behind the other wheel so as to form an inverted image of it that coincides with it. The same method can be used very nicely also for showing how a jet of water decomposes into drops.

EMSMANN observed a similar phenomenon with the familiar model used for explaining the earth's oblateness, which consists of two elastic brass rings, corresponding to two perpendicular meridians of the earth. On being revolved rapidly around an axis that represents the earth's axis, they are bent by the centrifugal forces into an elliptical form. The curved surface which they

[1] ROGET in *Phil. Transact.* 1825. I. 131. POGGENDORFFS *Ann.* V. 93.—PLATEAU, Ibid. XX. 319.—FARADAY, Ibid. XXII. 601.—EMSMANN, Ibid. LXIX. 326.

describe during their rapid motion shines out brightly owing to the light that is reflected by them, and dark lines are seen on this surface at the places where a front piece of the ring hides one behind it. The general principle of these phenomena has been enunciated by PLATEAU. When two illuminated curves traverse the field of view so rapidly as to leave behind them an apparently continuous illumination of the surface, there will be a dark line in this field of light connecting the points where the curves intersect each other if the light from one curve cannot pass through the other.

NEWTON[1] estimated the duration of the light impression as being equal to one second. Subsequently it was measured by various individuals by finding how long the impression lasted when a red-hot coal was revolved in a circle. SENGER[2] found the time to be a half of a second; D'ARCY,[3] eight sixtieths of a second; and CAVALLO,[4] one-tenth of a second. PARROT[5] found that the impression lasted for a shorter time in a bright room than in a dark one. Then followed PLATEAU's subsequent measurements[6] on the different durations of the impressions of different colours; and those of EMSMANN.[7]

MUSSCHENBROEK[8] mentions colour tops without naming any older observer. Special forms of such devices were described by E. G. FISCHER,[9] LÜDICKE,[10] and BUSOLT.[11]

The almost simultaneous invention of the stroboscopic discs by PLATEAU and STAMPFER at the end of the year 1832 has already been mentioned. The construction of the anorthoscope by PLATEAU[12] came in January 1836. The latter also developed the theory of the phenomena in much detail.

1704. I. NEWTON, *Optice.* Quaestio XVI.

1740. SEGNER, *De raritate luminis.* Gottingae 1740.

1760. MUSSCHENBROEK, *Introductio ad philos. natur.* §1820.

1765. D'ARCY, Sur la durée de la sensation de la vue. *Mém. de l'Acad. des Sc.* 1765. p. 450.

1795. T. CAVALLO, *Naturlehre,* translated by TROMMSDORF. III. 132.

1800 A. F. LÜDICKE, Beschreibung eines Schwungrades, die Verwandlung der Regenbogenfarben darzustellen. GILBERTS *Ann.* V. 272.

1810. Idem, Versuche über die Mischung prismatischer Farben. Ibid. XXXIV. 4. Beschreibung eines Chromoskops. Ibid. XXXVI. and LII.

1819. PARROT, *Entretiens sur la Physique.* Dorpat 1819–24. III. 235.

1825. ROGET in *Philosophical Transact.* 1825. I. 131; POGGENDORFFS *Ann.* V. 93. (Curves of wheel-spokes).

1827. E. G. FISCHER, *Lehrbuch der mechanischen Naturlehre.* Berlin II. 267. (Colour tops.) PARIS, Thaumatrop. POGGENDORFFS *Ann.* X. 480; *Edinb. Journ. of Sc.* VII. 87. TH. YOUNG, Optische Erscheinung bei einer schwingenden Saite. POGGENDORFFS *Ann.* X. 470–480.

---

[1] *Optice.* Quaestio XVI.

[2] *De raritate luminis.* Gott. 1740.

[3] *Mém. de Paris.* 1765. p. 450.

[4] *Naturlehre* translation of TROMMSDORF. III. 132.

[5] *Entretiens sur la Physique.* Dorpat 1819–24. III. 235.

[6] POGGENDORFFS *Ann.* XX. 304–324.

[7] POGGENDORFFS *Ann.* XCI. 611.

[8] *Introd. ad philos. natur.* §1820.

[9] *Lehrbuch der mechanischen Naturl.* Berlin 1827. II. 267.

[10] GILBERTS *Ann.* V. 272 and XXXIV. 4.

[11] POGGENDORRFS *Ann.* XXXII. 656.

[12] *Bull. de Brux.* 1836. III. 7. Idem in POGGENDORFFS *Ann.* XX. 319–543. XXXII. 646. XXXVII. 464. LXXVIII. 563. LXXIX. 269. LXXX. 150, 287.

1829. Plateau in Poggendorffs *Ann.* XX. 304–324, 543. (Different duration of colour impression; curves of wheel-spokes.) *Dissert. sur quelques propriétés des impressions produites par la lumière sur l'organe de vue.* Liège 1829.

1831. Faraday, On a peculiar class of optical deceptions. *Journ. of the Roy. Inst.* I.—Poggendorffs *Ann.* XXII. 601. (One toothed wheel seen through another; screw motion.)

1833. Plateau, *Correspond. math. et phys. de l'observat. de Bruxelles.* VII. 365.—Poggendorffs *Ann.* XXII. 647.—*Ann. de chim. et de phys.* LIII. 304.
Stampfer, *Die stroboskopischen Scheiben oder optische Zauberscheiben, deren Theorie und wissensch. Anwendung.* Vienna. Poggendorffs *Ann.* XXIX. 189. XXXII. *Jahrbuch d. polytechn. Inst. zu Wien.* Bd. XVIII.
Busolt, Farbenkreisel. Poggendorffs *Ann.* XXXII. 566

1834. Horner, Dädaleum. Poggendorffs *Ann.* XXXII. 650.—*Phil. Mag.* (3) IV. 36.
Talbot, *Phil. Mag.* Nov. 1834. p. 327, and IV. 113. (Rotating discs used for photometry.)

1835. Plateau in *Bull. de l'Acad. de Bruxelles.* 1835. p. 52.—Poggendorffs *Ann.* XXXV. 457–464. (Measurements of luminosity of intermittent light.)
Dove, *Über Diskontinuität des Leuchtens der Blitze.*
Addams, Optische Täuschung bei Betrachtung eines in Bewegung begriffenen Körpers. Poggendorffs *Ann.* XXXIV. 384; *Phil. Mag.* V. 373.

1836. Plateau, Anorthoskop. *Bull. de l'acad. de Bruxelles.* III. 7 and 364.—Poggendorffs *Ann.* XXXVII. 464.

1845. Emsmann, Optische Täuschung, welche sich an dem Abplattungsmodelle zeigt. Poggendorffs *Ann.* LXIV. 326.
Doppler in *Abhandl. der böhmischen Ges. d. Wiss.* V. Folge. Bd. 3.

1846. Dove, Über die Methoden aus Komplementärfarben Weiss darzustellen, und über die Erscheinungen, welche polarisiertes Licht zeigt, dessen Polarisationsebene gedreht wird. *Berl. Monatsber.* 1846. p. 70.—Poggendorffs *Ann.* LXXI. 97.—*Phil. Mag.* XXX. 465.—*Inst.* No. 712. p. 176.—*Arch. d. sc. ph. et nat.* V. 276.
Idem, Über ein optisches Verfahren die Umdrehungsgeschwindigkeit einer rotierenden Scheibe zu messen. *Berl. Monatsber.* 1847. p. 77.—Poggendorffs *Ann.* LXXI. 112.—*Inst.* No. 712, p. 177.
Müller, Anwendung der strobosk. Scheiben zur Versinnlichung der Wellenlehre. Poggendorffs *Ann.* LXVII. 271.

1849. Plateau, Sur de nouvelles applications curieuses de la persistance des impressions de la rétine. *Bull. de Brux.* XVI. I. 424, 588. II. 30, 254.—*Inst.* XVII. No. 818. p. 277. No. 830. p. 378. XVIII. No. 835 p. 5.—*Phil. Mag.* XXXVI. p. 434, 436.—Poggendorffs *Ann.* LXXVIII. 563. LXXIX. 269. LXXX. 150. 287.—Frorieps *Notizen.* X. 221. 325.

1850. J. Tyndall, Phenomena of water jet. (Illumination by electric sparks.) *Phil. Mag.* (4) I. 105.—Poggendorffs *Ann.* LXXXII. 294.—*Edinb. Journ.* L. 370.—*Inst.* No. 924. p. 303.
H. Buff, Einige Bemerkungen über die Erscheinung der Auflösung des flüssigen Strahls in Tropfen. Liebig *und* Wöhler LXXVIII. 162. (Illumination by intermittent light.)
Billet Sélis, Sur les moyens d'observer la constitution des veines liquides. *Ann. d. chim. et de phys.* (3) XXXI. 326.—Poggendorffs *Ann.* LXXXIII. 597.

1850. W. Swan, On the gradual production of luminous impressions on the eye and other phenomena of vision. Silliman *J.* (2) IX. 443.—*Proc. Edinb. Roy. Soc.* 1849. p. 230.
Stevelly, Attempt to explain the occasional distinct vision of rapidly revolving coloured sectors. Silliman *J.* (2) X. 401.—*Rep. of British Assoc.* 1850. 2. p. 21.
Sinsteden, Eine optische Stelle aus den Alten. Poggendorffs *Ann.* LXXXIV. 448.—*Cosmos.* I. 116.

1852. Montigny, Procédé pour rendre perceptibles et pour compter les vibrations d'une tige élastique. *Bull. de Brux.* XIX. 1. p. 227–250.—*Inst.* 1852. p. 216–220, 268.—Poggendorffs *Ann.* LXXXIX. 102–121.

1853. A. POPPE, Das verbesserte Interferenzoskop. POGGENDORFFS *Ann.* LXXXVIII. 223–230.
F. UCHATIUS, Apparat zur Darstellung beweglicher Bilder an der Wand. *Wiener Ber.* X. 482–485.
W. ROLLMANN, Über eine neue Anwendung der stroboskopischen Scheiben. POGGENDORFFS *Ann.* LXXXIX. 246–250.
J. PLATEAU, Sur le passage de Lucrèce où l'on a vu une description du fantascope. *Arch. d. sc. phys.* XX. 300–302.—*Cosmos.* I. 307–309. (Answer to SINSTEDEN.)
1854. EMSMANN, Über die Dauer des Lichteindrucks. POGGENDORFFS *Ann.* XCI. 611–618.—*Inst.* 1854. p. 276.
1855. LISSAJOUS, Note sur une moyen nouveau de mettre en évidence le mouvement vibratoire des corps. *C. R.* XLI. 93–94.—*Inst.* 1855. p. 245.—*Cosmos.* VII. 81–83.—*Arch. d. sc. phys.* XXX. 159–161.
Idem, Note sur une méthode nouvelle applicable à l'étude des mouvements vibratoires. *C. R.* XLI. 814–817.—*Cosmos.* VII. 608–609.—*Inst.* 1855. p. 402–403.
1856. Idem, Mémoire sur l'étude optique des mouvements vibratoires. *C. R.* XLIII. 973–976; XLIV. 727. XLV. 48–52.—*Inst.* 1856. p. 411. 1857. p. 237.—*Cosmos.* IX. 626–629. XI. 80–83, 110–112, 431–432.—*Ann. d. chim. et de phys.* (3) LI. 147–231.

Supplement by HELMHOLTZ *in the first edition.*

E. BRÜCKE has called attention to the fact that when discs like that represented in Fig. 40 are revolved at a certain rate, the intermediate rings look brighter than the inside and outside ones; and hence that for a certain frequency of alternation of white and black the total impression of light is not only greater than it is with less frequency when each colour comes into view by itself without being disturbed by the other, but is also greater than it is with greater frequency when white and black are fused into a uniform grey. He found that the effect was greatest for 17.5 stimuli per second, and that about twice as many were required to see a perfectly uniform grey.

When he looked at a disc, in which the white sectors were cut out, and which was covered with a red glass disc, the red got lighter for the same frequency that gave the maximum action of light effect. BRÜCKE thought this had something to do with the blending of the positive complementary after-image of red, mentioned in the next chapter (p. 251). Under the same conditions, spectral green becomes yellower, and spectral blue is not altered.

Evidently, we are concerned here with a complex antagonistic action between stimulation and fatigue of the retina. As often as the impression of white begins, at first the stimulation rises for a short time to a maximum, and thereafter dies down again, owing to the gradually increasing fatigue. Using white-black discs, the author has succeeded in developing after-images of the flicker, both with and without the interposition of the red glass. His experience is that the final persistent condition of fatigue is exactly the same for all parts of the disc, and that no trace of any difference can be detected in the after-image between the rings that flicker and those that do not;

notwithstanding the fact that the after-image was so distinct that the edge of the disc could be plainly recognized and the small knob that formed the end of the axle.

On the assumption that after each transit of a black sector this average state of fatigue recurs, as it finally persists in the after-image, it must be the first moments of the impact of white that make the deepest impression. If the effect is then interrupted, when it has reached its maximum, all the sectors of this set produce just this maximum impression. But with a smaller number of more persistent impressions the number of these maxima is smaller, and the duration of the gradually failing impression cannot make up for the diminution of their number. So far as the author is concerned, the flicker impression with a disc of this kind is not such that there is a greater brightness all over the flickering ring, for there are always dark places in the ring as a whole. But the white, where it is visible, appears to be relatively brightest and purest in the flickering rings, and for this reason it makes a comparatively strong impression on the eye. Anyhow the impression of a bright light does not cease all at once, to be succeeded abruptly by darkness.

When a revolving disc is viewed through a red glass, the complementary blue-green of the intrinsic light of the retina (see p. 242) is seen very distinctly on the black sectors just as it looks in the final after-image. On the flickering sectors, where the red appears in its maximum brightness and purity, it is certainly striking to see how in contrast with it the complementary blue-green is likewise more strongly forced on the attention, so that especially in indirect vision, these rings appear downright bluish on the red ground of the disc. But the author must differ from BRÜCKE's description about one matter; because he finds that the red is precisely more saturated and more brilliant in the ring between this flickering blue-green than in the other rings, when attention is directed to it. In this case it is as if two colours of opposite kind were seen superposed at the same place; and in the author's judgment it is the colour that is the most altered, even if it is less luminous, that is, the blue-green, that attracts attention most. Still it must be admitted that this whole theory of the fading of coloured light is complicated by too many unexplained phenomena for anyone at present to be able to elucidate the details.

PLATEAU's experiment mentioned on p. 208 has been repeated by A. FICK; who thinks that he has found small variations from the law there advanced, namely, that the impression of a rotating disc is the same as if the light from each ring were distributed uniformly over the whole ring.

1858. D. BREWSTER, On the duration of luminous impressions of certain points of the retina. *Athen.* 1858. II. 521.
1860. H. W. DOVE, Über einen besonderen Farbenkreisel des Herrn LOHMEIER in Hamburg. *Berl. Monatsber.* 1860. S. 491.
GOODCHILD, Trocheidoskop. DINGLER *J.* CLVII. 181–184.—*Pract. mechan. J.* 1860. April 4. (Colour discs used for contrast phenomena).
1862. F. ZÖLLNER, Über eine neue Art anorthoskopischer Zerrbilder. POGGENDORFFS *Ann.* CXVII. 477–484.
J. J. OPPEL, Vorläufige Notiz über eine eigentümliche Augentäuschung in Bezug auf Rotationsrichtungen. *Jahresber. d. Frankf. Vereins.* 1861–1862. S. 56–57.
D. BREWSTER, On the compensation of impressions moving over the retina. *Rep. of Brit. Assoc.* 1861 (2) p. 29.
1863. A. FICK, Über den zeitlichen Verlauf der Erregung in der Netzhaut. REICHERT und DU BOIS *Archiv.* 1863. S. 739–764.
1864. E. BRÜCKE, Über den Nutzeffekt intermittierender Netzhautreizungen. *Wiener Ber.* XLIX. 21. Jan. 1864.
AUBERT, *Physiologie der Netzhaut.* Breslau. S. 96–103.

§23. Variations of Sensitivity[1]

We have seen that after light acts on the retina, the nervous mechanism of vision continues in the state of excitation for a long time still. The best way to realize this persistence of the impression is to turn the eye to a perfectly dark field after it has been gazing at bright objects. But there is another very noticeable effect after bright light has been acting on any part of the retina. To new light falling on it from outside, this place now responds differently from those other parts of the retina that were not affected previously. In other words, the sensitivity of the visual organ to external stimuli has been altered here by the action of light.

The main thing, therefore, in this chapter will be to find out what sensations arise when a part of the retina that has been previously stimulated by bright light is acted on by other external light. However, it should be stated at once that some of the phenomena that are visible in an apparently dark field will have to be considered here also, because, as a matter of fact, there is no such thing as an absolutely dark visual field. Even when all external light is completely excluded, there still remains a certain feeble stimulation due to endogenous causes as manifested by the *light chaos* or *intrinsic light of the dark visual field* (§17). Now the sensitivity of the retina to these internal stimuli seems to be altered in the same way as its sensitivity to objective light; and hence part of our present subject is concerned with phenomena that occur in the dark visual field after the retina has entirely ceased to be stimulated. In this connection it may be added that in bright surroundings mere closing of the eyelids is not sufficient to

[1] Concerning the long-continued variations of sensitivity produced by adaptation to different intensities of light, see NAGEL's Appendix I, at end of this volume.—N.

exclude all objective light from the visual field, because it is easy to notice the added darkness that comes from squeezing the eyes tighter or from putting the hand in front of them. Indeed, in direct sunlight it is not even sufficient to put the hand in front of the eye, because a perceptible amount of red light comes through it still. Accordingly, what is to be understood here by a completely dark field is that of an absolutely dark room from which all traces of external light are excluded, or the field that exists in a lighted room when the eyes are closed and each of them covered tightly, but without pressure, by the palm of the hand or a dark bandage.

Moreover, the light that acts first on the retina and changes its sensitivity will be called hereafter the *primary light*, as distinguished from that which acts afterwards on the place in the retina that has been changed, and which may be called the *reacting light*;[1] because it is for us, so to speak, a reagent by which we test the sensitivity of the retina.

The diversity of the phenomena of this kind is indeed very great, and although quite a number of notable observers have investigated this field, many parts of it are still uncertain and incomplete. The difficulty about it is that at first every observer has to gain a certain practice in apprehending and judging accurately the phenomena encountered here; and in doing this generally these experiments soon prove to be so trying to the eyes that severe and dangerous ocular and nervous trouble may ensue if they are pursued too long. Hence most observers thus far have been able to establish themselves only a comparatively few facts and to make a few new discoveries. And every future observer who wishes to make experiments of this sort is advised not to do too many on any one day, and to take a long rest from such work the moment he notices slight pains in his eyes or head after making the experiments, or in general in looking at a bright light or vivid colours, or whenever the after-images begin to become more prominent and persistent than they are in the healthy eye.

There is the same distinction between *positive* and *negative after-images* as between positives and negatives in photography. *Positive* images are those in which the bright parts of the object are likewise bright and the dark portions dark. *Negative* images, on the contrary, are those in which the bright parts of the object appear dark and the dark bright.

The phenomena will be described at first simply with respect to luminosity, and without reference to the change of colour that in most instances accompanies change of brightness, being due probably to the

[1] ¶Or the "secondary stimulus." The primary stimulus is "the tuning light" (*das umstimmende Licht*) by which the retina is "tuned" for the "reacting light" (*das reagirende Licht*). (J. P. C. S.)

fact that the duration of the single stages of the phenomenon is different for different colours. In order to watch the normal course of after-images without being disturbed, the retina must first get rid of the after-images of previous luminous impressions; and for this purpose it is usually necessary and sufficient to sit a few minutes with the eyes tightly bandaged, until nothing is any longer to be seen in the dark visual field except the light chaos, whose characteristic patterns (bright clots, so to speak, separated by dark bands scattered like branches and network) we soon learn to recognize. When no traces of the outlines of external objects can be discerned any longer, and when they are not revealed by such exceedingly feeble illumination as may penetrate even through the closed lids, the eye is in condition to receive the impression.

Now if the eyes are directed for a short time towards a bright object, preferably a window, without having to be turned thither for the purpose, and then simply opened and closed, directly afterwards there will be left behind a positive image of the primary bright object, as was stated in the preceding chapter. The less the eyes have turned, the sharper and more distinct this image will be; and, according to the author's experience, it will be brightest provided the illumination of the retina by the primary light has not lasted more than about a third of a second. The experiments described in the previous chapter showed that the intensity of stimulation by light increases during the first moments of its action, but soon reaches its maximum. If the illumination continues for more than a third of a second, the intensity of the after-image, corresponding to the intensity of the residual stimulation of the nervous substance, diminishes rapidly. The probable reason for this will be shown later. Moreover, the greater the intensity of the primary light, the brighter is the positive after-image, and the longer it lasts. It should be noted here that degrees of brightness can often be discerned in the positive after-image that were not noticed in the direct observation, because the brightness was too intense. For instance, when a lamp with a round wick is quickly extinguished, by watching the flame vanish, we can see by the after-image that the flame was brighter at the edges than in the middle; although (see §21) it is hard to see this by looking directly at the flame itself. AUBERT noticed the same thing in the after-image of the electric spark. Viewed directly, the spark appeared to be a faded streak of light, but in the after-image it was sharply delineated. Incidentally, even with very moderately illuminated objects, for example, white paper bright enough for writing or reading, positive after-images can be obtained by the above method that can be recognized for about two seconds; whereas, on the other hand, the bright after-image of the sun often lasts several minutes.

For obtaining really beautiful positive after-images, the following additional rules should be observed. Both before and after they are developed, any movement of the eye or any sudden movement of the body must be carefully avoided, because under such circumstances they invariably vanish for a while. After having been tightly blindfolded for a long enough time, the observer, keeping his hands over them, should direct his eyes towards the object, being careful not to let them move; then he should quickly remove his hands and just as quickly replace them again. But this movement of the hands must be executed gently and lightly, without any strong exertion or shaking of the body. If this procedure has been well practised, the positive after-image can sometimes be seen so sharp and bright, that it gives the impression as if the hands over the eyes were transparent, and the actual objects were visible. There is time enough to notice in these after-images a number of individual circumstances besides, which there was not time to see during the actual observation. The dim surfaces disappear most rapidly, without altering their colour perceptibly. The brighter surfaces persist a longer time, during which their colours pass through bluish hues into a violet-pink, afterwards yellow-red. At the time when the brighter regions pass from blue into violet the design of the after-image often becomes quite indistinct, apparently because by that time the relative loss of light in the bright portions exceeds that in the fainter places, so that the brightness is more evenly distributed all over; and because, anyhow, as we shall see presently in the next chapter, nice discriminations cannot be made except when the state of stimulation of the retina is being changed; and the power of doing so is soon lost when the stimulation has reached a steady stage. Subsequently, the less brilliant objects in the positive after-images become quite dark, and the more brilliant ones, now coloured pink, alone survive for a longer time. The author was much impressed by the after-image of a bright carpet lighted by a sunbeam that came through the window and fell on it. A time came when the design on the carpet was perfectly distinct and everywhere equally bright, and when the spot of sunlight no longer attracted attention. Presently the design disappeared, and then the figure of the brighter spot came out again in pink-red light and remained to the last. Thus, perhaps also for certain degrees of illumination, the features of the image may become very indistinct in whole or in part, and afterwards more distinct; that is, apparently the image can almost disappear and then clear up again.

But by being keenly alert, it will be noticed that the ground of the image is distinctly brighter at the time when the character of it becomes confused than it is afterwards when the brightest places again appear outlined on a wholly black ground. In such cases, therefore, it is not that the luminous impression has vanished and returned, but simply that the contrasts of brightness are temporarily more subdued, and the power of perceiving them has to be restored by new changes of colouring and brightness of the after-image. Moreover, with images that comprised a variety of objects of different degrees of brightness, the author's invariable experience has been that the brighter a particular object was, the longer it took to disappear entirely from the positive image. The after-images which AUBERT got by illuminating the object by electric sparks were probably faint; and yet his experience was that the positive after-images persisted longer when the sparks were feeble than when they were strong.

On the other hand, if there has been any violent movement or pressure or shaking of the eye in uncovering and covering it, the result at first will be a confused light chaos, out of which then the after-image may gradually evolve. Similarly, an after-image already developed will be temporarily or entirely obliterated by pressure, movement, or trembling of the eye, or by external light.

Supposing that the external light had acted only a very short time, and was not excessively brilliant, and that the field of view has been kept entirely free from all traces of external light, usually the positive image will disappear without passing into a negative one.[1] But if while the positive after-image is still there, or just after it has disappeared, the eye is turned towards uniformly illuminated surfaces, or even with closed eyelids is directed to bright surroundings, a negative after-image appears. The stronger the positive after-image, the stronger also must be the reacting light to transform it into a negative one. There is always a certain intensity of the reacting light for which the positive image simply disappears without becoming negative. If the reacting light is more intense than this, a negative image comes out; if it is less intense, the image remains positive and simply becomes less distinct. Incidentally, as the intensity of the reacting light is increased, the distinctness of the after-image also becomes greater; until the intensity exceeds the degree that is best suited for discerning small fractional differences of luminosity, and then the image begins to get less distinct. In this way too after-images of fainter primary light can be obtained by using more intense reacting light, but they have

[1] Concerning the so-called PURKINJE after-image following brief stimulation by light, see pages 383, 444 at end of this volume.—N.

to be carefully watched, because they disappear very quickly. Moreover, after the positive after-image has disappeared the negative one remains visible on bright backgrounds a short time still; but it likewise gradually fades and disappears. In fact, when the field is perfectly dark, the effect may consist in a perceptible decrease of brightness of the intrinsic light of the retina. Usually then this intrinsic light itself in the immediate vicinity of the dark after-image looks a little brighter by contrast with it.

More intensity of the primary light imparts more clearness and longer duration to the negative after-image. Also when the primary source of light is a brilliant object, those parts of it of different objective intensities, but without any difference of luminosity so far as the sensation is concerned, can be differentiated in the after-image. After looking at the setting sun, the author has often noticed that objects which covered a part of the sun's disc could be distinctly recognized in the negative after-image, although on account of irradiation there was no trace of them to be seen in gazing directly at the sun. Even little things like branches and leaves of trees may subsequently become visible in this way. The sensitivity of those parts of the retina where the image of the sun itself was formed is, therefore subsequently more altered than that of the places affected by the blur circles and the diffused light; although there was no distinction between them in the original sensation. This is likewise the reason why after-images of the sun are usually bigger at first than the sun's disc, and become smaller afterwards; because at first there is joined with it an after-image of the blur circles on the outer edge of the sun, but it becomes negative more rapidly and then disappears before that of the centre of the sun, where the sun's full brightness has acted.

The influence of the duration of the primary illumination is not the same for the negative after-image as it is for the positive. That is to say, the intensity of the negative after-image increases with the duration of the illumination and appears only with longer duration to approach asymptotically a certain maximum. By long exposure to powerful illumination a permanent change in fact can be produced in the corresponding place of the retina, as happened to RITTER[1] after gazing steadily at the sun from ten to twenty minutes. To obtain distinct negative after-images, the primary illumination should be allowed, therefore, to continue a considerable time (about 5 or 10 seconds with moderate light). Then the positive after-image will be faint and will disappear quickly; but the negative will be clearer and will last longer. For example, after observing bright clouds through

[1] *Beiträge zur näheren Kenntnis des Galvanismus.* 1805. Bd. II. S. 175–181.

the window for a third of a second, the positive after-image fades away in about 12 seconds, and the negative image on a brighter ground in about 24 seconds. But on looking at the same object between 4 and 8 seconds, it was 8 minutes before the negative after-image disappeared. In this case the field of view was kept entirely dark, except that from time to time a little faint light was admitted through the closed lids, to test whether the after-image was still there. To keep the negative after-image quite sharply outlined, a definite point of the bright object ought to be sharply focused during the illumination. It is still easier in the negative after-image than in the more transitory positive one to recognize details afterwards that had escaped observation in direct vision. If two different points of the object have been fixated one after the other, two partly overlapping images will be noticed afterwards also. Thus, when the sun is in the field of view, and the eye is made to traverse the field rapidly, the entire path traced on the retina by the sun's image may be depicted also in the after-image. When the gaze has been fastened for a moment on single places in the field, there will be more intense round after-images of the sun at these points, and they will remain positive longer; and, when they have become negative, they will be darker and last longer. They are connected with each other by narrower faded bands, which are indeed at first bright also; but they soon become negatively darker, being more faintly outlined where the movement of the eye was more rapid. These bands are narrower than the sun's disc and faded along the edge, because only a chord of the round image of the sun has passed over the part of the retina corresponding to their edge, whereas the middle portion has been traversed by a diameter, and therefore the sunlight has acted longer on the latter.

After-images, whether positive or negative, follow the movements of the eye. Their apparent position in the field of view always corresponds to the place where an object would have to be whose image would fall on the part of the retina acted on by the primary light. Thus, if the yellow spot has been stimulated by bright light, no matter where the eye turns, the after-image will always be at the fixation point; and, when it is very active, it keeps finer objects from being seen. If there is a very vividly outlined after-image right by the fixation point, the spectator is easily tempted to try to fixate it. He turns his eye towards it, but apparently, like the *mouches volantes*, it always flees from the point of fixation toward the edge of the field. But if the spectator fixates a stationary point more to the outside, the after-images also stand still. Their movements depend always simply on that of the eye.

From the phenomena thus far described certain things can be inferred as to the state of the part of the retina and the corresponding part of the nervous mechanism of vision which have been stimulated by the primary light. In the first place we find that the state of stimulation persists there for a long time still after extinction of the primary light, as is indicated by the positive after-images; and, secondly, that the nervous substance in question is less sensitive to new reacting light falling on it than the rest of the retina that was not previously stimulated. Thus, *after light has acted on the eye,* (1) *stimulation keeps right on,* and (2) *the sensitivity to new stimuli is lowered.* That stimulation leaves behind a condition of lowered sensitivity for stimuli, is something that takes place also in the motor nerves and in other sensory nerves. Such a condition is called *fatigue.*[1]

With increased intensity of the reacting light the negative after-images go on getting more distinct, until the intensity reaches the point where small percentage reductions of it can be perceived best; and hence we infer that fatigue of the nervous substance affects the sensation of new light about in the same way as if the objective intensity of this light were reduced by a definite fraction of its amount. There is a lack of sufficient metrical data here, and hence in what follows all that will be attempted will be to indicate the general process of the intensity of the sensation of a fatigued portion of the retina considered as a function of the intensity of the reacting light. As long as the positive image still exists along with the negative, the stimulation of the retina is compounded of the stimulation due to the primary light which still continues and the stimulation due to the reacting light which has been diminished by fatigue. In this sense we can regard the brightness of the after-image as the sum of the brightness of the positive image and the brightness of the reacting light reduced by fatigue. Now if the diminution of the brightness of the reacting light is greater than the brightness of the positive image, the entire brightness of the after-image will be less than the brightness of the reacting light, as it looks to the unfatigued portions of the retina. And so the after-image will become negative. This is what happens when the reacting light is brighter. But when it is less bright, the brightness of the positive image is more than sufficient to make up for the loss by fatigue; and the image is positive.

Let H denote the apparent brightness of the reacting light in the unfatigued parts of the retina, and αH in the part that has been fatigued, where $\alpha < 1$; and let I denote the apparent brightness of the

[1] Concerning more modern investigations as to the sensitivity of the retina resulting from "adaptation" to darkness or to brightness. See NAGEL's Appendix I. A.—N.

positive image; then, according to the above, a must stay fairly constant when H varies. Supposing this to be the case, the brightness of the after-image will be $aH+I$, and that of the background on which it appears will be H. When

$$H = \frac{I}{1 - a}$$

then

$$I + aH = H ,$$

and the after-image, having the same brightness as the background, becomes invisible. When

$$H > \frac{I}{1 - a}$$

then

$$I + aH < H ;$$

that is, the after-image is negative. Finally, when

$$H < \frac{I}{1 - a}$$

the after-image is positive. If I is very small, the apparent brightness of the intrinsic light of the retina may be more than $\frac{I}{1-a}$; then the negative image will appear even in the darkest field. If, finally, the positive image has completely vanished, then H is the brightness of the ground, and aH that of the after-image. If with disappearing fatigue $1-a$ has become very small, a certain medium intensity of the reacting light will be necessary to make the difference appreciable. In the dark field it will not be seen then. At last when $1-a=0$, the after-image vanishes completely.

As to the negative images in the absolutely dark field, the evidence is that they arise by diminution of the intrinsic light of the retina. This intrinsic light, therefore, which we must attribute to the action of internal stimuli on the visual organ, is subject to the effects of fatigue, just as the impressions of external light are. That fatigue of the eye by stimulation affects its sensitivity to other stimuli, may also be shown, by the way, for electrical and mechanical stimulation of the retina. When a negative after-image has been developed, and then an ascending electric current is sent through the eye and optic nerve, the result being the bright bluish illumination of the field, the negative after-image will be reinforced by it; and when an image is in the act of changing from positive to negative, an ascending current will make it negative; and a descending current will make it positive. Thus when the eye is fatigued for light, it reacts more feebly to the electrical stimulus also. If colour sensations in the eye have been aroused by steadily maintained pressure, on releasing the pressure the images that persist still in the dark field can be made negative by letting light penetrate through the closed lids, or by looking at a lighted surface. Thus, fatigue due to stimulus by pressure also makes the eye less sensitive to light stimulus.

In cases of this sort where a fading after-image has been rendered visible for a moment by reacting light, occasionally directly afterwards a faint positive after-image again makes its appearance in the dark field. The inference here is that, while the stimulation in the fatigued area of the retina due to the reacting light is indeed weaker than in the unfatigued parts, it lasts longer. There is an analogy to this, by the way, in motor nerves; the contraction of a fatigued muscle being less powerful indeed but of longer duration than that of one that is not fatigued. This alternation between positive and negative images, which may sometimes occur, with hardly noticeable changes of illumination, from pinching the lids or moving the eyeball under the closed lids, perhaps too in consequence of subjective luminous phenomena produced by sudden pressure on the eyeball, has led some observers, PLATEAU in particular, to assume a spontaneous change in the condition of the nervous mechanism during the continuance of the after-effect. So far as this is concerned, the author shares FECHNER's opinion, that in most cases variation of illumination, movements of the eye or body, etc. are responsible for this change. But, of course, where two antagonistic influences are exactly evenly balanced, the least accessory circumstance may turn the scale one way or the other. It may be recalled that even the respiratory movements affect the intrinsic light of the retina. Sometimes too the images simply disappear, without changing into the opposite kind; indeed, as AUBERT aptly describes it, just like a moist spot drying up on a hot griddle. Incidentally, faint objective images sometimes vanish also in similar fashion when the gaze is riveted on an object, for example, in looking at a landscape by night. The impression the author gets is that it is impossible to compare the intensity of stimulation of different portions of the retina unless the stimulation changes from time to time. In case of objective images this can be accomplished every time by changing the point of fixation; but this cannot be done with subjective images. Reference will be made to the matter again in the theory of contrast. Incidentally, it appears that in trying to rivet the attention on images of this sort and to keep the eye from being distracted at all, there is the greatest feeling of exertion when the images are just fading out. After a while the strain ceases, and then the images come back. As to what internal change is involved here, the author has nothing to suggest.

The following phenomena, which are explained by the above principles, belong here also.

When a bright object, like a piece of white paper, is observed on a grey ground, and then the object is suddenly removed without changing the direction of the eye, a darker after-image of the white paper appears, as in the cases previously described. But if a little piece of black paper is observed on the grey ground, and this is pulled away, then a bright after-image appears. The place on the retina affected by the image of the white paper is more fatigued, that affected by the black image is less fatigued, than the rest of the retina on which the grey ground was imaged. Subsequently, the entire retina is acted on uniformly by the light from the grey ground, and therefore this light acts most strongly on the part of the retina where the black image was first, less strongly on that where the grey was before, and least strongly on that where the white image was. Now the experiment with the piece of black paper is important, because it shows that by prolonged observation of the grey ground the retina is fatigued where this light falls, and hence it becomes continually less and less sensitive to this light. Accordingly, when the black paper is removed, the light of the grey ground falls on an unfatigued part of the retina and produces there just the same impression which was made by the background at first. In the meantime, however, the portions of the retina on which the grey acts become fatigued, and so it looks much darker as compared with the fresh impression made on the unfatigued places of the retina. The difference between this experiment and the previous ones is that here the primary stimulus and the secondary stimulus are identical, both being the light of the grey ground. It shows that *external light of constant intensity acting uninterruptedly on the retina for a longer time arouses a sensation in it that continually gets weaker and weaker.* In fact, particuarly when the light is very weak, the intensity of the stimulation may fall off to such an extent as to be actually imperceptible. When night is approaching, suppose that a person gazes steadfastly at some dimly perceptible object without altering the direction of his eye; it soon vanishes completely, and it is not until he moves his eye away that the object will usually come in sight again as a negative after-image. This phenomenon is very striking at twilight on the sea especially when one tries to scan the horizon, because here the after-images of each part of the horizon are congruent with every other part; and, no matter where the look is specially directed, the after-image of the darker sea falls on sea and that of the brighter one on sky. If the eye is then elevated a little, a brighter streak appears on the lower part of the sky, bordered below by the edge of the sea, again becoming visible, and above by a line running parallel to this which goes through the new point of fixation. This streak is the negative after-image of the sea projected on the sky. If the eye is lowered instead of being elevated, a black streak appears, the negative after-image of the sky on the sea, bordered above by the horizon, and below by a line parallel to it. Thus, the horizon can be made visible by indirect vision, but it invariably vanishes on trying to look right at it.

Similar phenomena occur also in looking at a white or black square on a grey ground and then varying the point of fixation a little. Then the after-image of the paper does not completely cover the paper itself, and the brightness of the edges is changed. The portions of the after-image of the white paper that fall on the grey ground seem to be darker; and the portions of the after-image of the grey ground that overlap the white paper seem to be brighter. With black paper it is just the reverse. If the gaze has been riveted a long time on a certain point of the paper, and then if it is suddenly turned towards another point near by, the edges of the after-image will also come out distinctly, and the real nature of the object may be easily recognized. But if the point of fixation has been continually shifted, the after-images will be badly defined, and the bright ground in the region of the white paper seems merely to be a faint dark shade, the edge of the white paper being likewise a bright shade. A similar thing happens in looking a long time at a white square on a dark ground, and then, without changing the point of fixation, suddenly coming nearer the object, so as to increase its apparent size. In this case the edge of the square, except where it is still covered by the after-image of the previously observed pattern, appears to flare up brilliantly. On the other hand, by suddenly moving farther off, after having gazed steadily at the square for some time, it will appear on the dark ground surrounded by a darker border.

As to the peripheral portions of the retina, both PURKINJE and AUBERT noticed that the impression of bright objects disappears more quickly there than it does in the centre. Accordingly, this part of the retina seems to be fatigued much more rapidly. AUBERT found that the negative after-images on the periphery are not so intense as the central ones, but that otherwise they behave similarly. The author's experience shows that they are much more easily overlooked than central after-images, even on bright backgrounds; and that they cannot be readily noticed unless there is a change of the intensity of illumination.

Let us proceed now to the colour phenomena of after-images. After looking at coloured objects and observing the after-images on a wholly dark or white ground of different brightness, the image will be positive or negative depending on circumstances. In the beginning when the positive image is brightest, the colour is the same as that of the object; and the colour of the negative image, by the time it is fully and strongly developed, is complementary to that of the object.[1] The positive image is generally changed into the negative image by the interposition of pale or grey hues of another sort; and in fact, as a rule, the sequence of these colours is the same, no matter whether the transition takes place by gradual decline of the stimulation or by increase of the brightness of the ground.

The best way to develop positive images is by instantaneous action of the primary light. When a series of objects of various colours are seen by instantaneous exposure, the resultant positive after-image reveals the objects at first exactly in their natural colours. In most instances it disappears, a pink-red tinge spreading over it in which the colours that were there originally almost wholly fade out and are succeeded by faint yellowish-grey tones in which the positive image vanishes or is transformed into a dimly outlined negative after-image.

Negative after-images are obtained better by longer fixation at first. The way to see them is to place a bit of coloured paper on a grey ground, and fasten the attention on a special point of the coloured paper, and then suddenly pull the paper aside. A sharply defined negative after-image of the complementary colour will be seen then on the grey ground. Thus the after-image of red is blue-green; that of yellow is blue; that of green is pink-red, and *vice versa*. In general, what has been stated in connection with the after-images of white objects is true also of these after-images so far as duration and intensity are concerned.

Thus an eye which has been acted on by yellow light, say, is thereafter in a condition in which the blue components of white light affect it more than yellow does. Accordingly, the effect of fatiguing the retina is not uniformly extended to every kind of stimulation, but chiefly to stimulation similar to the primary stimulation. This fact is explained very simply by YOUNG's assumption of three different kinds of sensory nerves for the different colours. For since coloured light does not stimulate these three kinds of nerves all to the same

[1] ¶Consequently, we speak of the negative after-image as the "'complementary after-image" and the positive after-image as the "homochromatic after-image." (J. P. C. S.)

extent, different degrees of fatigue must also be the result of different degrees of stimulation. When the eye has been exposed to red, then the red-sensitive nerves are strongly stimulated and much fatigued; whereas the green-sensitive and violet-sensitive nerves are feebly stimulated and not much fatigued. If afterwards white light falls on the eye, the green-sensitive and violet-sensitive nerves will be relatively more affected by it than the red-sensitive nerves; and hence the impression of blue-green, which is complementary to red, will predominate in the sensation.

Corresponding results are obtained in observing negative after-images of coloured objects on coloured background. Invariably it is principally those constituents which were predominant in the colour of the primary object that disappear from the colour of the ground. Thus, a green object on yellow ground gives a red-yellow after-image; and on blue ground, a violet after-image. Considering yellow as composed of red and green and blue as composed of green and violet, and supposing then that the green is diminished in both of them owing to the effect of fatigue, we should expect that the after-image in the yellow will be nearer the red, and in the blue nearer the violet. Generally, the colour of the after-image is always between that of the ground and the colour complementary to that of the object; and in hue, but not in luminosity, may be regarded as being a mixture of the two.

The cases where the colour of the object is the same as that of the ground, or complementary to it, are of special interest. The best way of making observations in the first case is to place a black object on a coloured ground, and after gazing steadily for a while at a point on the edge of it, to remove it suddenly. Under these circumstances the portion of the ground that is seen near the black is to be regarded as the primary coloured object, and the entire coloured ground, after the black object has been removed, as the reacting light (or secondary stimulus). In this case a bright after-image of the black object will be obtained, in which the colour of the ground is not merely more intense but is also more saturated than in the region outside, so that as compared with the latter it seems to be mixed with a great deal of grey. With careful scrutiny, the development of dark and grey on the coloured ground can be noticed even before the black object is taken away. The effect is very striking at the latter instant, because then the colour at this place is seen in the way it affects the unfatigued eye at the first moment of beholding it. This development of grey in the ground does not occur simply with colours which are mixed with white, and in which it may be so noticeable that the hue of the ground disappears entirely, but even with homogeneous spectral colours and certain kinds of coloured glass, from which all foreign white light has been most carefully excluded. For instance, suppose a piece of red glass coloured with copper oxide, which transmits only red rays, is placed in front of the eyes, the head and the edge of the glass being enveloped in a dark cloth, so that nothing but red light can get to the eyes. Now if the observer looks through the glass at a white surface, and places a black object in front of the surface, and then suddenly removes it, the contrast between the red-grey ground and the saturated red of the after-image will be very distinctly manifested. The explanation of this phenomenon evidently is that the retina was fatigued for red by looking at the red ground, and hence this part of the retina is less sensitive than the unfatigued parts where the image of the black object was. If the red is also mixed with white, the sensitivity for red is lowered to a relatively greater extent than for the other colours in the white admixture, and hence the colour must become relatively whiter from fatigue of the retina; but as it also becomes less intense at the same time, it looks grey. However, this is not only the case with pale red, but also with perfectly pure red, and here the explanation becomes more doubtful. It might be natural to suppose that it was connected in some way with the "light-mist" of the dark visual field. If the eye is closed and completely darkened while the after-image is being developed, there is visible in the "light-mist" a clearly defined after-image of the ground in complementary colour, that is, blue-green in this particular case. The internal stimuli which arouse the sensation of the "light-mist" elicit simply the sensation of blue-green in the part of the retina that is fatigued for red, just as objective white light would do. Now if this sensation is compounded with that of objective red, then a pale (or grey) red must be the result of it, as is observed in the experiment.

Nevertheless, the author is not satisfied with this explanation; for the apparent intensity of the "light-mist" in front of the closed eye is certainly very slight. It is hard indeed to say exactly what its value is. However, the way the red turns grey can be observed even in very bright pure red light, for example, when white clouds illuminated by sunlight are viewed through red glass. YOUNG's hypothesis would afford the explanation in this case. As was explained above, we should have to suppose then that although the spectral colours highly stimulated only one or two kinds of nerves, at the same time they excited the others a little. This modification of the hypothesis was necessary to account for the change of hue in pure spectral colours with light of high intensity and for the results of mixing spectral colours. The same assumption is evidently adapted to explain the present phenomenon. If while pure red light stimulates the red-sensitive nerves excessively, it also excites the others a little; and if, owing to the strong stimulation, the sensitivity of the former fibres is more rapidly lowered than that of the latter, the colour impression must approach pale or grey red.

When the primary colour is complementary to the reacting colour of the ground, the latter colour comes out more saturated in the extension of the after-image than it does on the parts of the retina that have not been fatigued or on the parts that have been fatigued by the colour of the ground. If a blue-green object is placed on a red ground, and then removed after having been steadily fixated, a saturated red after-image appears, just as if a black object had been removed. But it is easy to realize that the colour in the after-image of a complementary object is even more saturated than in the after-image of a black object. The simplest way of doing it is to prepare an object which is part black and part coloured, for example, blue-green. Now place it on a complementary (red) ground and look at a point on the ground close to the border between black and blue-green; and then take the object away, and all over the after-image the colour of the ground will come out clearer than it was previously in the part of the ground that was uncovered. The after-image of the blue-green is somewhat darker than that of the black, but it is not as if the red were less luminous there; it is more as if the red in the after-image of the black were overspread by a pale mist, which does not interfere with the red in the after-image of the blue-green. Thus the after-image of red on red looks grey-red; of black on red white-red; and of blue-green on red, saturated red. These distinctions come out very plainly when the experiment is performed with all three shades side by side.

On the supposition that there is some white in the red of the ground, the result is easily explained. The eye is not at all fatigued by black. Its sensitiveness to the whitish red in the after-image of the ground is unaltered. The eye is fatigued for red by red. Its sensitiveness to red in the after-image is lowered, but its sensitiveness to the other constituents of white is practically undiminished, the sensation being that of faint whitish red (grey-red). On the other hand, blue-green renders the eye less sensitive to the non-red constituents of the ground light, and therefore leaves the red to stand out in the after-image with less foreign admixtures.

Similar experiments are just as successful with pure spectral colours. In the field of a telescope the author has projected separate portions of the spectrum, taking every possible precaution to get rid of the last traces of white light. The ground was so densely black that the outline of the diaphragm of the telescope could not even be distinguished on it, and all that could be seen were the cloudy patterns of the "light mist" of the eye itself. The eye was not exposed to any other light except that of a small portion of the spectrum. Upon this coloured field after-images of complementary spectral colours were projected. This was accomplished as follows. A little adjustable steel mirror was placed in front of the ocular at an angle of 45°. In it was reflected part of another very bright spectrum, suitably limited by a circular diaphragm. For this second spectrum such a high degree of purity is not necessary. The arrangements were such that the entire circle was all of one colour. When the little mirror in front of the ocular was drawn aside, the observer no longer saw the circle previously seen by reflection, but he looked through the telescope at the pure spectrum. On it appeared the after-image of the coloured circle. Exactly the same results were obtained here as in similar experiments with pigments. That is to say, the after-image of the complementary colours appeared to be more saturated as compared with the colour of the ground. The latter again seemed to be covered with a whitish "light-mist", which was drawn aside, as it were, where the after-image was, thus letting the colour of the ground come out in its greatest purity. The important conclusion from these experiments was that *the pure spectral colours, which are the most saturated objective colours in existence, still do not elicit in the unfatigued eye the most saturated colour sensation that it is possible to have.* This highest degree of saturation is obtained by making the eye insensitive to the complementary colours.

Here also it might be conjectured that the white sheen that spreads over the ground is the internal "light-mist," its disturbing factors having disappeared in the after-image. As a matter of fact, when the eye is directed on the dark ground beside the spectrum, a complementary after-image is seen. This explanation likewise seems unsatisfactory to the author because the phenomenon can be seen on very bright spectral colours as compared with which the apparent brightness of the "light-mist" is presumably utterly negligible. On YOUNG's theory, on the other hand, we would have here the pure colour sensations of the separate kinds of nerves, in contrast with which the spectral colours must always look somewhat paler, because, by the necessary modifications of that theory, each single kind of homogeneous light must stimulate more than just one single kind of nerve fibre by itself.

All these experiments on after-images of coloured objects on coloured ground may also be modified in various ways exactly as was done in the case of white objects; for instance, the point of fixation can be changed, or the object can be brought nearer the eye or moved farther away. Thus, suppose we have a blue disc on a yellow ground, and that after gazing steadfastly at a point of the disc, the point of fixation is altered; then the after-image of the blue disc will fall partly on the ground and partly on the disc; and the same way with the after-image of the ground. Where the after-image of the disc falls on the ground, the yellow looks more saturated; and likewise where the after-image of the ground falls on the disc, the blue looks more saturated. But where the after-image of the disc falls on the disc, and the after-image of the ground falls on the ground, the blue and yellow seem to be mixed with grey. The result of the other variations of these experiments can be understood without special explanation. Contrast phenomena are sometimes involved also. If a little bit of white paper on a red ground is fixated, and if the after-image is then projected on white, the after-image of the red ground will be blue-green, and that of the small white field will be red by contrast with the green, as will be shown in the next chapter. The best way to perform the experiment is to lay the coloured paper on a white sheet, and then holding the bit of white paper on the coloured paper with a forceps, to pull the coloured paper away. There is also a faint contrast colouring of this sort around the after-image of a coloured square on a white ground.

However, coloured after-images, usually with very diverse variations of colours, may be obtained also from white objects, not simply from coloured ones. These phenomena are commonly known as the *chromatic fading* of after-images. The sequence of the colours is different here, according to the duration and the intensity of the primary impression. The colour sequence after instantaneous observation the author finds is like that described by FECHNER[1] and SÉGUIN.[2] The original white passes quickly through greenish blue (green, SÉGUIN) into a beautiful indigo-blue, later into violet or pink-red. These colours are bright and clear. Then follows a dingy or grey orange, during which the positive after-image nearly always changes into a negative one, and in the negative image this orange often becomes a dingy yellow-green. After very brief action of the primary stimulus orange is usually the last colour, and the image fades out before it becomes negative. AUBERT also found the same colour sequence after looking at the rather bluish spark of a Leyden jar, except that the orange was not easy to make out on a dark ground; although both it and the subsequent green were very distinct on white. The image is surrounded by a yellow halo, perhaps the negative after-image of bluish light diffused in the eye by irregular refraction.

[1] POGG. *Ann.* L. 220.
[2] *Annales de chimie.* 3. Ser. XLI. 415–416.

The phenomena described so far have to do with the behaviour of the after-image in the perfectly dark field. As to the formation of negative after-images in this case, they seem to be merely nebulously sketched-in in the intrinsic light of the dark field. If, during the existence of an after-image of this sort, reacting light is gradually admitted by slowly withdrawing the hands or a dark cloth that has been over the eyes, it is generally noticed that the after-image in this case passes into the later stages of its colour development, and retreats again when the reacting light becomes fainter once more. For instance, if light is allowed to enter when the image is blue amid perfect darkness, it fades through pink-red into a negative yellow image. On again covering the eye quickly enough the blue reappears. If the image is pink-red amid perfect darkness, dim light makes it yellow-red, etc. After the positive after-image in the dark field has finally gone completely, a grey or green-grey negative after-image on a faintly illuminated ground may still be seen for a while longer; the brighter ground surrounding it, which corresponds to the unfatigued portions of the eye, now being pink-red.

In explanation of these phenomena PLATEAU supposed that the duration of the separate stages of the after-images may be different for different colours; and he tried to demonstrate this also directly by the experiments described in the previous chapter. A complete explanation would involve knowing perfectly not only the deportment of the residual stimulation but also that of the fatigue. Still some conclusions can be drawn from these experiments. Thus, in the perfectly dark visual field the first brightest stages of the phenomenon are fairly independent of the degree of fatigue, because this does not come into consideration until the brightness of the positive after-image can no longer be clearly distinguished from that of the intrinsic light. It is probable, therefore, that the green-blue, blue and pink-red phases are simply due to the residual stimulation, and that in the yellow and green phases, when the negative after-image develops, fatigue is a factor also. Accordingly, it is to be inferred that the residual stimulation for the three colours, red, green and violet, declines in some such way as that indicated in Fig. 50. The abscissae denote time, and the ordinates of the curves represent the intensity of stimulation. The unbroken line

Fig. 50.

corresponds to green, the dotted line to violet, and the broken line to red. The positive after-effect for all colours falls off continuously, the drop being greatest for red at first and afterwards least; the case being just the other way for green. For the intensities of colour sensation here represented, blue-green will predominate during the time from 0 to 1; blue at 1; violet at 2; purple at 3, which gradually changes more into red. However, owing to fatigue a greenish after-image is developed in the pale internal "light mist," and so this factor has to be considered as a matter of fact; and thus the fatigue for green, whose residual stimulation has disappeared most rapidly, appears finally to be the least. This green negative image, mixed with positive red, will give a yellow, which may appear brighter or darker than the ground according to the greater intensity of one colour or the other, and at last will change over to green, when the red also is extinguished. In PLATEAU's experiments on the duration of colour impressions, the same law of decline was obtained; those impressions which fall off most rapidly at first persisting longest ultimately in faint traces. The series of colour phenomena is quite different when fatigue has become more pronounced, as it is after prolonged action of white light or after the action of very intense light. FECHNER's observations indicate that with prolonged exposure to the stimulus of white the influence of fatigue is manifested by the white's becoming coloured even while it is still visible. Having kept his eyes shut a long time so as to get rid of the effects of previous impressions, he directed them at a white field on black paper illuminated by sunlight. During the first moments there was a sort of blinding, and he could not be certain about the presence or absence of colour. In fact, it seems that colour does not develop until after some time. But soon the paper was coloured distinctly *yellow*, then *blue-grey* or *blue* (without going through any transition stage of *green*, so far as could be detected in numerous experiments), the *red-violet* or *red*. The yellow phase was the shortest. The blue often lasted quite long before it changed to the next. No other colours succeeded red or red-violet, although the experiment was continued until the eye was under much strain. Even in diffused daylight he often perceived the given series of colours, but sometimes more distinctly than at others. The last two colourations could generally be recognized here better than the yellow. The phenomena were represented by FECHNER by three curves, but

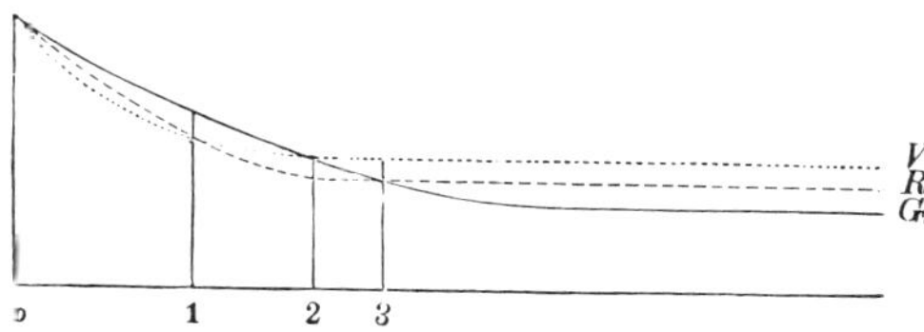

Fig. 51.

with different fundamental colours; as shown in Fig. 51, where again the abscissae are proportional to the time, and the ordinates indicate the intensity of stimulation of the retina in continuing to look at a white surface. The continuous curve represents green, the dotted curve red, and the broken curve violet. In the time from 0 to 1 the colour would be yellow-green; at time 1, pale green; at time 2, pale blue; at time 3, violet, and afterwards pink-red.

After longer and more intense action of primary white light the after-image on a wholly dark field shows the following colour series: *White, blue, green, red, blue*; and on white ground, ultimately *blue-green* and *yellow* besides. In the red the image becomes negative. SÉGUIN includes some intermediate stages in his description. The colours of the first series are for him *white, green, blue*; those of the second (negative) series being *yellow, red, violet, blue, green.* If the action of the white light has exceeded a certain time, this colour series is constant and is not changed any more by longer action. On briefer, but still not just instantaneous, exposure to a primary stimulus of white light that was distinctly tinged with yellow, the colour series was *yellow, blue, red-yellow*; then it became negative green. BRÜCKE gives: *Green, blue, violet, red*; after which the image is negative without distinct colour. Thus the blue phase seems to be always the first variation of the primary light impression; then follows a pink-red, red-yellow to green positive phase, depending on the duration of the primary impression.[1]

In these coloured after-images also, as was the rule with the others, the later phases of the after-image are produced by making the background whiter; whereas when the reacting light is reduced in intensity, the after-image may return through its earlier phases. Whenever the author has observed after-images of uniformly lighted surfaces, with the eye properly accommodated to see them, the alterations of colour were either simultaneously visible over the whole area of the after-image or they could be seen advancing from one side or the other, perhaps not very regularly. But after looking at the sun or some brilliant object of that sort, usually the changes of colour of the image proceed from the edge towards the centre. Irregularities of refraction, which, in the case of very bright objects, always scatter a great deal of light in the vicinity of the image, have something to do with these phenomena. But besides this, the eye is so blinded by the light that it is almost impossible to hold the accommodation steady and keep the

[1] For me and other deuteranopes (says NAGEL here) a deep blue after-image appears on looking at a dark ground after the action of bright white light. While this after-image lasts, all red, yellow and yellow-green colours (homogeneous lights as well as pigments) shade off to white or grey.—N.

eye fixed. The result is that the place on the retina where the image is formed of the central part of the sun is subjected more constantly and more intensely to the action of light than the part near the edge of the image. The sun itself is surrounded by the afterglow of the light diffusely scattered in the atmosphere and in the eye itself. When the eye has been in the dark and is then suddenly exposed to the sun for a moment, the contour of the sun's body can hardly be discerned in the dazzlingly luminous surface. Thus in these cases there is always a gradually diminishing action of light from the centre towards the edge, and the result is a correspondingly different deportment of the separate phases of the after-image. The more intense the action, the more slowly do the individual phases proceed on the whole; and hence at the edge of the after-image generally the earlier stages are seen gradually advancing towards the centre. Besides, owing to the slighter degree of fatigue, the sequence of colours out towards the edge is usually somewhat different from that in the centre. In keeping with this explanation, the after-image in its first stages has a greater extent than the apparent size of the sun, and it is natural to mistake the entire after-image for the image of the sun's disc alone, and to suppose that the different coloured rings developed there belong to it, when in fact they correspond to its halo. The best way of getting a well developed after-image of the sun is to take a very dark coloured glass (or a smoked glass or a pile of several glasses of complementary colours) and look at the sun through it. Then it will appear simply as a dimly visible disc of light. Now remove the glass for a moment, and close the eyes immediately. The action on them has been comparatively slight, and they have not had much time to change their adjustment. Meanwhile, however, the after-image develops very brilliantly. And in this case the author can also see a nucleus in the after-image which is uniformly coloured all over and is about the size of the apparent disc of the sun. Therefore, the deviations on the border may be attributed to errors of refraction in the eye.

Under these circumstances, in the environs of the sun's image the phases of the after-image due to white objects that have been seen for an instant pass rapidly in review: positive blue, pink-red passing through yellow into negative dark green; while the image of the sun itself in this *first phase* appears as a faded, more or less round *white spot*, which, at about the time when the ground has become pink-red, enters the *second phase* and is colored *bright blue*. The second phase usually passes rapidly into the *third*, by the blue becoming green first at the edge, then at the centre also, while a red-yellow border arises at the edge which is darker than the surrounding field, and on the outer edge of which, still in this same phase, a still darker blue-grey border is traced. If during this phase the eyes are directed to a white field, the positive green passes through violet into the negative blue-red of the following phase.

The *fourth phase* arises by the red of the border spreading itself over the centre of the image. The blue-grey border thereupon becomes broader and darker. The entire after-image now is darker than the surroundings. In contrast with it the latter appears whitish or greenish. This is the last negative green of the image of the sky. The after-images of any window-bars that happen to be in the field appear bright on it. If in this phase the eye is directed towards a white ground, the red passes into green-blue.

Finally, in the *fifth phase* the entire after-image takes on the blue colour of the previous border and disappears in the dark field generally in this stage of the blue, whereas on a white field it appears green-blue.

These are the phases as given by FECHNER. The author would like to add a *sixth phase*, in which there is nothing of the after-image any longer perceptible in the dark field, but perhaps still a yellow or brownish appearance on a white field. Finally, after quite a little time this disappears also. If during this time, and even still later, when the yellow appearance has gone, the eyes are exposed to white and then suddenly closed, a faint positive bluish after-image reappears, but it quickly fades out again. Then if the eyes are opened and exposed to white, the yellow after-image instantly reappears. In the writer's opinion the explanation of this effect is due to the fact already mentioned, that in a fatigued nerve the new stimulation dies out more slowly than in the surrounding unfatigued parts of the retina.

Incidentally, the way these after-images of intense light act does not seem to be essentially different for different persons when they are developed under the same conditions. At any rate the writer's observations in this connection are in accordance with those of FECHNER and SÉGUIN, as far as they go.

In this complicated colour sequence it may be conjectured that both the time needed for the impressions of the individual colours on the retina to disappear and for the perception of the intrinsic light may be altered by the ensuing fatigue; and since we have no exact knowledge of these relations and do not know how the various degrees of the fatigue itself die out in the separate colour sensations, it is impossible to give a complete explanation of the individual stages of this colour fading. The progress of fatigue and its effect on the progress of the stimulation for the separate purer colour impressions would first have to be determined and compared.

The phenomenon is certainly very much simpler in case of the fading of the after-image produced by the impression of saturated colours, but even then it is not entirely without colour changes. The main features of the phenomenon have already been described. First, there is a positive image in the same colour as the primary light, and afterwards a negative complementary image. But when the luminous impressions have been quite vivid, usually the transition from positive to negative does not occur by one image simply fading out and the other then becoming visible; but in this transition stage the colour changes by passing through whitish tones. In case there were no other colour in the field of vision but that of the primary light, the colours of the fading image always still seem to be tolerably saturated; and they have been so described by several observers, because there is nothing to compare them with in the dark visual field. But if at the moment of exposure the primary object displayed various colours of about the same luminosity, the after-images in the transition stage from positive to negative will exhibit much slighter differences of colour than the original colours; being all strongly mixed with the pink-red or yellowish white that appears also in the after-images of instantaneously exposed white objects. The after-image obtained by looking at a prismatic spectrum for a moment is particularly interesting in this connection. After the primary colours in the after-image have been visible for some seconds, and the faint colours at the ends of the spectrum have become entirely dark, the whole after-image changes into a whitish-red streak of the same form as the spectrum, in which colour differences are scarcely any longer indicated; except that the former yellow and orange tend towards blue somewhat, and here at the place where red was formerly its green-blue after-image is formed, being negative by this time. In order to tell where the previous colours had been, the writer made a black mark on the white screen where the spectrum was projected, parallel to the bands of colour which continued visible in the after-image. And thus he realized that the reddish-white after-image corresponds to the whole extent of the spectrum from orange to indigo. The same result is obtained by exposing coloured papers of nearly the same luminosity in sunlight, and then developing an after-image by taking a quick look at them.

The conclusion is that when coloured objects have been exposed to the eye for a moment, first, the predominating colour disappears in the after-image, and so the image becomes like that of a white object, usually the pink-red phase being specially prominent in this case. Then the complementary colour of the negative after-image develops gradually; but it may be visible even before the positive image has changed to negative, and thus it may look brighter than the dark ground. The author believes that the emergence of the complementary colour is due to the fact that by this time the positive image, having become faint and white, coincides with the negative and complementary image arising in the intrinsic light due to the fatigue of the eye. Obviously, such a coincidence between positive white and negative blue-green, as might be obtained by looking at red, may give a greenish-white positive image. These positive complementary images are mentioned by several observers.[1] When they are by themselves in the visual field or together with the primary colours only, the complementary colour appears to be tolerably saturated. But when they can be compared with after-images of other colours, the author's invariable experience has been that the complementary colour seemed to be strongly mixed with white or grey, as long as it was still brighter than the ground. It is in the negative after-image that it first becomes more saturated.

The way these phenomena would be explained by YOUNG's colour theory is thus: Each objective colour, even the most saturated, is subjectively mixed with white. The strong stimulation corresponding to the prevailing colour declines relatively faster than the weak stimulations due to the other colours contained in the white, and hence the resultant colour impression becomes weaker and at the same time approaches white. Then during the dimmer stages of the positive image, the negative image induced by fatigue finally gets the upper hand and exerts an appreciable influence by its colouration.

The fading that occurs after momentary exposure of the eye to single colours proceeds in a somewhat different manner, depending on their connection with the hues of the waning white. It is usually simplest with *green*, because its complementary pink-red is the same as the pink-red of the waning white. This hue is developed therefore in special intensity and beauty. *Greenish* blue passes through blue and violet, and *blue* through violet, into pink-red. In the latter case the succeeding phase of yellow comes out purer and more vivid, because it coincides with the complementary blue. In the case of the colours mentioned first, the green-blue and blue phase of the fading white preceding the pink-red phase may perhaps not be noticeable on account of their similarity to these colours themselves; but it seems to be the case with *yellow*, which passes through greenish white into violet, and with *red*. In the case of red there is more of a violet, subsequently grey-green colour, instead of pink-red. Incidentally, it disappears comparatively soonest of all. It has been mentioned that

[1] PURKINJE, *Zur Physiologie der Sinne*. II. 110. — FECHNER in POGGENDORFFS *Ann.* L. 213. — BRÜCKE, Untersuchungen über subjektive Farben, in the *Denkschr. der Akad. zu Wien*. Bd. III. S. 12. (See also NAGEL's Appendix I. B.—N.)

when there are no other colours in the visual field for comparison, the green stage frequently looks like a saturated green. AUBERT's experiments are also in essential agreement with these observations. He examined the electric spark through coloured glass. The only difference he got was that much adulterated yellow gave him still the yellow stage of the waning white after the violet, before it reached the negative blue. Usually too a corona of light formed, and ran through the stages more rapidly.

After longer or more intense action of primary coloured light, during the transition from the positive homochromatic image to the negative complementary one, some of the phases exhibited by white light are likewise noticeable at this time. In particular, the red border surrounded by the blue-grey one often appears. FECHNER has made experiments of the same sort by looking at the sun through combinations of various coloured media which transmitted only one or two colours of the spectrum. The author can supplement them by some observations of his own on prismatic colours, made by viewing a circular diaphragm illuminated by sunlight which had been transmitted through a prism. When the coloured light is so intense that it looks white or yellow, this effect remains also in the after-image at first, but gradually the characteristic colour is then developed clearly.

FECHNER obtained homogeneous *red* light by looking at the sun partly through a red glass and partly through a thick layer of litmus tincture. By direct observation it appeared yellow on account of its high intensity. The after-image too was yellow at first, the edge being red; and afterwards as the intensity diminished, it got red all over, and at the same time a dark blue-green border came into view. In this experiment when the field is dark, usually no distinct negative image develops. But on white ground the green-blue colour of the border becomes central. The author has noticed the same thing with prismatic red. In these experiments the transition from red to green-blue was made through violet. But after looking at a flame through a red glass for quite a while, the transition usually is made through a positive yellow-green followed by green-blue.

FECHNER obtained homogeneous *yellow* by combining two pale yellow glasses with a green glass and a pale red glass; so that except for a little green nothing but yellow was transmitted. The after-image looked yellow with red edge, the latter being surrounded by a dark blue-green ring. With a simple yellow glass, which transmitted red, yellow, green, and a trace of blue, there followed yellow, green, then blue-grey, with red-black encircling ring. With pure prismatic yellow, the author likewise observed the transition into green and the red-black ring. The green and red occur under the same conditions in the after-image of white. On the other hand, after having exposed his eye to a candle flame between 12 and 60 seconds, PURKINJE[1] saw the following colour series: brilliant white, yellow, red, blue, faint white, black.

FECHNER got a tolerably pure *green* mixed with yellow, by using a green glass with a bright blue glass and two bright yellow glasses. Through this combination the sun looked greenish-white; and so did the after-image, surrounded by a black-red ring. He got green, mixed with very little blue and yellow, with three green glasses and a yellow one. The sun looked almost white, the after-image being a little greenish with bluish-white border, and afterwards bluish-white with black-red ring around it, encircled for a while by a faint lilac sheen. With prismatic green the author got a green after-image bordered by blue; and on a white ground, dark purple bordered by yellow.

With a copper sulphate solution FECHNER got *blue* mixed with green. The sun looked white through it. At first the after-image was also white, and then blue. Then a positive green developed, surrounded by a negative red edge. With prismatic blue the author also got the purple border, but the surrounding area was complementary golden yellow.

With a thick layer of copper sulphate solution and ammonia and a piece of violet glass FECHNER got homogeneous *violet*. The sun appeared bluish-white. And so did the after-image at first; then it got dark violet, with a black-red ring around it, the ground beyond being greenish. The phenomenon disappeared before the dark red became central.

In all these cases where the border of the after-image begins to become negative, there is the same red border that occurs also with the after-images of white, as if the homogeneous colour were mixed with white; the fading phases of which become noticeable at the time when the positive after-effect of the principal colour is evenly balanced with the complementary negative one.

When the intensity of the primary stimulus of white or coloured light is low, or when it is moderate but very short-lived, positive images are left behind that fade out through very pale coloured whitish tones of indefinable hue, which may be changed in the most striking way by contrast; and this change is responsible for some very curious apparent conflicts in the results. When there are many different coloured objects in the visual field, the colour differences fade out in the after-image. The after-images obtained by AUBERT by illuminating coloured objects by the electric spark were apparently also of this sort. Thus, red squares on white were red to him in the after-image. On the

[1] *Beobachtungen und Versuche.* I. 100.

other hand, a broader red strip, cut from the same paper, with white squares on a white ground, gave green. The after-image of blue and yellow strips with black squares on a black ground was always yellow to him. On a white ground both strips gave blue after-images. What these differences are due to, is yet to be ascertained.

Other phenomena of the fading out of colour may be observed with colour tops with black and white sectors, by not letting them revolve so fast as to make a perfectly steady impression on the eye. If the top is made to spin slowly at first, and then gradually faster, while the eye watches it steadily without trying to follow the movement, it will be noticed that the white is coloured; being reddish on the front advancing edge, and bluish on the edge coming on behind. With fainter light the reddish hue tends more to red-yellow and the bluish to violet; but when the light is stronger, the first hue is more pink and the latter more green. With slower rotation the bluish hue at first is spread over a wider part of the white than the reddish. But when the speed is faster, red spreads out all over the white as pink-red, and green-blue extends over into the black sectors; and on the whole, violet appears then to predominate on the disc. With still greater speed, the different sectors can no longer be distinguished apart, and seem then to be sprinkled over with some fine particles, and little sparks flicker back and forth between violet-pink and green-grey. At last, when the speed is still further increased, the sparkling becomes fainter, and the grey mixture of white and black comes out more and more; being sometimes overspread by spots of violet-pink of variable size which are formed like the spots and streaks in watered silk.[1]

These various stages of the phenomenon may be seen side by side very nicely by using a disc with three concentric rings, like that in Fig. 52; in which the inside ring is composed of two, the middle ring of four, and the outside ring of eight equal sections, alternately white and black. When the disc is revolved at a certain speed, the inside field will look white with a greenish colouring, the middle field will be pink-red, and the outside field will exhibit the finely mottled flicker. With higher speed, the inside field shows the pink-red colouring and the middle field the finely mottled flicker, while the outside field gives a grey with violet tinge.

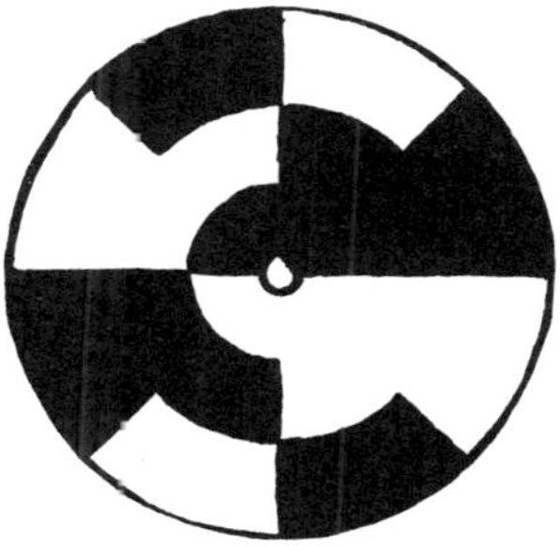

Fig. 52.

[1] See also BIDWELL, *Proc. Roy. Soc.* London LXI, 268–272, and v. KRIES, "Farbeninduktion durch weisses Light" in NAGELS *Handbuch d. Physiol. d. Menschen* Bd. III, S. 245.—N

It may be added that the ring where the pink-red is developed most purely always looks darker than an adjacent ring where the alternation is proceeding more slowly or more rapidly. The order of the colours, as they show up at first on the white areas of a disc of this kind cannot be recognized without some practice. It is easier to do it with a disc composed of two spirals of equal width, one black and the other white, as shown in Fig. 53. What this means is, that when a point on the retina is exposed to rapid alternations of whiteness and blackness, so that the stimulation is increased and lowered in quick succession, *the maximum stimulation does not occur at the same instant for all colours,* but the stimulation for red and violet begins sooner than that for green.[2]

Fig. 53.

These colour phenomena generally do not begin until some time after the observation has been in progress; and then they become gradually always more brilliant. Apparently, therefore, the eye has to be fatigued by the flicker to a certain degree before this result can be obtained. Moreover, other phenomena besides are connected with it which seem to proceed from an unequal sensitivity of different parts of the retina to this sort of stimulation. Thus, in the flickering light certain patterns become visible which are partly connected with definite places on the retina, that is, what is known as PURKINJE's *"shadow figure."*[1] When the revolution of the disc is so rapid that the single sectors can no longer be distinguished apart, the number of sectors appears to be greater, and they form, as it were, a lattice work of faintly outlined curved bars, whose meshes are longest in the direction of the radius of the disc. With increased speed of revolution the design becomes more delicate, like that of a piece of embroidery; and in that part of the flickering field corresponding to the yellow spot there is a peculiar round or oval pattern delineated by sharper contrasts of light and shade, which might possibly be compared with a rose of many petals, that are, however, more hexagonal in form. In its centre there is a dark point surrounded by a bright circle. The same figures can also be produced by turning with closed eyelids towards a bright light and moving the wide-spread fingers to and fro in front of the eye, thereby

[1] ¶On this point see also BIDWELL, *Proc. Roy. Soc.* London, 1896–7. LX, 368. (M.D.)

[2] *Beobachtungen und Versuche zur Physiologie der Sinne.* Bd. I. Prag 1823. S. 10.

exposing and shading it in rapid alternation. The whole point is to produce in some way this rapid change of light and shadow. In these figures PURKINJE makes a distinction between primary and secondary forms. The primary forms in his right eye are larger and smaller squares, changing from dark to bright like a checkerboard, which spread over the greater part of the visual field. It is only downwards from the centre that he sees a row of larger hexagons. Apparently, he observed only isolated features of the rosettes of the yellow spot, which are quite regular so far as the author is concerned. On the other hand, in the author's case the spots outside the centre are neither regular squares nor hexagons, but irregular, and increasing in size towards the periphery. PURKINJE got practically the same effect in his other (weak, left) eye also. The secondary forms were obtained especially when he turned his closed eyelids towards the sun. PURKINJE describes them as being eight-rayed stars and peculiar sharply broken spiral lines, which develop out of the primary patterns by displacement of the bright and dark squares; being, by the way, very changeable. He saw these secondary forms with either eye, merely symmetrically transposed in one eye as compared with the other.

On rotating discs these phenomena tend to disappear more and more as the speed of revolution is increased, the only traces that finally remain being the iridescent spots mentioned above. When the flicker is most vivid, sometimes the entire figure vanishes as you watch it steadily, and a dark red ground comes up behind it, which seems to be intersected by a lot of currents. This is the phenomenon in which VIERORDT[1] thought he recognized circulation. In the author's own case the image of this movement corresponds more to currents where there are no banks, continually changing their bed and shifting back and forth. Of course, it might be supposed that intermittent illumination makes the movement of the blood corpuscles visible, exactly as the movement and forms of drops can be exhibited in a jet of water. However, the author would not venture to give this explanation of the observations he himself has made.

When coloured light is made to alternate with black on the flickering discs, either by using coloured sectors or by looking at the black and white sectors through coloured glass, homogeneous colours show signs of colour fading even under these conditions. For instance, through a red glass that is opaque to everything but red, the advancing edge of the bright fields is orange, and the following edge is pink-red, corresponding to yellow and blue in white light. At the same time the black ground becomes covered with complementary green. The com-

[1] *Archiv für physiol. Heilkunde.* 1856. Heft II.

plementary colour[1] becomes even more distinct when one of the spiral bands on the disc is coloured and the other grey, and the top is stopped suddenly after it has been spinning some little time. This effect is obtained also under similar conditions when the disc consists of alternate white (or grey) and coloured sectors. SINSTEDEN[2] used for the same purpose an orange-red disc, with sectors cut out of it, which rotated over a white shaded one. When he stopped the upper disc, the lower one appeared vivid blue.

E. BRÜCKE also obtained similar results by setting a small black disc in vibration in front of a coloured glass plate. In this case the appearance in front of a green disc was especially curious, because the places, in front of which light and dark alternated, appeared pink-red; whereas the parts that were completely covered or uncovered were green.

The so-called *fluttering heart* is a characteristic phenomenon that probably belongs here.[3] On coloured sheets of stiff paper figures are made in some other vivid colour. Red and blue seem to work best. The colours must be very vivid and saturated. When one of these cards is moved laterally to and fro at a certain rate in front of the eye, the figures themselves seem to shift their positions with respect to the paper and dance about on it. Apparently, the explanation is that the luminous impression in the eye does not come and go with equal rapidity for the different colours, and so the blue apparently lags a little behind the red in the path traversed by the card. Something like it occurs also when the eye moves instead of the object. Thus, WHEATSTONE[4], BRÜCKE and E. DU BOIS-REYMOND,[5] looking over red and green carpet in gaslight, saw the pattern apparently move. BREWSTER states that the phenomenon occurs also when bright daylight enters an otherwise dark room through a small hole.

In this discussion so far the view has been adopted which is espoused especially by FECHNER; that is, that all phenomena of after-images depend partly on a persistent stimulation of the retina and partly on a lowered sensitivity to stimulus. As a matter of fact, if we adhere to our previous conception of stimulation and sensitivity, we must speak of stimulation as continuing to exist when we see a positive after-image in absolute darkness, and we must regard the sensitivity to stimulus as being lowered when the place in the eye

[1] DOVE in POGGENDORFFS *Ann.* LXXV. 526.
[2] Ibid. LXXXIV. 45.
[3] See page 446.
[4] *Inst.* No. 582. S. 75.
[5] *Die Fortschritte in der Physik im Jahre 1845*, reviewed by KARSTEN. I. 223.

where the negative after-image is developed is less sensitive to external light than the unfatigued retina. Thus the persistence of stimulation and the lowering of sensitivity is not an hypothesis but a direct consequence of facts. Moreover, these two circumstances suffice to explain by far the greatest number of the more obvious and constant phenomena in this region, especially the phenomena of the altered intensity of light and of positive homochromatic and negative complementary after-images. To bring the very intricate phenomena of colour waning of intense or persistent luminous impressions entirely within a simple scheme, would probably be a very difficult task indeed at present and would involve all sorts of arbitrary assumptions. However, we can inquire why these phenomena must be so variable. We really know neither the law by which a more or less advanced fatigue of the eye for a single colour disappears, nor how far the after-effect of light depends on fatigue. In FECHNER's opinion negative complementary images in the dark field are to be regarded as new modes of response to the internal stimuli of the retina. On the other hand, many physicists have looked on these images as effects of a new antagonistic activity of the retina. PLATEAU[1] in particular, has elaborated this view into a consistent theory. He demonstrated that the same complementary coloured images may be seen even without any external light at all; and as he was not then cognizant of the intrinsic light of the eye, he knew no other way of explaining the phenomenon except as a new antagonistic activity of the retina. Moreover, having noticed also subsequent variations of the positive and negative images, he advanced the proposition that whenever the retina was strongly stimulated by light, it experienced a series of oscillations, before it came to rest, which involved its passing successsively and alternately through opposite stages. These antagonistic conditions would correspond to the sensation of complementary colours. He connected this with certain contrast phenomena, which will be considered more directly in the next section; and assumed for the spatial extension of the impression a series of such oscillations. On the other hand, it must be remembered that negative complementary after-images do not consist in a lively activity of the retina, but, on the contrary, are rendered visible by being diminutions of the already existing sensation of intrinsic light; and that, moreover, as is almost always discovered on closer attention, those alternations between positive and negative images are dependent on external conditions, especially on slight changes in the illumination of the fundus of the eye. In the author's opinion it is very hazardous to develop a theory on the basis of such delicate and extremely un-

[1] *Ann de Chim et de Phys.* LIII. 386.—POGGENDORFS *Ann.* XXXII. 543.

steady phenomena as after-images are at the time of their conflict between positive and negative in the dark visual field, when the organ from which the light has long been excluded is in a state of greatly increased sensitivity, and when external influences that can be shown to be hardly perceptible are responsible for the transformation of the image. However, it is not surprising that under these circumstances we still do not know always what is the reason for the changes that take place. Incidentally, FECHNER has pointed out another difficulty in PLATEAU's theory. The latter has to assume that in the after-images complementary colours, being antagonistic activities of the retina, offset each other and induce darkness. For example, when there is a complementary after-image, the perception of the primary colour is affected. If the eye has been fatigued by green and red in succession, the after-image will be black. But how can this assertion be reconciled with the fact that sensations evoked simultaneously by objective complementary lights fuse into that of white, which is brighter than either of the two colours by itself?

BRÜCKE regards positive complementary after-images as incompatible with FECHNER's theory. In this connection the author pointed out above that the colouration of these images is in fact very pale; and that it is simply by contrast with the previously seen primary colour, and from not having other colours for comparison, that the complementary colour comes out so vividly. All one has to do is to look at two primary colours in quick succession, in order to see that in the last stages of the positive condition of their after-images there is just a faint suggestion of the complementary colours. And hence the author believes he may venture to consider these images as being a mixture of a positive pale after-image and a negative complementary one, and thus succeed in including these phenomena also in FECHNER's explanation. There is yet to be mentioned a puzzling phenomenon described by AUBERT in the case of after-images of objects which were illuminated by electric sparks. In this instance he saw, with black and red squares on a white ground, negative images apparently produced at the same time with the sparks. But with white squares on a black ground they failed to show up. Sometimes they seemed to be displaced with respect to the original object. They were succeeded by the homochromatic positive images. The after-images of coloured stripes on a white or black ground were always complementary and always brighter than the ground.

In this extremely perplexing region of the most manifold phenomena, the author believes it is best to be guided strictly by a theoretical opinion such as that of FECHNER's which easily explains the great mass of relevant phenomena, and which gives a good explanation especially of all those effects that are characterized by their energy, distinctness and constancy; even if we also find isolated and more transitory phenomena for which at present there is no perfectly satisfactory explanation. This is the case with the colour transformations that occur at the moment when the image changes from positive to negative, and when the antagonistic influences of the persistent stimulation and fatigue are found in a very unstable equilibrium. At present, the author does not know any phenomenon that is positively irreconcilable with FECHNER's principles of explanation.

Positive and negative after-images of windows were described by PEIRESC[1] in 1634. Afterwards the experiment became a kind of magician's trick. BONACURSIUS made a wager with the Jesuit, ATHAN. KIRCHER,[2] that he could make a person see just as well in the dark as in the light; and he won his bet by making KIRCHER look steadily at a drawing fastened in an opening of the window in a dark room. Then the room was made perfectly dark, and KIRCHER plainly saw the drawing again by looking at a piece of white paper held in his hand (which was not necessary). KIRCHER's explanation of it was that the eye sent the light out again that had been absorbed, and so illuminated the paper in front of him. MARIOTTE[3] repeated similar experiments. NEWTON was acquainted with images produced by intensely brilliant light and is said to have regarded them as being of a psychical nature,[4] because by fastening his attention on them, he could continue to elicit the after-images of the sun for quite a while. What led him to try these experiments was an inquiry from LOCKE, who had run across something about them in BOYLE's book *de coloribus* (published in English in 1663). A more complete theory of the phenomena was proposed by JURIN[5] in 1738, which was based on the assumption that as soon as a strong sensation ceased, another one was automatically aroused, which was in some measure a continuation of the original sensation and also something opposite to it. BUFFON[6] published minute descriptions of the phenomena which were afterwards used by Father SCHERFFER[7] as the material for the basis of his theory. As opposed to JURIN, he maintained that after-images (those that he had in mind were almost all negative) were due to lowered sensitivity of the fatigued retina. He made use of the same conception for explaining complementary colours, supporting his notion by NEWTON's rule for colour mixing. Another theory, which is rather artificial in some ways, and which suggests at once PLATEAU's oscillations, was proposed by GODART.[8] A great store of new observations especially with respect to coloured after-images was accumulated by different observers, among whom may be mentioned chiefly: DARWIN,[9] AEPINUS,[10] DE LA HIRE[11]

[1] *Vita.* pp. 175, 296.

[2] *Ars magna.* p. 162.

[3] MARIOTTE, *Oeuvres.* p. 318.

[4] D. BREWSTER, NEWTONS *Leben* (GOLDBERG's German edition) Leipzig 1833. S. 263.

[5] Essay on distinct and ind. vis. p. 170 in SMITH's *Optics*. Cambridge 1738.

[6] *Mém. de Paris.* 1743. p. 215.

[7] *Abhandlung von den zufälligen Farben.* Vienna 1765. — Latin edition, 1761; also in *Journal de Physique de* ROZIER. XXVI. 175 and 273. (1785).*

[8] *Journal de Physique.* 1776. VIII. 1 and 263.

[9] *Philos. Transact.* 1786. LXXVI. 313. — *Zoonomie* translated in German by BRANDIS. Hannover 1795. II. 387.

[10] *Journ. de Phys.* XXVI. 291. — *Novi Comment. Petrop.* X. 286.

[11] In PORTERFIELD *On the eye.* I. 343.

(on the coloured fading out of the sun's image), GERGONNE,[1] BROCKEDON[2] (who also tried to make use of them in a theory of esthetic colour harmony), LEHOT[3] (who directed special attention to the phenomena produced by suddenly varying the distance of the coloured field), GOETHE,[4] BEER[5] (on the way colours vanish when persons who have been operated on for cataract look far off), HIMLY and TROXLER,[6] PURKINJE,[7] OSANN,[8] SPLITTGERBER,[9] KNOCKENHAUER,[10] DOVE[11] (on subjective colours in case of objects in motion), SINSTEDEN,[12] SCORESBY,[13] GROVE[14] (on the reviving of after-images by alternate brightening and darkening of the field), SÉGUIN,[15] BRÜCKE[16] (who made numerous accurate observations on the fading of colours), and AUBERT[17] (on after-images produced by electric sparks).

Among the various efforts to coördinate this mass of material and give a consistent theory of the phenomena may be mentioned the attempt of PRIEUR DE LA CÔTE D'OR[18] to interpret them on the principle of contrast; and likewise BREWSTER's theory[19] that the complementary colour develops at the same time with the colour that the eye actually sees and tends to dim it. All these conflicting views were at last clearly brought out in the two comprehensive works published by PLATEAU[20] and FECHNER.[21] The arguments in favour of opponent activities of the retina were presented by PLATEAU in logical form. With extraordinary self-sacrifice FECHNER carried out an immense series of experiments in this field involving subjective measurements on himself. He gave the first satisfactory explanation of negative images on the principle of fatigue. The present state of the science is still essentially as indicated by these two works. Possibly, the conception of fatigue of the eye for a single colour would require to be more carefully defined. And this is done in YOUNG's colour theory. The author has tested it by experiments on the after-images of spectral colours,[22] the result being that he was particularly impressed by the great distinctness of positive after-images due to instantaneous action of light.

---

[1] *Journ. de Mathemat.* XXI. 291.

[2] *Quart. Journal of Sc.* No. XIV. 399; *Wiener Zeitschr.* VIII. 471.

[3] FECHNER, *Repertorium.* 1832. p. 229.

[4] *Farbenlehre.* I. 13, 20.

[5] *Das Auge oder Versuch das edelste Geschenk des Schöpfers zu erhalten.* S. 1–8.

[6] HIMLY, *Ophthalm. Bibl.* Bd. I. Stück 2. S. 1–20. Bd. II. St. 2. S. 40.

[7] *Beiträge.* I. 72, 96.

[8] POGG. *Ann.* XXXVII. 288.

[9] Ibid. LI. 587.

[10] Ibid. LIII. 346.

[11] Ibid. LXXI. 112. LXXV. 524, 526.

[12] Ibid. LXXXIV. 45.

[13] *Phil. Mag.* (4) VIII. 544. (1854.)

[14] *Phil. Mag.* (4) III. 435–436.

[15] *Ann. de Chimie et de Phys.* Ser. 3. XLI. 413–431.—*C. R.* XXXIII. 642. XXXIV. 767. XXXV. 476.

[16] *Denkschr. der k. k. Akad. zu Wien* III.—POGG. *Ann.* LXXXIV. 418.

[17] MOLESCHOTT, *Untersuchungen zur Naturl.* Bd. V. 279.

[18] *Ann. de Chimie.* LIV. p. 1.

[19] *Phil. Mag.* II. 89. IV. 354. — POGG. *Ann.* XXIX. LXI. 138.

[20] *Ann. de Chimie et de Phys.* 1833. LIII. 386; 1835. LVIII. 337.—POGG. *Ann.* XXXII. 543. Most complete in *Essai d'une Théorie génér. comprenant l'Ensemble des apparences visuelles, qui succèdent à la contemplation des objets colorés.* Bruxelles 1834.

[21] POGG. *Ann.* XLIV. 221, 513; XLV. 227; L. 193, 427.

[32] Presented before the *Niederrheinischen Gesellschaft für Natur und Heilkunde,* in Bonn, July 3, 1858, and in the *Naturforscherversammlung zu Karlsruhe.* Sep. 1858.

1634. PEIRESCII *Vita.* p. 175, 296.
1646. ATHAN. KIRCHER, *Ars magna.* p. 162.
1668. MARIOTTE, *Oeuvres.* p. 318.
1689. DE LA HIRE in PORTERFIELD, *On the eye.* I. 343.
I. NEWTON, Experiments on ocular spectra produced by the action of the sun's light on the retina. *Edinb. Journ. of Sc.* IV. 75; NEWTON'S *Leben* by BREWSTER, German trans. of GOLDBERG. Leipzig 1833. S. 263.
1738. JURIN, *Essay on distinct and indistinct vision.* p. 176. In SMITH'S *Optics.* Cambridge 1738.
1743. BUFFON, Dissertation sur les couleurs accidentelles. *Mém. de Paris.* 1743. p. 147.
1765. SCHERFFER, *Abhandlung von den zufälligen Farben.* Wien 1765; Latin ed. 1761.—*Journ. de Physique de* ROZIER. XXVI. 175 and 273.
AEPINUS de coloribus accidentalibus. *Nov. Com. Acad. Petr.* X. 282.—*Journal de Physique.* 1776. XXVI. 291.
1776. GODART, *Journ. de Physique.* VIII. 1 and 269.
1786. DARWIN, On the ocular spectra of light and colours. *Phil. Trans.* 1786. p. 313.—*Zoonomie,* German trans. by BRANDIS. Hannover 1795. II. 387.
1798. COMPARETTI, *Observationes dioptricae et anatomicae de coloribus apparentibus.* Patav. 1798.
1804. PRIEUR DE LA CÔTE D'OR, Bemerkungen über die Farben und einige besondere Erscheinungen derselben. *Ann. de Chim.* LIV. p. 1.—GILB. *Ann.* XXXI. 315.
1810. v. GOETHE, *Zur Farbenlehre.* I. 13, 20.
1817. SCHULZ, Über physiologische Farbenerscheinungen, insbesondere das phosphorische Augenlicht als Quelle derselben betrachtet. In GOETHE *für Naturwiss.* II. 20, 38.
1819. PURKINJE, *Beiträge zur Physiologie der Sinne.* I. 92.
1826. J. MÜLLER, *Zur vergl. Physiol. des Gesichtsinnes.* Coblenz. p. 401.
1830. LEHOT, *Annales des sciences d'observ. par* SAIGEY *et* RASPAIL. 1830. III, 3.—FRORIEPS *Notizen.* XXVIII. 177.—FECHNER *Repertorium.* 1832. p 229.
GERGONNE in his *Journ. de Mathem.* XXI. 291.
1833. BREWSTER in *Philos. Mag.* II. 89; IV. 354.—POGG. *Ann.* XXIX.
PLATEAU. *Ann. de chim. et de phys.* LIII. 386; LVIII. 337; POGG. *Ann.* XXXII. 543.—Most complete in *Essai d'une Théorie génér. comprenant l'ensemble des apparences visuelles, qui succèdent à la contemplation des objets colorés et de celles, qui accompagnent celle contemplation, c'est à dire la persistance des impressions de la rétine, les couleurs accidentelles, l'irradiation, les effets de la juxtaposition des couleurs, les ombres colorées.* Bruxelles 1834.
1836. OSANN, Über Ergänzungsfarben. POGG. *Ann.* XXXVII. 287.
1838. *G. TH. FECHNER, Über die subjektiven Komplementärfarben. POGG. *Ann.* XLIV. 221–245; 513–530.
Idem, Scheibe zur Ergänzung subjektiver Komplementärfarben. POGG. *Ann.* XLV. 227.
1840. *Idem, Über die subjektiven Nachbilder und Nebenbilder. POGG. *Ann.* L. 193–221, 427–465.
SPLITTGERBER in POGG. *Ann.* XL. 587.
D. BREWSTER in *Phil. Mag.* XXIII. 354.—POGG. *Ann.* LXI. 138.
1841. KNOCHENHAUER, Über Blendungsbilder. Ibid. LIII. 346.
1845. WHEATSTONE, Sur un effet singulier de juxtaposition de certaines couleurs dans des circonstances particulières. *Inst.* 1845. No 582. p. 75.
1848. H. W. DOVE, Über Scheiben zur Darstellung subjektiver Farben. POGG. *Ann.* LXXV. 526.
GRÜEL, Über einen Apparat für subjektive Farbenerscheinungen. POGG. *Ann.* LXXV. 524.
H. TAYLOR, On the apparent motion of the figures in certain patterns of blue and red worsted. *Phil. Mag.* XXXIII. 345.—FRORIEPS *Notizen.* IX. 33.—*Arch. d. sc. ph. et nat.* X. 304.

1850. J. M. SÉGUIN, Sur les couleurs accidentelles. *C. R.* XXXIII. 642. XXXIV. 767–768. XXXV. 476.—*Ann. de chim. et de phys.* (3) XLI. 413–431.—*Phil. Mag.* (4) III. 77.—SILLIMAN *J.* (2) XIII. 441.
SINSTEDEN, Über einen neuen Kreisel zur Darstellung subjektiver Komplementärfarben und eine eigentümliche Erscheinung, welche die Orangefarbe dabei zeigt. POGG. *Ann.* LXXXIV. 45.
E. BRÜCKE, Untersuchungen über subjektive Farben. POGG. *Ann.* LXXXIV. 418.—*Wiener Denkschr.* III. 95—*Arch. d. sc. phys. et nat.* XIX. 122.
1852. W. R. GROVE, On a mode of reviving dormant impressions on the retina. *Phil. Mag.* (4) III. 435–436.—*Inst.* 1852. p. 251–252.—*Arch. d. sc. phys. et nat.* XX. 227–228.—*Cosmos.* I. 237–238.
DOVE in POGG. *Ann.* LXXXV. 402. Explanation of "fluttering hearts."
1854. J. J. OPPEL, Über das Phänomen der flatternden Herzen. *Jahresber. des Frankfurter Vereins* 1853–1854. S. 50-54.—*Hallesche Zeitschr. für Naturwissenschaft.* V. 319.
W. SCORESBY, An inquiry into some of the circumstances and principles which regulate the production of pictures on the retina of the human eye with their measure and endurance, their colours and changes. *Phil. Mag.* (4) VII. 218–221; VIII. 544.—*Inst.* 1854. S. 154–156.—*Proc. of Roy. Soc.* VI. 380–383. VII. 117–122.—*Athen.* 1854. 1272.
1855. S. MARIANINI, Sur une manière de voir facilement les couleurs accidentelles. *Arch. d. sc. phys.* XXX. 235—*Cimento.* I. 165.
1856. SÉGUIN, Couleurs accidentelles. *Cosmos.* IX. 39.
VIERORDT, *Archiv für physiol. Heilk.* 1856. Heft 2.
1857. MELSENS, Recherches sur la persistance des impressions de la retine. *Bull. de Bruxelles.* (2) III. 214–252. *Cl. d. sc.* 1857. pp. 735–777.
1858. H. AUBERT, Über das Verhalten der Nachbilder auf den peripherischen Teilen der Netzhaut. MOLESCHOTTS *Untersuchungen zur Naturlehre.* IV. 215–239.
J. M. SÉGUIN, Notes sur les couleurs accidentelles. *C. R.* XLVII. 198–200.
HELMHOLTZ, Über Nachbilder. 34. *Vers. deutscher Naturf. in Karlsruhe.* S. 225.
H. AUBERT, Über das Verhalten der Nachbilder auf den peripherischen Teilen der Netzhaut, in MOLESCHOTTS *Unters. zur Naturlehre,* IV. 215.
1859. Idem, Über die durch den elektrischen Funken erzeugten Nachbilder. Ibid. V. 279.
H. AUBERT, Über die durch den elektrischen Funken erzeugten Nachbilder. MOLESCHOTTS *Untersuchungen.* V. 296–314.
1861. J. SMITH, On the chromoscope. *Rep. of Brit. Assoc.* 1860 (2), p. 65–66. Ibid. 1861. (2) 33.
1862. AUBERT, Untersuchungen über die Sinnestätigkeiten der Netzhaut. POGG. *Ann.* CXV. 87–116. CXVI. 249–278.
ROSE, Presentations of colour produced under novel conditions. *Rep. of Brit. Assoc.* 1861 (2) p. 33.
1864. AUBERT, *Physiologie der Netzhaut.* Breslau. S. 347-386.
1865. E. BRÜCKE, Über Ergänzungsfarben und Kontrastfarben. *Wiener Sitzungsber.* LI.

§24. Contrast[1]

In the previous chapter the question was, What effect is produced by seeing colours one *after* the other? But what we have to do now is to investigate the mutual influence of different luminosities and colours appearing together in the visual field *side by side* with each other.

The result of such a juxtaposition usually is that each portion of the visual field next a brighter one looks darker, and *vice versa*; and a

[1] Concerning the special relations of colour contrast in case of the so-called anomalous trichromats and certain dichromats, see v. KRIES's Appendix I. — N.

colour alongside another colour resembles more or less the complementary colour of the latter. The opposition thus manifested is implied in the term *contrast.* CHEVREUL draws a distinction between *simultaneous contrast,* as applied to the phenomena belonging here, and *successive contrast,* where two colours appear in succession upon the same retinal area.

However, cases also occur in which the colour of a part of the visual field is so altered by being adjacent to another colour that it becomes similar to the latter itself, and not to its complementary colour; and in these instances the term "contrast" might not seem to apply so directly, although, perhaps, as a matter of fact the alteration of one colour here is by a contrast with the complementary to another colour. So as to include these cases also, BRÜCKE calls the colour that is evoked by the action of one existing adjacent to it in the visual field, the *induced colour*; and the one that is responsible for the appearance of the other, the *inducing colour.* And so when the field, whose colour is altered, is itself coloured, we shall speak of this colour as the *reacting* colour, as formerly. The alteration of the reacting colour by the induced one leads to what may be called the *resulting* colour. In general, therefore, the idea of contrast is not directly appropriate except in the ordinary cases where the induced and inducing colours are complementary. But there are instances where the induced colour is identical with the inducing one.

The phenomena of successive contrast, which will be considered first, are easily comprehended from what has been stated in the previous chapter. After looking at a field of colour A and medium brightness, suppose the eye turns to look at another field of colour B. Then as a rule, the residual stimulation of the impression A will not be strong enough for a positive after-image to be projected on a second field of medium brightness; and so there will be a negative after-image of A upon the field B. Thus those parts of the colour B that are like A will be diluted. If B is of the same hue as A, it becomes whiter by contrast; if it is complementary, it becomes more saturated. If it lies on one side or the other of the colour circle between A and its complementary colour, it changes into an adjacent hue farther from A and nearer the complementary colour. Incidentally, the brighter A was, the darker B looks. Accordingly, this would be the general law of successive contrast, on the supposition that the luminosities of the two fields were such that only negative after-images could occur.

Even in comparing coloured areas with each other that lie side by side in the visual field, successive contrast, that is, contrast caused by after-images, is a very important factor, as any one can easily verify.

It has generally been supposed that in these cases it was simply a matter of simultaneous contrast, because hitherto in the theory of contrast little account has been taken of a certain characteristic of human vision. Under ordinary circumstances, we are accustomed to let our eyes roam slowly about over the visual field continuously, so that the point of fixation glides from one part of the observed object to another. This wandering of the eye occurs involuntarily, and we are so used to it that it requires extraordinary effort and attention to focus the gaze perfectly sharply on a definite point of the visual field even for 10 or 20 seconds. The moment we do it, unusual phenomena immediately take place. Sharply defined negative after-images of the objects develop, which coincide with the objects as long as the gaze is held steady, and hence cause the objects soon to get indistinct. The result is a feeling of not seeing and of having to strain the eyes, if we persist in trying to look at the fixed place; and the impulse to move the eye becomes more and more irresistible. The little deviations of its position are scarcely noticeable in the strain, but they are revealed by parts of the negative after-images flashing up on the edges of the objects, first on one side and then on the other. This wandering of the gaze serves to keep up on all parts of the retina a continual alternation between stronger and weaker stimulation, and between different colours, and is evidently of great significance for the normality and efficiency of the visual mechanism. For nothing affects the eye so much as frequent development of negative after-images caused by staring a long time at surfaces even only moderately illuminated. Strong negative after-images are, indeed, always an indication of a high degree of retinal fatigue.

Now let us consider what happens when the eye wanders in this way over a field where there are different colours or areas of different luminosity. If we observe a limited coloured field with the eye accurately focused on some point of it, a sharply defined after-image will be developed, which is therefore easily recognized. If two different points of the object in the same line of sight have been observed for a long time, two well defined after-images will be formed partly overlapping each other; but without special attention they are not now easily recognized as being copies of the object. But if the gaze has moved slowly over the object, without being held on any point, naturally the after-image will be simply a faded spot, and it is no longer so easy to recognize, although it is actually there for the attentive observer. Now if the look is transferred to an adjacent field of another colour, this colour of course will be altered by the influence of the after-image, exactly as if we had had these different colours one after the other in the field of vision. Accordingly, in a case like this, we do not have simultaneous contrast, at least not by itself; but we have here also successive contrast, and the phenomena are entirely, or in large part, identical with those described in the preceding chapter. In order to have simultaneous contrast alone, special pains must be taken to keep the fixation of the eye absolutely steady during the experiment.

Later we shall examine more carefully the phenomena of pure simultaneous contrast which continue during steady fixation of the eye. Now the phenomena will be described that belong partly to simultaneous, but mainly to successive contrast, as they are manifested under ordinary natural conditions of vision. The colour changes that occur in these circumstances are exactly the same as those already described for pure successive contrast. In general they are much more distinct and striking than those of pure simultaneous contrast; and when the two might cause different results, those of successive contrast invariably predominate in the natural use of the eye; and when both evoke the same effects, the alterations of colour always become much more considerable when the gaze ceases to be steady and begins to wander.

In general, contrast effects are promoted when the inducing colour is more intense than the reacting one, because then the after-images of the former are more vivid and more lasting. For example, if a small wafer of white paper is laid on a coloured sheet, this white will have the complementary colour. The colouring is more impressive, however, when grey is used instead of white; or even black, since in these subjective experiments all black is to be considered as a dark grey. However, as a rule, a medium grey is more satisfactory for the experiment than black. In such cases the contrast action may go so far that a tolerably vivid colour is reversed into the complementary. For example, if a small piece of orange-red paper (coloured with red lead) is laid on a red glass disc and held up against the bright sky, the reddish paper looks a vivid green-blue, that is, complementary to the colour of the red glass, being almost its own complementary colour too.

Moreover, it is conducive to have the inducing colour occupy a large part of the visual field, because then the various regions of the retina will be frequently and continuously stimulated by this colour and fatigued by it. The result is that the contrast colours are particularly vivid when the reacting colour occupies a small field surrounded by an extensive ground filled with the inducing colour. In this case, it is chiefly simply the colour of the small field that is altered, not that of the large field. But the contrast effects are not absent even when the two fields are of the same size; the influence then being a mutual one, and the colour of each being changed by that of the other.

Finally, the nearer together the inducing and reacting areas are in the visual field, the greater will be the contrast effect; because when the eye glides from one space over to the other, the after-image will be more strongly developed the sooner the gaze encounters the other field. This is shown very strikingly in the arrangement which CHEVREUL has selected for his experiments. From each of two colours, say, yellow and red, he cuts out two similar bands and places them side by side close to each other. Let us call them Y_1 and R_1. Then next the yellow band Y_1 he lays a second yellow band Y_2 at a little interval, and in the same way next the red band R_1 another one R_2. In this case the contrast action is not manifested anywhere except at the two middle bands Y_1 and R_1. The yellow of Y_1 becomes greenish by approaching blue-green that is complementary to R_1, and R_1 looks purple by being admixed with some indigo-blue that is complementary to Y_1. On the other hand, the two outside bands Y_2 and R_2 are not altered in appearance, so that there is a good opportunity of recognizing the contrast action. When the fields in contact are somewhat wider, this is also precisely why the contrast colouring is manifested particularly at the margins. Every time the eye sweeps from one field over A into the other field B, those parts of the retina that have just left the field A will be most fatigued by the colour A; and these are the places where the image of the edge of B falls now. Those parts of the retina which left A a little sooner and have already moved farther into the field B will be less fatigued; and hence for them the induced colour is not so strong. Consequently, every time the eye passes over to the field B, the marginal parts of B are most altered by contrast, and the parts farther from the edge less and less in proportion to their distance away. Thus, for instance, when a green and a blue field are in contact, the edge of the green looks a little more yellowish than the middle, and the edge of the blue a little more violet than its middle; because in the first case there is an admixture of yellow that is complementary to blue, and in the latter case an admixture of purple-red that is complementary to green. The play of after-images at the border of such surfaces can be watched very nicely by marking several points of fixation, and jerking the eye from one to the next, after holding it at each place for a brief time. It is easy to see then the well-defined after-images moving over on the other field. The earlier images, being shifted on ahead, will be paler, while the latest, lingering next the border, will be more intense.

If the question involves not difference of colour, but difference of luminosity, the reacting field will appear to be less bright when it is adjacent to an inducing field that is brighter than it; whereas next to a darker field, the luminosity of the reacting field will seem to be increased.

Incidentally, as compared with the methods of seeing negative images which were described in the preceding chapter, there are also other factors in these experiments that are conducive to eliciting the complementary colour. In general, a coloured object has to be deliberately focused for several seconds in order to obtain afterwards a distinct after-image that will persist for some time on a uniformly coloured ground. But in the experiments on contrast it appears that a tolerably cursory observation of one colour is sufficient to induce the complementary colour on the other field, and that this complementary colour is afterwards much more lasting than an after-image would be which was obtained under the same circumstances. In order to recognize an after-image on a uniformly coloured ground, it must be well developed and clearly outlined. It moves about as the eye moves, and so has to be perceived as any other subjective phenomenon. Ordinarily, we pay attention only to objective visual phenomena. But if a faded after-image covers a smaller coloured field, which has its own objective limitation and always appears under the influence of the after-image, this influence cannot be immediately separated in the perception from the other objective phenomena of the visual field, and hence it becomes much more easily an object of our attention. In Part III (Volume III) we shall have to study more closely this peculiarity of the way the attention is attracted.

In addition, the fatigue of the retina in these contrast phenomena is being always renewed, and so the effect is persistent; whereas in most methods of producing after-images it dies out pretty rapidly.

Let us turn now to the *phenomena of pure simultaneous contrast*. In order to recognize them positively as such, care must be taken in the arrangement of the experiments that no after-images can arise, and that the portion of the retina which is to perceive the induced colour has not been previously affected *en passant* by the image of the inducing field. As a rule, this can only be perfectly achieved by not letting the inducing colour be visible until after the eye has been focused on a definite point of the induced field. During the whole time of the experiment this point must be fixed steadily. If the inducing colour is not too intense or too saturated, all that is necessary then is for the eyes, which have been wandering about over dark, slightly coloured objects, or else have been closed, to be turned quickly towards the induced field and focused on a point there without letting them previously linger in the inducing field. In most cases this last method is sufficient, especially because the contrast phenomena of this group are most clearly exhibited precisely when the differences of colour between the inducing and induced fields are slight; whereas, conversely, the phenomena of successive contrast are promoted by strong antagonisms of colours and illumination.

In the author's opinion the phenomena belonging here are of an entirely different kind from those heretofore considered. In general, they may be characterized as cases in which it is not possible to make an exact estimate of the reacting colour by comparing it with other or inducing colours. Under such circumstances we are disposed to regard those differences which are distinctly and positively perceived in the observation as being greater than those which either stand out indistinctly or must be estimated by the aid of the memory. Doubtless, this is a general law in all our perceptions. By the side of a big fellow a man of medium size looks small, because at the moment we see clearly that there are larger men, but not that there are also smaller ones. The same man of medium size placed by the side of a small one will look large.

Now two colours or two luminosities can be compared most accurately when they are side by side in the visual field, with nothing but the difference between them to indicate their boundary. The farther they are apart, the harder it is to compare them. It is harder still when one of the colours has to be supplied by memory. Consequently, when a coloured field (the reacting field) is surrounded by another (the inducing field), the difference between the colour of the reacting field and that of the inducing field will be more distinctly perceived than the difference between that of the reacting field and other colours that are far away. The latter comparison becomes most difficult when the inducing field takes in the entire field of vision or at least most of it; and hence other colours will be perceived only by the peripheral parts of the retina, where colour discrimination is imperfect, or simply by the memory. In general, therefore, in accordance with the rule given above, the difference between the reacting field and the inducing field will appear to be too large as compared with the difference between the reacting field and other colours; and, in fact, the effect will be more decided in proportion as the inducing colour excludes all others from the visual field.

Moreover, we are more liable to err in estimating small differences than big ones; and, consequently, contrast phenomena are also relatively more distinct when the differences of illumination are slight than when they are considerable.

Finally, a difference appears bigger when it is the only thing that differentiates two adjacent surfaces than when it is merely one among several differences; and hence in general simultaneous contrast is more vivid when there is nothing between the induced field and the inducing field except the difference of colour.

Incidentally, there is one other point: the object must not continue to be focused too long. When fixation is long maintained, a series of phenomena occur resulting from fatigue of the eye that partly entail the opposite result from that of the original contrast.

Let us proceed now to the description of individual cases. The so-called *coloured shadows* are most conducive of all to vividness of contrast, because here the three mentioned conditions are generally fulfilled simultaneously. Among all contrast phenomena, therefore, coloured shadows have attracted most attention.

The easiest way to observe them is to illuminate a sheet of paper by weak daylight on one surface and by candle light on the other. Daylight or white light (reflected from a clouded sky or from any white surface lighted by the sun) or even moonlight is admitted through an aperture sufficiently small for the shadows cast by it to be distinct. Then an opaque object of any sort (a finger or a lead pencil) is placed on the paper. Two shadows will be perceived. The one that would be there if the candle were absent may be called the *daylight shadow*; and the one which depends on the presence of the candle, the *candle shadow*. The daylight shadow is illuminated by red-yellow candle light, but not by daylight. It appears in its objective colouration, namely, red-yellow. The candle shadow is illuminated by white daylight, but not by the red-yellow candle light. And thus while it is objectively white, it appears blue or complementary to the colour of the ground, which is a pale red-yellow, since the unshaded portions of the paper are simultaneously lighted by the white daylight and the red-yellow candle light. The colourations are most distinct when the intensities of the two sources are so equalized that both shadows are equally dark.

The blue in the candle shadow becomes more vivid when the eye is allowed to wander frequently over the red-yellow ground, but it also arises wholly without the assistance of after-images. Suppose a point a lying in the blue shadow is noted and marked; and an opaque screen is placed in front of the candle, so that for a while nothing but daylight falls on the paper, until the after-effect of the red-yellow light is completely gone, and the daylight again appears quite white. Now look at the point a, and take the screen from in front of the candle. Immediately the candle shadow becomes blue and stays blue, provided there has not been the slightest deviation in the gaze of the eye. Moreover, the contrast colour immediately appears in the shadow when the eyes are closed and covered for a while and then suddenly opened and turned towards the shadow.

Take a tube painted black inside, and adjust it so that on looking through it the eye sees only places on the paper that lie in the shadow of the candle light. If at first nothing but daylight falls on it, and then while the eye is looking through the tube the candle light is allowed to fall on it too, the observer will see nothing of the places illuminated by the candle light; he does not notice their presence at all, and the appearance of the regions of the paper which he sees through the tube remains unaltered. Incidentally, the objective colour of the paper in the shadow of the candle light is not changed. The reason for noting this fact here is because it was doubted by OSANN.

On the other hand, if the tube is held to the eye and directed so that a part of the field surveyed is illuminated by the red-yellow light of the candle, the shadow from the candle light becomes blue. When the blue has become real intense, let the tube again be pointed so that nothing but this subjective blue is in the visual field. The blue now persists, no matter whether the candle light is allowed to shine on the rest of the paper, or whether the candle is screened, which, of course, so far as the observer is concerned, amounts to the same thing; because under these conditions he is not aware of it at all. The blue colour in such case is so constant that OSANN has concluded from similar experiments that it is objective. This opinion is easily refuted at once by the fact that the blue colour persists even when the candle is extinguished. But at the moment when the black tube is removed from the eye, the subjective blue disappears also, because it is then recognized to be identical with the white that occupies the rest of the visual field. No experiment shows more impressively or more clearly the influence of judgment on our determination of colour. As the result of contrast, whether it be successive or simultaneous, once the judgment has been formed that the colour in the shadow of the candle light is blue, the colour continues to appear blue, although the circumstances that led to the decision may have ceased to exist; until the black tube is removed so as to enable us to make a new comparison with other colours and to form a different judgment in the light of new facts.

Instead of the red-yellow colour natural to candle light, other colours may be used too. The candle light may be coloured by interposing a piece of coloured glass in front of the candle, thus combining coloured candle light either with daylight or with uncoloured candle light. However, the phenomena are most brilliant when the experiment is conducted in a dark room, where coloured sunlight is admitted through an opening in the shutter covered with coloured glass, and white daylight through another small opening. In all these cases, whether the eye is held steady or not, the white light shows up in the colour complementary to the coloured light.

When the eye wanders, the complementary colour appears indeed even on absolutely black surfaces and on surfaces that are dimly illuminated by the prevalent colour. When the eye is kept steady, a dark area sometimes appears in the complementary colour and sometimes in the same colour. It is generally the first way by dim light, and the latter way by bright light. However, after somewhat prolonged fixation, it is always the same colour as that of the prevalent light, the complementary colour flashing up simply at the edges, owing to the unavoidable tiny fluctuations of the visual axis to and fro. As soon as the eye is allowed to wander, the complementary colour invariably comes out, or gets more brilliant in case it were dimly there before.

In fact, the complementary colour comes out when the light is made to go through two pieces of glass of the same colour, one of which, however, is not so highly coloured as the other; or when two pieces of the same kind of glass are used, provided some white light also is incident on one of them. In such cases, therefore, the hue of the paler shadow is thus converted into the opposite hue exactly.

The same contrast phenomena obtained with the coloured shadows invariably occur whenever most of the visual field is occupied by a predominating colour, or when a great part of the field is unilluminated, and there is in the illuminated portion a colour which is predominant in its extent and intensity.

Hold a little piece of white or grey paper in a short pair of nippers, or attach it to a wire and hold it directly in front of one eye; and close the other eye and look at it. Then insert behind it a large sheet of coloured paper or a large plate of coloured glass, so that most of the visual field is occupied by this coloured surface. As soon as this happens, the complementary colour appears on the little piece of paper. As a rule, the reacting white must not be too bright. If the experiment is performed in a room where the light comes from a lamp or from not too large an aperture in a window, the brightness of the white paper may easily be altered by letting the light fall on it more or less perpendicularly until the proper brightness is found. It is best to get a medium brightness of the white, which is about like that of the coloured ground. If the white is too bright or, on the other hand, too darkly shaded so that it begins to look black, the contrast colours will be less distinct or will not appear at all. The more of the visual field that is occupied by the coloured surface, the brighter the white can be made. By increasing the distance between the eye and the objects and thereby diminishing their apparent sizes, the induced colour will be found to get fainter or to disappear entirely. It likewise disappears with sustained fixation and becomes just like the inducing colour; all the more readily, the smaller the apparent size of the inducing field is, the more intensely it is illuminated, and the darker the induced field is. If the latter consists of a small black disc which is placed in front of a plate of coloured glass fastened in an opening of the shutter where the sky can be seen through it, the colour of the glass from the very start will frequently spread over the black disc, provided after-images are avoided. In this case the author's experience is that there is no difference in the various colours except that usually the commercial red glasses are darker than the yellow, green and blue glasses; and therefore more intensity of light, for instance, the light from sunlit clouds, is required for red glass to elicit the same colour at the start. In the case of the blue glasses, which also exhibit the phenomenon with tolerably dark colouring, it might be that the fluorescence of the crystalline lens and cornea also had something to do with the distribution of blue light over the dark disc. After brief fixation, the colour is always the same as that of the glass, and it is only at the edge of the black field that the complementary border appears, being due to the wavering of the visual axis of the eye.

Leaving out at first cases where the induced colour and the inducing colour are the same, we may express the main result of the preceding experiments thus. When a particular colour is made dominant in the visual field, a paler shade of the same hue will look white to us, and real white will seem to be the complementary colour. Thus the idea of what we mean by white is altered in this case. Now the sensation of white is not a simple sensation, but consists of the sensations of three fundamental colours compounded in a definite proportion. In a particular case, in order to recognize a given colour as being white, when it is impossible to compare it with something that is known to be white, we must again be able to recognize whether the relative intensities of the three fundamental colours of which it is composed have been altered or not. But, as we saw in §21, the comparison of the intensity of different sensations of colour is extremely uncertain and inaccurate. And, therefore, any determination of white based on such a comparison must be inaccurate too; and pretty considerable variations will be possible in our estimates of white on different occasions, as is actually found to be the case.

At the same time this explanation shows also why these caprices as to what is white do not ever go to the extent of making us take for white a colour that is saturated like the red in glasses that are stained with copper oxide, which transmit only light from the red end of the spectrum; even when we happen to be a long time in a place that gets all its light through a glass of this sort. As a matter of fact, in comparing a very bright red with a faint blue, there is no doubt about which is brighter. We decide positively about big differences, but not about little ones. If, therefore, homogeneous light is presented to the eye, and in it the sensation of the red fundamental colour is very intense as compared with the sensation of the other two fundamental constituents, the colour is pronounced to be red without deliberation. We do this even when the sensation of red has already been very much enfeebled by fatigue of the eye. It is true that a somewhat pale but still tolerably saturated red may be taken for white under such circumstances, as in the experiment described above, where a piece of paper coloured with red lead looked green in front of a highly illuminated red glass.

There is one other circumstance that keeps us from making too big an error in a case like this. This is the intrinsic light of the retina, which, when the eye has wandered about for some time, appears complementary to the prevailing colour and becomes noticeable in all perfectly dark places in the visual field. When we look steadily through a red glass, soon all perfectly dark objects appear to be a vivid green. Thus, alongside the red its complementary colour becomes visible, and we are thereby compelled to recognize red as red; we cannot confuse it with white. With dominant white illumination the mist looks white in the dark places, and for just this reason it requires careful attention to see it. Even in dim coloured light, for instance, the light of a lamp or candle, the intrinsic light of the retina becomes noticeable in this manner. All that is necessary is to hold a small black object, entirely unilluminated, in front of a white paper surface lighted by the candle, and let the eye wander over it and the paper surface; the indigo-blue sheen on the black, which is complementary to the red-yellow of the candle light, will then soon be perceived. White paper appears white just as well by candle light as by daylight. But if the paper is viewed through a tube blackened on the inside, which has only a small opening, and the appearance of the small part of the paper surface that is still seen is compared with the dark field, it is soon perceived that the former is red-yellow and the latter looks bluish; whereas by daylight there is no such difference. This is a means of recognizing the colour of the prevailing illumination even when there is no daylight for comparison. Consequently, too, the colour of the intrinsic light of the eye matches the white of daylight, and hence this white is of special significance for the eye still and is entitled to the name of white before all other whitish colours.

Of course, the intrinsic light of the eye is too feeble to be used in diffused coloured illumination for making comparisons in order to determine white exactly.

Hence, if there are a limited number of coloured objects in the field of vision, we are in a much better position for determining the relative differences between the various colours present and between each colour and their mean colour than the difference between this mean colour and white. Now, by the normal illumination of daylight, when a large variety of objects can be freely compared, the white of sunlight is the mean colour, from which the deviations of the other colours in the various directions of the colour chart are estimated. But if another colour A is predominant, so that the average of all colours seen at the same time resembles the colour A, we are inclined to use this average as the starting point of our temporary colour discriminations and to identify it with white.

In the author's opinion the characteristic thing about this interpretation of the phenomena is that, when after-images are avoided, a very weak colouration of the dominant light elicits quite as distinct contrast colourations as the most saturated. The weak red-yellow of candle light gives the coloured shadows a very intense blue. The author's experience is that there is no tendency for this blue to become more vivid and more distinct when the eye looks at it steadily, supposing an exceedingly red-yellow paper or red glass is used as background. But as soon as the eye is allowed to wander, the latter saturated colours certainly do give also much more saturated after-images than candle light.

The effect of slight differences is manifested in exceedingly striking fashion in a method devised by H. MEYER.[1] A sheet of nice white letter paper and one of coloured paper, green, say, are cut exactly the same size, and superposed so that one covers the other perfectly. A little piece of grey paper, just as dark as, or darker than, the green is inserted between them. Black or white paper is not so good. The green and grey underneath just manage to show through the outside white paper, and where the grey is now appears a very distinct and decided pink-red. With a different colour of the background the little piece of grey invariably shows through the white in the complementary colour. Frequently conditions are obtained that succeed in bringing out the complementary contrast colour more distinctly than the weak colour of the ground. The author's experience is not simply that in these cases the contrast colour is just as easily seen as if the background were a saturated colour; but that it seems to be easier to see it; for it took much practice and perseverance to succeed with the experiments on contrast colours in which the eye has to stare at a little piece of paper while a coloured sheet is shoved under it.

The two phenomena may be directly compared in the following manner. The sheet of red paper is covered with the translucent white paper, and a little piece of opaque white paper held by a forceps is laid

[1] POGGENDORFFS *Ann.* XCV. 170.

on the latter. The observer then looks steadily at the little piece of paper until it distinctly takes on the complementary colour; but not too long, otherwise the difference of colour is quickly obliterated by the after-images. Then he suddenly pulls the white letter paper away, and now he sees the little piece on the uncovered red paper. The complementary colouring is scarcely stronger than before, unless there was too much delay.

As a matter of fact, according to the above explanation of the vacillations as to what is meant by white, there is a certain limit always to the change which this conception can undergo. This limit has already been reached when the coloured ground is not very much saturated, and then, unless after-images are involved, it does not seem able to extend much farther. On the other hand, the nature of a colour can be determined much more positively by comparing it with a colour that is very much like it than by comparing it with a much more saturated colour. Moreover, two colours are easier to compare when they have the same luminosity than when their luminosities are very different. In the author's opinion this is the reason why contrast colouring is most positive where inducing and reacting colours are equally bright, and their difference is not one of luminosity but simply of colour.

This seems also to be the explanation of the following phenomenon. Holding a little piece of white paper with a forceps over an equally bright white ground, insert a coloured paper in between the two. If the new coloured ground is large enough, the little piece of paper shows up now on it in the complementary colour. Leave the coloured paper where it is for from two to four seconds, and then pull it away again, always taking care to look steadily at a point of the little piece of white paper. At this instant the latter will take on the same colour as that of the first inducing field just as distinctly and definitely as when it assumed the complementary colour before. Indeed, in all such cases, where the coloured ground was not very extensive, the way the same colour as that of the ground comes out will be even more distinct than the way the complementary colour came out before. In fact, when the coloured paper is removed, the white ground has a faint tinge of the complementary colour, and is nearly as bright as the little piece of white paper; and the effect of this is to promote the development of the contrast colour more than it was promoted by the more intense colouring of the coloured paper that was inserted under it at first. It is the same way when the ground and the little piece of paper are both black. In this case also the similar colouration that occurs when the coloured ground is removed is plainer than when it was inserted.

Of course, it is exactly the same way when the little piece of paper is removed along with the coloured ground, and then the after-images of both are projected on a white or black field. In the previous chapter the colour of the after-image of white in this case was explained as being a contrast colour; and now we see the justification for it.

Before leaving the cases of contrast where the induced colour constitutes the greatest part of the visual field, we must still consider the reason why the reacting field occasionally has the same colour as that of the inducing field. There are two conditions when this occurs: first, when the inducing field has a very great luminosity, and second, when the same point is fixated a long time.

When the inducing field has a very great luminosity, the author does not consider the appearance of the homonymous colouration in the reacting field as being a subjective phenomenon but as being caused by a scattering of objective light. Every transparent solid or liquid diffuses everywhere small quantities of light passing through it, and hence when much light traverses it, it appears to be dimly illuminated itself. The fact that this is also the case with the cornea and crystalline lens of the eye has already been stated (Vol. I, p. 193). Moreover, the entoptical objects in the vitreous humor will be recalled in this connection, because undoubtedly they must partially deflect the light in its passage through this medium. Light is reflected from the illuminated places of the retina to the other parts of the fundus. The effect of all this is that when a large amount of light penetrates the eye, invariably considerable quantities of it will be diffused over more or less of the fundus of the eye. This illumination by diffused light is manifested most distinctly in the second method of showing the vessels of the retina, by moving a candle flame to and fro below the eye, as described in §15, Vol. I. The shadows of the retinal vessels appear in the light mist, which in this case completely fills the fundus of the eye. Certainly, therefore, the illumination is an objective one and not simply a distribution of the sensitivity for light in the retina. It is easy to show by objective experiments with ordinary glass lenses that diffusely scattered light is always most in evidence in the vicinity of the regularly refracted beam of light, and that it gets less and less farther away from this beam. If sunlight passes through an opening in a black screen and falls on a distant lens, so that the image of the bright aperture is projected on a white screen, the little bright image will be surrounded by a white cloudy effect; which can also be seen when the image of the bright opening itself is allowed to pass close by the edge of the screen. That white cloudy effect, therefore, is no irradiation originating in the eye, but an objective phenomenon. It can be seen still better by making a small opening in the screen near the image of the bright aperture, but without letting it coincide with it. On looking at the lens through the opening in the screen, it will appear to be more brightly illuminated, the nearer the eye comes to the optical image of the source of light. A perfectly analogous phenomenon occurs in the eye. Look at a flame in front of a very dark field, for instance, in front of the open door of an absolutely dark room; it seems to be surrounded by a whitish cloud which is brightest in its immediate vicinity. This lustre can be noticed best by interposing a small opaque object between the eye and the flame, so that the latter is no longer visible. Instantly, the cloud of light in front of the background vanishes also, and the latter is seen in its characteristic black. If the light is coloured, then, of course, the diffused cloud of light is also of the same colour. In this case too the author believes that undoubtedly this light cloud comes from the diffusion of objective light, because the distribution of light is exactly the same as would be produced by a system of glass lenses under the same circumstances. But here, it is true, there is no proof by means of the shadows of the retinal vessels, like that which could be given in the case first mentioned. In the case of blue light we have to take into account also the white-bluish light diffused by the fluorescence of the lens, which likewise is scattered over the whole fundus of the eye. And so when a large amount of coloured light penetrates the eye, those parts of the retina where images of dark objects fall will invariably be feebly illuminated also by the dominant light, and this illumination will be greater in proportion as those places are nearer the images of bright surfaces. Besides, in the region of the dark image there is the internal stimulation of the nervous substance, which is responsible for the whitish intrinsic light of the retina. By itself this latter would appear by contrast complementary to the prevailing colour. But if much of the inducing light of the same colour is mixed with it, this colour will predominate from the start in the impression that is produced. And hence, as was noted above, a small black disc in front of coloured glass will show the complementary colour when the luminosity is low, and the same colour when the luminosity is high.

The second case, where the induced colour is the same as that of the inducing light, as is the case with long fixation, is explained by the gradual fading out of the images when the eye is held steady for a long time, as was described in the previous chapter. It was noticed there that if a place on the retina has been receiving the same impression of light continuously, the luminosity sensation gets weaker and weaker, and the colour becomes less and less saturated. However, this change of the impression is only noticed when comparisons are made with the impression produced by the same light on unfatigued areas of the retina. In this case, therefore, we cling to the judgment of the colour and luminosity that we formed at first glance.

If the surface on which the eye is focused contains portions of different relative luminosities, these differences gradually disappear as the impression gets weaker. Select some point on the surface for the point of fixation; but when the borders between bright and dark parts are faded, be careful about not obtaining too distinct after-images in consequence of slight movements of the eye. When the fixation is sharp and steady, differences of light that are often quite marked will fade out in from 10 to 20 seconds. The way this happens is at first by the brighter parts getting darker, and at the same time the darker parts getting brighter. It is striking too to watch here how sometimes a large mass changes into a faded dark spot, or a bright mass into a pale bright spot, as if the objects were painted with diluted colours and these ran together. Incidentally, the experiment is hard to perform in this way on account of the long steady fixation involving much strain. Every time the eye winks or moves ever so little, the image returns. It is much more convenient and satisfactory to use objects that have fixed positions on the retina itself, such as the retinal vessels. The methods of making the retinal vessels visible have been described in §15, Vol. I. What is common to all these methods consists in letting the shadows of the vessels fall in some unusual direction or in trying to prolong the umbrae of the shadows. But in this case it is also necessary to change continuously the direction of the light that casts the shadow, and only those vessels are visible whose shadows are shifted. As soon as the source of light is kept steady, the ramifications of the vessels disappear in a few seconds by becoming as bright as the rest of the visual field. They vanish more rapidly and more completely than the images of external objects that are hard to focus; and the weaker the illumination, the more quickly they disappear. The way to keep them longest is by concentrating sunlight with a lens on the external side of the sclerotica, because here the field is brightest.

Simple considerations easily show, by the way, that the disappearance of the retinal vessels is due to the same causes as the disappearance of all steadily fixated images, and that no special peculiarity of the parts of the retina behind the vessels is involved at all in this case. There is no ground for supposing that these places are endowed, say, with a higher sensitivity than the rest of the retina, and that, therefore, even if they are screened, the sensation there would be just as intense as elsewhere. For when the shadows are projected in an unusual direction, by illuminating a part of the sclerotica through the pupil or from outside so that it becomes a source of light for the fundus of the eye, the new parts of the retina where the shadows fall behave exactly in the same way as the places that are accustomed to them. The image disappears quickly on them too unless its location is varied, and the parts where the shadows usually fall cannot be distinguished at all by any continuously greater luminosity. Of course, bright gleams flash out spasmodically along with the shadow after it has stayed still a long time, and then begins to move again. But this happens just as well with lateral as with frontal illumination. In this case, therefore, it may be that the shaded parts of the retina recuperate, and when light falls on them again, they are more sensitive to it. But the after-effect of repose, as shown by the bright negative after-image of the shadow, does not last any longer than the after-image of dark outside objects. Undoubtedly, in the author's opinion, the rapid disappearance of the shadows of the vessels is exactly the same sort of thing as the disappearance of any objective image with moderate differences of luminosity which is steadily focused by the eye, except that in the former case the difficulties of fixation are absent.

Suppose now that a place A on the retina is continuously illuminated more highly than another place B; then, of course, since A is more fatigued than B, the initial difference of stimulation will be diminished to a certain extent, and so it gradually gets to be imperceptible and disappears wholly and entirely; possibly because it becomes really too weak to be perceived, or, as the writer is inclined to think, because our powers of discrimination for continuous nervous stimulations are much more imperfect than for varying stimulation. But since in these cases we stick to our judgment of the colour as we saw it first, and fail to notice the gradual change in it, the surfaces A and B in this experiment seem to us to get more alike, their average luminosity appearing to be about constant. As a general thing the brighter one A, in this case, becomes darker, and the darker B gets brighter. Thus, for example, when the writer stares a long time at a wall covered with silver-grey paper with dark grey leaves on it and some copper engravings, it looks as if it had a film of milk over it.

If there are different colours in the visual field, it is likewise only at the first moment that the impression of them is perfectly vivid. As we continue to look at them, all colours invariably grow darker and greyer, and therefore similar to one another. We notice that they do become similar, but we do not notice the change of the dominant colour; not accurately anyhow when there are no fresh impressions for comparisons; and so generally we consider it as not having changed.

Accordingly, after looking at a white field on red ground until the two colours become more and more alike, our judgment is that the white becomes red. The result is that every time the eye wanders to the border between the two fields, a green after-image flashes up on the white and a saturated red one on the red, the effect being augmented by contrast.

The tendency of the two colours to become like each other is very plainly manifested by looking at a small red field on broad white ground. FECHNER noticed that in this case too the white becomes reddish after a time, and uniformly so all over. Another small coloured field far off to one side has no influence on the progress of the phenomenon. But if the point of fixation is located on the border of two small fields of different colour, both lying on a white ground, according to FECHNER, the ground becomes coated over with the mixture of the two colours. And so this experiment shows that the colour perceived by the yellow spot is given a special preference, perhaps because this colour is judged most sharply and accurately, the colour sensation on the peripheral parts of the retina being much more imperfect.

In the cases heretofore considered where the inducing colour was supposed to occupy most of the visual field, or at least to dominate the others by its intensity and vividness, the contrast phenomena are very constant and distinct, and seem moreover to depend on no minor considerations. It is different when the field of the inducing colour is not so large and there may be a sufficient number of white and other objects besides near it on the border of the field of view. In this case the contrast effects throughout are not so constant any more, and will depend on many accessory conditions; which in the author's opinion are very important for the theory of these phenomena. If the field of vision outside the inducing and induced fields is dark, this does not matter much. But when the dark region comprises a very large part of the visual field, as is the case, for instance, in looking through a black tube, the intrinsic light of the retina seems to supply the lack of a white illumination, and the contrast phenomena become uncertain.

When a little piece of white, grey or black paper is laid on a coloured quarto or octavo leaf, and inspected from a distance of about a foot, as a rule, supposing the fixation is exact, no contrast colour is perceived except maybe some doubtful traces of it. However, if, as in MEYER's experiment above, the coloured octavo leaf is covered with a sheet of thin letter paper, it is remarkable how perfectly clear and constant the contrast colour comes out, in spite of the fact that the colour antagonisms are very greatly reduced by this method. Here also it is best for the little piece of paper to be grey and of about the same luminosity as the coloured paper.

The coloured paper covered by the letter paper makes a very faintly coloured whitish ground. Where the piece of grey paper is underneath, the objective colour of the upper paper is pure white. Now it might be supposed that by covering the objective white place with a white or bright grey bit of paper laid on top of the letter paper, it too would appear complementary to the ground. But, strange to say, that is not what happens. The little piece of paper exhibits its own objective colour, without contrast. Indeed, if a piece of paper is selected of exactly the same colour and brightness as the letter paper over the place where the grey is, and if this is inserted at the corresponding place of the letter paper, and now if the colours of the two places are accurately compared with each other, the contrast effect disappears even on the white place of the letter paper, where it was originally; and this place now looks white as long as the other little piece of paper is there beside it for comparison. Moreover, the contrast colour disappears also when the contour of the piece of grey paper underneath is traced on the letter paper in black lines. Thus the contrast colour continues only so long as there is no difference between the two fields except their difference of colour. The moment one field is outlined as a distinct body or by a definite contour, the effect disappears, or at least becomes very much more doubtful.

Secondly, the experiments with coloured shadows succeed even when a comparatively small part of the visual field is illuminated by coloured light; for instance, when a coloured plate is mounted vertically on a white sheet of paper, so that only part of the paper gets coloured light.

In the third place, contrast colours are brought out very beautifully also with a coloured field of moderate extent in the following method devised by RAGONA SCINA. In Fig. 54 *ab* and *ac* are two white paper surfaces, one horizontal, the other vertical; and *ad* is a coloured plate of glass inclined to the two paper surfaces at 45°; *e* and *f* are two black spots. An observer at *B*, looking down on the apparatus from above, sees the surface *ab* through the coloured glass, and the surface *ac* reflected in it. The image of *ac* coincides apparently with *ab*, and the image of the black spot *f* is at *g*, say, not far from the spot *e*. The light transmitted through the coloured glass is coloured; but the reflected light consists partly of pure white light that comes from the first surface, and partly of coloured light in comparatively small amounts that is reflected at the rear surface, or has been reflected several times inside the plate. Thus when

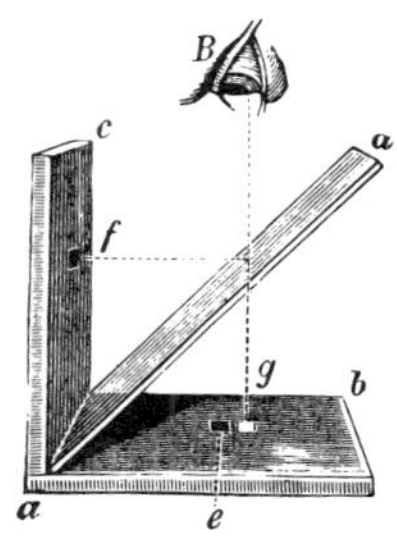

Fig. 54.

the colour of the plate is dark, the reflected light is almost white; anyhow it is much less coloured than the transmitted light. Consequently, the light the observer gets from the image of *f* at *g* is all transmitted or coloured light coming from *ab*; and the light from the bright ground is partly transmitted coloured light and partly reflected white light; whereas the light from the black spot *e* is all reflected white light. Now although this latter light is not altogether white, but always contains some of the coloured light of the glass itself, still by contrast with the colour of the ground it appears in the complementary colour. On the other hand, the spot *g*, of course, shows up in the saturated colour of the glass. For example, if the glass is green, then *e* appears pink-red, and *g* appears green.

Here too care must be taken not to have too much difference between the luminosity of *e* and that of the ground; and so with a coloured glass that transmits a lot of light the surface *ab* ought to be shaded by a white paper. Incidentally, the contrast colour of *e* is more distinct, when the spot *f* shows up there in the same colour as the ground than when it does not. Both spots are seen here under apparently the same conditions, and the contrast is heightened by comparing the way they look. Now if the observer will find a grey paper of exactly the same colour as that in which the spot *e* would appear to him without contrast, and hold a small piece of it over the coloured plate so as to hide half of the spot *e*, this little piece of paper will not appear in the complementary colour at all or even show any suspicion of it; and the moment the colour of the spot *e* is compared with it, and seen to be the same, the complementary colour of *e* disappears also and changes into a simple grey. This is exactly the same phenomenon as shown by the first method.

The following are some similar phenomena, which, it is true, exhibit only very small fields coloured by contrast, yet the effect is clear and vivid. Take a rather thick plate without much colour in it, such as ordinary greenish window glass, and notice the image of a bright white surface reflected in it. In this case the front surface of the plate reflects pure white light, and the rear surface greenish light, because the latter has been exposed to the absorption action of the glass. Now interpose a narrow black rod between the plate and the bright surface. There will be two images of it in the glass, one due to reflection at the front face of the plate, and the other due to reflection at the rear surface. Where the image in the front surface is seen, the eye still gets greenish light from the rear surface; and where the image in the rear surface lies, the eye still gets white light from the front surface. Hence the ground looks white and hardly at all greenish. The first image looks green; and the second image by contrast is very distinctly pink-red. The phenomenon is even more distinct when the rear surface of a coloured plate of glass of this sort is covered with tinfoil, and the after-images are observed at such oblique incidence that both of them seem to be equally intense.

The following experiment is similar. Place a coloured paper, green, say, on a white one (a grey one of the same luminosity is better). Near the edge where the green and white fields meet, make a small black spot on each of them, and place a crystal of Iceland spar over this place. Through the crystal all points of the base will be seen doubled. In the middle there will be a green-white strip, where the ordinary image of the white is covered by the extraordinary image of the green. It must be so arranged that one of the images of each of the two black spots will be seen in this strip. In the ordinary image of the black spot situated on the white, white is absent, but green is present; so the spot is green. In the extraordinary image of the black spot situated on the green, green is absent, but white is present; by contrast it appears a vivid pink-red.

In these last experiments the contrast action no longer depends simply on a definite distribution of colours in the field of vision. We have seen that this effect can be exactly the same with two different simple modifications of the experiment, and yet in the one case the contrast effect appears, in the other it does not. The moment the contrasting field was recognized as an independent body laid over the coloured ground, or was even divided off enough by something to indicate that it was a separate field, the contrast was absent. Accordingly, since the *judgment* of the position in space, i. e., of the corporeal independence of the object in question, is the decisive factor in the determination of the colour, the consequence is that the contrast colour here is not due to an act of sensation but to an act of judgment. The nature of this act of judgment by which we reach the perception of objects with definite characteristics will be more accurately described in Part III (Vol. III). As the acts of judgment here spoken of are always executed unconsciously and involuntarily, naturally it is often hard to determine what chain of impressions is responsible for the final result, and in the nature of the case very different circumstances may affect it. The author will endeavour to indicate here some of these conditions, as well as he has been able to ascertain them considering how new the subject is.

The experiments which have been described above have something in common which seems very much to support the occurrence of contrast action, although contrast can also occur without this condition. That is to say, in all of these cases a coloured illumination, or a transparent coloured veil, seems to be spread over the field. The immediate impression is not that this colouration is absent where there is white, that is, it is not just a mere substitution of the complementary colour of the ground in place of the white; but the idea seems to be that two new colours are substituted in the place of the white, namely, the colour of the ground and the complementary colour. The connection is clearest in the arrangement shown in Fig. 54, where the observer looks through the green glass inclined at an angle of 45°. He decides that the black spot on the horizontal surface is pink-red, but he also decides that this spot, as well as the entire surface with its pink-red colour is seen through the green glass, and that the green colour given by the glass extends uninterruptedly over the entire lower surface, and even over the dark spot. Thus he believes that he sees two colours together at this place, that is, green, which he attributes to the glass plate, and pink-red, which he attributes to the paper behind it; and the two of them together do, in fact, give the true colour of this place, that is, white. As a matter of fact, an object which, seen through a green glass, sends white light to the eye, as this spot does, would have to be pink-red. But when a white object of exactly the same appearance is placed above the plate of glass, every reason for resolving the colour of the object into two disappears; it looks white to us.

It is the same way when coloured surfaces are covered with translucent paper. If the ground is green, the paper itself seems to be greenish. Now if the substance of the paper extends without a perceptible break over grey underneath, the observer thinks he sees an object shining through the greenish paper; and an object of this kind must, on the other hand, be pink-red in order to give white light. But if the white place is outlined as an independent object, and there is lack of continuity between it and the greenish part of the surface, it is regarded as being a white object lying on this surface. In §20 above, it was stated that this sort of separation of two colours that are present in the same part of the visual field is a matter of judgment. We were confronted with this condition there as something that was an obstacle to the free realization of the sensation of a compound colour. A separation of this sort is a very frequent occurrence whenever the two colours are unevenly distributed. These phenomena were noticed first by VOLKMANN,[1] and he describes the effect by saying that we seem to see one colour *through* the other. In the author's opinion the faculty of making such a separation depends on the following circumstance. Colours have their greatest significance for us in so far as they are properties of bodies and can be used as marks of identification of bodies.

[1] MÜLLERS *Archiv für Anat. und Physiol.* 1838. S. 373.

Hence in our observations with the sense of vision we always start out by forming a judgment about the colours of bodies, eliminating the differences of illumination by which a body is revealed to us. In §20 it was noticed that in this sense we make a plain distinction between a dimly illuminated white surface and a highly illuminated grey one. Therefore, we have a certain difficulty about realizing that brightly lighted grey is the same as dimly lighted white. By some device the intense light must be confined strictly to the grey field, so that we cannot infer from the sense impression that the grey is more highly illuminated than the rest of the field of vision. It is then only that we recognize its identity with white. Just as we are accustomed and trained to form a judgment of colours of bodies by eliminating the different brightness of illumination by which we see them, we eliminate the colour of the illumination also. There is plenty of opportunity of investigating these same corporeal colours in sunshine outdoors, in the blue light of the clear sky, in the weak white light of the overcast sky, in the red-yellow light of the setting sun, and by red-yellow candle light. And besides all this there are the coloured reflections of surrounding bodies. In a shady forest the illumination is predominantly green. In rooms with coloured walls it is the same colour as the walls. We are never distinctly conscious of these latter variations of illumination, and yet they can be demonstrated often enough by the coloured shadows. By seeing objects of the same colour under these various illuminations, in spite of the difference of illumination, we learn to form a correct idea of the colours of bodies, that is, to judge how such a body would look in white light; and since we are interested only in the colour that the body retains permanently, we are not conscious at all of the separate sensations which contribute to form our judgment.

Thus too when we view an object through a coloured mantle, we are not embarrassed in deciding what colour belongs to the mantle and what to the object. We do the same thing in the experiments described above, even when the mantle over the object is not coloured at all; and it is this that causes, or at any rate promotes, the illusion into which we fall, and as a result of which we attribute a wrong colour to the body, complementary to that of the coloured part of the mantle.

But although we are trained to recognize correctly the colours of bodies in monochromatic light, our experience does not enable us to do so when two illuminations in different colours come from two different directions and from limited sources of light that cast sharp shadows. For in most of the cases of coloured illumination mentioned above, the coloured surfaces are very broad, and hence the coloured light is tolerably uniformly distributed over all sides of the observed object. Hence, with all coloured surfaces without distinction, wherever they are in the sphere of the coloured illumination, we get accustomed to subtracting the illuminating colour from them in order to find the colour of the object. We do the same thing with the coloured shadows where two coloured illuminations coalesce. Where candle light and daylight come together, the illumination of the ground is whitish red-yellow. This red-yellow of the illumination we subtract too from the colour of the shadow that gets no candle light at all, and consider it as blue, although it is white. How the idea is actually obtained that the coloured illumination is removed in these coloured shadows and also in the translucent paper cover over the objective white spot, can be seen especially when little irregularities of the paper make the illumination spotted; then the observer thinks he sees these little spots in the coloured light, although they are not there at all.

Some other illustrations of our faculty of distinguishing the colours apart of two objects placed one behind the other will also be added here. The first one is connected with VOLKMANN's experiment alluded to above. He held two small strips of coloured paper in front of his eye, one quite close and the other at the distance of distinct vision; and noticed that, instead of seeing the mixed colour, he saw one colour through the other. Hold a green veil close in front of the eyes, which is so highly illuminated that the entire field of view has a green tinge, whereas the pattern and creases in the veil are seen merely as a very faint blurred image. Then there will be no difficulty in recognizing correctly the colours of objects seen through the veil, although on the retina some of the green light of the veil is mixed in with all colours. It is even more striking still when presently the retina becomes fatigued for the green light; and then the objects seen through the veil will even be pink-red, although green light is mixed in their retinal images. The best way to see this is to close the left eye and look through the green veil with the other eye. Presently a white paper seen through the veil will look not simply *white* but even *reddish-white*. Then if the right eye is closed and the uncovered left eye opened, the paper will look *green* to this eye by contrast. When the eyes are opened alternately, the paper looks reddish with the right eye where the retinal image of the paper is greenish white; and, conversely, it looks greenish with the left eye where the retinal image is white.

The same result is obtained in the experiment described by SMITH of Fochabers (Scotland),[1] which was afterwards modified and theoretically explained by BRÜCKE.[2] When a bright flame is placed close by the side of the right eye, or when the eye is illuminated from the right

[1] *Edinb. Journ. of Science*, V. 52.—POGG. *Ann.* XXVII. 494.

[2] *Denkschr. der k. k. Akad. zu Wien.* III. Bd.—POGG. *Ann.* LXXXIV. 418.

side by the sun, so that no light goes directly into the pupil, the other eye meantime being shaded, white objects will look greenish to the right eye and reddish to the left eye. This is seen distinctly by opening the two eyes in succession, sometimes the right eye and sometimes the left eye; or by looking steadily with both eyes at a white sheet of paper and holding a little black rod vertically midway between the paper and the eyes. Then two images of the rod will be seen projected on the paper, one for each eye. The image on the left, where the surface of the paper is seen by the left eye, but not by the right eye, will look red, and the other image will look green. On the other hand, when a person looks steadily at a black plate and holds a white object in front of it some distance away, so that there are two images of it, the right image, which now is the one seen by the left eye, will be red, and the left image will be green. Thus, white looks greener to the eye that is illuminated from one side than it does to the eye that is not illuminated. Now under these circumstances, light penetrates through the sclerotica and eyelids into the illuminated eye, and this light is red, as we already know from previous experiments (Vol. I, p. 213). If sunlight is allowed to shine on the eye from one side, the red colour will be recognized on dark objects too. For example, on looking at a printed page, the black letters appear a beautiful red and the white paper green. This red light coming in from the side is diffused over most of the fundus of the eye, and the places on the retina of the illuminated eye where the image of a white object is formed are therefore simultaneously illuminated by white and red light, but the sensation is greenish white. The greenish colouring gets more and more distinct as the experiment goes on, because it depends on the eye's being fatigued for red. But with excessive red illumination of the retina the only way this can happen is by the illumination already diffused over the ground getting separated from the additional light coming from the objects; and thus this latter light looks greenish because the eye is fatigued for red. In contrast therewith pure white looks reddish in the eye that has not been affected.

Consider, moreover, the image of the wall-paper and of the ceiling of a room which is reflected in the highly polished surface of the top of a mahogany table. When the eye is accommodated for these images, the colours either look natural or, it may be, a little bluish, complementary to the colour of the table. On the other hand, when the eye is accommodated for the top of the table, the total light coming from it is overwhelmingly red-yellow. The author's experience in this case is that the complementary colouring of the images occurs especially when the reflected light of the object is feeble as compared with the illumination of the table. But if the light falls very obliquely so as to increase very much the intensity of the reflected light, and at the same time cause the grain of the wood to disappear, the images, on the contrary, will often look reddish, because then there is no inducement any more to complete the separation.

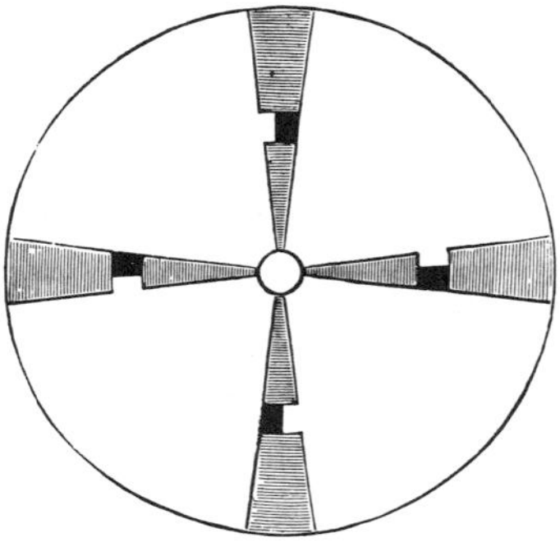

Fig. 55.

Although these circumstances that prompt us to effect a separation of white light into two portions are very conducive to the appearance of contrast, still they are not necessary. Similar contrast phenomena occur also in other cases, where a faint difference of colour is all that separates the induced from the inducing field. The effects are shown very beautifully on the colour top by inserting small coloured sectors on a white ground, so that the disc is like that shown in Fig. 55. Halfway from the centre the coloured sectors are interrupted by a piece composed of white and black. Thus when the disc revolves, naturally there should be a grey ring on a slightly tinted whitish ground. But as a matter of fact, this ring does not look grey. The colour is complementary to that of the sectors, and is most intense when its luminosity is the same as that of the ground or a little less. If the coloured sectors are wide, and consequently the colour of the ground too intense, the complementary colour of the ring will be fainter, or anyhow more doubtful, than when the ground is not so highly coloured. It is the same way when the grey ring is enclosed by two narrow black circles that sharply divide it from the ground. In the latter cases the contrast colouring is probably not entirely lacking, but it is connected with a considerable uncertainty of judgment as to the colour of the induced field; and by comparing it with a white field situated near the colour top, it is easy to reach the conclusion that the induced field is really white; whereas when the circles are absent, there is no doubt about the complementary contrast colour asserting itself in the perception. On the other hand, when a little bit of white paper is taken in a forceps and held over the greenish disc, absolutely no contrast colour whatever is seen on it, even when it is contrasted with the greenish field by no deep shadows. And when it is so turned towards the light that its luminosity is exactly the same as that of the grey zone, even the latter suddenly appears white in the vicinity of the little bit of paper, and like it; whereas the more distant parts of the ring generally continue to be coloured. If the grey zone is outlined by black lines, its colour in this experiment is recognized as pure grey all over. In this case it cannot be said that one colour was seen through the other. But in deciding as to the colour of the ring we start with the colour of the ground, and consider the colour of the ring as being a departure from the colour of the ground. When the two colours belong to two different bodies, there is no reason for connecting them together. We try rather to decide about the colour of each object independently of any accidental juxtaposition. But if a continuous flat surface, of the same structure and material all over, shows different colours in different places, the individual differences of these places being therefore in the colouring, necessarily in our judgment of them these different colours as such have to be connected and compared with one another. The result of this comparison, as experiment shows, is that the difference between the colours appears to be too great, whether it is because this difference, if it is the only one present and alone attracts the attention, makes a stronger impression than when it is one among several, and, therefore, is involuntarily considered as larger in the first case than in the second; or whether it is because in this case also the different colours of the surface are considered as being variations of the single fundamental colour of the surface, such as might be produced by shadows falling on it, by coloured reflections, or by being moistened by coloured fluids or sprinkled with coloured powders, etc. As a matter of fact, in order to produce an objectively white-grey spot on a greenish surface, a reddish pigment would have to be used.

Incidentally, it comes out plainly in the capricious results of these experiments, how hard it is for us to make accurate comparisons of luminosity and colour of two surfaces that are not directly in contact with each other and have no border between them. In the case of photometric methods we saw that the only certain and exact way of making the comparison was when there was nothing to distinguish the border between the two fields except difference of colour or illumination. The farther they are apart, the more inexact the comparison becomes; so that in such a case there is distinctly a wider latitude for the influence of accessory circumstances on our judgment of luminosity or colour. In the experiments which have been described the difference between the induced and inducing surfaces is brought out under the most favourable conditions; but the induced surface has to be compared with other surfaces lying off to the side in the visual field, so that this comparison can only be very imperfect.

This is shown still more plainly in the experiments now to be described, where the induced surface is in contact with two different colours on opposite sides. Then it will have the complementary colour on the corresponding edges. Or when the induced surface touches a darker surface on one edge and a brighter one on the other, the first edge will look brighter and the second edge darker. However, these contrast phenomena are likewise not distinct unless the only distinction between the inducing and induced fields is simply the difference of colour or luminosity, with no other border of any kind.

The experiments can readily be performed with transparent paper covers. Pieces of green and pink-red paper are fastened together so as to make a single sheet, half one colour and half the other. On the border line between the two colours a little strip of grey paper is attached; and over it all is laid a sheet of thin letter paper just large enough to cover it. The grey strip, where it touches the green, will now look pink-red, and where it touches pink-red it will look green. In the middle of it the two colours fuse into each other through an indefinite hue which perhaps is really grey, although it cannot be definitely recognized by us as such. The phenomenon is much more vivid when the length of the grey strip is oblique to the line of separation of the colours. Then the part of the grey that projects into the green may look just as vividly pink-red as the pink-red ground of the other side. The contrast colour is fainter, yet distinctly perceptible, when the middle longitudinal line of the grey strip is directly over the line of separation of the colours. Then the lateral edges of the grey appear coloured with a narrow border of complementary colour faded out towards the middle.

Similar effects are obtained by laying thin sheets of paper on top of each other step-fashion, so that the edge of each sheet is exposed in turn. If light is allowed to shine through a layer of paper whose thickness varies in this way, the objective brightness in each step will be constant of course, and yet each step will look darker at the edge where it touches the next brighter one, and brighter on the other edge where it touches the next darker one.

Fig. 56.

However, all these phenomena can be much more beautifully produced and delicately regulated on the colour top. The disc is made in black and white with sectors formed as shown in the accompanying Fig. 56. When it is set in revolution several concentric rings will be seen, the outer ones being always brighter than the next interior ones. Within every ring of this sort the angular width of the black segments is constant; and hence, with rapid rotation the luminosity is constant too. The luminosity varies simply from one ring to the other. And yet each ring looks brighter on the inside where it connects with the next darker ring; and darker on the outside where it joins the next brighter ring. If the differences of luminosity of the rings are very slight, it can scarcely be noticed sometimes that the inner rings are darker than the outer; and what the eye sees rather is simply the regular alternation of bright and dark at the edges.

If different colours are used instead of white and black, the colour of each ring will look different at the edges, although objectively the colour of each single ring is the same all over. Each element of the compound colour comes out more intensely on that edge of a ring where it joins another ring containing less of this colour. For instance, suppose that blue and yellow are mixed, and that blue predominates in the outer rings and yellow in the inner rings; then each ring will look yellow on the outside and blue on the inside. And if the differences of colour between the separate rings are on the whole very slight, here again we may get the illusion of the disappearance in the various rings of the differences of colour that are really there, and the alternating blue and yellow contrast colouring of the edges will seem to be applied on a uniformly coloured ground. It is very characteristic too that usually in these cases the compound colour does not make an impression, but what we seem rather to see are the two colours separately side by side and all in disorder.

These very striking contrast effects disappear, however, when the boundary between each pair of rings is outlined by fine black circles. Then each ring looks, as it really is, of the same luminosity and colour all over. Here too perfect continuity and uniformity of the parts of the different fields except as to colouring is again a decisive factor; and so here too we have to do with variations, not of sensation but of judgment. The differences of illumination of the various parts of the surface assume special importance again as individual perceptible differences; and since the differences between two elements of surface are plainer and more certainly detected when they are directly in contact than when they are farther apart, the differences of illumination along the edges of each pair of fields will be particularly forced on the attention; and because the perception of them is surest and most distinct, these differences look bigger than those between a pair of middle portions of two fields as to which the mind is more in doubt. In the experiments here described there was no sudden change of illumination in the middle of each field that could be perceived; and therefore the appearance was as if the colour of one edge had passed gradually through the middle of the field into that of the other. However, if a black mark is made in the middle of the induced field, or if a grey field, the two halves of which are unequally bright and separated by a clear line of division, is interposed between two coloured ones, the complementary colourings will extend from each side up to this boundary line and be separated by it. If the colour-differences between the induced and inducing fields are so marked that there is no doubt about detecting the difference everywhere, the contrast action disappears, or at any rate is much more uncertain: If there is some other delimitation of the induced field besides, the difference between its colouring and that of the inducing field is detected with far less certainty, and the contrast likewise disappears or is less pronounced.

Note (added by HELMHOLTZ to the first edition).—BURCKHARDT has devised a series of experiments on contrast colours in after-images which on the whole are extraordinarily vivid, because the conditions are especially conducive here to the production of contrast. The same cases are mentioned above on pages 244 and 278. The after-image of white surrounded by a monochromatic ground shows the same colour as this ground. If two different colours of equal extent come in contact with the white field, the after-image of the white will be a mixture of the two colours of the ground. If the after-image is projected on a coloured ground, the colour of this ground contributes in addition the colour that the after-image would show on white ground. The following experiment is very beautiful. Look steadily at a disc with two coloured sectors while it is standing still. Then suddenly begin to rotate it without moving the eyes. The after-image will be seen on the disc with the colouring of the sectors reversed.

In the theoretical explanation of contrast phenomena the earlier observers invariably assumed that the mode of reaction of the nerves, that is, the sensation, is altered at the induced places on the retina, and therefore that contrast phenomena belonged in some sense in the domain of sympathetic sensations (or synaesthesia). Many investigators have been inclined to explain irradiation in this way also. Undoubtedly, in a certain sense there is some justification for speaking of altered sensation, in the case of observations where no precise distinction is made between successive contrast and simultaneous contrast, and where, therefore, there might certainly be a modification of sensation due to after-images. Here, as far as possible, the author has endeavoured, methodically in every case, to make a distinction between successive contrast and simultaneous contrast; the result being that, wherever the inducing colour did not overshadow all others by its extent and luminosity, the occurrence of the contrast colour has been shown to be due to conditions which were established simply by the psychic activities by which it reaches visual perception. If the inducing field is supposed to be an independent body, usually the contrast colour does not come out so as to be perceived. The nature of the illusion of judgment that occurs in this case has already been indicated. Invariably we have to do with cases where a certain modicum of doubt exists as to the nature of the induced colour, because an exact comparison of it with white is not feasible; and where, therefore, our faculty of perception is influenced by subsidiary circumstances so as to misplace the colour in question first at one and then at the other limit of the interval in which the uncertainty exists. To those readers who as yet know little about the influence of psychic activities on our sense-perceptions it may perhaps seem incredible that through psychic activity a colour can appear in the visual field where there is none. The author must beg them to suspend judgment until they have become acquainted with the facts in Part III of this work, which will deal with the sense-perceptions. There they will find many examples of a similar kind. The present chapter has brought us already to the theory of the perceptions of vision; and it has been allowed to remain here in the theory of sensations, because heretofore contrast has always been considered as belonging here, and because the most ordinary phenomena in this region are of mixed nature.

Since most contrast phenomena are dependent on the extent of the uncertainty in the judgment of the intensity and quality of our visual sensations, practice in judging colours is bound to have a considerable influence on the appearance of contrast. An eye that is trained in estimating size, distance, etc., will be on its guard against many illusions into which an untrained eye will be betrayed, and it is the same way with determinations of colour; and hence the author's belief is that practised eyes generally see contrast less vividly than unpractised eyes. His experiments were easily verified for him by persons who were skilled in optical observations. On the other hand, in many books contrast phenomena are described in such fashion that he is compelled to suppose that many observers can see them much more easily and more frequently than he can do.

Whereas, owing to the dependence of the colouring on other circumstances which are simply matters of judgment, there can be no doubt as to the interpretation of contrast phenomena when the inducing field is circumscribed, the contrasts are much more constant when the inducing field is not circumscribed, and might therefore seem to imply rather that they are aroused by changes of the sensation itself. However, the conditions for reaching positive decisions as to the colour of the inducing field are evidently far more unfavourable still in these latter cases than they are in the former, simply because there is no other white with which to compare the colour of this field, or at any rate the comparison is much more restricted. Besides, although contrasts occur more constantly in an inducing field that is not circumscribed, at the same time so far as their intensity relations are concerned, they are perfectly analogous with those of the circumscribed field. In all these cases the contrast colour is evoked in full intensity even by a very low intensity of the inducing colour, and is but little augmented by increase of the latter. On the other hand, it may be distinctly augmented the moment the sensation is actually altered by after-images. It will finally be maintained in full intensity by the judgment when all other colours are removed from the visual field. And so the author does not doubt that when the inducing field is large, just as when it is small, the explanation of the phenomena must be that the contrast colour is determined simply by an exercise of judgment, although in the former cases he cannot yet give as satisfactory proof of this explanation.

LEONARDO DA VINCI was quite familiar with contrast phenomena. He says that of all colours of equal purity those are the most beautiful that are placed side by side with their opposites; that is, white with black, blue with yellow, red with green.[1] Later the contrast phenomena that especially attracted attention more than all others were coloured shadows. OTTO V. GUERICKE[2] knew about them and tried to utilize them to prove ARISTOTLE's statement, that blue could be obtained by mixing white and black. But more general attention was first directed to them by BUFFON.[3] However, his observations were merely occasional and always made at sunrise or sunset, when they were sometimes blue, sometimes green. Abbé MAZEAS[4] produced them by the light of the moon and of a candle. Moreover, he thought he was able to explain them as being due to diminution of the light. On the other hand, MELVILLE[5] and BOUGUER[6] tried to explain the phenomena on NEWTON's colour theory. The colours were supposed to be objective, because, in point of fact, blue shadows illuminated by the light of the blue sky are objectively blue in colour. BEGUELIN,[7] in particular, showed that the blue sky light is really the cause of blue shadows in many cases. The subjective nature of the colour of one of the shadows seems to have been discovered first by RUMFORD,[8] by observing it through a narrow tube. GOETHE,[9]

[1] *Trattato della pittura.* Kap. CC. — Coloured shadows in Chapters CLVI and CCCXXVIII.

[2] *Exper. Magdeb.* S. 142.

[3] *Mém. de l'Acad. de Paris.* 1743. p. 217.

[4] *Abh. der Akad. zu Berlin.* 1752.

[5] *Edinb. Essays.* Vol. II. p. 75.

[6] *Traité d'Optique.* p. 368.

[7] *Mém. de l'Acad. de Berlin.* 1767. p. 27.

[8] *Philos. Transact.* LXXXIV. 107; GRENS *Neues Journal der Physik.* II. 58.

[9] *Farbenlehre.* S. 27.

Grothuss,[1] Brandes,[2] and Tourtual[3] adopted the same view. On the other hand, other observers still contended for a long time for the objective nature of both shadow colours; for example, v. Paula Schrank[4] (who attributed the colour of the blue shadow to diffraction), Zschokke,[5] Osann,[6] and Pohlmann[7] (who adopted the view of Beguelin). But it was Fechner[8] chiefly that proved the subjective nature of these phenomena. Among other things he demonstrated also how the contrast colour once aroused might be maintained by an exercise of judgment; and, although he made a great many new observations, he did not venture to propose any theory of these phenomena. Plateau[9] included contrast phenomena in his theory of after-images; just as the change of the retina to the opposite state was a function of the time, it should likewise be a function of the (exciting) surface, the result being that right around the region of excitation the same phase occurs that is manifested in irradiation phenomena, and a little farther away the opposite phase that arouses contrast.

The explanation of contrast phenomena as being due to after-images had been proposed by Jurin,[10] and afterwards by Brandes. It was true for some of the phenomena, but not for all; and Fechner, in particular, showed that even without preceding fatigue of the retinal areas concerned, contrast colours could arise.

The modifications of individual colours by their juxtaposition to others were accurately described by Chevreul.[11] The complementary reflex images in plates of coloured glass were described by Brandes[12] and Osann. The best method of making this experiment was devised by Dove.[13] It was further modified by Ragona Scina.[14] The cases in which the induced field has the same colour as the inducing field were discovered by Fechner and Brücke.[15] H. Meyer[16] showed that a faint difference between the colours is more conducive than a big one. Incidentally, almost all the later observers adopted Plateau's view, that contrast is due to a change in sensation. The author himself has endeavoured in this article to separate the various concurrent causes more fully than has been done heretofore; and has taken pains to show that pure simultaneous contrast is due to a change, not of sensation, but of judgment.[17]

---

[1] Schweiggers *Beiträge zur Chemie und Physik.* III. 14.

[2] Gehlers *Neues Wörterbuch.* Art. Farbe.

[3] *Die Erscheinungen des Schattens.* Berlin 1830.

[4] *Münchener Denkschr.* 1811 and 1812, S. 293, and 1813, S. 5.

[5] *Unterhaltungsblätter für Natur- und Menschenkunde.* 1826. S. 49.

[6] Pogg. *Ann.* XXVII. 694; XXXVII. 287; XLII. 72.

[7] Ibid. XXXVII. 319–341.

[8] Ibid. XLIV. 221. L. 433.

[9] *Ann. de chim. et de phys.* LVIII. 339.—Pogg. *Ann.* XXXII. 543; XXXVIII. 626.

[10] *Essay on distinct and indistinct vision.* p. 170.

[11] *Mém. de l'Acad.* XI. 447–520.

[12] Gehlers *Neues Wörterbuch.* Art. Farbe. IV. 124.

[13] Poggendorffs *Ann.* XLV. 158.

[14] *Racc. fisico-chimici.* II. 207.

[15] *Denkschr. d. Wiener Akademie.* III. October 3, 1850.

[16] Poggendorffs *Ann.* XCV. 170.

[17] ¶Helmholtz's contention that errors of judgment are at the basis of our ideas of contrast (especially of simultaneous contrast and also of successive contrast in some cases) has been vigorously opposed by E. Hering and others. Hering (*Arch. f. d. ges. Zeit. f. Psychol. u. Physiol. d. Sinnesorg.* 1890. I. 18) has furnished striking evidence that the excitation of one region of the retina may modify the physiological state of contiguous

1651. LEONARDO DA VINCI (*1519), *Trattato della pittura.* Chapters CLVI,CC,CCCXXVIII.
1672. OTTO V. GUERICKE, *Experimenta nova, ut vocantur, Magdeburgica de vacuo spatio.* Amstelod. 1672. p. 142.
1738. JURIN, *Essay on distinct and indistinct vision.* p. 170.
1743. G. DE BUFFON, Sur les couleurs accidentelles. *Mém. de Paris.* 1743. p. 217.
1752. MAZEAS, *Mém. de l'Acad. de Berlin.* 1752.
1760. BOUGUER, *Traité d'optique sur la gradation de la lumière.* Paris 1760. p. 368.
MELVILLE, Observations on light and colours. Essays and observations. *Phys. and Litt.* Edinburgh II. 12 and 75.
1767. BEGUELIN, Mémoire sur les ombres colorées. *Mém. de l'Acad. de Berlin.* 1767. p. 27. 1783. p. 52.
1778. v. GLEICHEN alias RUSSWORM. Von den Farben des Schattens. *Act. Acad. Mogunt.* 1778. 308.
1782. H. F. T., *Observations sur les ombres colorées.* Paris 1782.
1783. FLAUGUERGUES, Sur les ombres colorées. *Mém. de Berlin.* 1783. p. 52.
OPOIX, *Journal de Physique.* 1783. Dec.
PETRINI, *Mem. di Mat. e di Fisica della Soc. Ital.* XIII. p. 11.
1787. CARVALHO E SAMPAGO, *Tratado das Colores.* Malta 1787.
1805. PRIEUR, Bemerkungen über die Farben und einige besondere Erscheinungen derselben. GILB. *Ann.* XXI. S. 315.—*Ann. de Chim.* LIV. p. 1.
HASSENFRATZ, Sur les ombres colorées. *Journ. de l'école polytech.* Cah. XI.
1810. v. GOETHE, *Zur Farbenlehre.* S. 27.
1811. GROTHUSS, Über die zufälligen Farben des Schattens. SCHWEIGGERS *Journal.* III. 14.
v. PAULA SCHRANK, Über die blauen Schatten. *Abh. d. Münchener Akad.* 1811. p. 293 and 1813. p. 57.

---

areas (theory of retinal induction). He holds this reciprocal relation of retinal areas to be of fundamental importance in contrast and does not accept the explanation of the phenomenon given by HELMHOLTZ. More recent experimental work in favour of HERING'S theory may be found in the publications of SHERRINGTON (*Journ. Physiol.* 1897. XXI. 33), BIDWELL (*Proc. Roy. Soc.* London 1901, LXVIII, 262) and BURCH (*Physiological Optics*, Oxford 1912); see also STARLING (*Human Physiology*, 1920, p. 573). The most recent paper (supporting HERING) is by C. v. HESS (*Arch. f. d. ges. Physiol.* 1920. 179. p. 50).

Although the experiments of these writers have been pretty generally accepted as furnishing convincing evidence that the view of HELMHOLTZ was wrong, the problem is not so simple as it seems. HELMHOLTZ certainly recognized this fact, and so did BIDWELL, who distinctly agrees (*loc. cit.*) with HELMHOLTZ that mental judgment is sometimes the sole cause of contrast phenomena. GREENWOOD (*Physiology of the Special Senses.* London 1910) has proposed an explanation of simultaneous contrast which helps to harmonize the two conflicting theories, and ALLEN'S recent paper (Reflex Visual Sensations and Color Contrast, *Journ. Op. Soc. Amer.* etc. 1923. VII. 913) is important. v. KRIES (*Allgemeine Sinnesphysiologie.* Leipzig 1923) in his recent critical analysis of the problem, has clearly set forth the complexity of contrast phenomena and pointed out the basis of the controversy. See also v. Kries's Note at end of this chapter.

For extensive discussions of contrast and related phenomena the reader is referred to articles by TSCHERMAK (*Ergeb. d. Physiol.* 1903. II. (2) 726), who gives a full bibliography to 1902; RIVERS (in SCHAFER'S *Textbook of Physiology.* 1900. II. p. 1060), and v. KRIES (*loc. cit.*). For literature since 1902 the *Zeitschrift für Psychologie und Physiologie des Sinnesorgane* (separated in 1906 into the *Zft. f. Sinnesphysiologie* and *Zft. f. Sinnespsychologie*) and the *Psychological Index* are indispensable. For reviews of literature see the *Psychological Review* up to Vol. X, the *Psychological Bulletin* and *The American Journal of Physiological Optics.* 1921. II. pp. 232, 316 (Reviews of progress of visual science, by TROLAND; see also his important monograph on "The present status of visual science." *Bull. Nat. Res. Counc.* 1922. Vol. 5. part 2. No. 27). — (M.D.)

1820. MUNCKE, Über subjektive Farben und gefärbte Schatten.—SCHWEIGGERS *Journ.* XXX. 47.

1826. ZSCHOKKE, *Die farbigen Schatten, ihr Entstehen und ihr Gesetz.* Aarau 1826.—*Unterhaltungsblätter für Natur- und Menschenkunde* 1826. S. 49.

1827. BRANDES, Art.: Farbe in GEHLERS *Neuem physik. Wörterbuch.* IV. 124.
TRESCHEL, *Biblioth. univers.* XXXII. 3.

1830. TOURTUAL, *Über die Erscheinungen des Schattens und deren physiologische Bedingungen, nebst Bemerkungen über die wechselseitigen Verhältnisse der Farben.* Berlin 1830.
LEHOT in: *Annales des sciences d'observation par* SAIGEY *et* RASPAIL. 1830. III. 3.—FRORIEPS *Notizen.* XXVIII. p. 177.

1832. OSANN, Vorrichtung zur Hervorbringung komplementärer Farben und Nachweis ihrer objektiven Natur. POGGENDORFFS *Ann.* XXVII. 694; XXXVII. 287; XLII. 72.
SMITH (of Fochabers) in *Edinb. Journ. of Science.* V. 52.
BREWSTER, Über den Versuch von SMITH in POGGENDORFFS *Ann.* XXVII. 494.
CHEVREUL, Sur l'influence, que deux couleurs peuvent avoir l'une sur l'autre, quand on les voit simultanément. *Mém. de l'Acad. de Paris.* XI. 1832. — *De la loi du contraste simultané des couleurs.* Strasbourg 1839.

1834. J. MÜLLER, in his *Archiv für Anat. und Physiol.* 1834. S. 144.—*Lehrbuch d. Physiol.* 2. Aufl. II. 372.
PLATEAU in: *Ann. de chim. et de phys.* LVIII. 339.—POGGENDORFFS *Ann.* XXXII. 543. XXXVIII. 626.

1836. POHLMANN, Theorie der farbigen Schatten. POGGENDORFFS *Ann.* XXXVII. 319–341.

1838. *FECHNER, Über die Frage, ob die sogenannten Farben durch den Kontrast objektiver Natur seien. POGGENDORFFS *Ann.* XLIV. 221–245.
DOVE, Über subjektive Komplementärfarben. POGGENDORFFS *Ann.* XLV. 158.

1840. *FECHNER, Tatsachen, welche bei einer Theorie der Farben durch den Kontrast zu berücksichtigen sind. POGGENDORFFS *Ann.* L. 433.

1847. D. RAGONA SCINA, Su taluni fenomeni che presentano i cristalli colorati. *Racc. fis. chim.* II. 207.

1851. E. BRÜCKE, Untersuchungen über subjektive Farben. *Wiener Denkschr.* III. 95.—POGGENDORFFS *Ann.* LXXXIV. 418.—*Arch. d sc. phys. et natur.* XIX. 122.

1852. A. BEER, Über das überzählige Rot im Farbenbogen der totalen Reflexion (Kontrastfarbe). POGGENDORFFS *Ann.* LXXXVII. 113–115. *Cosmos.* II. 95.

1855. H. MEYER, Über Kontrast- und Komplementärfarben. POGGENDORFFS *Ann.* XCV. 170–171.—*Ann. de chim.* (3) XLV. 507.—*Phil. Mag.* (4) IX. 547.

1858. CHEVREUL, Note sur quelques experiences de contraste simultané des couleurs. *C. R.* XLVII. 196–198.—DINGLER *J.* CXLIX. 435–436.

1859. NARDO, Nota sulle ombre colorate ottenute col solo concorso di luce bianche. *Cimento.* IX. 352-356.—*Atti dell' Istit. Veneto.* V —*Zeitschr. für Chemie.* 1860. pp. 18–20.
RAGONA, Su taluni fenomeni di colorazione soggetiva. *Atti dell' Acad. Palermit.* III.—*Zeitschr. für Chemie.* 1859. pp. 20–24.

1860. G. TH. FECHNER, Über die Konstrastempfindung. *Leipzig. Ber.* 1860. S. 71–145.
OSANN, Über Ergänzungsfarben. *Würzb. Zeitschr.* I. 61–77.
FECHNER, Einige Bemerkungen gegen die Abhandlung Prof. OSANNS über Ergänzungsfarben. *Leipz. Ber.* 1860. 146–165.
J. J. OPPEL, Über farbige Schatten bewirkt durch weisses Licht. *Jahresber. des Frankf. Vereins.* 1859—1860. S. 65—69.

1861. ROSSOLINI, Sulle ombre colorate. *Atti dell' Istit. Lombardo.* II, 318—321.

1862. H. AUBERT, Beiträge zur Physiologie der Netzhaut. *Abhandl. der schlesischen sellsch.* 1861 (1). S. 49–103. S. 344.
G. TH. FECHNER, Über den seitlichen Fenster- und Kerzenversuch. *Leipz. Ber.* 1862. S. 27–56.

1865. FR. BURCKHARDT, Die Kontrastfarben im Nachbilde. *Basler Verhandl.* 1865.

Note by v. Kries (prepared especially for the present English edition).[1]

It is well known that the theoretical explanation of simultaneous contrast, with respect to both luminosity and colour, has been a subject of particular controversy. As opposed to HELMHOLTZ's opinion of our having to do here invariably with what he calls "mental illusions", another view, due mainly to HERING, which has been prevalent for a long time, is that, by virtue of a reciprocal physiological action of adjacent portions of the visual organ, which follows simple laws, a change takes place in the sensation itself in the strict sense of the word. Thus, for example, the sensation at one place in the visual field might be darkened by high illumination at an adjacent place, or it might be shifted towards green by red illumination at an adjacent place, and so on. It is not possible to go into this subject here in detail, and the reader is referred therefore to other works where the writer has discussed it (NAGEL's *Handbuch der Physiologie*, III, pp. 232 foll.; also v. KRIES's *Allgemeine Sinnesphysiologie*, Leipzig, 1923, p. 261 and especially pp. 275 foll.). Perhaps here it will suffice simply to say a few words as to the present status of the question. To begin with, it may be regarded as certain that there actually is a reciprocal physiological action as HERING supposed, especially with respect to luminosity-contrast. However, it is another question as to whether this is the sole cause of contrast phenomena or whether conditions of another sort are likewise involved, especially conditions that are identical with, or at any rate not very far removed from, those which HELMHOLTZ inferred. At present the latter must be considered as being probably the case. All the facts that led HELMHOLTZ at the time to attribute these phenomena to mental illusions and that were adduced by him in support of this hypothesis are in its favour. Moreover, some very searching experiments on this point have been made very recently by JAENSCH and his pupils; the conclusion being that many, in fact most, contrast phenomena are *not* to be accounted for by a reciprocal physiological action such as HERING assumed. Consequently, instead of speaking of a simple change of sensations, JAENSCH employs the term "transformation" of luminosity or colour.[2] However, the idea that is meant to be

[1] ¶Concerning the above note on "Contrast," Professor v. KRIES writes (from Freiburg, January 6, 1924) that it had been originally intended to add an extensive article on this subject in the third edition, but that owing to the fatal illness of Professor NAGEL, who had been entrusted with this task, the plan had ultimately to be abandoned. Professor v. KRIES expressed the hope that some effort would be made to supply this deficiency in the English edition, and kindly offered to place the brief note above at the disposal of the editor to be utilized for that purpose in any way he deemed best. However, it was finally decided to insert here Professor v. KRIES's note just as it stands. (J. P. C. S.)

[2] R. E. JAENSCH and A. E. MÜLLER, Über die Wahrnehmung farbloser Helligkeiten und den Helligkeitskontrast. *Zft. f. Psychol.*, LXXXIII. p. 266. — O. KROH, Über Farbenkonstanz und Farben-Transformation. *Zft. f. Sinnesphysiologie* LII. p. 113.

conveyed here by the use of the term transformation amounts to saying that the connection between the sensation strictly so-called and the conceptions that are retained in the memory can be shifted; and so in any case it is not far removed from HELMHOLTZ's hypothesis of mental illusions.[1] —K.

§25. Various Subjective Phenomena

Some subjective phenomena of vision still remain to be described, the explanation of which is as yet unknown or at any rate is very doubtful, and which therefore could not be included in the preceding chapters.

1. *Phenomena of the yellow spot.* The yellow spot is a place on the retina which is distinguished in many respects. The peculiarities of its anatomical structure have been described in Vol. I, §4. Moreover, physiologically, it is characterized by the keenness of its perception of tiny images, wherein the fovea centralis far surpasses all other places of the retina. Its importance as point of fixation is a consequence of this extraordinary sensitivity. In Vol. I, §15, pp. 213-217, it was shown how the yellow spot could be made visible in the entoptical image. In this mode of observation it is characterized by the absence of vessels at its centre, and also by the shadows cast by the lateral slopes of the fovea in oblique illumination. With respect to the sensations of this place on the retina, we have already mentioned that when a current of electricity goes through the eye, the yellow spot is outlined sometimes as dark on a bright ground, sometimes as bright on a dark ground, depending on the direction of the current. Moreover, in intermittent light of moderate frequency the yellow spot appears as a peculiar star-shaped design in the iridescent patterns of the retina.

Another fact has now to be mentioned. In uniformly diffused illumination, especially in blue light, the yellow spot is characteristically outlined. Then different parts of the yellow spot, not always all at the same time, appear with different distinctness under different conditions. The centre of the yellow spot is the fovea centralis, and there the retina is very thin, transparent and without colour. According to KOELLIKER, its diameter is between 0.18 and 0.225 mm. Its distance from the posterior nodal point of the eye is 15 mm, and therefore on the average 75 times as great as its diameter. Hence, its apparent size in the field of view is that of a circle whose angular diameter is from 40 to 50 minutes of arc. Ordinarily, when visible, it looks like a regular well outlined circle. Surrounding the fovea centralis a dark halo is frequently observed, whose size about

[1] Concerning this, see v. KRIES, *Allgemeine Sinnesphysiologie* p. 139 and pp. 281–284.

corresponds to the place of the yellow spot where there are no vessels, as it looks when the vessels are made entoptically visible. The external boundary of this so-called *non-vascular* halo is indistinct. Its diameter, being about three times that of the fovea centralis, amounts therefore to something over two degrees. Sometimes its border is fairly circular in appearance, especially in dim light, and then again it is like a rhomb with its longer diagonal horizontal. It appears the latter way to the author, especially in good light. Anatomically, this place corresponds to the central, intensely yellow, part of the yellow spot. Its horizontal diameter as measured by H. MÜLLER in two eyes was 0.88 and 1.5 mm; and its vertical diameter 0.53 and 0.8 mm. Incidentally, the yellow colouring extends much farther yet, but it is weak and faded.

Finally, in good light the dark non-vascular halo is seen surrounded also by a bright halo, whose outer border is very indefinitely indicated, and which likewise looks to the author more rhomboidal than circular. Its two diameters are some three times as large as that of the dark non-vascular halo. An anatomically well-defined substratum of this place cannot be designated. The yellowish faded colouring of the outer parts of the yellow spot coincides to some extent with this bright halo. Still nothing can be said about the congruence of their areas, because the extent of the faint yellow colouring is too different in different eyes. Perhaps this outermost bright halo owes its origin simply to a contrast action also. It may be called LOEWE's ring after its discoverer, to whom it appeared to be circular in form.

LOEWE[1] discovered this ring by looking at a bright surface through a clear sea-green solution of chromium chloride. The ring appeared violet in comparison with the greenish ground surrounding the central darker halo, and so HAIDINGER compared it with an image of the iris surrounding the dark pupil. HAIDINGER showed that dichromatic means are not necessary for the production of the rings, and that they appear in the homogeneous blue of the prismatic spectrum, and also in mixed light with enough blue in it. In the latter case the differences of colour are different from the rest of the ground, depending on the quality of the colours admixed with the blue. This ring seems to appear with more distinctness to some eyes than to others; and in fact many eyes cannot see it at all. It is only with a certain medium brightness, about like that which is satisfactory for reading and writing, that the author can see it. By holding a blue glass in front of his eyes, and resting them a while by closing the lids, and then looking through the glass at a white paper surface, he can see distinctly the non-vascular corona as a rhombic shadowy spot surrounded by a rhombic bright blue strip, which is LOEWE's ring. With slightly more or less luminosity than this

[1] HAIDINGER in POGG. *Ann.* LXX. 403. LXXXVIII. 451.—*Wiener Sitzungsber.* IX. 240.

the ring looks smaller; and for any greater changes of luminosity all that the writer can see then is the dark non-vascular halo without a bright encircling border.

The dark non-vascular halo is the most constant feature of the phenomenon. Its behaviour was first accurately investigated by MAXWELL.[1] According to him, it always appears in blue, and not in other colours, when homogeneous light is used. Incidentally, it appears in mixed colours also, when they contain a great deal of blue; especially too in white, but faint. When the eye, after being rested, is turned towards a blue surface, it comes out, but soon vanishes again, more quickly in bright than in dim illumination. MAXWELL recommends placing blue and yellow glass or blue and yellow paper in front of the eye alternately. The spot appears in blue, and disappears in yellow. The author observes it most beautifully on the evening sky when the first stars begin to appear, and after having been outdoors for some time, so that the eyes are sufficiently rested. Closing them for a moment and then opening them towards the sky, the observer will see the non-vascular halo very distinctly for some time, often too the fovea centralis in its interior as a little brighter spot of pure blue, pretty sharply outlined. It is a singular thing here, as was noticed also by MAXWELL, that the luminous impression in the central places of the retina develops into sensation a moment later than it does in the peripheral parts. To show this, MAXWELL caused a series of dark bands to pass in front of a blue field at a certain rate. However, it may be plainly seen by simply opening the eyes. The darkness of the closed eyes plainly vanishes from the periphery of the visual field towards the centre, and the last remnant of it lingers as the MAXWELL spot. With certain degrees of brightness, particularly with that of the sky mentioned above when the first stars become visible, the phenomenon when the eyes are opened is more complex still. Thus while the darkness disappears from the periphery towards the centre in the manner described, we also see either the fovea centralis by itself or the entire MAXWELL spot flare up bright. Perhaps the bright flash precedes the dark phenomenon a little, but the time is so brief that they are apparently simultaneous, similar to what AUBERT noticed in after-images with illumination by electric sparks.

Sometimes, when the fovea centralis appears very distinct, the writer can detect in the non-vascular halo faint patterns of lines like the contours of a flower with many petals (for example, a georgina or dahlia). Perhaps they are indications of the same pattern that comes out more plainly in intermittent light.

[1] *Athenäum*. 1856. p. 1093.—*Edinb. Journ.* (2) IV. 337.—*Inst.* 1856. p. 424.—*Rep. of British Association*. 1856. II. 12.

Finally, the writer must mention also that often on getting up in the morning he has accidentally seen the MAXWELL spot, bright on a dark ground, in case the eye was first directed to a bright window with a broad luminous surface. So far he has not succeeded in eliciting the phenomenon deliberately. In this case it appears as a brilliantly bright circle of the size of the non-vascular halo, shaded off towards the edges and with indications of stellar design. This latter appearance suggests that when the eye is thoroughly rested and sensitive, the impression of light in the yellow spot persists longer than in the other parts of the retina; while, on the other hand, the action at the same place appears to begin later, as indicated by the phenomena obtained on opening the eye, which were described above. The reason why the highly pigmented part of the yellow spot looks dark on a blue field, is apparently on account of the absorption of blue light by the yellow pigment. Precisely those parts are yellow here that are directly in front of the elements or cones which are peculiarly sensitive to light. Incidentally, the explanation of why the spot is but faintly outlined subjectively, and is very transient, is the same as that of the fugitive appearance of the vascular figures. But the occasional bright flaring up of the yellow spot on opening the eye cannot yet be explained.[1]

The phenomena as described thus far relate to unpolarised light. But if the eye is directed to a field emitting polarised light, HAIDINGER's *polarisation brushes* appear at the point of fixation. For instance, they will be seen on looking through a NICOL prism at a well lighted sheet of white paper or at the surface of a bright cloud. The brushes are reproduced in Fig. 1, Plate III,[*] as they appear when the plane of polarisation is vertical. The brighter spots bounded by the two branches of a hyperbola look bluish on a white field. On the other hand, the dark brush separating them, which is narrowest in the centre and broader towards its ends, is yellowish in colour. When the NICOL is turned, the polarisation pattern turns through the same angles. BREWSTER noticed that the dark brush is much narrower at its centre when it is horizontal (that is, parallel to the line between the two eyes) than when it is vertical, as shown in the illustration; and the writer has verified this observation. Both MAXWELL and the writer find that the surface covered by the polarisation pattern seems to be the same

[1] The condition under which MAXWELL's spot is produced and in general the entoptical perceptions in the macular zone have recently been very thoroughly investigated and described by GULLSTRAND (GRÄFES *Arch. f. Ophthalmol.* LXII. 1905, 1; LXVI. 1907, 141) and by DIMMER (ibid. LXV. 1907, 486). A more detailed consideration of these investigations would lead us too far here. It may be simply stated that GULLSTRAND considers the yellow colouring of the macula to be a post mortem appearance.—N.

[*] Plate III faces page 384.

size as the non-vascular halo of the yellow spot. The edge of the fovea centralis about passes through the brightest part of the blue surfaces. BREWSTER estimates the diameter of the polarisation brushes as somewhat larger, namely 4°; and SILBERMANN gives 5°, the difference being perhaps due to the fact that they seem to be very much more distinct for some eyes than for others, and hence some persons perceive the faintest parts of the pattern to the farthest edge, while others do not. Thus, twelve years ago, immediately after HAIDINGER's discovery, the author took the greatest pains to perceive these brushes and could not do it at all. Recently, when he tried it again, he saw them the moment he looked through the NICOL prism. The centre of the dark brush is much darker in his left eye than in the other eye. It may be that the variable colouring of the yellow spot is responsible for it. Incidentally, when they are visible, they always quickly disappear again, as is the case with every subjective phenomenon connected with a structure of the retina. Then they reappear again when the polariser is turned through a right angle.

Individuals who perceive the brushes quite distinctly see them also in light that is only partially polarised, on brilliant surfaces, on the sky, etc.; and it enables them to tell equally well under any circumstances the direction of the plane of polarisation. However, STOKES showed that blue was the only one of the homogeneous colours that exhibited the polarisation brushes. In the less refrangible parts of the spectrum they do not show. In a blue field the bluish hyperbolic surfaces look bright, and the yellow brushes between them look dark, as, for example, in looking at a white surface through a strongly coloured blue glass in conjunction with the polariser. The writer fails to see the brushes, not merely in homogeneous green, yellow and red, but even in the mixed, tolerably saturated shades of these hues obtained with coloured glass. Consequently, even in white light the phenomenon depends on variations of the blue. At the place where the yellow brushes are there is no blue, and just for that reason they look yellow and darker.

When light is polarised by refraction, reflection or double refraction, all the colours are always almost equally affected by the polarisation. It is only when coloured light is absorbed in double refraction that light of certain colours may be polarised, while light of other colours is not. The most familiar example of this sort is the absorption by tourmalin, which is so often used for polarising light. This property, by the way, is very common among coloured crystals that are double refracting. It may be produced by colouring them artificially, and it depends on the fact that sometimes the ordinary ray is more absorbed (as in tourmalin), and sometimes the extraordinary ray (as in titanium dioxide and tin oxide). But most organic fibres and membranes are double refracting to a slight extent, both of them usually behaving like uniaxial crystals, with the axis in the fibres longitudinal, and in the membranes perpendicular to their surfaces. The phenomenon of the polarisation brushes may be explained by supposing that the yellow elements in the yellow spot are double refracting to a slight degree, and that the extraordinary ray of blue is more absorbed by them than the ordinary ray.

If blue light polarised in any way passes through a mass of fibres of this sort along the direction of their axes, it will be much absorbed. But if it traverses them at right angles, it will not be strongly absorbed unless it is polarised parallel to the fibres; being only feebly absorbed when the plane of polarisation is perpendicular to the direction of the fibres. Now in the yellow spot the so-called radial fibres of H. MÜLLER, which at other places of the retina are perpendicular to its surface, run diagonally, their posterior ends being nearer the fovea centralis.[1] In the fovea centralis the nuclear layers and the intermediate nuclear layer are either absent entirely or are anyhow very thin. On the other hand, in the region around the fovea centralis the inner nuclear layer and the intermediate nuclear layer are thicker than elsewhere. Something of the same sort is true with respect to the layer of ganglion cells, although even in the fovea centralis it is still three cells deep. Thus it seems as if the other elements belonging to the cones of the fovea centralis were heaped up in the region around this little pit and therefore the connecting fibres of nerves and tissue must run diagonally. At the edge of the fovea centralis, where the direction of the fibres is mainly diagonal towards its centre, according to the above supposition, light would be more strongly absorbed where the fibres were parallel to the plane of polarisation. Thus if the latter is vertical, darker places will be formed above and below the fovea centralis, and brighter places to the right and left. Likewise, the places where the fibres are not oblique to the surface of the retina, that is, in the fovea centralis itself, and out towards the outer border of the yellow spot, would have to be darker. Now the phenomenon of the polarisation brushes actually does agree with these consequences.

There are other opinions as to the origin of the polarisation brushes. One of these in particular, which was suggested by ERLACH and specially developed by JAMIN, has met with much favour. Their idea was that the brushes might be produced by multiple refractions at the boundaries of the ocular media. As a matter of fact, light polarised vertically and entering the eye from above or below would be more

[1] BERGMANN in HENLE und PFEUFFER *Zeitschr. für rat. Med.* (2) V. 245, (3) II. 83. — MAX SCHULTZE, *Observationes de Retinae structura penitiori*. Bonn 1859. p. 15.

strongly reflected, and less of it would get through, than in the case of the same kind of light entering the eye from one side; and hence the upper and lower quadrants of the visual field would have to be a little darker than the right and left. But if polarisation by refraction were responsible for the effect, in the first place the brushes would have to appear almost equally distinct in all homogeneous colours, whereas they appear distinct only in blue. In the second place, they would have to increase in intensity continuously out towards the edges of the visual field. On the contrary, they are restricted to a very small central part. In the third place, their centre would have to be where the optical axis meets the retina, and not at the point of fixation, these two places being different always. STOKES, BREWSTER and MAXWELL have all pointed out the insufficiency of this explanation, and the two latter have noted that the extent of the brushes agrees with that of the yellow spot. Various other explanations, not clearly worked out however, have been given by HAIDINGER and SILBERMANN.

According to HAIDINGER's description, there are also bright X-shaped lines in the blue field where LOEWE's ring is seen, but as yet they have not been observed by anybody else. The writer cannot see them.

1844. W. HAIDINGER, Über das direkte Erkennen des polarisierten Lichts. POGGENDORFFS *Ann.* LXIII. 29.

1846. Idem, Über komplementäre Farbeneindrücke bei Beobachtung der Lichtpolarisationsbüschel. POGGENDORFFS *Ann.* LXVII. 435.

Idem, Beobachtung der Lichtpolarisationsbüschel in geradlinig polarisiertem Lichte. POGGENDORFFS *Ann.* LXVIII. 73.

Idem, Beobachtung der Lichtpolarisationsbüschel auf Flächen, welche das Licht in zwei senkrecht aufeinander stehenden Richtungen polarisieren. POGGENDORFFS *Ann.* LXVIII. 305.

SILBERMANN, Essai d'explication des houppes ou aigrettes visibles a l'œil nu dans la lumière polarisée. *C. R.* XXIII. 624.—*Inst.* No. 665. p. 327.

1847. v. ERLACH, Mikroskopische Beobachtungen über organische Elementarteile bei polarisiertem Licht, in MÜLLERS *Archiv für Anat. und Physiol.* 1847. p. 313.

HAIDINGER, Helle Andreaskreuzlinien in der Sehachse. *Ber. d. Freunde der Naturwiss. in Wien.* II. 178.—POGGENDORFFS *Ann.* LXX. 403.

BOTZENHART, Polarisationsbüschel am Quartz. *Ber. d. Fr. d. N. W. in Wien* I. 82.

Idem, Sur une modification des houppes colorées de HAIDINGER. *C. R.* XXIV. 44.—*Inst.* No. 680. p. 11.—POGGENDORFFS *Ann.* LXX. 399.

1848. JAMIN, Sur les houppes colorées de HAIDINGER. *C. R.* XXVI. 197.—POGGENDORFFS *Ann.* LXXIV. 145.—*Inst.* No. 737. p. 53.

1850. D. BREWSTER, On the polarizing structure of the eye. SILLIMAN's *J.* (2) X. 394. *Rep. of British Assoc.* 1850. II. 5.—*Wiener Ber.* V. 442.

G. G. STOKES on HAIDINGER's brushes. SILLIMAN's *J* (2) X. 394.—*Rep. of British Assoc.* 1850. II. 20.

W. HAIDINGER, Das Interferenzschachbrettmuster und die Farbe der Polarisationsbüschel. *Wien. Ber.* VII. 389.—POGGENDORFFS *Ann.* LXXXV. 350.—*Cosmos. I.* 252, 454.

1852. Idem, Die LOEWEschen Ringe eine Beugungserscheinung. *Wien. Ber.* IX. 240–249. —POGGENDORFFS *Ann.* LXXXVIII. 451–461.

1854. W. Haidinger, Dauer des Eindrucks der Polarisationsbüschel auf der Netzhaut. *Wien. Ber.* XII. 678–680.—Poggendorffs *Ann.* XCIII. 318–320.
Idem, Beitrag zur Erklärung der Farben der Polarisationsbüschel durch Beugung. *Wien. Ber.* XII. 3–9.—Poggendorffs *Ann.* XCI. 291–601.
Idem, Einige neuere Ansichten über die Natur der Polarisationsbüschel. *Wien. Ber.* XII. 758–765.—Poggendorffs *Ann.* XCVI. 314–322.
Stokes, Über das optische Schachbrettmuster. *Wien. Ber.* XII. 670–677.—Poggendorffs *Ann.* XLVI. 305–313.

1856. J. C. Maxwell, On the unequal sensibility of the foramen centrale to light of different colours. *Athen.* 1856. p. 1093.—*Edinb. Journ.* (2) IV. 337.—*Inst.* 1856. p. 444.—*Rep. of Brit. Assoc.* 1856. II. 12.

1858. Power in *Phil. Mag.* (4) XVI. 69.

1859. Brewster in *C. R.* XLVIII. 614.

---

2. *Bright mobile points* appear in the field of view in looking steadily at a large uniformly illuminated surface like the sky or snow, especially during walking or other bodily exercise. At different places in the visual field little points leap up and run away quite rapidly in very different paths, which are generally not entirely straight. Then after little intervals new ones appear along the route one of them has blazed, and pursue the same path. Purkinje noticed that in gazing at a luminous area of limited dimensions, a window say, every point on the side away from the centre of the visual field draws a little shadow after it. As they seem to keep to fixed paths, many observers (J. Müller) have been disposed to regard them as having something to do with the circulation of the blood. But in the author's opinion these phenomena are much too sporadic to be taken for blood corpuscles. Their paths also are too far apart from each other, and their movements are too rapid, to correspond to a capillary network. If their appearance really has anything to do with the circulation of the blood, the most we could say about it would be that single lymph corpuscles, perhaps containing much fat, floating through larger vascular stems, may be manifested in this way. This phenomenon, by the way, seems to be readily seen by most people.[1]

[1] It is easy to recognize a connection between this phenomenon and the circulation of the blood, because in many persons the movement of the luminous corpuscles invariably shows a pulsating acceleration, and with all persons the movement can be easily retarded by a slight pressure on the temporal side of the eyeball and changed into a very distinct pulsating one. The little points are shoved forwards rhythmically with the pulse; with stronger pressure they move just a little with each beat of the pulse, and between times remain almost still. When the pressure is increased still more, the motion ceases, and at the very instant the images of the objects that are seen dissolve away.

Rood (Silliman's *Journ.* (2)XXX. 264 and 385)noticed long ago that the movement of the little points can be seen particularly plainly by looking through a blue glass. If the phenomenon is investigated when the field of view is illuminated by monochromatic light of various colours, it will be found that no trace of the corpuscles can be seen in red, yellow or green light and also in cyan-blue light; but that they are very plainly visible in indigo and violet light, and can be seen to a remarkable extent even in yellow-green light (between

Incidentally, the blood corpuscles are just large enough, if they were on the retina and did make an impression on it, to be recognized. On the average their diameter amounts to 0.0072 mm, and the size of the smallest perceptible distance is 0.005 mm (see §18). Various observers have also seen rows of advancing little globules and vaguer undulations and currents. The peculiar appearance of meandering currents, which occurs with intermittent light, and which is connected by VIERORDT with the circulation of blood in the choroid, has been alluded to above. Incidentally, something similar to this may be seen sometimes without intermittent light, by gazing at a bright surface, especially after the blood has been driven to the head by bending over. As soon as the retina is so far fatigued by the action of the light that the surface becomes dark, there appears, as it were, behind the vanishing bright surface a spotted reddish surface, with its spots sometimes in motion and sometimes still. STEINBACH and PURKINJE[1] have seen rows of little flowing globules, especially when mild pressure is exerted on the eye. The latter saw them first by observing the dark accommodation pattern, which in his case consisted of a central white circle surrounded by a brownish halo without any definite border. Alongside the white circle on the right and left, he saw two vertical lines of light, with rows of tiny balls moving along them, down on the right and up on the left. So far the author has not seen anything like this. During congestion in the cranial region, or when he bent over and then suddenly rose up, JOHANNES MÜLLER[2] saw something that looked like black bodies with tails to them, jumping and flying about in the most manifold directions. He compared it with the creepy sensations in the tactile nerves.

Sometimes, too, the writer has noticed a flicker on a wall roughly plastered with lime and very obliquely lighted by a small window, as of tiny objects in motion. The wall appears to be studded with a quan-

---

570 and 560$\mu\mu$). These are precisely the kinds of light that are most strongly absorbed by hemaglobin, and in view of this fact there can hardly be any doubt about its being a question here of images of the *red* corpuscles. The comparatively small number of corpuscles visible at one time might be due to the fact that, in the first place only those corpuscles in a perfectly definite plane of the retina (very close to the *membrana limitans interna*) are under favourable conditions for being seen, and, in the second place, even these are perhaps not easy to see unless they are oriented in a certain way in which they are particularly absorbent of light. Since the little points that move look *bright*, it might be supposed that this is due to spaces in the row of blood corpuscles that occupy a capillary.

Generally, the foveal region always remains free from this circulation effect. On this subject, see G. ABELSDORFF and W. NAGEL, Über die Wahrnehmung der Blutzirkulation in den Netzhautkapillaren. *Zft. f. Psychol. u. Physiol. der Sinnesorgane*, XXXIV. 291. 1904.—N.

[1] *Beobachtungen und Versuche*. I. 127.

[2] *Physiologie*. II. 390.

tity of small black irregular points. But these might perhaps have been after-images of the small points flaring up from unavoidable little perturbations of the eye.

Some other phenomena are described by PURKINJE that are produced by agitating the vascular system or by straining the eyes. His description reads[1]: "On a bright day when I have been out walking briskly for fifteen minutes or half an hour and then suddenly enter a dim room, or at least one very much darkened, a dull light waves and flickers in the field of vision, like the expiring flame of alcohol spilled on a horizontal surface, or like a place rubbed with phosphorus faintly flickering in the dark. On looking more closely, I notice that the flickering haze consists of innumerable extremely tiny luminous points moving amongst each other along various lines, piling up now here, now there, and forming vaguely bounded spots which are again torn asunder to be reassembled elsewhere. Each point leaves a trail of light behind it as it moves, and these trails form manifold intersecting nets and little stars. A large tract in the interior of the visual field teems with this sort of thing and interferes with distinct vision. The appearance is more like the swarming of the motes of sunbeams than anything else."

When he covered his right eye and looked at a bright surface with his weak farsighted eye, he got the same effect; and likewise when the pressure on his left eye was gradually increased more and more. The points appear more vivid with the eye open than with it closed, especially when it is turned towards a distant place that is not entirely dark. Thus, external light is requisite for the effect.[2]

If he had been running or if he pressed on his eye or coughed violently, he would see pulsating spheres on the surface of the bright sky. There would be a pair of them on the right side of the visual field, a row of them below, and three on the left side. The point of fixation pulsated too; and there were grey bands also, partly circular vascular bands surrounding the point of fixation and partly radial bands.[3]

3. *Figures, which become visible when the retina is uniformly illuminated.* PURKINJE[4] noticed that when he gazed steadily at a large rather brilliant surface (for instance, at a uniformly clouded sky or at a candle flame close by), in a few seconds luminous points kept on leaping up in the middle of the visual field; which without having changed their position quickly disappeared again, leaving black points behind, which likewise quickly vanished. If, while the luminous

[1] *Beobachtungen und Versuche* I. 63.

[2] See footnote on page 8 of this volume.—N.

[3] *Beobachtungen und Versuche.* I. 134.

[4] Ibid. I. 67.

points were leaping out, he turned his eye towards a very dark place, or closed it, the phenomenon proceeded in the same manner, only in a diminished light; as if the points had been simply kindled by the preliminary gazing and then glowed by themselves alone. The writer likewise has frequently seen luminous points of this sort by accident, which could not be after-images, because there were no corresponding small bright objects in the field of vision, and which left dark after-images behind them. Usually, however, there was only one at a time, and rarely a repetition of it.

Here too we must mention PURKINJE's figure of the spider web,[1] consisting of luminous reddish lines on a red ground, imitating the web of the garden spider, sometimes more simply, sometimes in a more complex way. In order to make the figure come out well, PURKINJE laid down in such a position that the rays of the rising sun were bound to fall on his eyelids. On waking, he saw the figure behind the closed lids.

PURKINJE's work generally is exceedingly full of subjective observations of a similar kind, and will be for a long time still a rich mine for work in this region. But many of the phenomena which he describes have not been verified by other observers; and concerning them there is for the time being a question as to whether they were not due to individual peculiarities of his organ of vision.[2]

Note by W. Nagel

The appearance of flicker or flare, which occurs particularly in the dark visual field, and the causes of which are unknown, also deserves to be mentioned. The flicker observed in pathological conditions, which lasts from a quarter to a half hour and which is accompanied or followed by headaches of the nature of migraine, is best known. The flicker is an alternation between bright and dark sensation, and spreads during the attack over constantly increasing areas of the visual field. At the same time these areas become wholly or partially insensitive to external light stimuli (so-called flicker scotoma). Closely related qualitatively to this flicker phenomenon is a physiological one, which can occur in perfectly normal conditions, and which in the case of many observers, including the writer, occurs very frequently without any special conditions to account for it. The writer notices it only in the dark, generally when there is, or has been, some slight pressure on the eye, but also without such pressure. The frequency of the intermission

[1] Ibid. II. 87.

[2] See also the phenomena in No. XXII of the first volume, and in Nos. IV, V, XV of the second volume of his observations and experiments.

appears to be very regular, probably just as in pathological flicker scotoma, about 10 to 15 oscillations per second, according to the writer's estimate. However, though the pathological flicker phenomenon usually continues a long time in a uniform way, the physiological type invariably occurs periodically in the writer's experience; being perceptible for one or two seconds, rising in intensity during this time and then falling off again. This is succeeded by a pause from three to five times as long without any flicker, after which the process is repeated.

As to the causes of this phenomenon, as also of the flicker in migraine, positive opinions can hardly be ventured at present.—N.

Appendix by W. Nagel.

Adaptation, Twilight Vision, and the Duplicity Theory.[1]

A. The Adaptation of the Eye for Different Intensities of Light.

1. Dark Adaptation

What is meant by the *adaptation of the eye*[2] is an adjustment of the degree of sensitivity of the retina for different intensities of illumination. By virtue of this capacity, the eye is enabled to go on seeing in very dim light, and, on the other hand, to bear the brightness of the sunshine on a summer day without being blinded by it or harmed. When little or no light falls on the retina, thanks to the process of adaptation, its sensitivity becomes far and away higher, as compared with the stimulus, than that of an eye that has been gazing for a long time at a bright surface, like the daylight sky for instance. This change which takes place with complete, or almost complete, exclusion of light is called the *dark adaptation* of the eye. The reverse process of becoming accustomed to bright light is called *light adaptation*. However, these terms are employed also to describe the states of the eye resulting from long-continued darkness in one case and from long-continued exposure to bright light in the other case. Experiment shows, as might have been inferred in advance, that the change is a limited one, and reaches a fairly stationary stage after a certain time.[3] When outside light is shut out entirely, the sensitivity of the eye increases for several hours; whereas the reverse process of light adaptation by exposure to bright light is completely accomplished in a few minutes. These changes and the limitations of the capacity of adaptation will be treated more in detail presently.

[1] ¶In connection with the subjects treated in this Appendix, the following comparatively recent works may be consulted: W. DE W. ABNEY, *Researches in colour vision.* 1913; and J. H. PARSONS, *An introduction to the study of colour vision.* 1915. See also: L. T. TROLAND, The progress of visual science in 1920. *Amer. Journ. Physiol. Optics* II. 1921, 316–391. Idem, Brilliance and chroma in relation to zone theories of vision. *Jour. Opt. Soc. Amer.* VI. 1922, 3–26. — Idem, The present status of visual science, *Bull. Nat. Research Counc.* V. (part 2), 1922, 1–120. — Report of committee on colorimetry for 1920-21. *Jour. Opt. Soc. Amer.* etc. VI. 1922, 527–596. (H.L.)

[2] For literature prior to 1902, consult A. TSCHERMAK, Die Hell-Dunkeladaptation des Auges, usw.; *Ergebnisse der Physiologie.* I. 2, S. 695. 1902.

[3] ¶The terms *photopia* and *scotopia* are also convenient for describing the states of light adaptation and dark adaptation, respectively. Thus also we may speak of the photopic eye or the scotopic eye. (J. P. C. S.)

Undoubtedly, it would be taking a rather one-sided view to estimate the state of adaptation of an eye merely by the threshold stimulus at the moment, that is, by the minimum effectual intensity of stimulus; because it is by no means certain in advance that the sensitivity to stimuli that are appreciably above threshold value must always proceed parallel to "threshold sensitivity." In other words, it is doubtful, for example, whether we have the right to suppose that in the case of an eye *A*, whose threshold stimulus is a hundred times lower than that of another eye *B*, a definite "super-liminal" intensity of light must be reduced to a hundredth of its value in order that *A* and *B* shall have equal sensations of subjective brightness. Such comparisons are of course only possible between the two eyes of the same observer, and for this very reason cannot be made very precisely. Moreover, the moment a stimulus appreciably above threshold value acts on the retina, its state of adaptation is changed extraordinarily quickly; and therefore, such comparisons between the sensitivity of two eyes can be made only at one instant. Under such circumstances, the agreement may be considered sufficiently accurate when the sensitivity measured for threshold values and the susceptibility to stimulation as determined binocularly for greater than threshold values turn out to be of the same order of magnitude. Now according to the writer's observations, this is always the case; and his impression is that the values of the sensitivity as determined by the two methods will be found to be more and more nearly in agreement, the more pains are taken to avoid the sources of error above mentioned; and that consequently also the determination of the state of adaptation by the threshold values is justifiable.

The first systematic investigations of dark-adaptation were made by H. AUBERT,[1] who likewise proposed the word "adaptation". He took the trouble of determining the threshold stimulus of the organ of vision after it had been kept in darkness for various lengths of time. After coming into the dark room, where the experiments were conducted, the threshold was found to be lower. The reciprocals of the threshold values afforded a measure of the degree of sensitivity of the eye for the time being, which could be exhibited in the form of a curve, where the abscissae denoted the number of minutes spent in darkness and the ordinates the corresponding values of the sensitivity. Many such curves showing the process of dark adaptation have been published since the time of AUBERT. His method of making the measurements was to send a current of electricity from a constant cell through a platinum wire until its glow was just visible in the dark room. The length of wire traversed by the current could be varied, and the longer it was, the fainter it shone. In a few minutes after entering the dark room, AUBERT found a rapid rise of sensitivity to light, and that the sensitivity continued to increase more slowly thereafter on remaining longer in the dark. After two hours in darkness the sensitivity had increased to about 35 times its original value according to AUBERT's calculation.

[1] *Physiologie der Netzhaut.* Breslau 1865. Among older works on adaptation the following are important still: CHARPENTIER, Expériences sur la marche de l'adaptation rétinienne. *Archives d'ophtal.* VI. 1887. TREITEL, Über das Verhalten der normalen Adaptation. v. GRAEFE's *Arch. f. Ophthalm.* 1887.

More recent investigations indicate a much greater increase of the sensitivity. In these determinations of relative values of sensitivity, naturally, much depends on the conditions under which the observations are made, and above all on the starting point of the series of determinations, that is, on the state of adaptation of the observer's eye when the dark room is entered. Apparently, AUBERT began his experiments with his eye in a medium state of adaptation, but we do not know definitely what it was. But for other reasons, which will be apparent later, the results obtained by his method of measuring sensitivity could not be of general value.

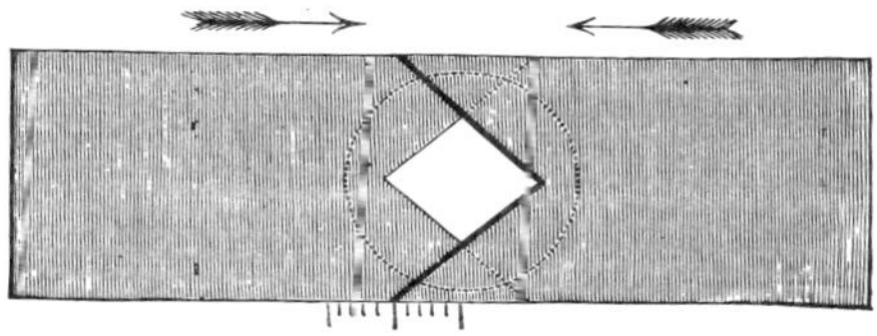

Fig. 57.

Oculists usually determine the capacity of an eye for adaptation by means of an instrument designed by FOERSTER,[1] called a *photometer* or *photoptometer.* It consists of a wooden box blackened on the inside, with two apertures on one side where the patient puts his eyes. On the opposite interior wall there is a white sheet of paper with broad black bands or other

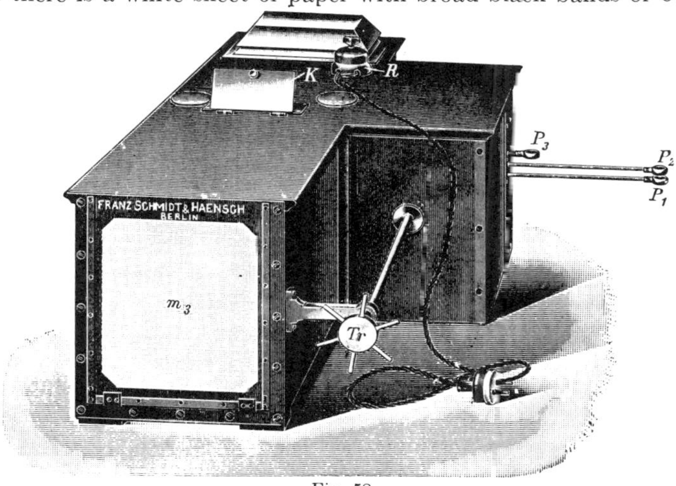

Fig. 58.

figures on it. This is the object to be observed. It is illuminated by a little window 5 cm square, in the same side of the box where the eye-holes are. The window is closed with translucent white paper or milk-glass; and on the outside there is a little lantern with a candle in

[1] R. FOERSTER, *Über Hemeralopie und die Anwendung eines Photometers in der Ophthalmologie.* Breslau 1875.

it which shines on the window. The amount of light from the window that falls on the object is regulated by a square diaphragm, as represented in Fig. 57. The size of the opening in square millimetres can be read off on a scale.

A normal eye, which has been previously dark-adapted from 15 to 20 minutes, with an aperture of 2 square millimetres, can distinguish the bright from the dark parts of the object on the back of the box. However, in this case the assumption is that the eye was not completely light-adapted before the dark adaptation was begun, but that it was in the medium state of adaptation such as exists in a moderately bright room.

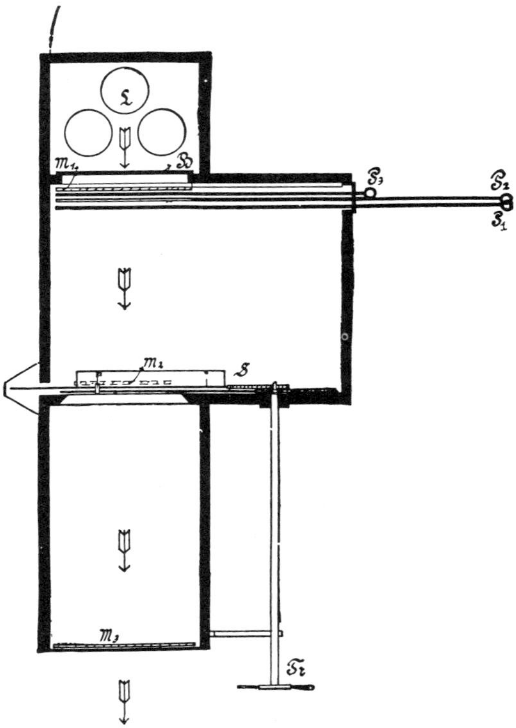

Fig. 59.

In Foerster's instrument, the amount of variation in the size of the diaphragm is far too little for measuring the threshold value of the eye for each state of adaptation. Thus, while the apparatus is well suited for ophthalmological practice and for finding gross irregularities in the power of adaptation, it is not designed for more accurate measurements.

For such purposes the writer devised an apparatus called an *adaptometer*.[1] Its construction is shown in Figs. 58 and 59.

On the front side of a wooden box 80 cm long there is a milk-glass plate m_3 which acts as object. The source of light is situated on the opposite side. It consists of three incandescent lamps each equivalent to 25 candles. A bright blue glass plate is interposed in the path of the light to make the lamp light as pure white

[1] W. Nagel, Zwei Apparate für die augenärztliche Funktionsprüfung usw. *Zeitschr. f. Augenheilk.* XVII. 1907.

as possible. There are two contrivances for reducing the intensity of the light. Just in front of the lamps three plates are inserted, made of metal and provided with holes of such size that each of the discs reduces the amount of light falling on it by one twentieth. Thus with two screens the intensity is diminished to 1/400; and with all three to 1/8000. Moreover, in the middle of the box there is a square diaphragm as in FOERSTER's instrument (see Fig. 57), which is intended for finer adjustments of the degree of intensity of light. It has a milk-glass plate behind it, and its size can be varied from 1 square millimetre to 10000. Thus the intensity of light falling on the front side of the milk-glass plate can be diminished from its maximum to $\frac{1}{8000} \cdot \frac{1}{10000} = \frac{1}{80000000}$ and the interval from 1 to $\frac{1}{8000000}$ can be evaluated with sufficient accuracy.

A similar apparatus has been described by H. PIPER[1]. By means of contrivances of this kind the complete course of dark adaptation can be followed. If comparisons are to be made of the powers of adaptation of different individuals with healthy eyes, and eventually with diseased eyes also, so as to study the process of adaptation under different conditions, it is a good plan to use a definite and constant size of field, because, as will be seen, the threshold stimulus is dependent on the size of the retinal area stimulated. The writer has chosen for this purpose a circular field of 10°. The way this is obtained is by putting a circular diaphragm 10 cm in diameter in front of the milk-glass plate of the adaptometer, and placing the eye 57 cm away from the apparatus. This is a convenient distance because the observer can then reach the handle of the instrument and adjust the intensity of the light for himself. If the problem is simply to determine the threshold for a particular degree of adaptation, the gaze ought not to be fixed, but the eye should be allowed to roam about in the field, so that the image of the disc of the adaptometer falls on various portions of the retina in succession. On the other hand, if the object of the experiment is to test the sensitivity of a particular part of the retina, a fixation mark is provided for the eye in the form of a little dark red point. If this mark emits a pure red light, it is more easily seen in the fovea centralis than in the peripheral parts of the retina, and hence the gaze is attracted by it. A contrivance designed for this purpose has been described by the writer elsewhere.

To reach the threshold of the light sense when the eye is in a good state of light adaptation, it is necessary to get fairly close to the

[1] H. PIPER, Zur messenden Untersuchung und zur Theorie der Hell-Dunkeladaptation. *Klin. Monatsbl. f. Augenheilk.* XLV. 357. 1907.

highest intensity of light that this apparatus will afford. All three of the screen plates must be removed.

To bring about such a state of "good light adaptation", all that is necessary is to stay outdoors 20 or 30 minutes on a bright sunny day, looking mainly at bright objects that are, however, not particularly dazzling. After such preparation the threshold stimulus for the majority of eyes is practically the same. It amounts approximately to an intensity of illumination on the adaptometer disc equal to one metre-candle; in other words, a white surface subtending an angle of 10° will just be visible when it is illuminated by a standard candle at a distance of one metre.

In order to compare the sensitivity in the state of dark adaptation with this, the observer retires to a perfectly dark room and stays there for an hour, or his eyes are tightly blindfolded for that length of time. For eyes thus dark-adapted an illumination of one metre-candle is dazzlingly bright. To find the threshold value, the illumination must be diminished to between fifty and one hundred and fifty thousandths. In other words the sensitivity to light has been increased anywhere from fifty to a hundred and fifty thousand times. The increase of sensitivity after dark adaptation for one hour in the case of an eye that was previously in a state of "good light adaptation" may be called the "*amplitude of adaptation*" of the eye in question. Its numerical value is the difference between the threshold of the light-adapted eye and that of the dark-adapted eye. As a matter of fact, dark adaptation takes longer than an hour to be complete, as was shown by AUBERT; and as has been confirmed many times by tests made by PIPER and the writer. After dark adaptation of both eyes for eight hours, PIPER[1] obtained twice as high a value of the sensitivity to light as he got after dark adaptation for one hour. The writer has kept one eye closed for sixteen hours, without being able to decide whether the limit of the process of dark adaptation was reached. Still by the end of the second hour further increase of sensitivity is very slight. In the experiment just mentioned, the monocular amplitude of adaptation amounted to 270,000.

When the weather is not very bright, the degree of light adaptation is not generally reached for which the threshold is one metre-candle. Moreover, it takes a good deal of preparation and practice to make very quickly the first measurement of sensitivity after going from the bright region to the dark. When the day is a little cloudy, even after being outdoors, the sensitivity is found to be from ten to twenty times higher than it is with complete light adaptation. If the state of light

[1] H. PIPER, Über Dunkeladaptation. *Zeitschr. f. Psychol. u. Physiol. der Sinnesorgane.* XXXI. 1903. 161–214.

adaptation attained in cloudy daylight is compared with the value of the sensitivity after dark adaptation for one hour, the latter is found to be only between three and eight thousand times more If the experiment is started with the state of adaptation that is reached after being in a room a long time in daylight, the sensitivity is found to be one thousandth of that in case of dark adaptation for one hour. Even on sunshiny days, complete light adaptation is not developed indoors at longest, but the sensitivity is about one or two hundred times more than that of an eye adapted to the brightness outdoors.

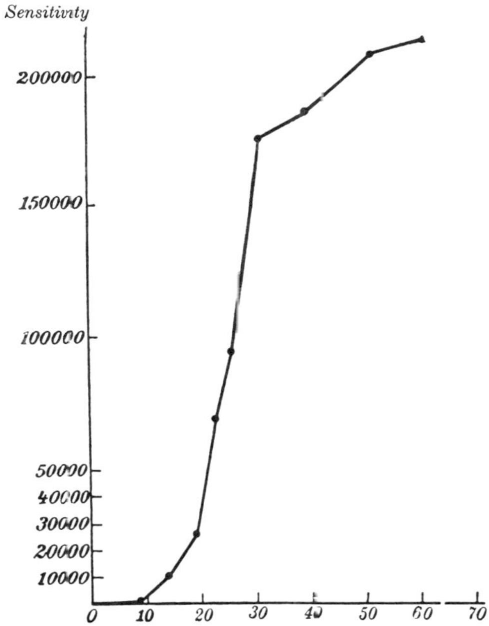

Fig. 60.

In pathological cases, in what is known as *night blindness* or *hemeralopia*, which is a symptom of many eye diseases, the amplitude of adaptation is very considerably diminished. MESSMER,[1] who investigated a number of patients afflicted with night blindness, measured amplitudes of adaptation with the adaptometer which were as low as 125, 25, and even 14. In slight cases he found values of 1250 and 1666; whereas for normal observers the amplitudes were between 3332 and 10415. Similar results were obtained by HEINRICHSDORFF and others.[2]

It should be particularly noted that these measurements of sensitivity to light are for eyes showing normal pupil reaction. In order to compare the values of light adaptation with those obtained after long dark adaptation, to be really accurate, the diameter of the pupil in each measurement must be known; otherwise, the actual retinal sensitivity under the different conditions cannot be compared. However, since all the measurements, including those

[1] MESSMER, Über die Dunkeladaptation bei Hemeralopie. *Zeitschr. f. Sinnesphysiol.* XLII. 83. 1907.

[2] ¶F. BEST, Über Nachblindheit. *Arch. f. Ophthalm.* XCVII. 1916, 168–197. — W. DE W. ABNEY, Two cases of congenital night blindness. *Proc. Roy. Soc.*, 90 B. 1917. 69–74. (H.L.)

made after light adaptation, are carried out in a dark room, the size of the pupil is of less influence than it would be under different conditions. The pupil always takes a certain time after being in the dark to reach the equilibrium condition corresponding to the lack of light, but this had practically been reached already before the measurements here described were made.

With the aid of the adaptometer or some similar instrument, the progress of dark adaptation through its various phases may be more accurately followed, by determining the threshold value at regular intervals, say, every five minutes, after entering the dark room. The results obtained by plotting the reciprocals of these values as ordinates and the numbers of minutes spent in darkness as abscissae, will be "adaptation curves" similar to those constructed by AUBERT and other investigators. An example of the typical progress of dark adaptation in a normal eye is given in Table I; and the results are exhibited graphically in Fig. 60. As a matter of fact, on account of the wide range of variation of the sensitivity, the ordinates have to be represented on an enormous scale, in order to get a true picture.

Table I

Increase of Sensitivity in the Dark

Minutes in Darkness	Values of the Sensitivity (1 unit sensitivity being that for which the intensity of 1 metre-candle is the threshold)
0.5	20
4	75
9	1850
14	10400
19	26000
23	69500
26	94700
31	174000
39	195000
51	208000
61	215000

The steepest part of the curve is comprised between the 19th and 31st minutes, the ascent being more gradual thereafter. This is typical of the normal curve of adaptation, the ordinates indicating the absolute degree of sensitivity at the time.

It is not without interest to consider the same process in another way, by exhibiting the *relative* increase of sensitivity. The table shows that between the 4th and 9th minutes the sensitivity is increased about 25-fold; whereas in the same length of time, between the 9th and 14th minutes, the increase is not more than 5.6-fold, and in the twelve minutes between the 19th and 31st minutes the increase is 6.7-fold. By plotting the logarithms of the values of the sensitivity as ordinates the *relative* increase in this sense will be exhibited. A definite increase in the height of the ordinate of the curve will mean a rise of sensitivity in a definite ratio (that is, its multiplication by a certain coefficient), and the steepest parts of the curve will be at those times when the relative increase of sensitivity to light was most rapid. In this way curves like that shown in Fig. 61 will be obtained. It is steeper at the beginning of the hour of dark adaptation than it is in the middle. It is usually steepest between the second and eighth minutes, just where the curve in Fig. 60 is almost horizontal.

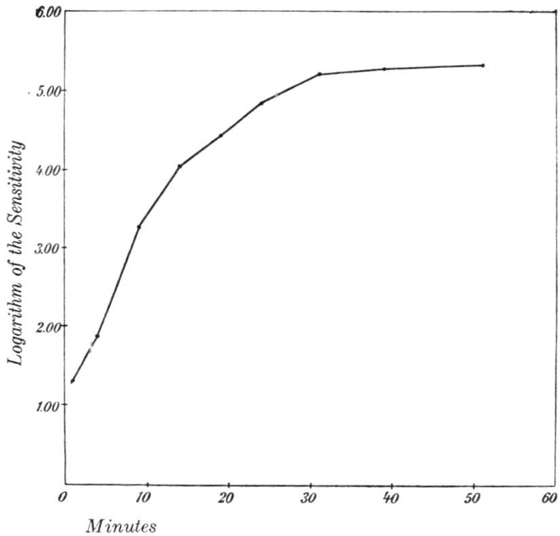

Fig. 61.

Comparing the adaptation of a large number of persons with normal eyes, we shall find that it proceeds very much the same way in all of them, the only variation being in the final limit attained. But in those diseases of the eye, where night blindness occurs (see p. 319), the sensitivity at the end of an hour is not only apt to be much less than in the case of the normal eye, but the form of the curve of adaptation may be abnormal also. It is very common to find an unusually flat stretch of curve for the first quarter of an hour, that is, a retarded increase; although at the end of the hour the final sensitivity may be similar to the normal. Such persons are much annoyed by not being able to get accustomed to the change from a bright to a dimly lighted chamber as quickly as normal individuals can do. At first they see practically nothing. In the more severe cases the sensitivity increases but little in the course of hours, and naturally such individuals are still more helpless in twilight.

For further information concerning adaptation in pathological cases, the reader may consult the works of HEINRICHSDORFF[1], MESSMER[2], LOHMAN[3], and HORN.[4]

According to the investigations of PIPER (*loc. cit.*) and WÖLFFLIN[5], there is no particular connection between the amplitude of adaptation or the course of adaptation and the time of life. Moreover, TSCHERMAK's statement[6], that special characteristics of dark adaptation are peculiar to different types of colour vision, has not been confirmed. The fact is rather that the capacity for adaptation has nothing to do with the type of colour system; and in this respect both normal and anomalous trichromats do not differ from the partially colour-blind (dichromats). Even the so-called achromats, or totally colour-blind, are not abnormal in this matter. In the case of a totally colour-blind girl, MAY[7] and the writer found the amplitude of adaptation below the average, but still within normal limits (around 5000).

Among other agencies that may modify the course of dark adaptation, the action of the nerve poisons, strychnin and brucin, should be mentioned. Injected under the skin, both of these poisons, according to the experiments of DRESER[8] and WÖLFFLIN[9], produce a distinct increase of one-fifth or one-fourth in the amplitude of adaptation. The rate of adaptation appeared to be increased also. WÖLFFLIN used doses of between two and five thousandths of a gramme of strychnin, and two hundredths of a gramme of brucin. FILEHNE[10] claims that santonin tends to retard adaptation; but this is a mistake, as has been shown by KNIES[11] and WÖLFFLIN (*loc. cit.*). In conjunction with VAUGHAN[12], the

[1] HEINRICHSDORFF, Die Störungen der Adaptation und des Gesichtsfeldes bei Hemeralopie. GRAEFES *Arch. f. Ophthalm.* CX. 405. 1905.

[2] MESSMER, Über die Dunkeladaptation bei Hemeralopie. *Zeitschr. f. Sinnephysiol.* XLII. 83. 1907.

[3] LOHMANN, Untersuchungen über die Adaptation usw. *Habilitationsschrift*. MÜNCHEN und GRAEFES *Archiv f. Ophthalm.* 65, 1907.

[4] HORN, Über Dunkeladaptation bei Augenhintergrundserkrankungen. *Dissertation*, Tübingen 1907.

[5] WÖLFFLIN, Der Einfluss des Lebensalters auf den Lichtsinn bei dunkeladaptiertem Auge. GRAEFES *Arch. f. Ophthalm.* 61, 524. 1905.

[6] TSCHERMAK, *Über physiologische und pathologische Anpassung des Auges.* Leipzig (VEIT u. Co. 1900); also: *Ergebnisse der Physiologie*. I. 2. S. 700. 1902.

[7] MAY, Ein Fall totaler Farbenblindheit. *Zeitschr. f. Sinnesphysiol.* XLII. 69. 1907.

[8] DRESER, Über die Beeinflussung des Lichtsinns durch Strychnin. *Arch. f. exp. Pathol. u. Pharm.* XXXIII.

[9] WÖLFFLIN, Über die Beeinflussung der Dunkeladaptation durch künstliche Mittel; GRAEFES *Arch. für Ophthalm.* 65.302.1907. See also SINGER, Brucin und seine Einwirkung auf das normale Auge. Ibid., 50.

[10] FILEHNE, Über die Wirkung des Santonins usw. PFLÜGERS *Arch. f. d. ges. Physiol.* LXXX. 96. 1900.

[11] M. KNIES, Über die Farbenstörung durch Santonin. *Arch. f. Augenheilk.* XXXVII.

[12] C. L. VAUGHAN, Einige Bemerkungen über die Wirkung des Santonins auf die Farbenempfindungen. *Zeitschr. f. Sinnesphysiol.* XLI. 399. 1906.

writer has made a large number of experiments with santonin poisoning, large doses and small doses, without discovering any effect whatever so far as adaptation is concerned.

In an indirect way the mydriatics have an effect. For example, when the pupil is dilated by atropin, abnormally large amounts of light can enter the eye and produce a state of blindness or glare. The writer has occasionally exposed one of his eyes, with the pupil dilated, for about a half hour to the glare of the sun in the street, and then found that adaptation was much retarded. The first or flat part of the sensitivity curve, as seen in PIPER's curves (also in the curve shown in Fig. 60), extends over 20 minutes, instead of from 8 to 10 minutes. Thereafter the sensitivity increased rapidly and reached its normal value in from 70 to 80 minutes.

AUBERT noticed that the progress of dark adaptation was not only not impeded by flashes of light considerably above the threshold intensity, but that, in fact, it was aided in some measure. This observation has since been confirmed many times in the writer's laboratory. That the instantaneous light of a match reflected from dark walls into the observer's eye lowers the threshold of the light sense considerably, as much perhaps as one-third, can easily be verified. This effect does not wear off for several minutes. It is particularly distinct when the observer has reached a practically stationary condition of adaptation after having been in the dark for an hour.

In peculiar contrast to this, is the fact that after being in the dark for a very long time, a sensitivity is attained which is diminished by stimuli that are close to threshold value. For example, the above mentioned maximum sensitivity attained by the eye after 16 hours of darkness, in which an illumination of 1/270,000 metre-candle on the adaptometer field was found to be the threshold value, is merely transitory. A few observations in which the eye was exposed to stimuli not more than three times as strong as the liminal value sufficed to lower the sensitivity to about one-half. The writer, having repeatedly made similar observations, can state that the increase of sensitivity which takes place after an hour in darkness is of a much more transient nature than that which occurs during the first hour.

It has never been certainly demonstrated that the state of adaptation of one eye has any influence on that of the other. The dark adaptation in each eye seems to be independent. It may be that occasionally with just one eye dark-adapted one gets the impression of not attaining the same degree of sensitivity in monocular observation as when both eyes are dark-adapted; but this is partly due to subjective luminous appearances that are apt to occur when the state of adaptation is not the same in both eyes. It may be partly due also to the fact that it is more difficult than one might suppose to bandage an eye so tightly that not a glimmer of light can be detected after long adaptation. In investigations of this nature, it must be remembered that in the dark-adapted state, the threshold stimulus for monocular vision is found to be higher than for binocular vision (see below, page 339).

So far as the final degree of sensitivity is concerned, it is of no importance whether the eyes are free during dark adaptation or whether they are under more or less pressure from being bandaged.[1] Even the passage of an electric current through the eye was found by the writer and his assistants to have no effect on the threshold stimulus for light. Accordingly, it has not been shown that there is any change of excitability of an electrotonic nature.[2]

2. Light Adaptation

The decrease of sensitivity to light connected with the process of *light adaptation* cannot be so conveniently followed over a long interval as the reverse process, because precise sensitivity determinations can only be made by threshold methods and in a dark room. A quantitative investigation of the decrease of sensitivity in light adaptation can therefore only be carried out, by first finding the threshold value after being in the dark for a long time, then exposing the eye to a brighter light for a definite time, then suddenly cutting off this light adaptation, darkening the room again, and quickly making another threshold measurement, before the dark adaptation has had time to be appreciable. In this manner it has been possible to determine how much the sensitivity to light is lowered by a definite intensity of illumination of known duration. W. Lohmann[3] has taken no little trouble to make a large number of measurements of this sort. He varied systematically the intensity of the light that was used in the light adaptation, and also the duration of the exposure of the eye. Lohmann performed the experiment above described, first, with a light adaptation lasting for one minute; then, after re-inducing the state of dark adaptation, he used light adaptations of 2 minutes, 3 minutes, 6 minutes, etc. In this way he got an idea of the behaviour of the process of light adaptation with a given intensity of illumination. The starting point of his experiments was invariably a state of dark adaptation produced by

[1] Nagel, Einige Beobachtungen über die Wirkung des Druckes und des galvanischen Stromes auf das dunkeladaptierte Auge. *Zeitschr. f. Psychol. und Physiol. d. Sinnesorgane.* XXXIV. 285. 1904.

[2] ¶See S. Hecht. The dark adaptation of the human eye. *Jour. Gen. Physiol.* II. 1920. 499. 517; also, The Nature of foveal dark adaptation. Ibid. IV. 1921. 113-141. (H. L.)

[3] W. Lohmann, Über Helladaptation. *Zeitschr. f. Sinnesphysiol.* XLI. 290 1906.

being in the dark for 45 minutes. Adaptation in broad daylight proceeds with extraordinary rapidity, as has been stated, and probably may be completed at the end of a minute under some circumstances. And so LOHMANN generally used moderate intensities, by which the process could be followed more closely. In observations of this kind it must be kept in mind that after extinguishing the light used for light adaptation, the sensitivity begins to rise again immediately, faster, indeed, in proportion as the light adaptation was more moderate. Consequently, the threshold determination must be made as quickly as possible after extinguishing the light. But it must not be done too quickly, because during the first few seconds in the dark observation is extremely difficult on account of successive after-images.

For these reasons those measurements of LOHMANN's are mentioned first in which the threshold determination was made 10 seconds after extinguishing the light. As a matter of fact, the sensitivity will have increased a little in this time, to a different extent also for different periods of light adaptation. But this is an error that can hardly be avoided. Incidentally, it does not diminish the value of LOHMANN's results. The difficulty of the task compels us here to put up with approximate determinations.

Table II

Decrease of Sensitivity to Light during Light Adaptation under different degrees of Illumination and for different durations of Light Adaptation.

Intensity of illumination of the white surface at which the subject gazes for being light-adapted	Duration of Light Adaptation						
	⅓ min.	⅔ min.	1 min.	2 min.	3 min.	6 min.	10 min.
	Sensitivity Value						
5 metre-candles	23000	17500	10400	8130	5200	3470	3000
25 metre-candles	9950	7440	5200	3360	2740	2040	1450
50 metre-candles	5800	3700	3250	2600	2038	1600	1130
moderate daylight	435	230	200	115	87	48	40

Table II is based on LOHMANN's results Instead of using his threshold value, the sensitivity values here have been reduced to unit sensitivity (threshold at one metre-candle). The results are exhibited in Fig. 62 in the form of a curve. We see how the sensitivity falls off abruptly during the first minute. When the adaptation is for diffusely reflected daylight, the sensitivity is enormously reduced in 20 seconds. The ordinate which would indicate the degree of sensitivity before beginning the light adaptation would have to be about five times as high as the tallest ordinate of the curves shown in the diagram.

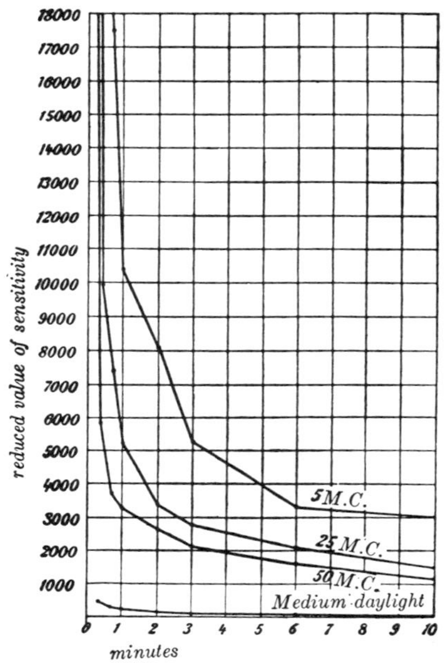

Fig. 62.

After the first precipitous drop, the descent of the curve becomes much more gradual. Light adaptation therefore continues beyond the first two minutes, but very slowly. But for all that, even in adaptation to daylight, the lowering of sensitivity can be traced for at least ten minutes. From the experiences of daily life it would be natural to suppose that, especially under the action of light of low intensity, the sensitivity would quickly reach a definite level, as would be indicated by horizontal procedure of the curve. But, as LOHMANN's tests show, this is not so. For illuminations of 25 and 50 metre-candles, LOHMANN continued the experiments farther still. The results are given in Table III, the sensitivities being referred to unit sensitivity, as in the preceding table.

Table III

Decrease of Sensitivity for long-continued Light Adaptation

Duration of Light Adaptation in minutes	10	15	29	34	39	49	60	70	79	80	99	109	110
25 metre-candles.......	1450	1000		250		125	95	62		54			54
50 metre-candles.......	1130	312	104		46		36		28		25	24	

We see from the table and from Fig. 63 that the sensitivity continues to fall off after the first ten minutes, and it is more than a half hour before adaptation reaches a kind of level. This level, by the way, is curiously different in height according to the intensity of the light used for adaptation. For obvious reasons the scale of coördinates is not the same in Figs. 62 and 63, but since in both cases the ordinates denote the values of the sensitivity based on the same unit, the curves may readily be compared with each other and likewise with the curve of dark adaptation shown in Fig. 60.

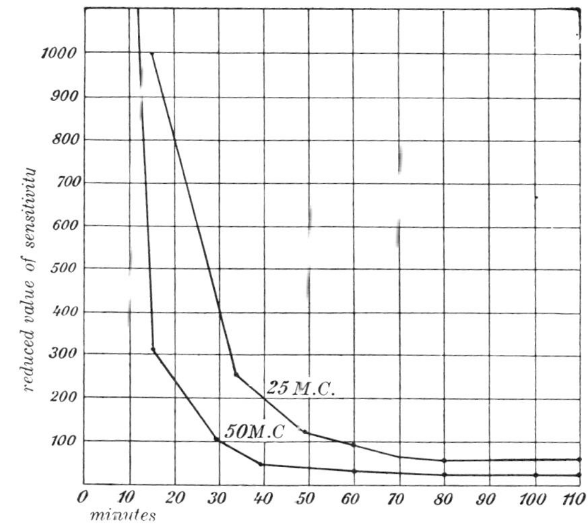

Fig. 63.

The fact that moderate intensities of light are slow in nullifying the effect of dark adaptation, may easily be verified. If one eye is tightly bandaged for half an hour and the other exposed to diffused daylight for the same length of time, on removing the bandage, the sensitivity of the eye that was closed sinks rapidly down to a certain point; but in a dimly lighted room continues then for a half hour or an hour distinctly higher than that of the other eye which has not been closed. The difference can be perceived by alternately opening and closing the eyes.

The basis of all the quantitative results on light adaptation which have been given above is the measurement of the liminal value made ten seconds after extinguishing the light used for light adaptation. Another set of LOHMANN's experiments affords a picture of the rise of sensitivity during the first minutes in darkness, that is, of the after-effects of light adaptation during the early period of dark adaptation. Without going into the method of these experiments, the way the dark adaptation proceeds, after previous light adaptation with an intensity of illumination of 75 metre-candles, is exhibited in Fig. 64. Having been previously thoroughly dark-adapted, the observer gazed at a white surface illuminated by 75 metre-candles, for one, two, three, six or ten minutes. The gradual renewal of sensitivity during the first four minutes in darkness is then determined. The first measurement is made after ten seconds. In the diagram the numbers of seconds reckoned from the completion of light adaptation are plotted as abscissae, and the values of the sensitivity as ordinates. These values are relative, but they may be compared, approximately at least, with those of the foregoing curves of adaptation, by noting that the highest ordinate in the figure corresponds to a sensitivity value of about 16000 on the basis of the unit of sensitivity in the previous diagrams. Fig. 64 brings out clearly how, after brief light adaptation lasting only one minute, the sensitivity rises again more quickly than it does after the illumination of 75 metre-candles has acted a longer time. If the light adaptation is produced with lower intensities, the curves get steeper still and start off with higher ordinates.

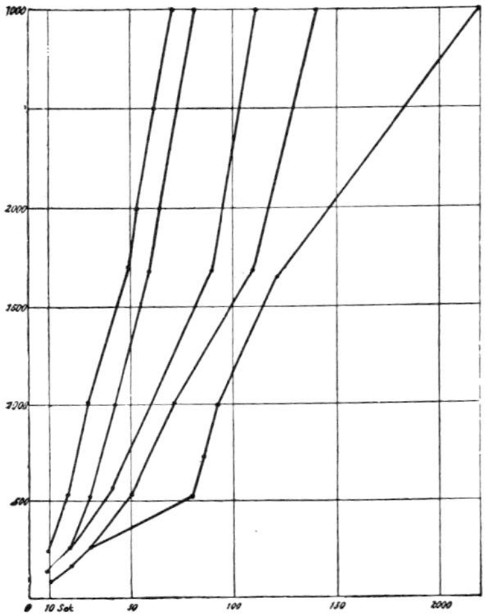

Fig. 64.

When we consider the rapidity with which the sensitivity of a retina adapted to moderately strong light starts to rise as soon as the light is extinguished and also on suddenly changing into a dimly lighted room, it is easy to explain the following observation, which may be regarded as a special kind of after-image. When one comes from the street into a comparatively dark vestibule and looks steadily for some seconds at an object with strong contrasts of brightness (for example, a bright wall with an inscription in large letters), a very vivid after-image is obtained, which is always negative. This is because the adaptative increase of sensitivity has taken place more quickly in the parts of the retina where the dark letters are imaged than in the more highly illuminated portions. Thus a stage of adaptation at which the sensitivity is rapidly rising is a prerequisite. Subjectively, the intensity of illumination may seem to be just as bright as it was in the experiment in the dark vestibule, and yet if the observer is in a more or less stationary state of adaptation, he will try in vain to produce negative after-images of considerable intensity, because in such a case much stronger stimuli will be needed for that.

At the suggestion of the writer, Mr. H. J. Watt[1] carried out an investigation on the production and behaviour of negative after-images, comparing the results in the case of a light-adapted eye and the case of a thoroughly dark-adapted eye, the subjective brightness being the same for both eyes. No marked difference could be detected in the two cases. It is rather just at that particular stage when the sensitivity is rapidly rising, at the beginning of dark adaptation, that the conditions are peculiarly favourable to the production of those characteristic negative after-images.

[1] H. J. Watt, Über die Nachbilder subjektiv gleich heller aber objektiv verschieden stark beleuchteter Flächen. *Zeitschr. f. Sinnesphysiol.* XLI. 312. 1906.

3. Local Variations in Retinal Sensitivity

When a person in a state of medium adaptation enters a dark room where there is a very faintly luminous object, it can be perceived better with the peripheral parts of the retina than with the central parts, as may be easily verified. If the luminous object is small, subtending a visual angle of from 1° to 2°, this difference between peripheral and central vision persists even after long dark adaptation; in fact, it becomes more pronounced. But with objects whose apparent size is 10° or more (for instance, the disc of the adaptometer), the difference between the centre of the retina and the periphery disappears more and more with protracted stay in the dark, and after an hour's adaptation can scarcely be detected. More exact investigations enable us to explain this phenomenon as follows.

The centre of the retina, the so-called fovea centralis, behaves quite differently from the rest of the retina so far as the process of adaptation is concerned. Indeed, it shares with the rest of the retina the very rapid rise of sensitivity that occurs in the first two or three minutes of darkness, but after that it remains stationary[1]; whereas in the other parts of the retina it is then that the sensitivity to light has its biggest increase. Thus the foveal region behaves quite analogously to the total retina of an eye afflicted with a bad case of hemeralopia, so that we can use v. KRIES's expression and speak of a "physiological hemeralopia" of the fovea.

This explains why in dark adaptation *small* objects are more easily seen by the peripheral retina than by the fovea. The reason why, on going into a dark room, the periphery is better suited for seeing faint objects (even those that are too big for the fovea) than the central part, is different. Even in a bright room the light that reaches the extreme peripheral parts of the retina in the equatorial region is, as a rule, essentially less than that that goes to the centre of the fundus of the eye. The periphery therefore is in a condition more nearly like that of dark adaptation, and hence in dim light it is from the very start better adapted for perceiving feeble light stimuli.

Systematic investigations of the distribution of the sensitivity to light in the dark-adapted retina have been carried out under the direction of v. KRIES.[2] The eye of the observer fixated a tiny luminous point. At the same time a small white object, whose illumination can

[1] ¶See S. HECHT, *loc. cit.* (H. L.)

[2] J. v. KRIES, Über die absolute Empfindlichkeit der verschiedenen Netzhautteile im dunkeladaptierten Auge. *Zft. f. Psychol. u. Physiol. d. Sinnesorgane.* XV.327. 1897. Among earlier investigations should be mentioned especially those of GUILLERY, in *Zft. f. Psychol.* usw. XII. 261, 1896; XIII. 187. 1897; PFLÜGERS *Arch. f. d. ges. Physiol.* 66. 401. 1897.

be gradually regulated at will, was presented to the eye at various lateral distances from the point of fixation. In this way the sensitivity of the separate parts of the retina could be compared. The results of these measurements for a small region of the central part of the retina, extending 4° to right and left of the point of fixation, are exhibited in Table IV; and graphically in Fig. 65. In these experiments the angular diameter of the object was 0.35°, and it was illuminated by bluish white light.

Table IV

Sensitivity (relative value)	Temporal distance in degrees	Nasal distance in degrees	Width of blind area in degrees
1	1.07	0.85	1.92
1.78	1.22	1.06	2.28
7.12	1.70	1.36	3.08
16.02	2.3	1.92	4.22
28.48	3.0	2.50	5.58
44.50	3.75	3.33	7.08
64.08	4.04	4.04	8.08

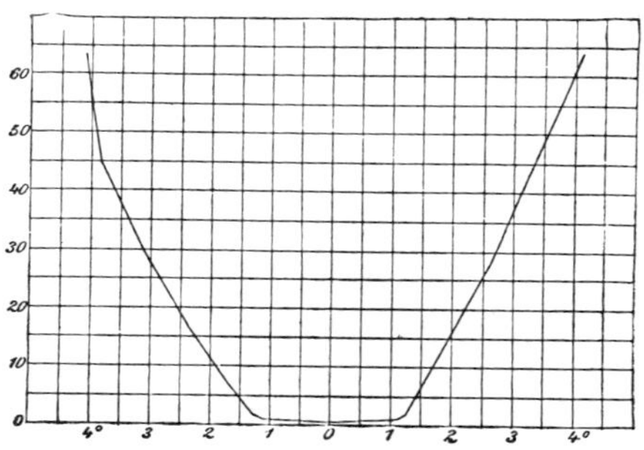

Fig. 65.

It is evident how the sensitivity begins to rise rapidly a little more than 1° away from the point of fixation; and how the lowest sensitivity is in the foveal field over an angular diameter of about 2°. Still farther towards the periphery the sensitivity to light in the uniformly dark-adapted eye goes on increasing, and, according to the investigations made by v. KRIES and his students, reaches a maximum at an angular distance from the fovea of from 10° to 20°. In some experiments made by BREUER and the writer together, taking the sensitivity at 6° from the centre as unity, the sensitivity at 12° on the temporal side was found to be 1.38, and at 18° 1.64; on the nasal side at 12° it was 1.54, but at 18° it fell to 1.37. Numerical comparisons of the sensitivities in the foveal and extra-foveal regions, as will be explained presently, are of no value except for a particular quality of light. For an eccentricity of 10°, in case of good dark adaptation, PERTZ (see v. KRIES, *loc cit.*) found the sensitivity for a blue light (subtending an angle of 8.6′) to be about 1400 times more than the foveal sensitivity. Using mixed white light, subtending an angle of half a degree, after an hour's adaptation, the writer found the sensitivity in the most sensitive part of the periphery to be in round numbers a thousand times greater than it was in the fovea under the same conditions. In this connection, it should be noted that the distribution of sensitivity in the light-adapted eye is entirely different, the maximum sensitivity to light in this case being in the fovea centralis. This is easy to prove by having a small point-source of light of intensity just a little above the foveal threshold somewhere in a dark room. If the eye is thoroughly light-adapted, it is very difficult to enter the room and find this spot of light right away. It is almost by accident that it is seen, and then it is surprisingly distinct, although it may be quite easily lost again. On the other hand, a dark-adapted eye will have no difficulty in discovering the luminous spot immediately, because the peripheral sensitivity is greater than that in the fovea.

Accurate measurements of the decrease of sensitivity from the fovea to the periphery of the light-adapted eye, with due consideration for the state of adaptation, cannot be made except with coloured lights. However, in the observations on the distribution of sensitivity to coloured lights in the light-adapted eye, as carried out at the writer's suggestion by C. L. Vaughan and A. Boltunow[1], the decrease of sensitivity out towards the periphery was found to be exactly of the same nature for red, green, and blue lights; and, perhaps, therefore, the quantitative results of these observers may be applied without hesitation to all light of any kind at all. What they found was that the sensitivity 10° to one side of the centre of the fovea was not more than one-fourth of its value in the fovea; at 20° away it was one-tenth; and at 35° away one-fortieth.

The relative central scotoma of the dark-adapted eye. As has been already stated, light stimuli, whose intensities are below the threshold value of the fovea cannot be seen in a central region of about two degrees, but are visible immediately as soon as they fall on the retina outside this area after it has been made more sensitive by being dark-adapted. The region within which stimuli of still lower intensity cannot be perceived is correspondingly wider. The dimensions of this "blind region" are given in the fourth column in Table IV for the intensity of light employed in that case.

[1] Vaughan u. Boltunow, Über die Verteilung der Empfindlichkeit für farbige Lichter im helladaptierten Auge. *Zeitschr. f. Sinnesphysiol.* XLII. 1. 1907. See also Guillery, *Zeitschr. f. Psychol. u. Physiol. d. Sinnesorg.* XII. 261. 1896; XIII. 189. 1897.

In ophthalmology those parts of the visual field from which no sensation of light can be aroused in the eye are called *scotomata*. The scotoma is said to be *absolute* when no light at this place, however brilliant, can arouse a sensation in the eye; and it is said to be *relative* when the part of the retina concerned requires a much stronger stimulus for excitation than the surrounding parts. In this sense there is a relative scotoma in the visual field of the dark-adapted eye at the place corresponding to the fovea centralis.

The existence of this scotoma can be very easily verified as soon as dark adaptation has proceeded far enough. Astronomers all know that a faint star can be seen better by not looking straight at it, but a little to one side of it, in other words, by letting the image fall near the fovea.[1] The best way to see this is to consider a cluster of stars, like the Pleiades, in which there are some bright stars and some fainter ones side by side. Looking directly at this constellation, four stars, or five at most, will be visible; but looking a little to one side, straightway a whole lot of fainter stars come into view. Gazing at the sky with its myriad stars on a clear night, one will be surprised how many of them become invisible when he stares straight at them. It is not so easy for some persons to be convinced of the existence of this relative central scotoma, because there is some difficulty in looking straight at an object which disappears as soon as it is fixated. R. SIMON[2] investigated the relations of fixation in the dark-adapted eye. The subject was told to look at a luminous point whose light-intensity was below the foveal threshold; and by locating exactly where the blind place was, the point on the retina was found which was focused under these conditions by a dark-adapted eye. CHRISTINE LADD-FRANKLIN[3] had found that, in fixating a point of light so dim as to be below the foveal threshold in intensity, individual observers looked in different directions. SIMON corroborated this. In his own case this spot was above the fovea on the temporal side in his right eye, and right above the fovea in his left eye.

It is an interesting fact that this fixation place varies under certain circumstances even in the same individual. Thus, from the tendency to fixation, a luminous point whose intensity is near the foveal threshold is focused on the retina at a place very close to the fovea. The dimmer the light, the farther away is the part of the retina that is used in looking at it. The change in the subjective brightness of the

[1] ¶ARAGO said: "Pour apercevoir un objet très peu lumineux, il faut ne pas le regarder." (J. P. C. S.)

[2] R. SIMON, Über Fixation im Dämmerungssehen. *Zeitschr. f. Psychol. u. Physiol. d. Sinnesorg.* XXXVI. 186.

[3] See A. KÖNIG, *Ges. Abhand.* S. 353.

point, induced by progressive dark adaptation, proceeds in the same direction as the change in the objective intensity of the light stimulus. Thus, with a luminous point having a certain intensity below the foveal threshold, SIMON found that, after 10 minutes dark adaptation, his point of fixation was about 2° on one side of the fovea; after 10 minutes more about 1.5°; and at the end of an hour only about 1°. Thus, perhaps, for every eye there is a fixed *direction* in which it turns in the tendency to fixate when the light-stimulus is below the foveal threshold, although the distance from the fovea of the point used for fixation may be variable. What is the direction from the fovea of the point on the retina that is selected for the fixation of dim lights, probably depends essentially on muscular and dioptric conditions, and to a less extent on the fact that certain parts of the retina immediately surrounding the fovea are prominent as being sensitive and are preferred, therefore, for fixation.

In general, we have no notion of the focusing of a point that is outside the fovea, that is, of an unusual point for fixation. For this reason many observers are sceptical at first when the attempt is made to demonstrate foveal hemeralopia to them. They are persuaded that a dim object has to be fixated in the same way as a bright one; although as a matter of fact they do not look straight at the dim object, and that is how they see it. With many persons the experiment succeeds better when they look first in any other direction, and then suddenly direct the gaze on the faint spot that is to be descried. Then immediately it vanishes for a time, until by slight involuntary ocular movements the image is again shifted to a more sensitive part of the retina near the fovea. The following experiment is exceedingly instructive. A small round white paper disc between 5 and 6 cm in diameter is fastened on the black outside wall of a box, which has a tiny electric light inside. A little hole is made in the centre of the disc, and a piece of red paper inserted behind it, so that the light from the lamp shines through it. The white paper disc is illuminated from the front, on the side where the observer stands, by a small lamp whose intensity can be regulated. This may be a little gas jet or an electric lamp in a milk-glass lantern with an iris diaphragm. The observer, having been previously dark-adapted, stands about 3m away and looks at the little red spot, at the same time regulating the illumination of the white disc, until it ceases to be visible to him. Then if he turns his eye the least bit so as to make the image of the disc fall outside the fovea, it straightway reappears, dark grey at first, becoming bright white when the eccentricity is increased.

Anyone who has had some practice in this experiment with this fixation-point, can usually perform it afterwards with the same result when the red point is extinguished. Thus all that is needed is practice in learning how to direct the eye in a dark room towards a dim object that disappears when it is fixated. The reason why it seemed worth while to describe this experiment in detail is because competent observers have been heard to say, either that there is no such thing as a central scotoma, or that it is merely a trivial difference of sensitivity between the fovea and the rest of the retina. The experiment which has just been described has invariably converted such sceptics.

4. Increase of Sensitivity to Light in the Fovea Centralis in Darkness

Although the fovea is hemeralopic in comparison with the retinal periphery, it is not entirely devoid of the power of adaptation.[1] Undoubtedly, it is very much less than it is in the other parts of the retina. A measurement of the amplitude of adaptation is beset with difficulties, at least in the later stages, where the sensitivity of the periphery increases faster than that of the fovea; the result being, therefore, that it requires much practice to succeed with the foveal fixation of the stimulus-light used for the measurement. The determinations are easier in the first stage of the dark adaptation that is preceded by complete light adaptation. Then for a couple of minutes still the fovea is the most sensitive place of the retina and therefore easily holds the fixation steady. One circumstance, indeed, that makes the experiment more difficult is that at this stage of complete light adaptation, as has been already stated, an object, whose visual angle is not greater than that of the fovea and whose intensity is close to the foveal threshold, cannot be easily found except by accident, so to speak. In this case every observer is in a position similar to that of a patient afflicted with so-called *retinitis pigmentosa* whose field of vision is restricted to the region of the fovea. The vision is like that produced by looking through a narrow tube. Accordingly, in the experiments made by Schaefer and the writer[2] for measuring the foveal adaptation, the plan was adopted of making the brightness of the test-object distinctly beyond the threshold at first, so that the observer, coming into the dark room, could easily locate it. The field was then quickly darkened by an assistant until it was just on the border of visibility. During the first minute the sensitivity was found to increase about five-fold. However, there was some uncertainty about these determinations. But after the first minute in the dark, the increase of sensitivity can be measured quite accurately. It amounts to about four-fold and lasts from 5 to 8 minutes. In all, therefore, from the

[1] ¶See S. Hecht, *loc. cit.* (H. L.)

[2] *Zeitschr. f. Psychol.* XXXIV. 1904. S. 271.

very beginning of darkness, foveal sensitivity increases about 20 times, which is certainly not an extraordinarily higher value. The writer has not detected any increase of foveal sensitivity beyond the tenth minute.

It is important to realize that foveal adaptation proceeds in exactly the same way and at the same rate for lights of different colours or wave-lengths. Thus, lights that look equally bright to the fovea of the light-adapted eye, will appear the same way when the fovea has been dark-adapted for any length of time. The so-called PURKINJE phenomenon, which will have to be considered presently, does not, therefore, exist in pure foveal vision.

Another thing to be emphasized is that, *generally, in order to demonstrate that there is an increase of sensitivity in the fovea, it is necessary to begin with very thorough light adaptation (looking at the bright sky).* No satisfactory measurements of adaptation like those described can be obtained by merely changing to the dark room from a room illuminated by ordinary daylight. The sensitivity may probably rise a little during the first seconds, and that will be all.

TSCHERMAK[1] has stated that dark adaptation, and particularly the PURKINJE phenomenon, occurs in the fovea too, provided light is excluded for a sufficient length of time. And so he assumes a much slower adaptation in the fovea than in the periphery. Some other writers make the same statement, but the discrepancy between this result and that given above is due to using test-objects that are so large that their retinal images extend beyond the rod-free foveal region. In the immediate vicinity of the fovea, where the rods are all isolated (as to the significance of this, see below, page 344), as a matter of fact, the adaptative increase of sensitivity in darkness is so minute that it takes hours of darkness to detect it at all. But, of course, this is something entirely different from the increase of sensitivity which takes place in the first few minutes of darkness, and which occurs in the fovea as well as in the rest of the retina.

5. Relations between Sensitivity to Light and the Stimulated Area of the Retina

In his experiments in the dark room, AUBERT[2] noticed that it took less illumination to see large areas than small luminous objects of the same intensity. Many investigators since then have studied this matter, in hopes of ascertaining the laws and have sought to find the conditions governing the relations between sensitivity to light and the superficial dimensions of the source of the stimulus. Among the earlier writers may be mentioned TREITEL[3], RICCO[4], and CHARPENTIER.[5] In very recent years the problem has been worked out in the

[1] A. TSCHERMAK, PFLÜGERS *Arch. f. d. ges. Physiol.* LXX. 1898. S. 297.

[2] H. AUBERT, *Physiologie der Netzhaut.* Breslau 1865.

[3] TREITEL, Über das Verhalten der normalen Adaptation. v. GRAEFES *Arch. f. Ophthalmologie.* 1887.

[4] RICCO, Relazione fra il minimo angolo visuale e l'intensita luminosa. *Annali d'Ottalmol.* VI.

[5] CHARPENTIER, *C. R.* XCI. 995. 1880; and *Arch. d'opht.* II. 487. 1882.

writer's laboratory from various angles by PIPER[1], LOESER[2], HENIUS[3] and FUJITA.[4]

The relations between sensitivity to light and the area of the surface are different in the two stages of adaptation of the eye; and they are different also in the fovea and in the periphery. The rule formulated by RICCO and confirmed by LOESER, namely, that the product of the area of the retinal image and the intensity of the light is constant for threshold stimuli, or, in other words, that the sensitivity is proportional to the area of the image, is true for purely foveal vision. In the case of a luminous circular field the law may also be stated as follows: The product of the angular diameter of the field and the square-root of the threshold intensity is constant. That this law is approximately true, at any rate, is shown by the following example taken from a set of experiments by LOESER.

Table V
(Distance of object from observer is 8 metres.)

Diameter of object D, in mm	Square-root of light intensity expressed by the diameter of a stop, J	Visual angle $\frac{D}{E}$	Product $\frac{D.J}{E}$
20	0.87	2.5	2.18
14	1.27	1.75	2.22
8.5	2.4	1.06	2.5
5	3.45	0.63	2.26

For the periphery of the retina, the relations are different. The following table, taken from PIPER's results, refers to the stage of dark adaptation.

Table VI

Area	$\sqrt{\text{Area}}$ or angular size	Threshold value	Relative stimulation value	Product of angular size and threshold value
1	1	10	1	10.0
10	3.15	2.94	3.4	9.3
25	5	1.96	5.1	9.8
100	10	1.02	9.8	10.2

[1] H. PIPER, Über die Abhängigkeit des Reizwertes leuchtender Objekte von ihrer Flächen- bzw. Winkelgrösse. *Zeitschr. f. Psych. u. Physiol. d. Sinnesorg.* XXXII. 98.

[2] L. LOESER, Über die Beziehungen zwischen Flächengrösse und Reizwert leuchtender Objekte bei fovealer Beobachtung. *Beiträge z. Augenheilk. Festschrift für* J. HIRSCHBERG 1905.

[3] K. HENIUS, Die Abhängigkeit der Lichtempfindlichkeit von der Flächengrösse usw. *Zeitschr. f. Sinnesphysiol.* XLIII. 99. 1909.

[4] T. FUJITA, Versuche über die Lichtempfindlichkeit der Netzhautperipherie unter verschiedenen Umständen. Ibid. XLIII. 243. 1909.

It is evident at once that there is no question here of any proportionality between area and stimulus value, and so, of course, the product of area and threshold value cannot be constant. But there does exist an almost complete proportionality between the stimulus and the square-root of the area (or the visual angle of the retinal image, on the supposition that the objects concerned are geometrically similar). The threshold values are given in this table in terms of a unit that enables us to see easily the connection between the value of the stimulus and the visual angle.

PIPER has shown that, over a wide range, the stimulus value of a luminous surface, at the measured threshold value, is independent of the form of the surface. As a result of his observations, he formulated the following statement: For the dark-adapted peripheral retina, the value of the stimulus of a luminous surface is proportional to the square-root of the area of the retinal image; or in other words, the product of the threshold value by the square-root of the area of the retinal image is a constant.

As shown by the latest investigations of HENIUS and FUJITA, this law is not absolutely true unless PIPER's experimental conditions are observed, that is, unless the light is mixed white light and the angular size of the object is not more than 10°. The departures from the rule are more marked when still larger surfaces are used, as in shown in Table VII (taken from HENIUS's results). The values of the stimulus are computed (like those in Table VI) in terms of a unit defined to be the stimulus due to an object which subtends an angle of 1°. The place of the retina used for observation was 10° above the fovea.

Table VII

1	2	3	4	5
Area	$\sqrt{\text{Area}}$ or diameter of the round object in degrees	Threshold value	Product of 2 and 3	Stimulus value (Sensitivity)
1	1	454	454	1.0
4	2	260	520	1.7
9	3	127	381	3.5
16	4	94	376	4.8
25	5	79	395	5.7
49	7	60	420	7.5
100	10	46	460	9.8
225	15	35	525	13.0
400	20	26	520	17.5
625	25	25	625	18.1
900	30	21	630	21.6

From 10° on, therefore, the sensitivity, as expressed in terms of the value of the stimulus, rises distinctly more slowly as the size of the angle increases. This increase is much less still when the stimulating light is red; even for small angles. Furthermore, in fields smaller than 1° the variations from PIPER's rule are considerable, as FUJITA has shown.

But in the case of vision with light-adapted eye, the relations are essentially different. For luminous areas which subtend an angle that exceeds the foveal visual angle of from 1.5° to 2°, no clear connection has been found between the threshold stimulus and the size of the angle. This is shown distinctly in Table VIII taken from FUJITA's results.

Table VIII

White Light Stimulation, 30° from fovea, Light Adaptation

Visual Angle	Sensitivity (as reciprocal of threshold value)
1°	25
2°	40
3°	50
5°	50
10°	50
15°	50
20°	53
30°	53

However, the last two values indicate a small increase as compared with the preceding ones; but it is extremely likely that a slight admixture of unavoidable dark adaptation is responsible for this. But the falling off of the sensitivity for angles below 3° cannot be accounted for by sources of error. There is a connection here between sensitivity and size of field, which comes out still more clearly in FUJITA's experiments in which values of the stimulus were compared for angles less than 1° (at 10° from the centre, using white light and light adaptation).

The nature of the dependence between the value of the stimulus and the size of the field in the case of these small areas (Table IX) conforms

Table IX

Visual Angle of object	Relative Area of object	Threshold value	Relative sensitivity
1.0°	1.00	2300	1.00
0.5°	0.25	4500	0.51
0.25°	0.06	10000	0.23

very closely to PIPER's rule for large surfaces in dark adaptation. On account of the difficulty of making these measurements, they are to be considered only as being relatively accurate. It is certain, however, that for small objects subtending angles of 2° or less the value of the threshold stimulus of the luminous surface decreases with the visual angle, but not so fast as to obey RICCO's law for the fovea.

In consequence of the fact that the connection between threshold sensitivity and size of object is distinctly different in light and dark adaptation, particularly for surfaces of medium size subtending visual angles between 5° and 15°, the amplitude of adaptation must be different for luminous areas of different size. In fact, it must increase with the size of the field. TREITEL was aware of this fact. By systematic investigation of the process of adaptation with objects of different sizes, PIPER was able to get considerable differences in the rise of the curves of adaptation.[1]

6. Binocular Stimulus Summation[2]

In determining the threshold of the light sensation, it is not immaterial whether the observation is made with one eye or with both eyes. The truth is, in the state of dark adaptation the threshold in binocular vision is distinctly below what it is in monocular vision; being about half as high, according to PIPER's measurements.[3] This result has been confirmed in the writer's laboratory by numerous other observers. It was corroborated also by W. LOHMANN.[4] Testing a number of persons and comparing the results, he found unequivocally that the increase of sensitivity due to the participation of the other eye did not occur to the same extent with everybody. The differences were most marked in the case of subjects who had a squint. This may be the reason why some observers (WÖLFFLIN[5], for example) have found practically the same values of the thresholds of monocular and binocular vision. The writer had no trouble in verifying the fact of binocular summation of stimulus in his own case and in that of numerous other observers.

[1] ¶See H. PIÉRON, De la variation de l'énergie liminaire en fonction de la surface retinienne excitée pour la vision periphérique. (Cones et batonnets.) *Compt. rend. soc. de biol.*, LXXXIII. 1921, 753. Also, Des principes physiologiques qui doivent présider à toute étude de la lumière. *Rev. gén. des sci.*, XXXI. 620 and 656. (H. L.)

[2] ¶W. DE W. ABNEY and W. WATSON, The threshold of vision for different coloured lights. *Phil. Trans. Roy. Soc.*, 216 A. 1916. 91–142 (see page 109). — P. REEVES, Effect of size of stimulus and exposure time on retinal threshold. *Astrophys. Jour.*, XLVII. 1918. 141–146. (H. L.)

[3] H. PIPER, Über das Helligkeitsverhältnis monokular und binokular ausgelöster Lichtempfindungen. *Zeitschr. f. Psychol. u. Physiol. d. Sinnesorg.* XXXII. 161.

[4] W. LOHMANN, Untersuchungen über Adaptation und ihre Bedeutung für Erkrankungen des Augenhintergrundes. v. GRAEFES *Arch. f. Ophthal.* LXV. 1907.

[5] E. WÖLFFLIN, Der Einfluss des Lebensalters auf den Lichtsinn bei dunkeladaptiertem Auge. Ibid. 61. 1905.

Fechner[1] stated that a bright surface viewed with both eyes does not look brighter than it does in monocular vision. Helmholtz corroborated this, and it is a well known fact so far as vision in a bright place is concerned. The only effect of closing one eye may be to make a faint shadow come over the surface.

In performing this experiment, supposing that the initial adaptation is of that moderate kind customary indoors, it should be borne in mind that a certain degree of dark adaptation takes place very quickly in the eye that is closed or screened with the hand. When the eye is uncovered, it is not as if the free eye were reinforced by a second eye of equal sensitivity, but by an eye which at the instant when it begins to take part has increased sensitivity. As a matter of fact, when light adaptation is very good, and one of the eyes has been closed for a very short time, there is generally no appreciable difference of brightness between monocular and binocular vision.

That there is a real difference between the pair of eyes for photopia and scotopia so far as the summation of binocular stimuli is concerned, has been proved by Piper. His method consisted in determining the monocular and binocular thresholds alternately throughout the entire test of the dark adaptation. The results of one of these tests, in which the writer acted as observer, are given in Table X; and graphically exhibited in Fig. 66. The experiment was made before the construction of the adaptometer; consequently, the absolute sensitivity values are not directly comparable with those given in other places, but are represented in terms of an arbitrary unit.

Table X

Binocular		Right Eye		Left Eye	
Time in min.	Sensitivity	Time in min.	Sensitivity	Time in min.	Sensitivity
0	86	½	111	1	111
3½	272	4½	498	5	498
8½	2724	9½	2914	10½	3419
14½	11815	15½	13521	16	14516
20½	41649	21½	27778	22½	22957
27½	65746	28½	38447	30	33058
37	81632	39½	40000	40½	36982
52½	97656	56	40000	57	41649
59	97656				

A distinct separation between the curves for monocular and binocular vision is not manifest until after 14 minutes. Investigating various soldiers, who were employed as normal controls in an investigation of hemeralopia, Messmer (*loc. cit.*) got quite similar results.

[1] Fechner, Über einige Verhältnisse des binokularen Sehens. *Abhandl. der sächs. Gesellsch. d. Wissensch.* VII. 1860. S. 423. See also Bd. III. S. 424.

When the observations are made with both eyes, one of which is light-adapted and the other dark-adapted, the light-adapted eye behaves as if it were shut; that is, in this case the threshold is found to have the same value in both monocular and binocular vision. PIPER concludes correctly that this affords further proof of the independence of the process of adaptation in the two eyes. Extending his investigations further, he was able to show that even with stimuli considerably above the threshold, a binocular summation of stimulus occurred, provided both eyes were dark-adapted. In his method of experiment the intensity of one of two white surfaces could be regulated at will, and

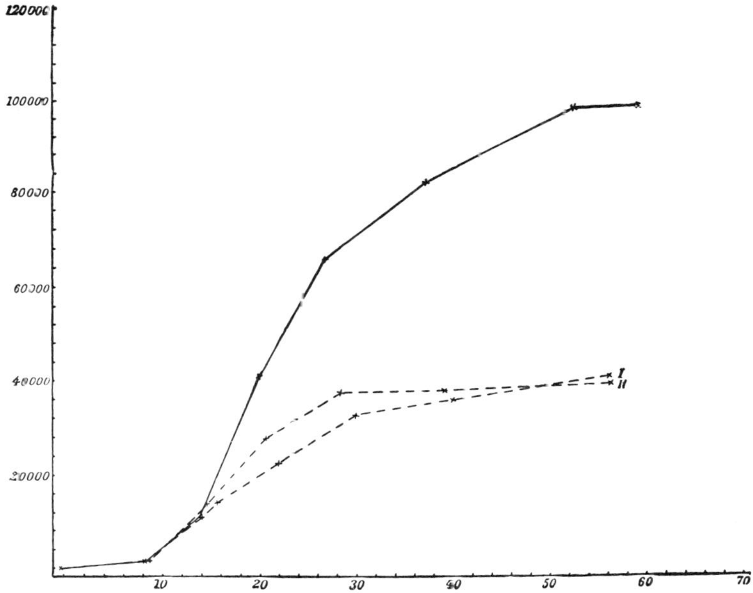

Fig. 66.

this surface was exposed to one eye only, while the other could be seen by both eyes. This enabled him to make a comparison between monocular vision and binocular vision. In order to obtain the same apparent brightness with both methods of observation, it was found that the intensity of illumination of the field seen with one eye must be from 1.6 to 1.7 times that of the other. With very low intensities, near the threshold in fact, the value is approximately double.

Comparing these results on binocular vision with those in which the determination of the threshold was made for retinal areas of different sizes, we find a remarkable analogy, as follows: In complete light adaptation of the eye the stimulating values of the quantities of light that fall on different parts of the retina are not to be added together when the stimulated area exceeds a certain size amounting to a few degrees of visual angle; and, furthermore, when both eyes are light-adapted, the stimulations due to the quantities of light reaching the two eyes are not to be added together. But in the state of dark adaptation the summation occurs distinctly and regularly in both cases. The possible experimental explanations of these facts will be discussed in another place.

7. The Least Amount of Energy Needed for Stimulation[1]

The absolute measure of the least amount of energy that is needed to stimulate the organ of vision has been determined very recently by various observers. Thus, WIEN[2] has computed the energy that reaches the eye every second from the most faintly visible stars as being 4×10^{-8} erg. A thorough study of the subject was made by v. KRIES[3] with special reference to the quality of the stimulating light and the region of the retina stimulated. The measurements by Dr. EYSTER were made on the most sensitive part of the retina, with blue-green light of wave-length $507\mu\mu$, which is the part of the spectrum for which the ratio between visibility and energy has its greatest value, according to LANGLEY's measurements. With the most thorough dark adaptation, and under the most favourable conditions, for utilizing the energy, as to both area and exposure, EYSTER obtained values of from 1.3 to 2.6×10^{-10} erg. For continuously visible light the value found was 5.6×10^{-10} erg per second. DR. BOSWELL[4] made similar determinations for purely foveal vision. As stimulus he used the longest waves in the spectrum for which KÖNIG has found a fairly uniform ratio between brightness and energy, that is, sodium light ($589\mu\mu$); because, for reasons to be mentioned presently, foveal fixation is easier with this light than with light of shorter wave-lengths. BOSWELL's observations were made in a state of slight dark

[1] ¶P. REEVES, The minimum radiation visually perceptible. *Astrophys. Jour.*, XLVI. 1917. 167-174. — P. L. DUNOÜY, Energy and vision. *Jour. Gen. Physiol.* III. 743–764. — W. W. COBLENTZ and W. B. EMERSON, Relative sensibility of the average eye to light of different colors and some practical applications to radiation problems. *Bull. Bur. Stand.* 1917 sci. papers, No. 303, 167–236. — H. PIÉRON, De la variation de l'énergie liminaire en fonction de la durée d'excitation pour la vision peripherique. *Compt. rend. Acad. des Sci.*, CLXX. 1921. 1203–1206. (H.L.)

[2] WIEN, Über die Messung von Tonstärken. *Diss.* Berlin 1888.

[3] J. v. KRIES, Über die zur Erregung des Sehorgans erforderlichen Energiemengen. Nach Beobachtungen von Herrn Dr. EYSTER mitgeteilt. *Zeitschr. f. Sinnesphysiol.* XLI. 373. 1906 (and *Ges. Abhandl. z. Physiol. d. Gesichtsempfindungen*, 3, 4.)

[4] F. P. BOSWELL, Über die zur Erregung des Sehorgans in der Fovea erforderlichen Energiemengen. Ibid. XLII. 299. 1907.

adaptation, perhaps, therefore, under conditions in which the foveal sensitivity had already reached its highest value; which, according to the results given above as found by SCHAEFER and the writer, occurs after 10 minutes of dark adaptation with previous highest light adaptation. BOSWELL found the just effective energy for short exposure, taking the average of numerous measurements, to be 31.6×10^{-10} erg. For long exposure from 16 to 20 times as much energy per second is necessary.

It is extremely remarkable that the amount of energy necessary to stimulate the peripheral retina is only about 15 times as much as is required to stimulate the fovea. Much greater differences might be expected on account of the considerable differences between the fovea and the periphery in the perception of very faint light. However, it must be remembered that the superiority of the periphery is particularly evident in long continued observations of large objects. When the values of the stimuli given by EYSTER and BOSWELL for long exposures are compared, it appears that the periphery under these conditions is about a hundred times superior to the fovea. We saw above, that for simple threshold determination with objects 1° in diameter, the eye being free to look in a chosen direction for as long a time as desired, the periphery of the dark-adapted eye is about a thousand times more sensitive than the fovea. Considering also the fact that in threshold determinations with the adaptometer, where the eye is free to move, the observational conditions are more favourable than in measurements with a spectrum apparatus and an ocular slit, and that this circumstance is more evident in eccentric vision than in foveal, we need not wonder any more at the comparatively slight difference between the stimulation of fovea and periphery as found in v. KRIES's laboratory. This is even more readily appreciated in the light of BOSWELL's comparative results on energy values when other lights besides yellow were used. It was found that the stimulating value of green light was more than twice that of sodium light.

B. Duplicity Theory and Twilight Vision.

1. The Duplicity Theory

Aside from the quantitative changes that are the result of the variation of sensitivity, a series of qualitative changes of vision are also associated with adaptation. These are manifested especially with respect to colour vision and in some other connections also (time relations, etc.). Some of them have indeed been mentioned already in the preceding section, particularly those phenomena which have recently led more and more to a very definite way of regarding the contrivance and function of the visual organ. Inasmuch as the actual facts of observation can be much more easily described in terms of the assumptions of this theory, the fundamental ideas will be stated here.

In the first edition of the "Physiologischen Optik" HELMHOLTZ had expressed the opinion that it was not likely that the rods were also sensitive to light in addition to the cones (see page 31). In the second edition he accepted the hypothesis of H. MÜLLER and KOELLIKER, which is that the rods also are concerned in sensation of light (see page 30), although he conjectured that their rôle in the localization of the sensations was different from that of the cones. In this connection no allusion was made to the hypothesis which had been advanced by MAX SCHULTZE[1] in 1866. The fundamental ideas of this theory have since been accepted by many writers.[2] According to SCHULTZE, all that the rods do is to mediate simple sensation of light without colour discrimination, the cones being the apparatus of colour perception. SCHULTZE's view received its chief support from the decrease of the colour sense out towards the periphery of the retina; which has its structural correlations in the preponderance of rods over cones in the peripheral parts. It is supported also by comparative anatomy. In animals that either live exclusively in dark places (caves, etc.) or are not active and alert except in dim light, SCHULTZE found the rods developed in much greater numbers than in animals that do not shun bright light. Indeed, in the case of many nocturnal animals he could find no cones at all, and nothing but rods. On the other hand, in the case of animals that live much in bright sunlight, such as lizards and snakes, he found only cones and no rods.

The foundation of SCHULTZE's theory cannot be said today to be perfectly valid. Peripheral colour blindness is explained in a partly different way from SCHULTZE's hypothesis. And, moreover, with respect to the data of comparative anatomy, it is certainly not true that nocturnal animals have rods only. According to more recent anatomical investigations, there are no mammals and birds without cones. Still the general fact is correct that the rod mechanism, as compared with the cone mechanism, is much more highly developed in nocturnal and twilight animals than in those whose habits are distinctly diurnal. The difference, however, is not so much in the number as in the length of the rods and the amount of visual purple they contain.

The theory of the functional difference between the rods and cones

[1] *Arch. f. mikroskop. Ana* . II. 1866. S. 247–261.

[2] For the literature on the subject see TSCHERMAK, *Ergebnisse der Physiologie* I. 2. 1902.

was put on a much firmer basis by the researches of PARINAUD[1] and v. KRIES.[2] Independently of each other, and proceeding along different lines, they arrived at a complete confirmation of MAX SCHULTZE's hypothesis and shaped it into a reasonable theory. By referring to it as the *duplicity theory* (*Duplizitätstheorie*), which is the name given to it by v. KRIES, the implication is that there is not simply a morphological duality of the elements of the retinal neuro-epithelium, but a corresponding duality of function as well, and that to a certain extent there are two kinds of vision. One kind is that which is active when the eyes are light-adapted and stimulated by strong light—*Tagessehen* (or *daylight vision, photopia*), as v. KRIES designates it. Opposed to it is the so-called *Dämmerungssehen* (or *twilight vision, scotopia*), when the eye is dark-adapted and the light stimulus is weak. On the duplicity theory the organ for daylight vision is the "daylight mechanism" or brightness-mechanism represented by the totality of the cones; the "twilight mechanism" or darkness-mechanism being constituted by the rods along with the visual purple absorbed in their outer segments.

In agreement with each other and with SCHULTZE, PARINAUD and v. KRIES assumed that only one quality of light sensation can be mediated by the rods. Thus, to a certain extent the twilight mechanism must be considered as being totally colour-blind, whereas the daylight mechanism is *farbentüchtig* or capable of discriminating colours.[3]

The function of the daylight mechanism, pure and simple, is exemplified in vision with the foveal region of the retina where there are no rods. But the function of the twilight mechanism cannot be isolated so simply. According to the theoretical assumptions, under ordinary circumstances in not too strong light, rods and cones function together simultaneously. But the rods are supposed to have a much greater capacity for dark adaptation. Thus, with low intensity of illumination, the stimulus may be sufficient to excite the rod or twilight mechanism, without being adequate to stimulate the cone mechanism. And hence below a certain limit of intensity, whatever possibility of vision there may be is to be considered as being due entirely to the mediation of the rods. As to whether the rods continue to function along with the cones

[1] PARINAUD, *Arch. gén. de méd.* April 1881; *C. R. Aug.* 1881, 286; *C. R.* Nov. 1884; *Annal. d'ocul.* 1894. CXII. 228; *Arch. d'ophthal.* XVI. 87. 1896.

[2] J. v. KRIES, Ber. naturf. Ges. Freiburg, IX. 1894. Über die Funktion d. Netzhautstabchen. *Zft. f Psychol.* IX. 81. 1895; GRAEFES *Arch. f. Opth.* XLII. 95. 1896.

[3] ¶"Broadly speaking vision with the dark-adapted eye, i. e., scotopic vision, is monochromatic or tone-free. Vision with the light-adapted eye, i. e., photopic vision, is polychromatic or toned. In the former the threshold stimulus intensity is low; in the latter relatively high." J. H. PARSONS, *loc. cit.* p. 203. (J. P. C. S.)

at high intensities of light, the two kinds of visual epithelium being therefore united for seeing by very bright light, the duplicity theory does not definitely attempt to say. During morning and evening twilight, and in dim light generally, the functions of the two mechanisms are interlinked in a complicated fashion, as will be described in the following sections.

The foundation for the duplicity theory is primarily in the comparison between foveal vision and peripheral rod vision.

In the preceding chapter of this Appendix it was shown (p. 329) that the absolute sensitivity in the centre of the retina at the fovea is not anywhere nearly so high as it is in the parts of the retina to one side of the fovea, say, 10° from it.

On the supposition that the angular diameter of an object which sends out white light is about one degree, and that the eye is thoroughly dark-adapted, the intensity of the light required in order for the object to be just visible when it is fixated directly must be in round numbers one thousand times greater than when the eye views the object with the most sensitive parts of the periphery of the retina; as was stated above. In light adaptation, on the contrary, the sensitivity in the fovea is somewhat greater than in the periphery. The simplest explanation of the superior sensitivity of the dark-adapted periphery of the retina is by assuming that the sensitivity of the rods, which are absent in the fovea but present in the periphery along with the cones, increases in darkness to a much greater degree than that of the cones; and that, therefore, with decreasing illumination, as for example, in the evening twilight, the rods more and more take over the rôle of receptors; until finally, for a certain degree of darkness, the intensity of light is no longer sufficient to stimulate the central part of the retina where there are no rods. Thus the condition is brought about which is characteristic of twilight vision, and in which there is a deficiency of function, a "scotoma", in the centre of the visual field.

2. Quality of the Light Sensation in Twilight Vision

Another striking characteristic of twilight vision is the lack of all colour discrimination. The eye is totally colour-blind, as can be verified without difficulty. The observation may be made at night in any room, with a suitable source of light and a device for regulating its intensity, so that it can be made sometimes dark and sometimes bright. As soon as it is possible to distinguish colour, it is a sure sign that the intensity is already above the threshold of the fovea centralis. Suppose, for example, there is a piece of red paper in the dark room, and that at a certain degree of illumination the red colour can be discerned. Under these circumstances, a little bit of this paper subtending an angle of one or two degrees cannot be made to "disappear" in the foveal region of the retina; as can be done so easily with a white object and sufficiently low degree of illumination, by suddenly looking straight at it (see page 333). The intensity is above the threshold of the foveal sensitivity.

The experiment turns out the same way when the colour of the paper is orange or brown, and to some extent also even when it is green. But in the case of blue objects there are difficulties with experiments of this kind. Many persons will say that a blue paper looks still distinctly blue, although the intensity of illumination of the surface is certainly below the foveal threshold; that is, although a small piece of the blue object of proper size disappears when it is focused in the fovea. It is a fact also that the sky at night when not absolutely devoid of light is frequently a beautiful dark blue[1], and a landscape in moonlight may look as if suffused in a lustre of bluish white.

These observations show that the quality of the light sensation in twilight vision is not always colourless, that is, white or grey, but that, at least under certain circumstances, it may be bluish. This is not a contradiction of the existence of total colour blindness in twilight vision; because that expression does not imply the inability of seeing light of any kind of colour at all, but merely the impossibility of distinguishing colours as qualities that are different one from another. When a person looks through a piece of glass which is transparent only to rays of some one particular colour, a dark red ruby glass, for example, he is practically colour blind. Objects seen through such a glass all appear in various gradations from bright red to black. The colouring in this case is all so intense that there is never any doubt even for a moment as to whether everything is coloured or colourless. It is a different matter when the piece of glass is lightly coloured, and particularly if the glass is blue. Seen through blue goggles everything white looks bright blue at first glance. (Other rays besides the blue rays come through these goggles, and hence red and green objects do not look much changed and often appear almost in their natural colours.) However, after wearing the goggles some hours, the blue disappears more and more from the visual field, and then white objects cease to look blue, and begin to look white. This is particularly the case when the goggles are made like snow-spectacles so as to prevent side light from entering the eye; and consequently comparisons cannot be made with parts of the field that do not radiate blue light.

In a case like this, in large portions of the field in a comparatively short time the specific coloured character of a sensation of blue that is

[1] Dove, Poggendorffs *Ann.* LXXXV. 1851. 397. (See also above, p. 226).

not a very saturated one may disappear and give way to a sensation of colourlessness. And so also in twilight vision something analogous may occur, particularly as the stimulus here is feeble anyhow, and simultaneous contrast between different coloured parts of the field is absent entirely.

On the other hand, it might be supposed that the coloured quality of twilight vision would be particularly manifest when there was a possibility of comparing such a sensation with the sensation aroused in the light-adapted eye under the conditions of daylight vision. The writer has made observations of this sort in the following manner. The observer places his head in front of the open side of a box divided in two parts by a vertical partition, so that each eye looks into one half. The back of the box is made of milk-glass, and behind it there is an annex for holding the contrivances for regulating the illumination of the two halves of the milk-glass plate. There are two iris diaphragms, in which little pieces of milk-glass are inserted. Coloured glasses can be placed over these diaphragms also. In front of each of the apertures a special source of light can be adjusted. By binocular comparison the observer can decide whether the two fields on the inside of the box are equal in luminosity and in colour.

For one of the fields, say, the one on the right, the intensity of the light is so regulated that, although it is below the foveal threshold, it looks as bright as possible to a *thoroughly dark-adapted eye*. (The best way to get this adjustment is by covering the field with black paper with a hole in it subtending an angle of between 3° and 4°. If the piece of milk-glass as seen through this hole when the dark-adapted eye looks straight at it can be made to disappear with certainty, the intensity of the light is far enough below the foveal threshold.)

The observer's right eye is thoroughly dark-adapted by blindfolding it tightly for one hour; the other eye meantime being kept as fully light-adapted as possible. A very bright source of light must be adjusted in front of the left iris diaphragm. Immediately after going into the dark room the luminosity of the two fields is compared, the right field being viewed with the right eye, and the left field with the left eye, the eyes being closed alternately. If the correct intensities of light have been selected, the left field therefore being illuminated about a thousand times more intensely than the right one, the difference of colour in the two fields is manifested in the most striking way. The left field, viewed by the photopic eye, looks generally distinctly yellow-red alongside the greenish blue right half as seen by the scotopic eye. To obtain equality of the fields, a highly coloured blue filter must be interposed in front of the left source of light.

The technical difficulties of this method are so very great that the writer has not yet succeeded in making an accurate determination of the colour of twilight vision in terms of the wave-length of a definite colour of the spectrum. But by a different method v. KRIES and the writer together[1] have obtained a basis for a determination of this sort. In experiments which will be described below in another connection matches were made, for the eye of a colour-blind person (namely, the writer)[2], between homogeneous lights and a mixture of spectrum red and blue (using a field considerably larger than the foveal region). For reasons that cannot be fully understood until we get further on in the subject, a colour match of this sort does not generally continue valid when the lights in the two halves of the match are reduced in the same proportion (for example, by narrowing the width of the ocular slit) so as to approach the conditions of twilight vision. The colour match previously made under the conditions of full daylight vision is therefore incorrect from two points of view, first with respect to luminosity, and then also with respect to hue. Thus, the nearer the conditions are to those of twilight vision, the more the specific colour of twilight vision, the cyan-blue mentioned above, blends for the observer into the colour of the field; and indeed, as we shall see, this takes place to an unequal extent for the different colours of the spectrum. Thus depending on the wave-length of the homogeneous light used in making the colour match, this homogeneous light, or the red-blue mixture, will change *more* towards blue, when the intensity of the entire field is reduced. If the mixture were made up of red of $670\mu\mu$ and blue-violet of $435\mu\mu$, what was found was, that a homogeneous light of wave-length $495\mu\mu$ (which is colourless to the colour-blind and corresponds therefore to his so-called "neutral point" in the spectrum[3]) appeared brighter and *bluer* than the mixture when the intensity of the whole colour match was reduced. The same behaviour was observed for all kinds of light up to about $485\mu\mu$. But if the wave-length of the homogeneous light is shorter than 480, the mixture was found to be *bluer* than the homogeneous light when the intensity was lowered. Thus between these limits, 480 and $485\mu\mu$, there is a homogeneous light which for the dichromat does not change its hue in the transition from daylight vision to twilight vision. v. KRIES and NAGEL called this place the "invariable point"[1] in the spectrum and assumed for reasons which will not be discussed here, that it was situated nearer

[1] J. v. KRIES and W. NAGEL, Einfluss von Lichtstärke und adaptation auf das Sehen des Dichromaten. *Zeitschr. f. Psychol. u. Physiol d. Sinnesorg.* XII. 29.

[2] ¶NAGEL was a deuteranope. (J. P. C. S.)

[3] ¶The *neutral point* of the dichromatic spectrum is the spectrum colour that matches white for this type of colour blindness. (J. P. C. S.)

the upper limit, 485$\mu\mu$. Accordingly, it may be conjectured that the quality of the sensation in twilight vision is similar to that which is aroused by this light in daylight vision. This, however, is not pure white, but very distinctly blue. Recent researches[1], which will be taken up later, point to the same conclusion. The statements made here are applicable strictly to colour-blind persons, but they are doubtless true also to a great extent with respect to so-called normal colour vision; because, as numerous observations have shown, the twilight mechanism in the eye of the normal individual and in that of the colour-blind person seem to function in identically the same way.

Incidentally, it is quite conceivable that the light sensations that occur under the conditions of pure twilight vision have a certain range of fluctuation as to their quality, varying from absolute colourlessness to a cyan-blue of no little saturation.

Perhaps this may be connected with a previous colour modulation (*Umstimmung*) of the visual organ. But the writer, judging by his own observations, does not believe this is the case. On the contrary, according to his experience, the blue hue of twilight vision comes out most distinctly right after long dark adaptation, where there cannot be any question of colour adaptation, the indication being rather that the eye must be "neutral" in HERING's use of this term.

The writer would like to guard against what seems to him to be the mistake of using such observations as the above as the origin or basis of any theory as to the rods in the retina being the anatomical substratum of the blue sensation. A conclusive argument against such a view is the fact that, while the peculiar characteristics of twilight vision are lacking entirely in the fovea centralis, still it is undoubtedly capable of mediating the blue sensation. It is only under very special conditions, when the two mechanisms of daylight vision and twilight vision operate together, that it is possible for the blue sensation arising in the twilight mechanism to be blended with colour vision to any noticeable extent.

3. Twilight Values of Pure Homogeneous Kinds of Light

Rays of different wave-lengths originating from one coherent spectrum have unequal stimulating values for the retina operating under the conditions of twilight vision (state of dark adaptation, with intensity of light below the foveal threshold). As the effects of all wave-lengths are qualitatively of the same kind, that is, as no colour

[1] W. NAGEL, Farbenumstimmung beim Dichromaten. *Zeitschr. f. Sinnesphysiol.* 44. 1909.

sensations are aroused whether the waves are long or medium or short, it is possible to make exact matches and quantitative measurements of the effects of individual wave-lengths without much difficulty.

The method employed in investigating spectral light is different from that used in studying more mixed light like that reflected by the colours of pigments. The nature of the reactions to homogeneous kinds of light is naturally of chief interest; and consequently this matter will be considered first.[1] The investigation may be made by means of HELMHOLTZ's spectrophotometer to be described hereafter. Another apparatus is needed in conjunction therewith to enable the experimenter to adjust a comparison light anywhere alongside the various homogeneous kinds of light in the spectrum. It must be possible to regulate the intensity of this comparison light over a wide range, and it should be big enough to subtend in the field of view an angle of from 5° to 10°. Looking through an ocular slit in HELMHOLTZ's instrument, what the observer sees is a circular field of the prescribed size divided by a vertical diameter. One of the semicircles is completely illuminated by light from the comparison lamp, while the other is illuminated in turn by the different kinds of homogeneous light in the spectrum. A homogeneous light, say cyan-blue, is generally used also for the comparison light. It is better not to regulate the luminosity of this half of the field, at least not entirely, by changing the width of the slit, because the range is not sufficient for the necessary changes. An adjustable pair of NICOL prisms is the best contrivance for this purpose. The intensity of the light in the other semicircle is not deliberately changed at all, the width of the slit being the same for each of the various kinds of spectral light that are admitted through it.

In order to make the observations under the conditions of twilight vision, the observer must keep his eyes thoroughly dark-adapted. The intensity of the entire field should be so low that all colour differences vanish; and therefore either very faint sources of light should be used, or merely a small fraction of the radiation of a lamp that is steady enough. This can be accomplished by placing the lamp far away and letting it shine on a sheet of magnesium oxide or white baryta paper held at a suitable angle in front of the collimator slit. Another convenience about this arrangement is that both slits can be illuminated by a single source of light.

[1] See F. HILLEBRAND, Über die spezifische Helligkeit der Farben, mit Vorbemerkung von E. HERING. *Sitz. Ber. K. Akad. Wien. Mathem. naturw. Kl.* XCVIII. III. 1889.—J. v. KRIES u. W. NAGEL, Über den Einfluss von Lichtstärke und Adaptation auf das Sehen des Dichromaten (Grünblinden), *Zeitschr. f. Psychol. u. Physiol. d. Sinnesorgane.* XII. 45; also in: *Abhandlungen zur Physiol. d. Gesichtsempfindungen aus dem physiol. Institut zu Freiburg i. Br.* Herausgeg. von J. v. KRIES. Heft 1.

The stimulus value for the dark-adapted eye of a particular homogeneous kind of radiation may be expressed simply in terms of the value of the intensity of the comparison light when there is a perfect match between the two halves of the field. The relative stimulus values obtained by systematic investigations with different kinds of spectral light may be termed (see v. KRIES and NAGEL, *loc. cit.*) the *twilight values* (or twilight valences), and can be exhibited in the form of a curve.

Table XI (according to v. KRIES and NAGEL)
Twilight Values of Homogeneous Kinds of Light

Wave-length	Twilight value in arbitrary units
670.8$\mu\mu$	?
656	19.3
642	36
628	110
615	254
603	276
591	599
582	1276
571	2061
561	2477
552	2930
544	3027
536	2820
525	2055
515	1576
505	1015
496	697
488	486
480	318
469	263
460.8	146
448	46
436	17

The typical distribution of the twilight values in the dispersion spectrum of gaslight for the writer's eye is given in Table XI. The relations are exhibited by the curve in Fig. 67. The maximum is in the green at 544 (for other measurements made by the writer it was at 536$\mu\mu$). The curve falls rapidly on both sides, but most steeply towards the red, while the descent from the blue-green to the violet is very gradual. On the red end near the border between the orange and red, the values are already almost too small to be measured; and in the red from 670$\mu\mu$ on, while traces of twilight vision can still be detected, they cannot be numerically measured; being dependent besides on the almost unavoidable admixture of diffused light in the instrument.

The statement that this is a typical curve means that it not only represents the case of persons with the same kind of colour vision, but cases of so-called normal colour vision and of various congenital types of colour blindness as well. It applies likewise, as has been shown by the writer, to anomalous trichromats to be described later.

In his experiments on the luminosities of the spectral colours for various absolute intensities, A. König (*Ges. Abhandl.* p. 144, and Beiträge zur Psychologie und Physiologie der Sinnesorgane, *Festschrift zu* Helmholtz 70 *Geburtstag*. 1891. p. 309) used for the lowest degree of intensity one that was probably not far from the conditions required for pure twilight vision. On this occasion he got the important result that the great differences in the distribution of spectral brightness, that exist between normal, red-blind, green-blind and totally colour-blind persons when the intensity of the light used is relatively high, disappeared almost entirely when the lowest degree of intensity was employed. More recent investigations have completely corroborated this result and shown that it is valid also for the anomalous trichromats, as above stated.

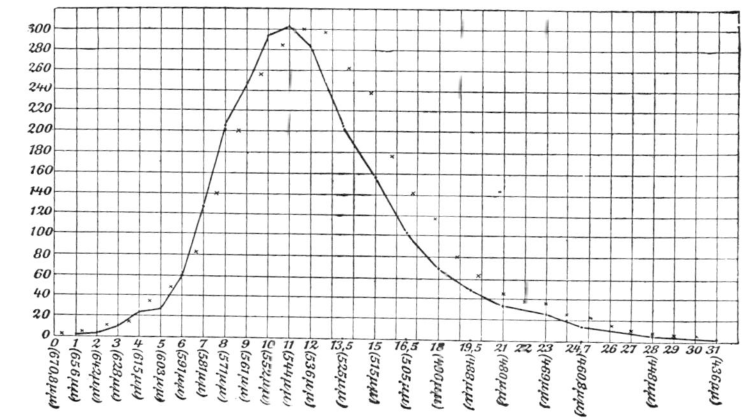

Fig. 67.—Distribution of the twilight values in the dispersion spectrum of gaslight (according to Nagel). The crosses indicate the results of previous, perhaps rather less accurate determinations.

In repeated parallel experiments with persons of different types of colour vision the same result has been always verified again. *There is no known form of anomalous colour vision for which any noticeable deviation has been found in the distribution of twilight values over the spectrum.* The only cases that have not yet been tested are those of pure tritanopia or violet blindness.

A. Tschermak[1] has stated over and over again that the adaptative ability of the eye for long and short wave-lengths for the two principal types of colour

[1] A. Tschermak, *Über physiologische und pathologische Anpassung des Auges.* Leipzig 1900.—Beobachtungen über die relative Farbenblindheit im indirekten Sehen. Pflügers *Arch. f. d. ges. Physiol.* LXXXII. 559. 1900.—Die Helldunkeladaptation des Auges und die Funktion der Stäbchen und Zapfen. *Ergebnisse d. Physiol.* I, 2. 703 and 747. 1902.

blindness (red blindness and green blindness) is regularly different from the two types of colour vision that are presumed to correspond to them. But no proof of the truth of this assertion has ever been given or even so much as attempted.

Evidently, these determinations of the twilight values always apply simply to a particular spectrum, and if the observations are made with an interference spectrum, the curves will have a different form. Moreover, in using a definite apparatus the distribution of the twilight values will be different according to the nature of the source of light. An AUER lamp containing a lot of green light will give a different curve from that of the reddish incandescent electric light.

The most complete determinations of this sort, so far as perfecting the method is concerned, are perhaps those made by SCHATERNIKOFF.[1] The values which he obtains for gaslight are very close to those of the writer. For direct sunlight and reflected skylight the values are given in Table XII and the curve shown in Fig. 68. As was to be

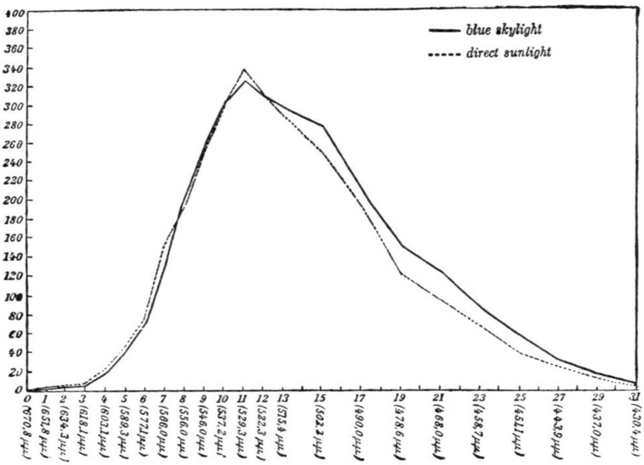

Fig. 68.—Distribution of the twilight values in the dispersion spectrum of sunlight and blue skylight (according to SCHATERNIKOFF).

expected, the peak of the curve (at $529.3\mu\mu$) is a little nearer the blue end than it is in gaslight (537.2). The values for blue skylight are somewhat higher in the blue-green and blue than for direct sunlight.

Even without any spectrum apparatus, an approximate idea of the distribution of twilight values may be obtained, which brings out especially the considerable differences of the luminosities of the colours in the daylight

[1] M. SCHATERNIKOFF, Neue Bestimmungen über die Verteilung der Dämmerungswerte im Dispersionsspektrum des Gas- und Sonnenlichts. *Zeitschr. f. Psychol. und Physiol. der Sinnesorgane* XXIX. 255. Also, *Abhandlungen zur Physiologie der Gesichtsempfindungen,* herausgeg. von J. v. KRIES. Heft 2. 1902.

Table XII (according to SCHATERNIKOFF)

Wave-length	Twilight Values for	
	blue skylight	direct sunlight
670.8	7.7	5.9
651.8	12.5	10.5
634.3	22.2	33.3
618.1	70.7	86.3
603.1	189	214.4
589.3	411	459
577.1	725	752
566.4	1369	1535
556.0	2019	1933
546.0	2578	2546
537.2	3000	3000
529.3	3213	3353
522.3	3060	3067
515.4	2959	2833
502.2	2758	2460
490.0	2067	1935
478.6	1497	1205
468.0	1224	945
458.7	830	658
451.1	580	399
443.8	299	212
437.0	160	112
430.4	69	46

vision and twilight vision. With a good assortment of saturated coloured papers of about post-card size, they can be arranged under very dim illumination in a graded series from the brightest to the darkest. Thorough dark adaptation is necessary and also an illumination correctly adjusted until the colours of the papers are no longer apparent. Afterwards when bright illumination is turned on, it is evident that the order of arrangement is absolutely wrong for daylight vision. Brilliant red papers will be found at the darkest end of the row, and the orange, say, near a dark blue that in daylight vision looks very much less luminous than the orange. On the other hand, the greens are placed near the bright end of the row.

More accurate experiments, in which also numerical values for the differences of luminosity are obtained, can be made by using the colour top. A coloured disc of sufficient size, 18 or 20 cm in diameter, is set to spinning. On it are placed two smaller discs, one black and the other white, which can be shifted over each other, The illumination is lowered until a thoroughly dark-adapted eye can no longer distinguish any colours, and the ratio between the black and white sectors is so adjusted that the grey of the mixture and the dim colourless light of the larger disc appear to have exactly the same luminosity. While this can be quite accurately done, of course no precise twilight values can be obtained in this way, because the colours on the disc are not homogeneous.

The twilight value curves give a picture of the distribution of the luminosity in the spectrum as it must look to a thoroughly dark-adapted observer viewing an entire spectrum under the conditions of twilight vision; that is, at an intensity for which colours are no longer discernible. Observations of a dim, colourless spectrum of this sort were made first by HILLEBRAND[1] in HERING's laboratory; the result being as a matter of fact that under these circumstances the spectrum had a colourless grey appearance. Its brightest place was where pure green or somewhat yellowish green is seen in a bright spectrum. There was nothing visible at all at the places where it would otherwise be red; but towards the violet side the colourless spectrum extended to the end without being noticeably shortened.

The most striking way of showing the changes that take place in the spectrum is to project it on a screen in a dark room and then cut down the intensity by means of a pair of NICOL prisms interposed in the path of the light. If the intensity is not very great to begin with, an episcotister[2] transmitting not more than one three-hundredth of the total amount of light, may be used to produce the necessary dimness. Of course, in order to see the spectrum without any colour in it and yet not too faint, the spectators should previously stay quite a while in the dark or at least in a dim room. In Figs. 1 and 2, Plate II,[*] the writer has endeavoured to exhibit the appearance of the spectrum in twilight vision as contrasted with the coloured spectrum of higher intensity.

The characteristics of twilight vision, as described here, enable us also to explain (as was done first by LUMMER[3]) an interesting phenomenon which was noted by H. FR. WEBER.[4] He observed that when a body is heated to incandescence in the dark, usually it does not look *red hot* at first, but grey, "dull cloudy grey." This *grey glow* appears at temperatures around 400°C. With rise of temperature there is the appearance of a yellowish-grey radiation, the yellow-green rays being particularly prominent under spectroscopic examination. According to DRAPER[5] it is not until a temperature of 525° is reached that the red glow appears. Although the radiation consists mainly of waves of the lengths of red light, its intensity at 400° is not high enough to stim-

[1] F. HILLEBRAND, Über die spezifische Helligkeit der Farben. *Sitz.-Ber. Wien. Akad.* XCVIII, Abt. 3. 1889. S. 70.

[2] ¶An instrument with rotating sectors for measuring changes in the intensity of the transmitted light, devised by AUBERT. See footnote in Vol. III, towards end of §32. (J. P. C. S.)

[3] O. LUMMER, Grauglut und Rotglut. *Verh. d. D. Physik. Ges.* XVI. 121. 1897; WIEDEM. *Ann.* LXII. 14. 1897.

[4] H. FR. WEBER, Die Entwicklung der Lichtemission glühender fester Körper. *Sitz.-Ber. Akad.* Berlin 1887. 9. June. WIEDEM. *Ann.* XXXII. 256. 1887

[5] *Amer. Journ. of Sc.* (2) IV. 1847; *Phil. Mag.* (3) XXX. 1847; *Scient. Memoirs* p. 33. London 1878.

[*]Plate II faces page 352.

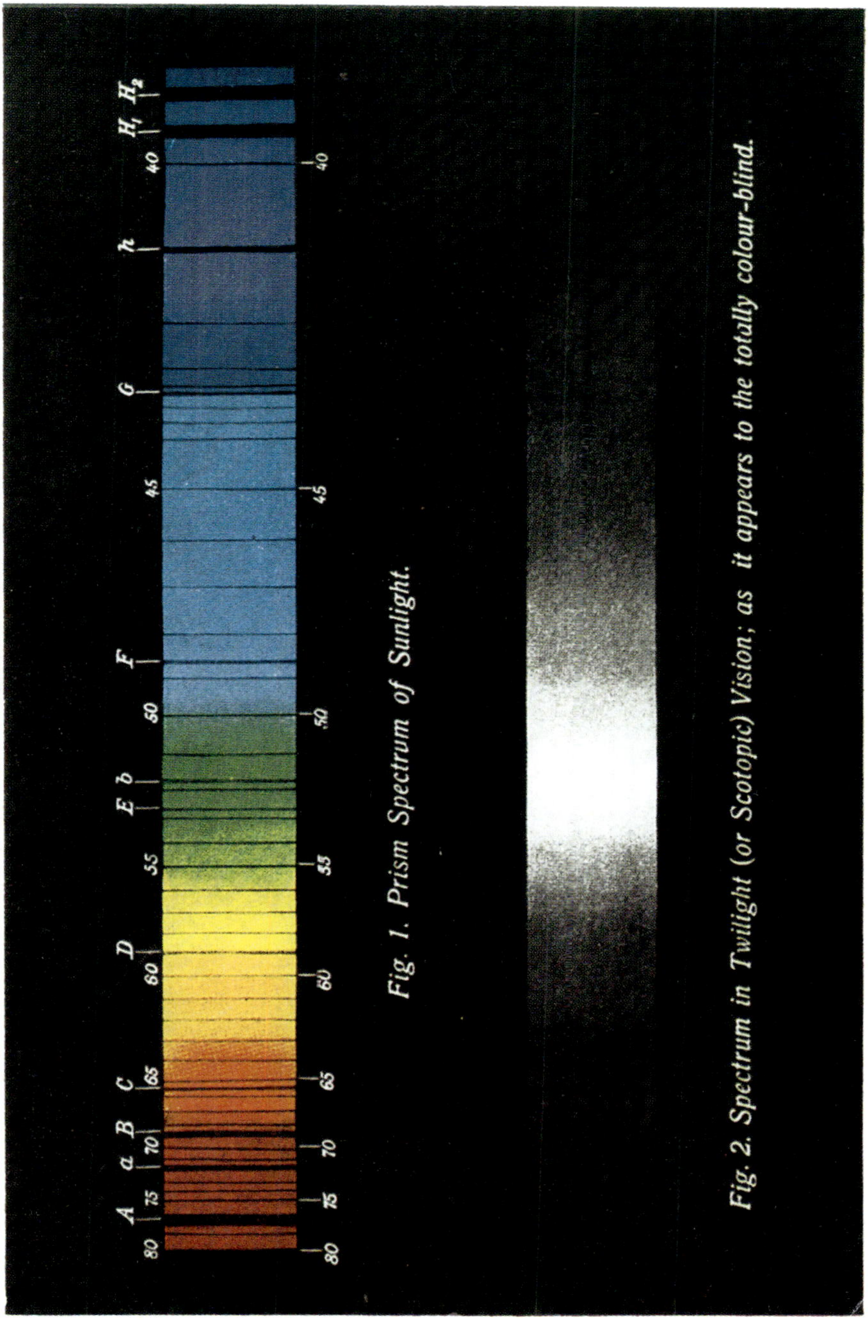

Fig. 1. Prism Spectrum of Sunlight.

Fig. 2. Spectrum in Twilight (or Scotopic) Vision; as it appears to the totally colour-blind.

PLATE II

ulate the colour sensitive mechanism, and yet, owing to the presence of rays of shorter wave-lengths, is sufficient to stimulate the twilight mechanism, which therefore mediates colourless light sensation, as it always does when it operates alone. Thus, as the temperature rises from 400° to 525°, the red portion of the total radiation is not increased in anything like the same proportion. On the contrary, with rising temperature, it is the middle and short-wave radiation that increases. But in consequence of the dark adaptation, the sensitivity of the eye is far better adapted to the medium waves (from yellow to green) than to the long red waves. It hardly needs to be said that a light-adapted observer, entering a dark room, cannot perceive the grey glow, and the first thing of the kind that he can see will be the red glow. He notices it about at the same moment as an observer who has been thoroughly dark-adapted beforehand, and who has been able to see the grey long before.

Two observers, one fully dark-adapted, the other light-adapted, do not obtain the threshold for the red colour at exactly the same temperature of incandescence, the scotopic eye being a little superior to the photopic eye in this respect. The explanation is given by some experiments which the writer got Mr. BOSWELL[1] to make. The threshold for the appearance of colour on a portion of a surface was determined, first, when the surface was illuminated simply by the coloured light, and second, when it was illuminated at the same time by a certain amount of mixed light. The regular result was that admixture of a slight amount of white light lifts a coloured light over the threshold when it is by itself below the threshold of visibility; the light appearing then in its specific colour. However, admixture of any light of appreciable twilight value, green light, for example, acts in the same way as the addition of white light. Moreover, by this addition a red light, say, that is below the threshold can be lifted above it. The eye being dark-adapted, naturally the green light acts longest on the colour-blind twilight mechanism before it has any effect on the mechanism that is sensitive to colour; and hence the complementary relation between green and red is not manifested at all, but simply the colourless part of the sensation, which to a certain extent paves the way for the perception of the coloured light, that is, the red light in this case. Obviously, this enabling influence can be exerted only by light stimuli whose intensities are themselves not far above the limit of perceptibility of the scotopic eye.

4. The Purkinje Phenomenon

We have seen that in the state of pure twilight vision the retina is not sensitive at all to the long red waves above 670 and is just sensitive to the shorter red waves. Under such circumstances, therefore, objects that are deep red must look perfectly black. In dim twilight the cap of a German infantryman, which is blue with a red band, does not look different from that of a sanitary officer with its dark blue band.

[1] F. P. BOSWELL, Über den Einfluss des Sättigungsgrades auf die Schwellenwerte der Farben. *Zeitschrift f. Sinnesphysiologie.* XLI, 364. 1906.

This lack of sensitivity to red light in the twilight mechanism is noticeable, however, not only in the case of pure isolated twilight vision, where all colouration disappears, but to a certain extent also in the much more frequent and complicated case when the two mechanisms of twilight vision and daylight vision both operate at the same time. This condition is very pronounced early in the morning at break of day. As a result of the preceding long spell of darkness, the sensitivity of the twilight mechanism that is characteristic of dark adaptation remains enhanced. The dim light is not yet sufficient to overcome dark adaptation and produce light adaptation. On the other hand, however, the light may already be strong enough to stimulate the daylight mechanism and therewith to mediate the sensation of colour also. In most of the retina we must suppose the elements of twilight vision and daylight vision are so interlaced that in observing coloured areas of even moderate dimensions the functions of the two mechanisms are combined. We are accustomed to the function of the daylight mechanism; it is normal colour vision as it occurs in bright light. To that is now added twilight vision, with the peculiarities described above. The most remarkable one is the high stimulating action of the green and cyan-blue rays of medium wave-length, and the ineffectiveness of the red rays. In observing green and blue objects, the colourless or bluish sensation of twilight is associated with the specific colour sensations, and the colours are seen as if white or bluish white were actually mixed with them, and so they look both *brighter* and more *unsaturated* than when the bright colours with the same objective intensity of illumination are viewed by an eye that is not dark-adapted. The contrast between such objects and those which emit or reflect red light only is that, so far as the latter are concerned, there is no difference between the behaviour of the scotopic eye and that of the photopic eye. For the twilight mechanism these rays have no stimulating value, and so nothing is superadded to the effect produced on the daylight mechanism.

The most striking consequence of this special position of the long waves in the combination of daylight vision and twilight vision must be, therefore, that red objects, and even orange ones, appear to be *relatively* darker than blue and green ones. This is familiar to everybody and is known as the PURKINJE phenomenon. The appearance of colours in the morning twilight was described by him as follows[1]: "Blue is what I saw first. The shades of red that are usually brightest in daylight, namely, carmine, vermilion, and orange, are for a long time the darkest, and not to be compared to their ordinary brightness."

[1] PURKINJE, *Neue Beiträge zur Kenntnis des Sehens in subjektiver Hinsicht.* Berlin 1825, S. 109–110.

The reason why PURKINJE speaks here of blue as being the brightest was perhaps due to an accidental choice of the coloured objects, among which the green must have been particularly dark. Thus in case of a spectrum that is dim but not too dark to be void of colour to a dark-adapted eye, the brightest part is not the blue, but the green, which of course looks whitish in the faint light that is there. The blue, violet, and yellow are also whitish and very unsaturated, all the more so, the more completely the eye is dark-adapted. The red alone is very dark and remains a saturated colour, not whitish. For a certain particular (quite weak) intensity of illumination of the spectrum and with a very thorough dark adaptation, almost the whole spectrum can be made to look as if tinged with bright blue, bordered only on one side by deep red.

It is especially instructive to make these observations with only one eye dark-adapted. After staying in a fairly bright room with one eye tightly blindfolded, the observer enters the dark room where the faint spectrum is projected and looks at it first with the photopic eye and then with the scotopic eye. In the latter case it appears as described above, but of course for the photopic eye it lacks the brilliancy and falling off in saturation in the portions corresponding to the medium and short wave-lengths, and the entire spectrum looks very dark but saturated everywhere; the brightest place being in the yellow, although under such conditions it looks more brown.

Perfectly analogous observations may be made also with sets of coloured papers, bundles of wool, etc., but in such cases, owing to the lower saturation of the pigment colours, the characteristic difference in the individual colours is not so distinct.

As is evident from what has been said, two factors are necessary for the production of the PURKINJE phenomenon, namely, dark adaptation and low intensity of the stimulating light. HERING[1] emphasized the importance of the state of adaptation, while HELMHOLTZ and A. KÖNIG[2] paid more attention to the intensity-ratios. The two agencies must act together; and hence it is going too far to say, as HERING does (*loc. cit.*, page 542), that KÖNIG has written about the PURKINJE phenomenon "without being aware of the main characteristic of it." The point that needs rather to be emphasized is that during most of the day there exists already a state of adaptation in which the PURKINJE phenomenon takes place, certainly always indoors. Under these conditions all that is necessary to produce it is

[1] E. HERING, Über das sogenannte PURKINJEsche Phänomen. PFLÜGERS *Arch. f. d. ges. Physiol.* **60**, 519.

[2] A. KÖNIG, Über den Helligkeitswert der Spektralfarben bei verschiedener absoluter Intensität. *Gesammelte Abhandlungen zur physiol. Optik.* Leipzig 1903, No. 20, S. 144.

to change the intensity. Special measures are needed, as stated on page 318, to throw out the twilight mechanism for a time, and then it succeeds only for a brief space. By staying quite a long while in a room with ordinary moderate illumination and then suddenly lowering the illumination considerably, all the conditions are present for obtaining the desired effect. If two large pieces of red and blue (or green) paper have been previously selected, so that in full daylight the red looks a little brighter, in the reduced light of the dark room the red piece appears darker at once. Of course, the change of luminosity is much increased by longer dark adaptation, but in the condition of the eye, which may be described as the usual one, the PURKINJE phenomenon can be induced by simply changing the intensity of the light.

A particularly clear demonstration of the phenomenon can be made by projecting sufficiently large patches of red and blue light on a white wall; with some device (like an iris diaphragm, pair of NICOL prisms, or a rotating sector) interposed in the path of the light for reducing the illumination. The two coloured fields can then be shown in strong and weak illumination in rapid alternation; and every time they are darkened, the luminosity of the red falls off relatively much more than that of the blue. In this experiment also the effect is better when the observer has spent some time in the dark or in a dim room, because then the difference in the appearance of the pair of colours in strong and weak light will be more striking. But even with the state of adaptation produced in a moderately bright room, the phenomenon is quite distinct.

In discussing the matters alluded to here, E. HERING[1] makes a distinction between two kinds of adaptations, which he calls *Daueradaptation* (or "time adaptation") produced by long continued exclusion of light, and *Momentanadaptation* (or "instantaneous adaptation"), which occurs the instant darkness begins. These terms are used also by HERING's pupil, TSCHERMAK. But there is no exact definition of what is meant by "instantaneous adaptation," and consequently it is hard to say exactly what those writers have in mind when they speak of a process taking place at the instant darkness occurs. It is a known fact that the sudden darkening of the entire visual field has an appreciable effect on the appearance of a small bright object, because it modifies the light sensation both qualitatively and quantitatively. But there is a sharp distinction between these effects and those of "time adaptation." If, prior to the beginning of "instantaneous adaptation," the eye is *thoroughly light-adapted*, the PURKINJE phenomenon never occurs. The latter effect is associated rather with the existence of some moderate degree at least of "time-adaptation", such as occurs naturally in "indoor adaptation." *Any matches of luminosity or colour as made by the eye in the state of maximum light adaptation are never invalidated by what* HERING *calls*

[1] E. HERING, Über das sog. PURKINJEsche Phänomen. PFLÜGERS *Arch. f. d. ges. Physiol.* LX. 1895.

"instantaneous adaptation," that is, by a sudden proportional dimming of both halves of the match or the total field. In the very nature of what we mean here by dark adaptation, and what HERING calls "time adaptation," the most important factor is the functioning of a new stimulus receptor, working according to different laws from that of the daylight mechanism. We shall find a series of other differences in the way these two mechanisms operate besides those that have already been mentioned. But it must be pointed out here with special emphasis that the process of dark adaptation in the sense intended here is not equivalent to the idea of the simple increase of sensitivity of the retina in darkness; that is, to a purely quantitative change of sensitivity; but that a very fundamental quantitative variation of the retinal function, most plainly indicated by the occurrence of the PURKINJE phenomenon, is an integral permanent part of the concept of "dark adaptation." The term "adaptation of the retina" was used first by AUBERT. Doubtless, he was not then aware of the close connection between the quantitative and the qualitative variations of the sensitivity of the retina in darkness. But the parts of the process of adaptation which he studied are just those that are inseparable from the characteristic conditions of PURKINJE's phenomenon. It is justifiable, therefore, to apply the term "dark adaptation" to the whole complicated process.

What HERING may have meant by "instantaneous adaptation" is not related to those matters at all, or very distantly anyhow; so that the use of this expression results simply in confusion. If the PURKINJE phenomenon is observed immediately by an eye when it is suddenly darkened after having been in a state of medium adaptation (or indoor adaptation, as we called it above), then (to stress this point again) this is the direct consequence of the darkening and the direct consequence of the fact that a *moderate time adaptation* had already occurred previously, but it does not indicate that the instantaneous darkening has induced a process in the eye that is essentially like that caused by long continued darkness.

5. Absence of the Purkinje Phenomenon in the Fovea Centralis

In a central region of the retina corresponding to a visual angle of about 1° 30′ the PURKINJE phenomenon cannot be elicited by any means whatever.[1] This area, as will be seen presently, corresponds pretty nearly to the fovea in the most central part of the retina where there are no rods at all. The anatomy of the fovea is extremely hard to determine. It is easy to see why this is the case in the light of what G. FRITSCH[2] tells us concerning the extraordinary diversity in the morphological formation of the centre of the retina. However, what seems to be universal in all normal eyes is a central region where the visual epithelium is composed of nothing but cones, without any rods in between them. In line with this definite anatomical result there is the physiological fact, that there is a central place where the PURKINJE phenomenon cannot occur, and which, if it is capable of adaptation at all, certainly does not show any of the characteristics of it that have

[1] ¶See L. T. TROLAND, Apparent brightness; its conditions and properties. *Trans. Illum. Engineer. Soc.* IX. 1916, 947–966. (H. L.)

[2] G. FRITSCH, *Über Bau und Bedeutung der Area centralis des Menschen.* Berlin 1908.

been described above.[1] Thus what is intended here in speaking of the fovea is always the region that is devoid of rods.[2]

It is extremely easy to verify the absence of the PURKINJE phenomenon in this region. Perhaps the most striking way of doing this is by the method employed first by O. LUMMER.[3] Two large pieces of coloured paper or cloth, from a half to three quarters of a square metre in size, one red and the other blue or green, are attached to the wall of the dark room, touching each other along a straight line. The colours are so selected that by bright gaslight or Argand burner the red looks to the normal eye distinctly brighter than the blue. The observer, supposed to be sufficiently dark-adapted, stands about five or six metres away and looks at the coloured areas. If the illumination is considerably lowered, the PURKINJE phenomenon appears immediately most distinctly, the red becoming deep black red and the blue bright bluish white. With further decrease of illumination still, the colour of the red surface entirely disappears and looks then dark grey or even black. But for the purpose of the present experiment the illumination should be lowered just far enough for the red still to show colour distinctly. The coloured field is then covered by a black curtain leaving just a small circular area exposed, half of it being red and the other half blue. This exposed area should subtend an angle of about one degree from where the observer stands. *As long as the observer continues to look straight at this little spot, even with the dim illumination, the red sems to him to be brighter than the blue; in other words, the* PURKINJE *phenomenon is absent.* But if he looks ever so little away from this spot, the blue comes out with a whitish glow distinctly bright. As soon as the curtain is raised again, immediately the marked difference of brightness in the large field is seen. This alternation can be repeated as often as desired.

The only precaution to be taken is not to lower the illumination until the red disappears. The small red field readily attracts the attention and so insures the formation of the image in the fovea. Otherwise, an inexperienced observer might very easily let his eye wander and thus obtain a "parafoveal" image on parts of the retina where there are rods. Dichromats and anomalous trichromats are more apt to fail with this experiment than those with normal colour vision, because the dim red is not such a striking colour to them and does not hold their attention. Of course, the experiment can also be carried out in another way by using a lantern and glass filters to project

[1] ¶See S. HECHT, *loc. cit.* (H. L.)

[2] ¶W. DE W. ABNEY and W. WATSON, *loc. cit.* (H. L.)

[3] O. LUMMER, Experimentelles über das Sehen im Dunkeln und Hellen. *Verhandl. d. Deutschen physik. Gesellschaft.* VI. No. 2, 1904.

a coloured field on a white wall. A pair of NICOL prisms is employed for reducing the illumination, and the field is contracted by the insertion of a diaphragm. A similar effect can be obtained very well also by using the spectrophotometer and pure homogeneous colours; provided the instrument will give a field of at least 4° in extent (from 10° to 12° is better) which can be conveniently stopped down.

These methods are somewhat unsatisfactory because the two halves of the field are illuminated by light that cannot be compared *exactly* as to brightness on account of the difference of colour. The red has to be such that with brilliant illumination it is undoubtedly brighter than the blue, and, on the other hand, the dark adaptation must be sufficient.

The experiment can be performed in a much more conclusive way on a dichromat, especially if he is what is called green-blind. For with a subject of this sort it is possible to find two kinds of light that match perfectly as to colour and yet show a very considerable difference in their twilight values. Red of wave-length $670\mu\mu$ or longer and green of wave-length $545\mu\mu$ are two such radiations. In HELMHOLTZ's spectrophotometer these colours can be thrown on a 7° field, and can be made to match perfectly by the colour-blind eye thoroughly light-adapted. Now if dark adaptation is produced (or has been produced already by bandaging one eye), and the intensity of both halves of the field at the same time is considerably lowered (say, to one fortieth), the match is gone completely, the green showing up bright and whitish. If now the visual field is gradually contracted by means of an iris diaphragm, the difference between its two halves is very slight even at a diameter of 3° and disappears entirely at about 1° 30′. The eye may have been dark-adapted for any length of time, and any total intensity of the field may be chosen; but the colour match remains good under all circumstances, provided the eye looks steadily at the centre of the field. If the line of fixation deviates as much as one or two degrees, it is enough to make the whitish glow appear again in the green.

Very similar tests can be made by making a match between blue-green, about $500\mu\mu$ and a purple mixture (say, $650+460\mu\mu$). The homogeneous (blue-green) side of this match has the greater twilight value, and therefore on a large field looks brighter and whiter to the scotopic eye. The twilight value of the homogeneous light is about 25 times that of the mixed light. In the previous red-green match the ratio between the twilight values (on the supposition that there are no red waves shorter than $670\mu\mu$) is at least 1000:1.

It is not possible for observers with normal colour vision to make true colour matches when the difference of twilight values of the two halves of the field is even nearly so much as the above. The very highest ratio is 6:1. This is the case with matches between different spectral white mixtures, one half of which is white obtained by mixing red and blue-green, the other being a mixture of yellow and indigo. It was by using such white mixtures that it was first ascertained that the matches did not last when the intensity of illumination was varied. This result was obtained independently and almost at the same time by CHRISTINE LADD-FRANKLIN[1] and H. EBBINGHAUS.[2] Their tests were made with colour-blind persons. A. TSCHERMAK[3] extended these experiments by testing persons with normal colour vision; and according to him colour matches of this kind were not valid for the foveal region of the retina, supposing dark adaptation had been sufficiently long continued. But under the given condition v. KRIES and the writer[4] were not able to find a single instance where a match that had been made at high intensities was not valid in the fovea of the scotopic eye.

KOSTER[5] and SHERMAN[6] have likewise maintained the existence of the PURKINJE phenomenon in the fovea, but there are considerable possibilities of error in the methods they used. The question as to whether there are any traces of the phenomenon in the fovea and what would be the significance of it, will be considered later. It will be sufficient to say here that in the writer's opinion it has not been proved that the PURKINJE phenomenon occurs in the part of the retina where there are no rods.

The size of the central area of the retina where the PURKINJE *phenomenon does not occur* can be determined with some degree of accuracy by making as perfect a spectral match as can be between two halves of the field whose twilight values are as far apart as possible, first, with a light-adapted eye, and then with the eye dark-adapted. If the field is greater than 2°, the dark-adapted eye detects immediately the inadequacy of the match. By means of an iris diaphragm the size of the field is now quickly diminished (to between 1° and 2°) until the two halves again match. A measurement of this sort cannot be

[1] C. LADD-FRANKLIN, On theories of light sensation. *Mind.* (N. S.) II. No. 8, 473–490, 1893.

[2] H. EBBINGHAUS, Theorie des Farbensehens. *Zeitschr. f. Psychol. u. Physiol. der Sinnesorgane* V. 145–238, 1893.

[3] A. TSCHERMAK, Über die Bedeutung der Lichtstärke und des Zustandes des Sehorgans für farblose optische Gleichungen. PFLÜGERS *Arch. f. d. ges. Physiologie.* LXX, 297–328, 1898.

[4] J. v. KRIES and W. NAGEL, *Zeitschr. für Psychol. und Physiol. der Sinnesorgane.* XXIII. S. 162.

[5] W. KOSTER, Untersuchungen zur Lehre vom Farbensinn. GRÄFES *Arch. f. Ophth.* XIV. 1895 and *Arch. d'opht.* XV. 1895.

[6] F. D. SHERMAN, Über das PURKINJEsche Phänomen im Zentrum der Netzhaut. WUNDTS *philosoph. Studien* XIII. 1898.

made very accurately, because naturally the effect of the slightest movements of the eye, involving displacements of the image on the central region of the retina, will be to make the size of the field seem to be too small.

Very much more accurate results were obtained by v. KRIES and the writer[1] by a method for the details of which the original publication must be consulted. In the case of the writer's right eye (after being dark-adapted by bandaging it tightly for many hours) the PURKINJE phenomenon was found to be absent in a region of the retina whose horizontal extent was 107′. In the left eye it was 88′. In the right eye the vertical extent was 81′. In the writer's case the point used for fixation is not in the centre of this region, but a little to one side. In the right eye the region in question extends from the point of fixation more to the temporal side, and in the left eye more to the nasal side; that is, it is not symmetrical in the two eyes. The longer the dark adaptation, the sharper was the line of demarcation between the central region where the PURKINJE phenomenon did not occur and the surrounding retina where it was manifested.

6. The so-called Colourless Interval

In recent literature on the visual sensations there is a lot of discussion of "the colourless interval."[2] It comprises a certain range of low degrees of intensity within which the stimulating light is visible, but the quality of the sensation is too indefinite for any colouration to be perceived. The actual existence of such an interval is implied in what has been already stated, at least for the state of dark adaptation and more extensive areas of the retina. Indeed the possibility of seeing a dimly illuminated spectrum without colour when the eye is dark-adapted depends on it. Another way of expressing this fact is by saying that coloured lights whose intensity is increased from zero intensity cross first the "absolute" (BUTZ[3]) or "general" (v. KRIES[4]) threshold, but need to be at higher intensity before crossing the "specific" threshold where they are discerned as coloured.

The question as to whether *all* lights that appear coloured at high intensity exhibit a colourless interval, has been variously answered.

[1] J. v. KRIES and W. NAGEL, Weitere Mitteilungen über die funktionelle Sonderstellung des Netzhautzentrums. *Zeitschr. f. Psychol. und Physiol. der Sinnesorgane* XXIII. 167–186.

[2] ¶See W. DE W. ABNEY, *Researches in colour vision* (1913); and ABNEY and WATSON, *loc. cit.* (H. L.)

See also H. D. PARSONS, *loc. cit.* p. 60. (J. P. C. S.)

[3] Untersuchungen über die physiologischen Funktionen der Peripherie der Netzhaut. *Dissertation*, Dorpat 1883.

[4] NAGELS *Handbuch der Physiologie des Menschen.* Bd III. S. 19. 1905.

PARINAUD states that red is seen as red as soon as it rises above the threshold; and also KÖNIG and v. KRIES are inclined to think that the general and specific thresholds for homogeneous red are identical. But CHARPENTIER, KOSTER, HERING, and TSCHERMAK maintain the existence of a colourless interval for red light also. This difference of opinion is easily accounted for by taking into consideration the distribution of the twilight values in the spectrum, as given above (p. 352). Beginning at the border between red and orange, say, from wave-length $650\mu\mu$ on downwards, homogeneous kinds of light have measurable twilight values, which means, in other words, that they must show a colourless interval. But for wave-length 670 the twilight value is already almost too small to be measured, and the almost imperceptible real effect on the twilight mechanism of the eye would perhaps disappear entirely if filters were used to exclude from the field every vestige of light of wave-length shorter than $670\mu\mu$. Even in the most accurate measurements hitherto this has not been done, owing to the technical difficulties. The twilight value of light of greater wave-length still, say between 680 and 700, definitely vanishes, and hence there cannot be a colourless interval for these radiations. Now red light of long wave-length by itself is never obtained with pigments, and it is only by special precautions that it can be obtained with light filters. Owing to this circumstance, an observer can easily be made to believe that there is a colourless interval for red light also. But from what has been said this is true only for the shortest red waves on the border of orange, and even here there is just a faint trace of this effect.

Opinion is also divided as to whether the colourless interval can be observed in the centre of the retina where there are no rods. Here again CHARPENTIER, KOSTER, and TSCHERMAK are among those who believe this to be the case; while PARINAUD, KÖNIG, and v. KRIES state that spectral lights are coloured as soon as they rise above threshold value (with the possible exception of greenish yellow of $580\mu\mu$, according to KÖNIG). The writer himself is a dichromat, and of course in his case there is no chance of observing a colourless interval except with the brilliant hues of red, orange, blue and violet. With fields of appropriate size for being imaged in the region of the fovea centralis (that is, with visual angle of about one degree), lights of this kind above the threshold always look coloured to dichromats. The writer finds the same thing with persons who are not colour-blind, although in their case there is also no colourless interval for green. He has not made any tests with yellow light.

In observations of this nature the same difficulties are encountered as in deciding whether or not the PURKINJE phenomenon occurs in the fovea. The trouble lies in the foveal fixation of the coloured field. Without having a point of fixation to rivet the observer's attention, there is much risk of his first seeing the image of the coloured field as formed on the parts of the retina outside the fovea coming over the threshold, and of course the image then will be colourless (unless the light happens to be red of short wave-length).

From the theoretical standpoint it is of slight interest whether, under any conditions at all, some indication of a colourless interval likewise in the foveal part of the retina could be shown for homogeneous spectral lights. What is more important and more certain is the fact that it can simply be a question of some vestiges of such an effect, not to be compared with the pronounced phenomenon that is observed in the dark-adapted peripheral retina.

For larger areas of the retina that have been completely light-adapted, the colours or at any rate red, orange, green, blue and violet appear immediately at the specific threshold. It is easy to verify this, because in this case there is no necessity of local fixation. Of course, such fields cannot be observed with the entire retina, because it is practically impossible to bring the whole retina to a state of complete light adaptation.

Suppose a person, after having been thoroughly light-adapted, enters a dark room where there is a blue or orange coloured surface of suitable size subtending an angle of from 10° to 20°. If the illumination is steadily increased from sub-liminal to super-liminal values, the deeply saturated colour will come out from the blackness. This observation can be repeated several times during the first minutes of darkness. But then a change occurs quite suddenly. The threshold is lowered and at the same time the colour becomes less saturated; and after a quarter of an hour the sensation produced by sub-liminal stimuli is no longer that of the specific colour but that of the vague almost colourless super-liminal hue. There are also other changes. During the first few minutes if the coloured surface is visible at all, it is sharply outlined; and if, for example, it is a square, it is immediately recognized as being such. In the later stage of the adaptation the borders of the surface that is illuminated just above the threshold are vague and confused, and the longer the dark adaptation continues the more hazy the outline becomes. *During the first minutes after light adaptation vision takes place by means of the daylight mechanism, enabling us to discriminate colour and to have keen perception of form, but requiring comparatively strong light stimuli; but during the later stages, it takes place by means of the colour-blind twilight mechanism in which recognition of form is vague, but which is extremely sensitive to light. It is only when the sensitivity of the twilight mechanism has been so increased during the process of adaptation that it exceeds that of the daylight mechanism for the light used in the test, that the conditions are produced for the appearance of the colourless interval.*

In this connection, it should be recalled that the "grey glow" of incandescent bodies cannot be seen unless the eye is dark-adapted. For the photopic eye the glow is red to begin with. Anyhow the grey glow is never seen in the centre of the retina.

7. Capacity of the Retina for Space and Time Discriminations, in Daylight Vision and Twilight Vision

In ordinary daylight vision the ability to make space discriminations, or what is usually meant by the "visual acuity", is greatest in the fovea by far. Even a few degrees to one side of the point of fixation it is quite a good deal less (see page 34). The approximate form of a curve constructed according to DOR's measurements[1] is shown in Fig. 69, where the ordinates indicate the visual acuity and the abscissae denote the distances in degrees from the point of fixation designated by F. According to BURCHARDT[2], the visual acuity at places from 15′ to 20′ away from the point of fixation was equal to the maximum value, but at a distance of 30′ it was only 80 or 50 percent. of the maximum.

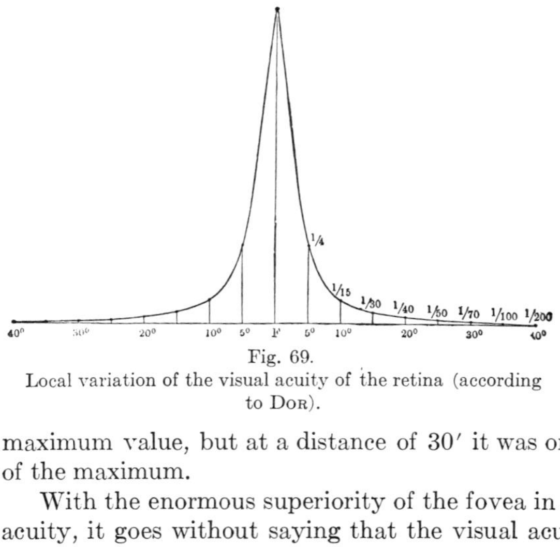

Fig. 69.
Local variation of the visual acuity of the retina (according to DOR).

With the enormous superiority of the fovea in the matter of visual acuity, it goes without saying that the visual acuity in pure twilight vision, where the fovea is excluded as a result of "physiological hemeralopia", will under all circumstances remain far below the maximum acuity of daylight vision. Indeed, strictly speaking, visual acuity in bright daylight cannot be compared with that in twilight vision, because it is dependent on the luminosity of the test chart; in other words, the visual acuity diminishes within certain limits as the

[1] *Archiv. f. Ophthalm.* XIX. 3. S. 321. 1873.

[2] BURCHARDT, *Internationale Probe zur Bestimmung der Sehschärfe.* Berlin 1893.

luminosity decreases. And even when the eye is thoroughly dark-adapted, the subjective brightness can never be so great as it is in daylight vision in bright light, without its rising above the threshold of foveal sensitivity and therewith crossing the border between twilight vision and daylight vision. However, a comparison may be made between the visual acuity of a scotopic eye and that of a photopic eye by making observations on the two eyes, either when the objective intensity of illumination is the same for both eyes in different states of adaptation, or when the subjective brightness is the same. In the latter case the objective illumination of the test chart must be much smaller for the scotopic than for the photopic eye. Experiments of this kind were made by BLOOM and GARTEN.[1] When the illumination is very low, their results are that the visual acuity of the dark-adapted periphery is somewhat higher than that of the photopic eye for the same illumination; but that when the illumination is slightly increased, the visual acuity of the photopic eye is the superior of the two. For approximately the same subjective brightness, these authors found the visual acuity of the scotopic eye was invariably less than that of the photopic eye; but v. KRIES[2] found it to be the same in the periphery for both conditions of adaptation. The essential point is that, whatever the differences may be, they are anyhow insignificant as compared with the differences between the absolute maxima of visual acuity of the photopic eye and the scotopic eye.

The quantitative connection between visual acuity and the intensity of illumination is shown by the measurements made by A. KÖNIG.[3] The test-objects used in these experiments were SNELLEN'S hook-shaped characters, painted black on a white background. The state of adaptation always corresponded to the intensity; that is, the more complete the dark adaptation was, the weaker was the illumination. The results are shown in Table XIII. The figures in the first column give the intensity of illumination B, the unit being the intensity of a HEFNER lamp one metre away. The third column gives the values of the visual acuity S expressed in terms of the ordinary SNELLEN unit. It is evident from the table that a curve plotted by taking the intensities B as abscissae and visual acuities S as ordinates would not enable us to get a very good idea of the connection between these magnitudes,

[1] S. BLOOM and S. GARTEN, Vergleichende Untersuchungen der Sehschärfe des hell- und dunkeladaptierten Auges. PFLÜGERS *Arch. f. d. ges. Physiol.* LXXII. 372, 1898.

[2] J. v. KRIES, Über die Abhängigkeit centraler und peripherer Sehschärfe von der Lichtstärke. *Zentralbl. f. Physiol.* VIII. 694, 1895. See also: BUTTMANN, Untersuchungen über Sehschärfe. *Diss.* Freiburg i. B. 1906.

[3] A. KÖNIG, Die Abhängigkeit der Sehschärfe von der Beleuchtungsintensität. *Sitz. Ber. Akad. Wissensch. Berlin.* 13. May 1897. S. 559–575; and *Gesammelte Abhandlungen* S. 378.

Table XIII (results of A. König)

B	log B	S	B	log B	S
0.00036	0.56—4	0.031	1.03	0.01	0.692
0.00037	0.57—4	0.046	1.16	0.06	0.564
0.00037	0.57—4	0.038	1.19	0.08	0.596
0.00038	0.57—4	0.038	1.38	0.14	0.615
0.00063	0.58—4	0.046	1.38	0.14	0.577
0.00087	0.80—4	0.055	1.76	0.24	0.692
0.0013	0.94—4	0.055	2.14	0.33	0.744
0.0022	0.11—3	0.077	2.20	0.34	0.795
0.0023	0.35—3	0.062	2.20	0.34	0.795
0.0034	0.36—3	0.092	2.28	0.36	0.923
0.0034	0.53—3	0.062	2.37	0.37	0.744
0.0043	0.53—3	0.062	2.97	0.47	0.667
0.0048	0.63—3	0.077	3.95	0.60	0.769
0.0080	0.68—3	0.088	4.64	0.67	0.654
0.0086	0.90—3	0.088	6.06	0.78	0.846
0.0091	0.93—3	0.096	6.81	0.83	1.115
0.0096	0.96—3	0.092	6.81	0.83	1.038
0.010	0.98—3	0.096	9.97	1.00	1.115
0.012	0.00—2	0.092	12.88	1.11	0.866
0.013	0.08—2	0.088	13.6	1.13	1.000
0.032	0.12—2	0.123	13.6	1.13	0.982
0.035	0.51—2	0.123	13.6	1.13	0.963
0.035	0.54—2	0.123	13.6	1.13	0.872
0.037	0.54—2	0.132	14.1	1.15	1.192
0.051	0.56—2	0.154	15.7	1.19	1.054
0.068	0.71—2	0.154	16.0	1.20	1.192
0.069	0.83—2	0.176	20.6	1.31	1.100
0.11	0.84—2	0.185	26.1	1.42	1.093
0.12	0.04—1	0.242	26.7	1.43	1.154
0.13	0.09—1	0.205	28.3	1.45	1.262
0.15	0.13—1	0.231	47.5	1.68	1.308
0.15	0.16—1	0.205	51.2	1.71	1.430
0.15	0.16—1	0.231	54.5	1.74	1.169
0.16	0.18—1	0.200	80.0	1.90	1.313
0.18	0.19—1	0.246	94.5	1.98	1.458
0.20	0.27—1	0.231	119.—	2.08	1.540
0.22	0.30—1	0.308	123.—	2.09	1.437
0.22	0.33—1	0.262	123.—	2.09	1.283
0.22	0.34—1	0.308	168.—	2.23	1.471
0.22	0.34—1	0.277	264.—	2.42	1.568
0.24	0.38—1	0.269	264.—	2.42	1.556
0.28	0.45—1	0.286	316.—	2.50	1.600
0.29	0.46—1	0.308	494.—	2.69	1.667
0.34	0.53—1	0.346	645.—	2.81	1.683
0.35	0.55—1	0.320	794.—	2.90	1.723
0.36	0.56—1	0.338	824.—	2.92	1.662
0.36	0.56—1	0.359	878.—	2.94	1.700
0.41	0.61—1	0.461	1042.—	3.02	1.744
0.41	0.61—1	0.374	1082.—	3.03	1.631
0.44	0.64—1	0.410	1261.—	3.10	1.643
0.44	0.64—1	0.400	1975.—	3.30	1.667
0.50	0.70—1	0.462	2346.—	3.37	1.703
0.52	0.71—1	0.462	2500.—	3.40	1.708
0.67	0.83—1	0.423	3511.—	3.55	1.700
0.67	0.83—1	0.523	7413.—	3.87	1.651
0.88	0.94—1	0.558	7413.—	3.87	1.708
0.91	0.96—1	0.615	7900.—	3.90	1.683
0.95	0.98—1	0.538	14040.—	4.15	1.708
0.99	0.99—1	0.500	31590.—	4.50	1.733
1.00	0.00	0.596	64480.—	4.81	1.750
1.03	0.01	0.558			

because if the high values of B are to be indicated clearly, the smaller values would all be crowded very close together. And so König employed the convenient method of plotting the logarithms of the intensity as abscissae, which shows the connection between the intensity of illumination and the visual acuity in an extremely simple fashion.

The curve (Fig. 70) is composed of two tolerably straight portions which meet each other at an obtuse angle. The break in the curve occurs at an intensity of 0.1 metre-candle. Thus, as the luminosity increases, visual acuity increases at first in proportion to the logarithm of the luminosity, and then for a short interval the relation is more complicated; but soon after the increase becomes again proportional to the luminosity, but the ratio between the two is now much larger, as shown by the steep portion of the curve. In the equation $S = a \log B$ the factor a is about 10 times as large for the steeper gradient of the curve.

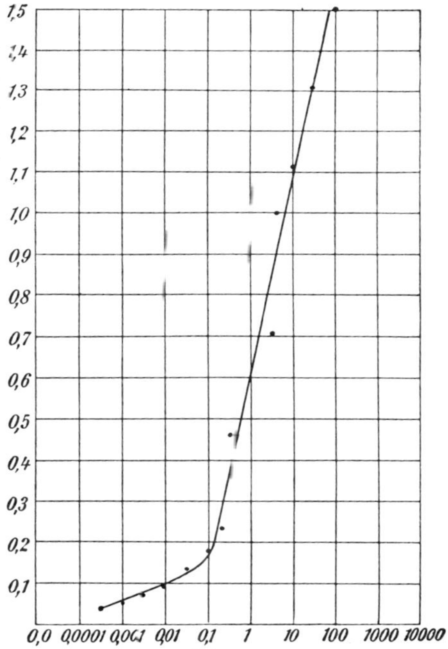

Fig. 70.

Functional connection between visual acuity and the logarithm of the intensity of illumination.

A white surface illuminated by one-tenth of a metre-candle is not far from the threshold of daylight vision. Thus the portion of the curve that lies to the left of the abscissa 0.1 corresponds about to the range of twilight vision. On the other hand, it was stated above (page 318) that a white surface illuminated by one metre-candle or more was above the threshold of an eye that has been light-adapted by bright light. This part of the curve therefore lies definitely within the region of daylight vision. As to the intervening segment comprised between intensities of one-tenth and one metre-candle, it may be considered as representing the case of daylight vision that has not been dulled by very high intensities of light. Pertz[1] found the threshold of

[1] In the article by v. Kries "über die absolute empfindlichkeit usw." *Zeitschr. f. Psychol. u Physiol. der Sinnesorgane* XV. 1897.

foveal vision to be the luminosity of a magnesium oxide surface illuminated by 0.033 metre-candle. At this intensity the curve, as shown in Fig. 70, begins to bend into the steeper branch, and here for the first time the daylight mechanism is appreciably involved, until when the intensity gets to be about ten times as great, it is the sole factor in the power of the retina to make space-discriminations. Thus it may be said that at the border between twilight vision and daylight vision there is some sudden change in the relations between intensity of illuminations and visual acuity, the daylight mechanism being dependent on the intensity of illumination in a much greater proportion.

Likewise there are marked differences between the photopic mechanism and the scotopic mechanism with respect to *the duration and course of the process of stimulation.* It is true we are not yet in a position to describe exhaustively the course of the stimulation in the two cases. However, it may be stated that the sensational response to a short-lived stimulus is more sluggish with the scotopic mechanism than with the photopic mechanism. This is shown most distinctly by rapidly intermittent stimulation; for example, by interposing a rotating sectored disc between the source of light and the eye, or simpler still, a rotating disc with alternating black and white sectors. As the speed of rotation, slow to begin with, is gradually increased, the sectors, which at first can clearly be distinguished, become more and more indistinct, the alternation from bright to dark being perceived merely as so-called "flicker", until finally at a certain speed the disc appears a uniformly steady grey. Knowing the angular speed and the number of sectors, the number of interruptions can be calculated at which flicker just ceases. It is true it is not altogether easy to determine this limit, because the direction of fixation must be kept constant, and the observation has to be made with a visual angle that it is not too large, say, from 3° to 5°. For the latter purpose, the greater part of the rotating disc can be covered with an opaque surface with an aperture of the correct size.[1]

[1] ¶For recent accounts of flicker the following may be consulted:

H. BENDER, Untersuchungen am LUMMER-PRINGSHEIMschen Spectralflickerphotometer. *Ann. d. Physik.* XLV. 1914. 105–132.—W. W. COBLENTZ and W. B. EMERSON, The relative sensibility of the average eye to light of different colors, and some practical applications to radiation problems. *Bull. Bur. Stand.* Sci. paper No. 303. 1917. 167–236.—E. C. CRITTENDEN and F. K. RICHTMYER, An average eye for heterochromatic photometry and a comparison of a flicker and equality of brightness photometer. *Bull. Bur. Stand.* XIV. 1918. 87–114. — U. EBBECKE, Über das Augenblicksehen II. Über das Sehen im Flimmerlicht. PFLÜGERS *Arch.*, 1920. CLXXXV. 181–195.— Idem, Über das Sehen im Flimmerlicht, PFLÜGERS *Arch* CLXXXV. 1921. 196–223. — C. E. FERREE and G. RAND, Flicker photometry. I. The theory of flicker photometry. II. Comparative studies of equality of brightness and flicker photometry with special reference to the lag of visual sensation. *Trans. Illum. Engin. Soc.* 1922. 50 pages. — H. E. IVES, Studies in the

HELMHOLTZ states (page 213) that with a disc of alternate black and white sectors of equal width illuminated by the strongest lamplight, the flicker ceases for him when the stimulus from a single sector lasts for one forty-eighth of a second; and under the illumination of the full moon, when the stimulus lasts one twentieth of a second.

This relation between the brightness of the light and the number of interruptions at which flicker is abolished is easily explained. The black portions of a colour-top of this sort cannot be considered as being absolutely black. Some light is reflected from them, although of course much less than that reflected from the white portions. When the illumination is diminished, the observer gets the impression of a stationary disc with greater diminution in the brightness of the white, which gradually changes to dark grey becoming more and more like the black. This colour acting on the eye alternately with the black fuses with it more easily, that is, for fewer alternations, than a pure white would do. The careful measurements made by SCHATERNIKOFF show that the *fusion frequency*, as v. KRIES calls the number of revolutions necessary to make flicker disappear, depends not only on the absolute objective intensity of the light, but also on the subjective brightness of the white as determined by the state of adaptation. As long as the conditions of twilight vision are completely or at least very approximately maintained, the fusion frequency increases with increasing dark adaptation, since the same effect for the brightness of the white is produced in this way as would be produced by objective increase in the intensity of the light.

SCHATERNIKOFF also compared the fusion frequency of the photopic eye with that of the scotopic eye, his method being similar to that used by BLOOM and GARTEN (*loc. cit.*) in their visual acuity tests; that is, by obtaining a brightness that was subjectively the same for the eye in both states of adaptation, the objective illumination being therefore very different for the two cases. Even under these conditions the photopic eye requires a greater frequency than the scotopic eye, and hence the decrease of the fusion frequency when the illumination is low cannot be attributed simply to the decrease of illumination of the

photometry of lights of different colors. *Phil. Mag.* 6th Ser., XXIV. 1912. 149–188; 744-751; 853–863. —H. E. IVES and E. F. KINGSBURY, The theory of the flicker photometer. *Phil. Mag.* 6th ser. XXVIII. ,1914. 708–728; XXXI. 1916. 290–322.—H. E. IVES, A polarization flicker photometer and some data of theoretical bearing obtained with it. *Phil. Mag.* 6th Ser. 1917. XXXIII. 360–380. — Idem, Hue difference and flicker photometer speed. *Phil. Mag.*, 6th Ser. XXXIV. 1917. 99–112. — Idem, Critical frequency relations in scotopic vision. *Jour. Optical Soc Amer.*, VI. 1922, 254–268.—Idem, A theory of intermittent vision. *Jour. Optical Soc. Amer.* VI. 1922. 343–361. — E. THUERMEL, Das LUMMER-PRINGSHEIMsche Spectral-Flickerphotometer als optisches Pyrometer. *Ann. d. Physik*, XXXIII. 1910. 1139–1160. (H. L.)

white, but we must assume that the operation of the mechanism of vision depends on other factors besides. SCHATERNIKOFF found that the fusion frequencies of the photopic eye and the scotopic eye were in the ratio of 5 to 3. For as exact a match as possible, as to both brightness and colour, the writer has obtained bigger differences, with his eyes in the two states of adaptation. Discs like those described by HELMHOLTZ in connection with Fig. 42 are convenient for observations of this kind. Experimenting with this apparatus, with one eye thoroughly dark-adapted and the other light-adapted, the writer found that the speed of rotation could be so regulated that for the photopic eye there is no flicker at all in the outside ring, little in the middle, and distinct flicker in the central ring; whereas the entire disc was free from flicker so far as the scotopic eye was concerned. The brightness was the same under the two conditions.

Without paying special attention to the state of adaptation, PORTER[1] compared the fusion frequencies for a series of different intensities of illumination from the highest to the lowest, and obtained the remarkable result that the relation between the frequency and the intensity of illumination can be expressed by a curve composed of two approximately straight portions meeting each other at an angle, similar to the curve in Fig. 70 which represents KÖNIG's measurements of visual acuity. In each of the straight portions the fusion frequency increases in proportion to the logarithm of the intensity, the factor of proportionality, however, being different for the two branches.

The similarity between the two curves, as v. KRIES[2] has pointed out, becomes still more perfect when the intensity is noted at which the bend in the curve occurs. This intensity, at which there is suddenly a new relation between intensity of light and spatial and temporal power of discrimination, turns out to be in fact practically the same in both cases, being the illumination of a HEFNER lamp at a distance of one-tenth of a metre in KÖNIG's experiments and that of a standard candle at a distance of one-eighth of a metre in PORTER's experiments.[3]

Aside from the fusion of periodic luminous stimuli, the time relations in a single brief stimulation are likewise of interest in connection with the duplicity theory. These quite complicated phenomena are to be described hereafter. It will appear then that although it is not yet possible to give a complete explanation of them, still they present a number of characteristics indicating with much probability a difference

[1] *Proceedings of the Royal Society*, London, LXX. 313.

[2] J. v. KRIES, Über die Wahrnehmung des Flimmerns durch normale und durch total farbenblinde Personen. *Zeitschr. f. Psych. u. Physiol. d. Sinnesorg.* XXXII. 113; and *Abhandlungen zur Physiol. d. Gesichtsempfindungen.* Drittes Heft 1908.

[3] ¶See J. H. PARSONS, *loc. cit.*, p. 208. (J. P. C. S.)

of behaviour between the daylight mechanism and the twilight mechanism, and consequently are very easy to comprehend on the basis of the duplicity theory.

8. Total Colour Blindness Considered as being Twilight Vision Alone

At the time the first edition of this work appeared only two kinds of colour blindness were known, namely, red blindness and green blindness, as HELMHOLTZ distinguished them; both afterwards included by HOLMGREN under the name of "partial colour blindness". Each of these types was capable of making certain qualitative colour distinctions. More modern investigations of the vision of such persons will be described below. Here however something should be said about what is now known as *total colour blindness*, a much rarer type discovered about 1880, in which there are no qualitative differences in the appearance of different colours, and which has therefore been called *achromatopia* or achromatic vision. Persons who suffer from this anomaly distinguish merely unequal shades of brightness in objects of different colours, the quality of the light sensation always remaining the same.[1]

In the case of certain totally colour blind individuals, HERING and HILLEBRAND found that the distribution of luminosity in the spectrum was the same as for the thoroughly dark-adapted eye in the state of twilight vision. Further researches conducted by KÖNIG, v. KRIES and others showed that the vision of persons who see with this kind of luminosity distribution has a number of other common propensities also, and that this anomaly, being apparently always congenital, is characteristic of a well-defined class, and may be described therefore as "typical, congenital total colour blindness". At the same time it should be mentioned that there are also achromatopes with a totally different luminosity distribution in the spectrum. In these cases it is nearly always possible to trace the origin of the anomaly to some injury of the optic nerve; and, besides, the other peculiarities of congenital total colour blindness which will be described below are not manifested.

In achromatic vision the spectrum looks like a shaded surface without any differences of colour, a conception of which an achromatope has no notion (see the illustration on Plate II).* To him the greatest luminosity (the source being gaslight) is where the green is, between 530 and 540$\mu\mu$. Red light of longer wave-lengths than the line C is not visible to him at all, but towards the violet end the visible spectrum in his case extends about as far as it does for a normal eye.

[1] ¶That is, they see only shades of grey. (J. P. C. S.)
*Plate II faces page 352.

But whereas in the normal eye this peculiar distribution of luminosity does not occur unless the intensity of the entire spectrum remains below the threshold of colour vision, that is, does not occur except in pure twilight vision, almost the same distribution of luminosity persists for the totally colour-blind whether the illumination of the spectrum is low or high. If the intensity of illumination of a spectrum is gradually increased from the lowest degrees, the instant it begins to appear coloured to an observer with normal colour vision, the only change whatever that will be apparent to an observer who is totally colour-blind will be a mere increase of the total luminosity.

The exact agreement between the luminosities of colours for the achromatopes and a dark-adapted person with normal vision may readily be verified by making twilight matches with the colour-top and especially with the spectrophotometer also. By the same process as described on page 317 for finding the twilight values of the normal eye, these values may be obtained also for the totally colour-blind individual, and not merely when the intensity of illumination of the spectrum is low, but also when it is high. There is an upper limit due to the sensation of glare that is very annoying to colour-blind persons. Curves showing the relative stimulating value in the spectrum of a totally colour-blind person have been obtained by KÖNIG[1], v. KRIES[2]

Table XIV

Wave-length	Twilight Value for W. NAGEL	Luminosity Value for for the totally colour-blind H.
628 $\mu\mu$	3.1	—
615	—	10.0
603	10.0	13.2
595	—	26.1
591	25.9	—
580	45.8	30.0
570	57.8	52.7
561	65.0	—
553	76.0	81.8
545	88.1	84.9
535	86.3	88.1
528	81.6	78.3
520	70.2	—
514	61.5	64.5
500	39.0	43.8
488	24.5	33.7
473	13.5	18.8
457	6.2	11.5
446	2.5	—

[1] *Zeitschr. für Psychologie und Physiol. der Sinnesorgane* IV, S. 241. 1892.

[2] Ibid., XIII. S. 293.

and the writer. The results of the investigation of a girl sixteen years old (Miss H.), as made by Dr. MAY[1] and the writer together, are exhibited in Table XIV and Fig. 71. The luminosity values for Miss H. and the twilight values in the case of the writer (which agree with those that are typically normal) are plotted here side by side. The colour-blind subject made the observations on the HELMHOLTZ spectrophotometer in the bright room with moderate intensities of the spectrum that were not dazzling for her. The writer's observations were

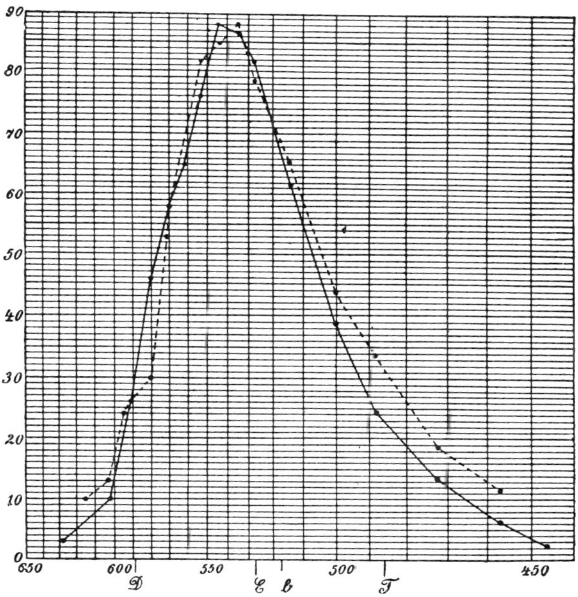

FIG. 71.—Distribution of luminosity for the totally colour-blind - - - - - - and twilight values for a deuteranope———, in the prismatic spectrum of the NERNST light.

made in the dark room after being dark-adapted with eyes bandaged for an hour. The results are for the spectrum of the NERNST lamp. The agreement between the two sets of observations is very obvious, and even better than in a previous comparison made between the writer and another colour-blind girl, which was reported by v. KRIES.[2] Perhaps, the main reason for this is that meanwhile the writer himself had had more practice in making twilight matches, which are not altogether easy to make. In both cases there is a distinct difference in the green and blue between the values of the colour-blind observer

[1] *Zeitschr. für Psychologie und Physiol. der. Sinnesorgane* XLII. S. 69. 1908.
[2] Ibid., XIII. S. 293.

and those of the writer, which can hardly be accounted for as being due to errors in the settings. At these places the twilight values are below the luminosity values for the totally colour-blind subject. The reason for this discrepancy will be discussed later, but here it may be simply mentioned that the difference would not exist, or at any rate would be very much less, if the totally colour-blind observer had also made the observations after previous long-continued dark adaptation, which in the present case was purposely avoided.

The peculiar luminosity distribution in the spectrum of the totally colour-blind eye, which coincides with that of the normal scotopic eye, together with the connection to be explained presently between the curve of twilight values and the curve showing the absorption of energy by solutions of visual purple, led A. König to make the assumption, that in case of eyes that are totally colour-blind vision is performed entirely by the mediation of the rods in the retina that contain visual purple, the cones being lacking or not functioning. Further study of the vision of achromatopes has confirmed this assumption in a very remarkable manner. The vision of these folks exhibits a number of marked peculiarities which are united to form a characteristic symptom-complex.

From the sensitiveness of visual purple to light, it would be natural to expect that persons dependent on this visual substance would see badly when the light was very bright, and that ultimately indeed they would become temporarily blind. This is really the case. Achromatopes are all, to a greater or less degree, light-shy (or photophobic) and averse to looking at a bright light. Their pupils are usually very small, the eyelids almost closed, and the head bent forward in a characteristic way whenever bright light comes from above. Totally colour-blind persons may be readily recognized by this attitude and by their timid glance. Dark glasses give them much relief. The photophobia of such persons as have come under the writer's observations is the effect of a different cause from that of individuals who are light-shy as a result of injury to the eye. Light is not painful or disagreeable to achromatopes; they are simply aware that they do not see well when they look at anything bright. Miss H., the totally colour-blind girl mentioned above, who was most amiable about acting as a test patient in numerous experiments made by Dr. May and the writer, was easily induced to look at an extremely bright light with one eye, the other being blindfolded. It gave her no pain, but in a few minutes she was practically blind and unable at any rate to see luminous bright fields in the spectrophotometer. Nor did she complain when a large area of her retina was brightly illuminated by the Thorner ophthalmoscope, although she was not able to see with this eye for a quarter of an hour afterwards.

Naturally, it was a matter of special interest to see what was the behaviour of the central part of the retina in the case of achromatopes; because here in the region of the fovea there is no visual purple, and there are no rods; and here also the characteristic phenomena of twilight vision are absent.[1] Various possibilities may be conjectured as to the behaviour of this part of the retina of totally colour-blind persons. Thus, the cones might be lacking entirely and their places occupied by rods; or instead of the cones there might be special structures different from the rods, but not able to function or at least not sensitive to light. Lastly, the vacancies due to the absence of cones might be only partly filled by rods. No matter which of these possibilities proves to be the case, the fovea must be either a completely blind spot, or a place which is stimulated by light exactly in the same way as the scotopic eye in the case of twilight vision. In the latter case, especially, a visual acuity is to be expected that continues below that of the normal fovea, having about the same range of values as are found in pure twilight vision under the most favourable circumstances. Now this is actually what is found to be the case, as v. KRIES has showed.[2] In the most favourable conditions the visual acuity of achromatopes may be as high as one-sixth or one-tenth, values even a little higher (as much as one-fourth) being obtained in sporadic cases. Of special interest in this connection is A. KÖNIG's result which indicates that as the illumination is increased the visual acuity of the totally colour-blind eye increases by a different rule from that of the normal eye. There is no bend in the curve (Fig. 70) at the threshold of daylight vision, but even for intensities of illumination higher than 0.1 HEFNER-candle the totally colour-blind curve is fairly straight and continues as a prolongation of the initial, less steep portion. A maximum is then soon attained, where glare interferes with perception of form. These results are entirely in harmony with our previous assumption that for intensities of illumination higher than a tenth of a metre-candle the visual acuity is controlled by the daylight mechanism represented by the cones, this mechanism being lacking in the totally colour-blind.

In some totally colour-blind people a spot has been found in the centre of the retina that is perfectly insensitive to light, that is, a blind spot or "central scotoma"; but in the case of other totally colour-blind persons it is not certain that there is a central scotoma. A. KÖNIG considered it as a particularly impressive proof of the correctness of his view concerning the function of the rods, that there was a central scotoma in the case of achromatopia which he investigated. On the other hand, not being able to find a defective spot of this sort, HER-

[1] See W. DE W. ABNEY and W. WATSON, *loc. cit.* (H. L.)

[2] *Zentralblatt für Physiologie,* 1894. S. 694.

ING and HESS used it as an argument to throw doubt not only on KÖNIG's theory but also on the duplicity theory in the form given it by v. KRIES. But v. KRIES, not having found a scotoma in a careful investigation which he made, has pointed out that this particular matter is of no importance for or against the duplicity theory. At present we still know nothing at all certain as to the original cause of total colour blindness, and therefore cannot assert that, because there is lack of cone *function,* the central part of the retina, which is normally crowded with nothing but cones, must be without any functional elements at all. We might just as readily assume that rods have taken the place of the cones. Consequently, it may be considered that the cases with scotoma (GRUNERT lists 10) and those without (GRUNERT lists 8) represent two different modes of origin of total colour blindness in fetal life. There is, however, still another possibility, namely, that even in those cases of total colour blindness in which no scotoma has been found as yet, there may be one present corresponding to the fovea, only it has escaped observation.

The establishment of a scotoma as large as the rod-free region, that is, having an angular diameter of from 1° to 1° 30′, is not easy even when the place in the retina that has become insensitive to light by some pathological process lies to one side of the fovea; but it is far more difficult still to confirm the existence of real central scotoma. The chief difficulty about finding a small defect like this in the visual field consists in maintaining the direction of fixation, and this condition cannot be fulfilled unless foveal fixation is possible. The writer failed to find a central scotoma in the case of Miss H., although a careful search was made; but in spite of this negative result, it is quite possible that it was there, and was simply hard to locate.

Nearly all achromatopes exhibit what is called nystagmos, that is, a restless, frequently very rapid movement of the eyes first one way and then another. Watching the eyes of a patient of this kind, the observer will notice particularly active nystagmos when he tries to rivet his gaze on an object, that is, fixate it. The subject simply cannot do it. This gives a peculiar expression to the eyes of an achromatope in addition to the other characteristics that are the result of photophobia. The tendency to perfect coördination of the movements of the two eyes and for permanent binocular vision is absent or perhaps weak; and the result is a strange sort of disconnection in the movements of both eyes, as was particularly noticeable in the case of Miss H. A permanent squint is sometimes the consequence. Strabismus divergens is reported in so many cases that its occurrence in connection with total colour blindness cannot be an accident. When the tendency to binocular vision is absent, as it is just before going to sleep, or when the eyes are closed, the axes of the two eyes are usually divergent. Apparently when there is a slight divergence the external muscles of the eye are most relaxed; and therefore it might be expected that, when the tendency to binocular fixation is absent from some defect of foveal vision, there is likewise a disposition to let the eyes diverge.

The writer desires to emphasize the fact that the existence of nystagmos and strabismus, or, to put it another way, the lack of a *definite* place of fixation in the centre of the retina, does not necessarily involve the assumption of a central scotoma in place of the fovea. It is just as satisfactory an explanation to assume that the foveal region was occupied by elements with rod-like functions. The absence of a place quite specially adapted for keen vision is sufficient by itself to explain to a certain extent why the eye is so restless in a case of this kind. The explanation is made still clearer by supposing that the rod mechanism is more quickly fatigued than the cone mechanism in the fovea; an assumption which has much to support it.

A characteristic distinction between daylight vision and twilight vision, as we saw above, is the dissimilar reaction to short-lived stimuli recurring in quick succession. In this respect also the vision of the totally colour-blind is similar to twilight vision. As stated above, flicker of a rotating disc with alternate black and white sectors ceases in twilight vision for a much slower speed of rotation than in daylight vision with equal subjective brightness. At the suggestion of v. KRIES, UHTHOFF made special determinations on some totally colour-blind patients whom he had under examination, and found that for even higher intensities of illumination that are above the threshold of daylight vision for the normal eye the totally colour-blind cease to see flicker at a speed of rotation of the disc that is not high enough to make flicker disappear for a person with normal vision. This result was completely confirmed by the writer in tests made with the colour-blind girl above mentioned.

Accordingly here also there is perfect analogy between the vision of the achromatope and twilight vision and a further support for the assumption, that the totally colour-blind person sees regularly only by means of the elements of the retina that under other circumstances mediate twilight vision.

9. Night Blindness as Functional Abeyance of the Rods[1]

PARINAUD regarded the relation between the light sense and colour sense in so-called night blindness or hemeralopia as one of the most essential supports of the duplicity theory. More recent investigations

[1] ¶W. DE W. ABNEY, Two cases of congenital night blindness. *Proc. Roy. Soc.* 90. B. 1916. 69–74. (H. L.)

have completely confirmed this view. Night blindness may be considered as a sort of obverse condition to total colour blindness. In the latter, according to the duplicity theory, the rod mechanism is isolated, the function of the cones being in abeyance. Conversely, in night blindness the function of the cones is found to be more or less intact, the rod-mechanism being seriously impaired. It is true that in those cases in which the rod mechanism can be considered as completely lacking, the cone mechanism has perhaps always been considerably impaired too, so that it is not the case of a retina where the rods alone have become incapable of functioning. Still there is a decided approximation to this state of affairs.

In night blindness the faculty of adaptation is much affected, and dark adaptation is so much retarded that the sensitivity usually reached in a half hour will not be attained until after several hours in darkness; or the amplitude of adaptation will be more or less circumscribed, and only a moderate increase of sensitivity to light will be experienced after a long period of darkness. In higher degrees of night blindness both the rate of adaptation and the amplitude are invariably diminished. As was mentioned on page 319, the sensitivity may be enormously reduced. In this case the foveal threshold in the state of light adaptation may be entirely or almost entirely normal; which is an important fact from the theoretical standpoint. With thorough light adaptation and in bright daylight an individual with a moderate degree of night blindness may see practically as well as one with normal vision. On the other hand, with high degrees there is as a rule a certain amount of amblyopia, which, considering the origin of the disease and the processes going on in the retina and choroid, is not surprising. No serious defect of colour vision is manifested unless there is some congenital, typical anomaly. The only thing that is occasionally remarkable is a lowered sensitivity to blue light, particularly in foveal vision. Small, dark blue objects, like corn-flowers that grow in grain-fields, are not seen as blue but simply as "dark". This is similar to the condition in the normal eye when it has been blinded by very bright light.

The reaction of the night-blind to the longer waves of red light is particularly remarkable and important theoretically. After dark adaptation lasting a quarter of an hour, more or less, his sensitivity to an area illuminated by composite white light which extends past the foveal region will be much less than that of a person with normal vision dark-adapted to the same degree; but the threshold for red light in cases of slight night blindness is not at all higher than it is for persons with normal vision, and very little higher in cases of moderate night blindness.

Connected with this is the fact that the PURKINJE phenomenon (see p. 357) is not nearly so distinct in night blindness as in normal vision. Suppose that red and green surfaces are arranged side by side in the dark room, the colours being so selected that they look about equally bright in daylight vision both to a night-blind individual and to a person with normal vision; and then let both observers enter the dimly lighted dark room together. After a few minutes the green will look decidedly brighter to the person with normal vision than the red, before any sign of this difference is apparent to the night-blind. He will not be aware of the phenomenon until much later. Measurement shows that after staying in the dark for about half an hour the increase of brightness of the light of shorter wave-length is from 10 to 100 times greater for the normal eye than for the night-blind eye. In far advanced cases of retinal pigment atrophy, which always accompanies night blindness, twilight vision may be completely destroyed. Under these conditions the visual field is considerably contracted in size, and the PURKINJE phenomenon cannot generally be evoked in the part of the central field that is left. For such patients the brightness relation of a pair of colours is just the same in weak illumination and after long dark adaptation as in bright daylight.

Concerning the time relations of the light sensation in night blindness little is known as yet. It is especially not known whether the "fusion frequency" for regularly intermittent light stimuli (page 373) varies with the state of adaptation in the case of night-blind persons in the same way as it does in normal vision; or whether the entire retina in their case behaves in the same way as only the central region does under normal circumstances. It is very probably that the latter is the case in the earlier stages of dark adaptation, but that in the later stages the part of the twilight mechanism which still continues to function asserts itself, as shown by the more extended course of the process of stimulation and consequent lower fusion frequency. This seems to be the case for normal vision on the basis of SCHATERNIKOFF's experiments mentioned above (p. 373).

The so-called PURKINJE after-image, to be mentioned presently, which apparently represents a peculiarity of rod stimulation, was not obtained by the night-blind patient examined by v. KRIES. Moreover, certain night-blind individuals, to whom the writer tried to show the after-image, could not see it at all even under the most advantageous conditions. In milder cases of this malady the PURKINJE after-image is not absent; still the writer has the impression that it is not so easy to perceive as it is for observers with normal eyes. These observations, as we see, are in good agreement with the duplicity theory and with PARINAUD's way of regarding night blindness. Further experiments would be of interest especially in those cases in which one eye is distinctly hemeralopic, the other eye not being yet so or only in slight measure. Of course, in this case the plan would be to make the subjective brightness of the light-stimuli equal for both eyes.

Although the facts here presented as to the light-sense and colour-sense of the night-blind point to the correctness of PARINAUD's hypothesis, besides many other facts concerning the pathogenesis of the condition of hemeralopia, which cannot be discussed here, but which point in the same direction; HESS[1] has come out recently as vigorously objecting to the use of the results found in night blindness as arguments in favour of the duplicity theory. He maintains that the night-blind persons examined by him were able to perceive the PURKINJE phenomenon; that after dark adaptation they exhibited less sensitivity to light in the centre of the retina than in the periphery; that red pigments in dim light were visible to them without colour, etc. These results are in keeping with some which the writer could instance concerning various night-blind subjects. On the basis of such observations, HESS concludes that PARINAUD's hypothesis of the origin of hemeralopia as being due to the disappearance of visual purple is thus upset. But the hypothesis does not imply that *everybody* that is night-blind is *entirely* without visual purple and the twilight mechanism in v. KRIES's sense. Nobody seriously thinks this. Night blindness is a symptom of a number of ocular diseases and occurs in the most various degrees. In many cases it is progressive, and consequently it is not surprising that in mild and medium degrees of it the effects mentioned by HESS should have been obtained. These effects are explained by v. KRIES, PARINAUD, the writer and others as being the expression of a participation of the twilight mechanism in vision. The process of dark adaptation by which the twilight mechanism is gradually inserted along with the daylight mechanism and made to function is merely accomplished far more slowly in the night-blind patient than in the case of a person with normal vision; and the PURKINJE phenomenon takes a longer time to occur and eventually is fainter, frequently with almost no traces of it. Moreover, in night blindness the sensitivity to light in darkness does not *cease* increasing, but it merely proceeds more slowly and to a less extent. All that HESS's experiments show is that certain qualitative changes in colour vision that go hand in hand with adaptation may occur in spite of the existence of night blindness. This has never been denied. The views which are here advocated would not be affected even if it can be shown (which incidentally HESS has not done) that the cone mechanism in night blindness is sometimes or always impaired

[1] C. HESS, Untersuchungen über Hemeralopie. *Arch. f. Augenheilk.* LXII. 1908.

also. The outstanding result shown by the symptoms is the *severe damage* of the twilight mechanism compared with which any lesion of the daylight mechanism is quite secondary. Even if it can be inferred from PARINAUD's statements that he meant to assert the occurrence of *damage* to the rod mechanism entirely by itself, it would perhaps be hard to gainsay such a positive statement as this without additional evidence against it.

10. Assumptions of the Duplicity Theory

The experimental observations described in the preceding sections indicate how under the influence of long-continued exclusion from light the retina undergoes fundamental changes in its mode of function. They show also that in the most striking and significant of these changes the centre of the retina, the fovea centralis, does not participate. As a matter of fact, when an eye which has been previously exposed to bright illumination is plunged suddenly in darkness, the sensitivity to light does increase even in the centre of the retina; but, as was stated above, this increase, even under the best circumstances, that is, after previous very thorough light adaptation, amounts only to a small fraction of the increase of sensitivity that takes place in the periphery; and, besides, it disappears in a few minutes, being succeeded by an approximately stationary condition. For the same size of luminous object, the periphery of the retina shows an increase of sensitivity thousands of times greater, lasting from a half hour to an hour, and not complete then. But the most remarkable thing is the difference in the reaction to radiations of different wave-lengths. In the fovea the change of sensitivity is of the same degree for all kinds of light; in the periphery the differences between light of long waves and light of medium and short waves are found to be enormous. The process in the fovea gives the impression of being a simple recovery process, such as takes place in quite similar fashion in other sense organs, the ear, for example. On the other hand, the adaptation of the periphery of the retina to different degrees of intensity is connected with striking qualitative changes in the mode of reaction. The latter, together with the important quantitative changes of excitability, find their most natural explanation on the basis of the theory of the double functions of the rods and cones, as formulated by v. KRIES.[1] His own words are as follows: "Accordingly, we should ascribe to the rods the property of being able to undergo very extensive changes of adaptation, which however arouse merely colourless sensations of brightness; and, lastly, the property of being affected by different kinds of light in just the same proportions as correspond to the distribution of brightness

[1] NAGELS *Handbuch der Physiologie des Menschen.* Bd III. S. 185.

in the spectrum as it occurs in twilight vision. On the other hand, the cones are to be considered as having comparatively little power of adaptation and as being able to discriminate colour in the centre of the retina and its immediate vicinity, although everywhere they are of such nature as to be strongly stimulated even by light of long wave-lengths, and hence by their activity the relations of brightness occur that are peculiar to daylight vision and favourable to light of long wave-lengths."

According to what was stated on page 347, there might perhaps have been included in these propositions the conjecture that the sensation mediated by the rods may vary, according to circumstances, between being colourless or bluish.

On the basis of the determinations of the fusion frequency with rotating discs, we must, moreover, assume, with v. KRIES, that the process of stimulation in the twilight mechanism, presumably therefore in the rods, is longer and more drawn out. In the matter of space discrimination, the parts of the retina where the rods are, never attain anything like the same capacity as the fovea where there are nothing but cones. Finally, the experiments of PIPER and HENIUS (see page 336 above), which shed light on the difference in the relation existing between the sensitivity and the stimulated retinal area in daylight vision and in twilight vision, enable us to realize that a real summation of stimulations is also a very much more important factor in the twilight mechanism than it is in the daylight mechanism.

The question may now be considered whether these differences in the excitability of the rods and cones as postulated by the duplicity theory can be reconciled with the known anatomical, physical and chemical characteristics of the photo-sensitive elements, and to what extent this is possible. Here the anatomical results obtained by RAMON Y CAJAL[1] with respect to the nervous connections of the rods and cones agree very well with the observed physiological facts. So far as the cones are concerned, especially those in the foveal region, at present it is quite generally supposed that each is connected with a single fibre; and that conduction in the retina and optic nerve is isolated to a certain degree at least. On the other hand, the idea is that the rods are collected in groups by cross connections in the retina. According to the views developed by HELMHOLTZ (see page 35), the isolated relation of the cones must be very favourable for nicer space discrimination, whereas the connections of the rods must be comparatively unfavourable. On the other hand, the arrangement of the

[1] See R. GREEFF, Die mikroskopische Anatomie des Sehnerven und der Netzhaut. GRAEFE-SÄMISCHS *Handbuch d. Augenheilk.* 2. Aufl. I. (5) 1901.

rods in groups must be conducive to the integration of separate sources of stimulation. Thus if, say, ten rods are connected together in such fashion that their impulses must all be conducted along a single nerve fibre, the stimulation of two rods in this group will not produce light sensations involving space-discrimination, but rather merely a single sensation so far as the sense of space is concerned, only it will be stronger than if just one rod had been stimulated. Subjectively, the result of this dispersion of a light stimulus over several rods in a group is seen in augmentation of brightness.

Of course, this does not imply that there is no integration whatever of separate stimuli in the cones, for example, in the foveal region. Evidently, the question here is rather one of quantitative differences that are certainly quite considerable in amount.

In trying to explain the inferior power of the twilight mechanism for temporal discrimination, we are on less firm ground. More complicated modes of connection in the retina might be imagined, involving the insertion of several neurones in the transmission system of the twilight mechanism. But even if a marked difference in this respect could not be shown between the routes of conduction from rods and cones, it would not involve any difficulty. For both the rate of reaction and the conductivity in the case of different kinds of tissues, even those particularly distinguished for excitability, exhibit differences that exceed those in question here.

A question of special interest is, what is the reason for the difference of excitability between the daylight mechanism and the twilight mechanism for light of long wave-lengths, and what is it that is mainly responsible for the unequal distribution of the stimulus values in the spectrum for the two mechanisms. There is a possibility here that the stimulus values, like photo-chemical actions, depend on the wave-lengths in some unknown way. To a certain extent this is undoubtedly true. We do not know at all why the so-called infra-red rays do not stimulate the retina of the human eye, although they distinctly stimulate some lower organisms; and we are equally ignorant as to why just those rays of wave-lengths $590-600\mu\mu$ produce the brightest sensation of light in the foveal cone-mechanism, although there is no known absorbent material in the cones that absorbs especially these particular waves, and although the maximum energy of spectral radiations is not in this part of the spectrum but in the green.[1] And yet no sooner had the duplicity theory been formulated, and a claim entered

[1] ¶See F. Weigert, Ein photochemische Modell der Retina, Pflügers *Arch.* CXC. 1921. 177–197. Also, Über die Photochemie der Retina. *Zft. f. Elektrochem.* XXI-XXII. 481–487. Also, Zur physikalischen Chemie des Farbensehens. I. Über die Lichtempfindlichkeit der Farbstoffe. *Zft. f. physikalische Chem.* C. 1922. 531–565. (H. L.)

on behalf of the rods as organs of twilight vision, than attention was directed to the very remarkable analogies between rod vision and the chemical action of light on visual purple. KÜHNE's researches had already shown that visual purple was only slightly affected by yellow light and hardly at all by red light. The more precise quantitative relations have been given in a previous section. Here it is sufficient to to say that the results obtained by KÖTTGEN and ABELSDORFF as to the *amounts of energy absorbed* by visual purple and TRENDELENBERG's determinations of the *bleaching values* or chemical effects produced in the visual purple indicate that both of these magnitudes depend on the wave-length. Now these relations are found to accord in most striking manner with the way the *twilight values* depend on wave-length, as can be seen in the graphical representation in Fig. 72.[1]

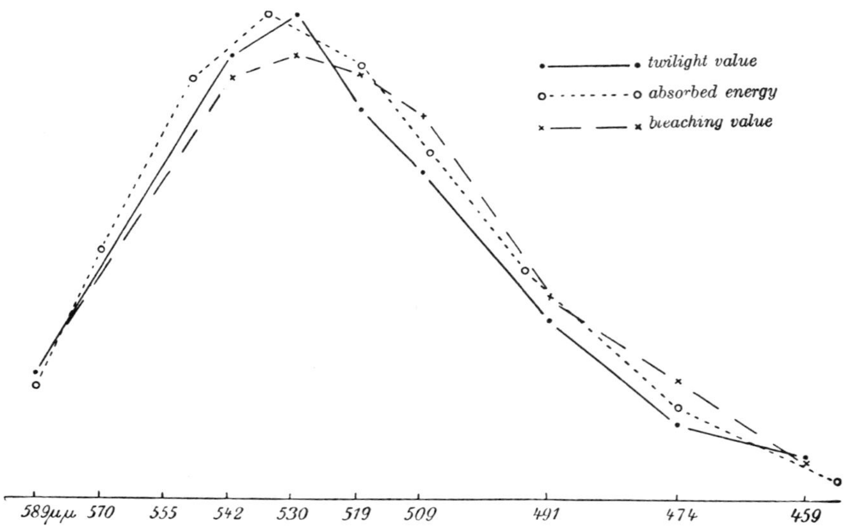

Fig. 72.

This cannot be purely accidental. It would seem rather to indicate beyond doubt that visual purple is of the greatest importance for twilight vision, and especially that its regeneration in darkness is the most fundamental thing in dark adaptation. Just what its special part is, and how it and its decomposition are related to the whole process of the stimulation of the eye by light, are questions that cannot be answered at present.

[1] ¶See S. HECHT and R. E. WILLIAMS, The visibility of monochromatic radiation and the absorption spectrum of visual purple. *Jour. Gen. Physiol.*, V. 1922. 1–34.—See also §18A in this volume, and literature there cited. (H. L.)

Plate III

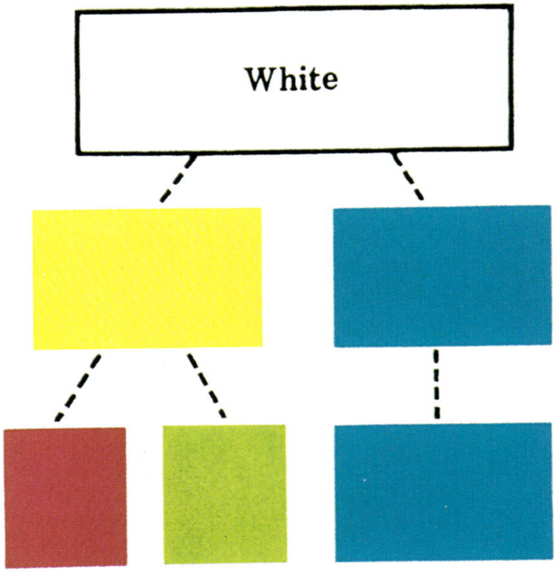

Fig. 83.—Stages 1, 2 and 3 of the actual development of the colour sense.

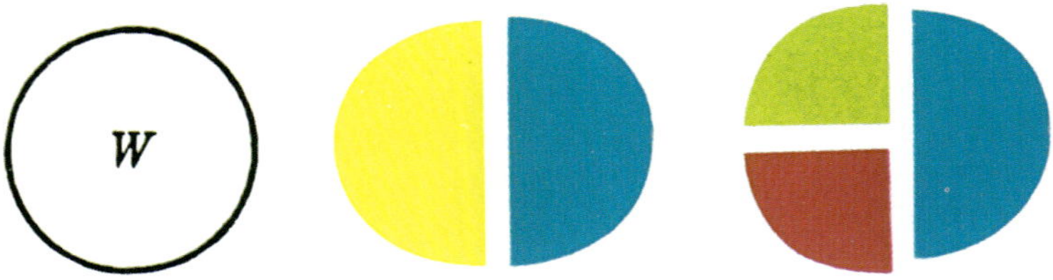

Fig. 84.—The cleavage products in the three stages of the colour sense. This diagram does not represent the entire light-sensitive molecule, but only the specific cleavage products which, according to the LADD-FRANKLIN theory, constitute the several nerve-excitants for the colour sensations (See WOODWORTH'S *Psychology*). For other diagrams, see *Psychological Review*, 23:247, 1916; *Zeitschrift f. Psychologie*, Bd. 6, etc.

The close connection between the optical properties of visual purple and the stimulating effect of spectral lights, which is proved to exist by the researches alluded to above, comes out still more clearly in a series of experiments carried out by STEGMANN under v. KRIES's direction. Visual purple, as is well known, occurs in the outer segments of the rods, colouring them in their entire length, and hence it must be in a layer of some little thickness. Light which traverses the layer with its purple contents will undergo a partial absorption in the layers which it encounters first, and in fact those rays will be most absorbed, and consequently most enfeebled, that have the strongest stimulating actions on the rods, namely, the green rays. The more concentrated the "solution" of visual purple is in the layer of rods, the greater the absorption will be. Thus, the ratio between the stimulating effects of two kinds of light, one of which (say, green) was strongly absorbed, while the other (say, orange) was not much absorbed, would have to depend to a certain extent on the amount (or "concentration") of visual purple in the layer of rods. Now we know it to be a fact that the accumulation of visual purple in the retina is notably increased by long-continued exclusion of light; and therefore matches made under the conditions of twilight vision between two kinds of light like those mentioned above will generally not be independent of the duration of the previous dark adaptation. As a matter of fact, STEGMANN and the writer, working together, found that there was no doubt about it.

If, for example, after being dark-adapted for a period of five or ten minutes, "twilight matches" were made between spectral orange and blue-green (both of which would, of course, look colourless), and if then this match was considered by the same eye after it had been dark-adapted for a much longer time, it ceased to be valid: the light of longer wave-length (orange) looked the brighter of the two, and it was necessary to reduce its intensity to three-fourths of what it had been. Accordingly, this is a change in the opposite sense from that which would correspond to the PURKINJE phenomenon; but it occurs in the direction that was to be expected if the increased accumulation of visual purple has any influence on the stimulating effect. The observation also shows that the place of action of light in the rods cannot be, or at least cannot be entirely, in the inner segment, or at the border between the inner and outer segments, but must be farther outwards. The farther outwards in the rod the place of stimulation is, the more the absorptive action of the visual purple must influence the threshold value. On the other hand, it does not follow from the experiment just described that the place of stimulation is only at the outermost end of the rod. It may just as well take place in the entire outer segment.

On the assumption that typical congenital total colour blindness (achromatopia) is to be considered as vision in which the cones have no share, it is natural to expect in the case of such persons that the difference between matches made with retina poor in visual purple and with retina rich in visual purple will be still more distinctly marked. An achromat can make matches in the spectroscope between orange and blue green in a bright room, whereas observers with normal vision must be already dark-adapted for some time before they can obtain anything like exact matches. In that state of "indoor adaptation", which is sufficient for making observations in the case of achromats, the amount of visual purple to start with is decidedly less.

With the totally colour-blind girl mentioned previously, the writer has made experiments of this sort with the HELMHOLTZ spectrophotometer. One eye was tightly bandaged for a long time, the other being kept always in the state of medium light adaptation. She then made matches with the photopic eye between orange-yellow ($600\mu\mu$) and green-blue ($490\mu\mu$). The width of the slit for the green-blue was 0.44 mm on the average (0.49–0.40). The observations then were made with the eye that had been bandaged for an hour, and for this purpose the intensities of the two fields were considerably lowered, in proportion, of course, by means of an episcotister. The first settings of the slit after removing the bandage amounted to 0.85 and 0.80 mm, but as the experiment continued, these values were lowered to an average of 0.6 mm. After this eye had become light-adapted, the values for it were also between 0.4 and 0.44, whereas for the other eye, that had meanwhile been blindfolded, the values increased to between 0.55 and 0.6.

The fact that the twilight value for blue-green light and, for that matter, for every green and blue, is found to be lower when the eye has been dark-adapted for a long time than when it was darkened simply long enough to make the matches, gives the explanation of the marked separation in the blue between the two curves in Fig. 71. One of these curves represents the twilight values of a normal eye, and the other shows the distribution of spectral brightness for a totally colour-blind person. By long exclusion of light, the observer who was not colour-blind made his retina far richer in visual purple than that of the totally colour-blind girl could possibly have been, because her settings were made in a bright room. Owing to the greater amount of visual purple, the twilight values obtained for green and blue lights, which are strongly absorbed, turn out to be lower than in the case of the totally colour-blind observer.

Comparative physiological investigations on animals whose retinas contain very different amounts of visual purple are also of much interest in the present connection. As was stated above, M. SCHULTZE's argument that animals that are accustomed to live in darkness should have rods only and no cones, has not been altogether verified. But it is a fact that in distinctly nocturnal animals like owls the rods are unusually large and numerous and also particularly rich in visual purple; whereas the retinas of diurnal birds of prey contain little visual purple or none at all.

Functional differences were established first by ABELSDORFF, in the case of the reaction of the pupil under the influence of light of different colours.[1] In diurnal animals he found the greatest pupillary contraction for those spectral rays that also look brightest to the light-adapted eye, that is, at the border of yellow and orange; whereas in nocturnal animals the maximum effect of pupillary contraction appears to be distinctly shifted towards the green, completely corresponding to the subjective impression of brightness of the scotopic eye with its large amount of visual purple.

Quantitative comparisons of photo-electrical reactions for light of different wave-lengths can be made far more accurately than measurements of the "pupillomotor" effect. In a manner similar to that used by earlier investigators (see page 56), HIMSTEDT and the writer together studied the electrical reactions in the frog's eye to light of various colours, the state of adaptation being taken into consideration for the first time. The distribution of electrical reaction in the spectrum was first found for the eye when it was as thoroughly dark-adapted as possible, and then for eyes which had been previously exposed to bright daylight. As might be expected from what is known about the effect of adaptation on the human eye, higher intensities of light were needed to get measurable electrical effects with the photopic eye, whereas very slight intensities were adequate for the scotopic eye. Taking account of the high sensitivity in the state of dark adaptation, the investigators tried not to disturb it any more than was absolutely necessary and therefore used very small amounts of light for producing stimulation. But the connection between the effect and the wave-length varies also with the adaptation. In the dark-adapted eye the maximum effect is shifted considerably towards the violet end of the spectrum.[2]

At the writer's suggestion, H. PIPER measured the stimulating effect of coloured lights on other animals. Some of them were warm-

[1] ¶See H. LAURENS, The pupillomotor effects of wave lengths of equal energy content. *Amer. Jour. Physiol.* LXIV. 1923. 97–119. (H. L.)

[2] ¶See E. L. CHAFFEE, W. T. BOVIE and A. HAMPSON, The electrical response of the retina under stimulation by light. *Jour. Opt. Soc. Amer.* VII. 1923. 1–44.—Additional more modern literature on photo-electrical and photo-mechanical effects will be found at end of §18A. (H. L.)

blooded. Invertebrates were represented by the cuttle fish. Here again the comparative observations on diurnal and nocturnal birds proved to be particularly interesting. The results in these latter cases are shown in Fig. 73. Some of the diurnal birds were hawks, chickens and pigeons. Several species of owls were among the nocturnal birds. The difference in the two groups comes out clearly in the figure, especially the entirely different positions of the peaks of the curves. In the individual groups there is quite good agreement; only one curve, that of the mouse-hawk, being different from that of the other diurnal birds by being less steep in its descent towards the violet end of the spectrum.

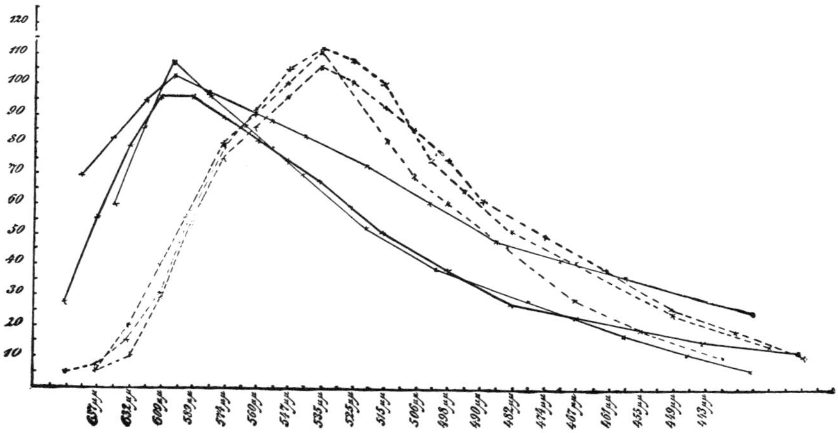

Fig. 73.

These facts certainly are strong indications that in these animals also the cones are the organs of daylight vision and the rods of twilight vision; and the probability is that, especially in their reactions to light of different kinds they behave much in the same way as the corresponding structures in the human eye.[1]

Finally, stimulation of the retina by light is accompanied by certain distinct structural changes, which like the photo-chemical and photo-electrical effects, are unquestionably dependent on adaptation, that is, on whether the retina is illuminated or unilluminated. These

[1] However, there is still much here that needs explanation. Thus, it is singular that Piper failed to find variations in the distribution of stimulations as a result of adaptation in the case of rabbits, cats, and even dogs. In Hess's experiments with chickens the values of the relative brightness were found to be shifted by adaptation more towards the violet end of the spectrum; but this is not so surprising because the retina of a chicken's eye is perhaps not entirely without visual purple.

changes are the migration of the pigment in the pigment epithelium of the retina and the contractions of the cones. It is natural to treat these two processes together, as HERZOG and EXNER and JANUSCHKE have done. According to these authors, we have here a contrivance for throwing in and out the two combined mechanisms in the retina, that is, for throwing in the twilight mechanism and throwing out the daylight mechanism. In dim light or complete darkness the pigment returns into the cells leaving the intervals between the individual rods free. It might be that under these conditions, when a small amount of light comes through the pupil, the portion of it that falls on a single rod is greater (owing to diffusion of light all over the retina) than if the rods were isolated from one another by a thick mantle of pigment. In this case the only light that can get to the rod will be light that traverses the retina in the longitudinal direction of the rod. The supposition is that the cones, which in twilight vision are mostly pushed forward between the rods, remain out of action, because the small amount of light acting on them is not yet above the threshold of stimulation for the cone mechanism.

When, on the other hand, the pigment migrates forward in bright light, that is, under the conditions of daylight vision, an exceedingly large amount of light is prevented from reaching the rod mechanism which has been rendered highly photo-sensitive by accumulation of visual purple in the dark. The result is that the adaptation to the increased intensity of light is more gradual, and meantime the less photo-sensitive cones can emerge from the mantle of pigment to be freely exposed to the stronger stimulus.

While this view of the photo-mechanical processes in the retina is certainly attractive in some ways, still it must be admitted that there are numerous arguments against it that cannot be treated lightly. The writer did not accept it in the summary of the objective changes that take place in the retina as set forth in the *Handbuch der Physiologie des Menschen*, Bd. III, S. 92; nor can he accept it now, especially since GARTEN in his very thorough review of this subject in the third volume of the second edition of GRAEFE-SÄMISCHS *Handbuch der Augenheilkunde* has not been able to adduce any arguments that are favourable to the HERZOG-EXNER hypothesis. The writer still thinks that the fact that the pigment and cone movements could be established very easily in cold-blooded animals and birds, but not in man and the higher mammals, is comparatively the least weighty of all the arguments against it. The mere fact that these processes have not been demonstrated in the human eye is no proof that they do not occur. It may be that they go on more rapidly than in cold-blooded animals. HESS tried in vain to find pigment migrations in monkeys; but GARTEN, employing a more delicate method, did succeed in detecting it to a slight extent. But even if it should be overwhelmingly proved that the photo-tropic reactions of the pigment epithelium do not occur in the human eye, that would be no argument against explaining the processes in the eyes of frogs, fishes, and birds in the above manner, and assuming that the same purpose, namely, adaptation of the eye to different intensities of light, is achieved in the eye of the mammal in a different way. The act of accommodation is performed by different classes of animals in fundamentally different ways.

But there are other considerations against this hypothesis of an in-and-out mechanism. For instance it is not easy to see what is the purpose of the elongation of the cones in darkness. Why do not all of the cones remain immobile on the membrana limitans, as some of them do? What is the object of the pigment in the region of the human eye where there are no rods? How are we to explain the fact that over a considerable portion of the retina in the case of animals with a tapetum, like the dog and cat, there is no pigment at all?

Without being able to answer such questions as these, it is impossible to speak yet of a complete theory of the photo-mechanical processes in the retina. It is very probable that the process of light and dark adaptation is not *rigidly* connected with the existence of photo-tropic movements of retinal elements, but that these changes may accompany adaptation, the most essential basis of it being the formation and bleaching of visual purple.

It is not probable that the pigment as such has anything to do with the formation of visual purple, because the latter occurs in large amounts even in albinos who have no pigment and in the parts of the retina of the cat where there is no pigment. The purpose of the pigment is probably purely optical and consists in the prevention of lateral diffusion of light in the layer of rods. But, as has been said, it is still obscure why with one set of animals this protective mechanism exhibits such plain reactions to light, and yet does not do so with other animals.

Appendix by v. Kries

I. Normal and Anomalous Colour Systems

1. Laws of Mixture of Light

The views developed by HELMHOLTZ as to the nature and origin of the visual sensations and the fundamental mechanism of the organ of vision have recently undergone a certain modification, as was explained in the preceding Appendix; because it has been necessary to assume that twilight vision is a special mode of vision, presumably having to do also with a separate and distinct part of the organ as a whole. Most of HELMHOLTZ's investigations were concerned with the other mode of vision known as *daylight vision*, that is, the vision of the thoroughly light-adapted eye, which is probably confined to a small region belonging to the fovea centralis. Here also later observations have considerably advanced our knowledge. This has been due mainly to improved methods. Thus whereas formerly the laws of light-mixture were merely qualitative, these laws have now been determined quantitatively for normal and anomalous vision.

It would take too much space to enumerate here all the many devices which have been designed in the last ten years for studying the mixture of spectral lights. It must suffice to say that, in so far as they are intended for subjective observation, they are almost all based on a principle which (as well as the writer can ascertain) was introduced first by MAXWELL.[1] Objective spectra are not used in this method, but their real images are focused directly in front of the observer's eye, so that any particular region can be separated and inspected through a slit in the ocular. Accordingly, the observer does not see the spectrum with its hues blending into each other from place to place; but what he sees is the surface of the object-glass illuminated all over by light of the same wave-length and appearing therefore as a uniform field.[2]

[1] *Philos. Transactions* CL. p. 57, 1860.

[2] ¶"By far the best method" is HELMHOLTZ's "colour-mixing apparatus," (presumably the same instrument as is described in the text), which "is, in effect, a double spectroscope; there are two collimator-tubes which throw light, after it has passed through a prism, into the two halves of a single telescope. The eye-piece of the telescope has been removed, and a plate carrying a narrow slit put its in place; the effect of this is that an eye looking through the telescope sees, not a narrow image of the collimator-slit, but the whole surface of the prism lighted up by homogeneous light." CHRISTINE LADD-FRANKLIN, Article on Vision in BALDWIN's *Dict. of Philos. and Psychol.* (J. P. C. S.)

Here we shall describe merely a little more in detail the apparatus which was described by HELMHOLTZ himself in the second edition of this work. Except for many special modifications, it is the same instrument as that used so much by KÖNIG and by NAGEL and the writer. It is shown in plan in Fig. 74.

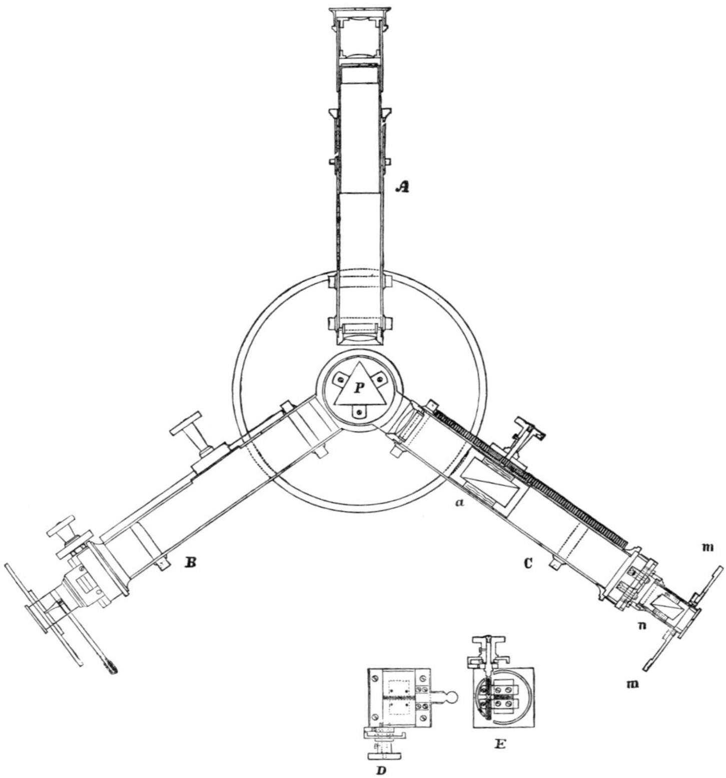

Fig. 74.

"A large equilateral prism is mounted at P and rigidly connected with the massive central base of the instrument. B and C are two collimator tubes that can be turned around a vertical axis which is underneath the prism. The two kinds of light that are to be mixed emerge from these tubes and fall on the prism. The light from B leaves the prism on its right face and that from C on its left face. Thus the observer whose eye is at A sees the light from the right collimator through the left side of the prism, and *vice versa.* For observations of colour mixture the ocular of the telescope is removed, and all that remains in its place is a narrow rectangular slit at the focus of the object-glass. The width of the slit can be adjusted by screws. At the ends of the tubes *B* and *C* there are two other finer slits, whose widths can be read on the heads of micrometer screws for adjusting them. The main thing about adjusting the width of each of these slits is to keep the middle line of the slit fixed, so as to alter the brightness of the image without changing its colour; and hence by turning the same screw the two edges of the slit are shifted equally in opposite directions. The details of the slits are shown in the inserts at *D* and *E*. To unite two colours, one from each of the collimators *B* and *C*, each tube contains a double refracting calc-spar prism (*a* in Fig. 74) so combined with a glass prism that there is no deviation of the mean direction in which the two images are seen. After traversing these prisms, the rays proceed as if they came from two separate images of each slit; these images being farther apart, the farther the prism is from the slit. Each of the calc-spar prisms can be shifted back and forth by means of a rack and pinion. For this purpose a piece is cut out of the side of each of the tubes *B* and *C* to let the prism *a* and its screw be free to move one way or the other. The two bundles of rays, however, that correspond to these two images, are polarised at right angles to each other. If, therefore, a rotatable NICOL prism is inserted in front of each slit, the relations between the two beams of light can be varied at will. One image can be extinguished entirely, while the other attains its greatest intensity. The amount of rotation is measured by means of two graduated circles *mm*, fastened to the pieces of tube that hold the NICOL prisms.

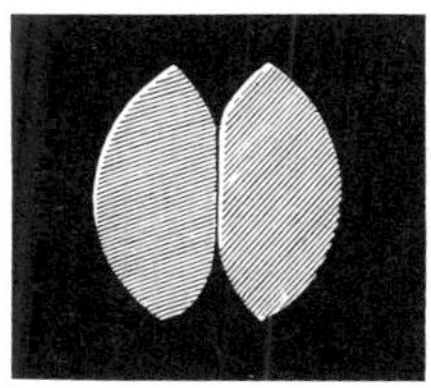

Fig. 75.

"Each of the two slit images is developed in a spectrum by the prism *P*. But depending on the apparent separation of the images, the two spectra are more or less shifted towards each other so that different pairs of colours can be made to overlap. Finally, a real image of these two pairs of spectra is produced in the plane of the slit at the focus of the object-glass of the telescope. Thus on each side of the prism *P* light of both colours meets here in the slit. Looking towards the prism through the slit, the observer sees therefore a field of the form shown in Fig. 75. The narrow dividing line corresponds to the front edge of the prism *P*, the round edges on the sides to the circumferences of the object-glasses of *B* and *C*. The colour and brightness of the two fields can thus be compared; and the observer can try to make them match, if they do not do so.

"The two colours in each colour mixture can be both made to vary towards red or towards violet by turning the corresponding tube *B* or *C*. On the other hand, by shifting forward the calc-spar prism *a*, one of the colours will change more towards red and the other towards violet. One intensity of the components may be varied in a way that can be accurately measured by turning the NICOL prism *n*. The relative brightness of the two mixtures is adjusted by altering the width of the slit in *B* or *C*.

"For determining the wave-lengths, the ocular lens is inserted in *A*. Then when sunlight is used, the FRAUNHOFER lines of the four spectra will be visible in the ocular slit. Individual lines can be isolated, and since the wave-lengths of even the finer lines are accurately known, it is possible to determine the wave-lengths of the four average colours of the slit."

The apparatus designed by v. FREY and v. KRIES (*Arch. f. Anat. u. Physiologie.* Physiol. Abt. 1881; page 336) is adapted also for scientific observations and measurements; as is also an instrument described by ASHER (*Verhandlungen der Deutschen Physikal. Gesellschaft.* V. 1903) and one mentioned by HERING (PFLÜGERS *Arch.* LIV. 1893. 312), which, so far as the writer knows, has never been described any more fully.[1]

Various instruments have been designed for demonstrations of colour-mixing effects: ZOTH (PFLÜGERS *Arch.* LXX. 1898. 1), SCHENCK (*Sitzungber. d. Marburger Gesellschaft z. Bef. d. ges. Naturw.* 1907), BASLER (PLFÜGERS' *Arch.* 116. 1907. 628), SAMOJLOFF (*Zeitschr. f. Physiol. d. Sinnesorgane.* XLIII. 1909. 237), and KROGH (*Scandinav. Arch. f. Physiol.* XVIII. 1906. 320). The writer has described a simple apparatus designed for laboratory instruction: (*Zeitschr. f. Psychologie u. Physiol. d. Sinnesorgane,* XLIII, 1919. 59). Some instruments intended for purposes of practical investigation will be described later.

In describing the results of these investigations, it is best to consider trichromats and dichromats together. Two facts stand out prominently as being of more general significance and having wide bearing. The first of these is that typical dichromats (for whom according to HELMHOLTZ's way of looking at it the colour triangle would collapse into a straight line) fall into two distinctly separated types. The vision of these two types presents differences which certainly cannot be due to purely physical causes, for example, to more or less yellow colouration in the ocular media; and yet in each type separately there are individual distinctions which are not very great and are doubtless physical in their origin.

The other important fact is that colour matches made by persons with normal colour vision are valid for both types of dichromats.

[1] ¶S. GARTEN, HERINGS Farbenmischapparat für spektrale Lichter. *Zft. f. Biol.* LXXII. 1920. 89–100. (H. L.)

Objectively unlike mixtures of light which appear to be alike to persons with normal colour vision appear alike also to dichromats. This is true except for the presumably physical differences just mentioned, such differences as are to be found also among persons with normal colour vision.

The first of these facts amounts to nothing more than what was already known before, for example, that in the case of some dichromats the red end of the spectrum is abbreviated. But the fact comes out far more distinctly in the difference of brightness between two kinds of light belonging to the half of the spectrum that corresponds to the long waves. These may be so chosen that all typical dichromats make correct matches with them for certain definite ratios of intensity; although these ratios have to be entirely different for the two classes of dichromats. In these tests the writer used red light of wave-length 671$\mu\mu$ (lithium line) and yellow light of wave-length 589$\mu\mu$. The following table[1] exhibits the results obtained with twenty dichromats.

Amount of red that it takes to match a given yellow

1. W.N.	36.5	F.	214
	36.3	V.	213
	36.3	M. Sc.	211
	33.5	E. J.	205
	38.4	H.	196
2. L. V.	37.3	E. I.	198
3. A. V.	37.0	E. II	210
4. Sc. En.	37.0	K.	200
5. O.N.	37.8	W.	210
6. K.Th.	37.0	B.	203
7. H.Th.	36.9	Th.	225
8. O.Th.	38.0		
9. F.	40.0		

It is obvious here that, while the two types are sharply distinguished from each other, there is very close agreement between the individuals of each class.

A detailed and complete determination of the mode of vision of the individual dichromat can be obtained by making a set of systematic observations on colour mixture, that is, by what the writer speaks of as "calibrating a spectrum" (*Eichung eines Spektrums*). The way this is done is to illuminate one of the fields of the apparatus by a mixture of light of short wave-length and light of long wave-length, the other field getting from its collimator a pure spectral light. This latter field is illuminated in turn by a series of pure spectral lights, and the mixture of red and blue light that looks like it in each instance is

[1] v. Kries *Zeitschrift für Psychologie etc.*, XIII. S. 925

obtained in the adjoining field. Thus for every part of the spectrum the amounts of red and blue are found which are in the mixture that matches this place. The red values of two dichromats, one of each type, are exhibited in Fig. 76, and the blue values in Fig. 77.[1] It will be observed that while the red values are quite different for the two types, the blue values are not characteristically unlike.

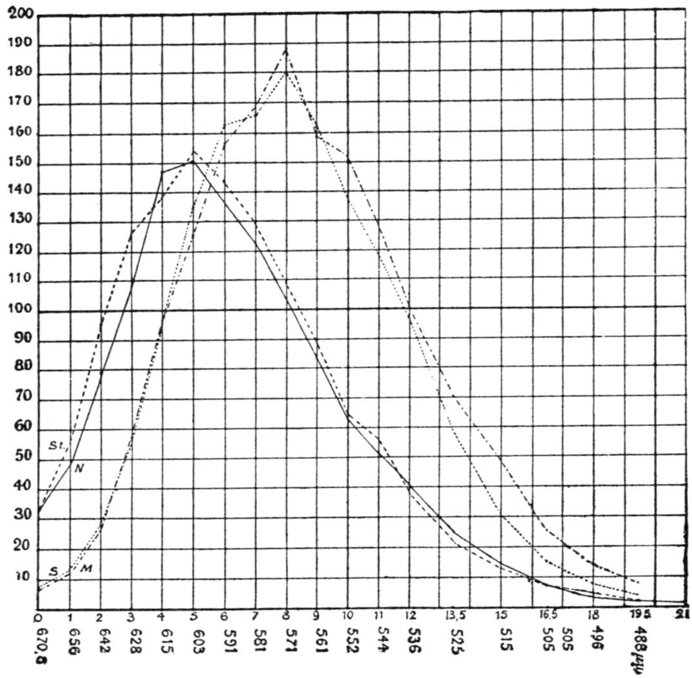

Fig. 76.

The second fact mentioned above might be proved by calculation, in a perfectly exact and general way, provided the spectrum could be "calibrated" in a similar way for an eye with normal colour vision. So far this has been done only for the less refrangible half of the spectrum. In this region a trichromat can match a given colour by a mixture of two others, one on either side of it. This part of the spectrum, therefore, can be "calibrated" by using two standard lights; and then we are in a position to calculate immediately the distribution

[1] See numerical data, *loc. cit.* p. 252 and in NAGEL's *Handbuch d. Physiol.* III. page 153.

of the stimulus values for dichromats, knowing that for one class of dichromats the stimulus value of yellow-green light is about 20 times that of red light and for the other class about twice that of red light.

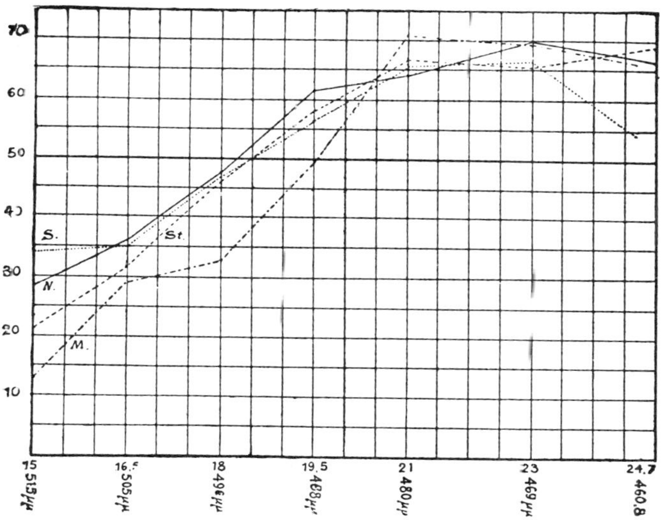

Fig. 77.

The calculated distribution agrees very satisfactorily with that found for dichromats, as the following table shows.

Wave-length of the Homogeneous Light	Amounts of Lights 670.8 and 552μμ in the mixture		Stimulus Value for the eye of the green-blind		Stimulus Value for the eye of the red-blind	
	670.8	552	Calculated	Observed	Calculated	Observed
0 (670.8)	88.5	—	33	33	4.9	4.9
3 (628 μμ)	251	10.0	106	107	28.8	38.5
4 (615 μμ)	276	27	126	147	54.2	63
5 (603 μμ)	270	49	145	151	86	84
6 (591 μμ)	202	67	135	137	108	105
7 (581 μμ)	123	76	114	124	117	113
8 (571 μμ)	73	91	110	103	137	126
9 (561 μμ)	21	80	76	82	111	106
10 (552 μμ)	—	71	64	64	101	101

Thus it may be generally stated as a result of calculation, invariably verified in individual cases by experimental test, that colour matches made by persons with normal colour vision are in fact valid for both kinds of dichromats.[1]

As was stated, the entire spectrum would have to be "calibrated" for persons with normal colour vision in order to make a similar comparison throughout the whole range. Three calibrating lights would be needed for this purpose and therefore much more complicated apparatus. Observations of this kind have not yet been carried out. We are certainly therefore not in a position to prove the above law systematically. And yet it is easy to prove it by means of numerous and varied individual experiments, and there are no exceptions to it, so far as the writer's experience goes. The changes that have to be made in any colour match made by a person with normal colour vision to make it right for a dichromat are usually very slight and such as occur with that particular type, being probably due to physical causes.

The connection thus established and experimentally verified between the two types of dichromats and between both of them and normal trichromats is identical with what HELMHOLTZ had already supposed to be probably the case. In order to have a name for it that expresses simply the experimental results without involving any theoretical consequences, the writer speaks of the two colour systems of dichromats as being *reduction forms* of normal vision. At any rate the simplest and most natural explanation of these abnormalities is to regard them as being *deficiency effects*, in the sense that each of these types lacks one of the component factors of the visual organ as assumed in the HELMHOLTZ theory. This was just the idea that was conveyed by the terms red-blind and green-blind that were formerly in use. And yet these expressions have been the source of much misunderstanding, for which the words themselves are perhaps partly responsible. As a matter of fact, green-blind persons are not really blind to green light, nor red-blind persons to red light; nor can it be assumed that the former lack the sensation of green and the latter the sensation of red[2]. In order to have brief descriptive terms for the relation that has been found to exist here, without expressing any theoretical bias, the writer suggests the names *protanopes* and *deuteranopes* to describe the two kinds of dichromats, that is, persons who lack the first component or the second component, respectively, of the normal visual organ.

[1] A very simple proof of this statement is that matches of pure yellow with a red-green mixture are not the same for both kinds of dichromats unless the relative amounts of red and green are such that the mixture has the same colour as pure yellow for persons of normal vision (see *Zft. f. Psychologie*, etc., XIII. 277.)

[2] Moreover, simply because the dichromat is supposed not to have some one of the three components of the organ of vision, we have no right at all to infer that ordinary heterogeneous daylight which has no colour for a normal eye looks blue-green to one kind of dichromat and purple to the other kind.

It may, therefore, be considered as established that colour matches which are valid for both kinds of dichromatic vision are valid also for trichromatic vision. Starting from this assumption, we can exhibit the laws of colour mixture for trichromats in the form of a colour chart which will not only express the qualitative relations, as was done formerly (see page 139), but will agree also with the quantitative results of observation. On the basis of the measurements of dichromats mentioned above, the writer has constructed a colour chart of this sort.[1]

Moreover, we obtain here something definite as to the way in which the separate components of the organ of vision are affected by light of different kinds. The term *valence* proposed by HERING is usually employed to denote the intensity of the action of a given light on a part of the visual organ.[2] Here, on the other hand, the term *calibration value* (*Eichwerte*) will be used for the amount of each of the three calibration lights in the mixture that looks the same as the given homogeneous light. *Calibration curves* and *valence curves* are curves for which the ordinate at any point represents one or the other of these two magnitudes as a function of the wave-length in the case of a definite spectrum.

Now a simple consideration shows[3] that the blue valences, that is, the stimulus values for the blue component substance of the eye, depend on the wave-length of the light in the same way as the observed blue calibration values do; whereas the valences of the first and second component substances may be any linear functions of the observed amounts of light of long wave-length.

Owing to the frequency of the occurrence of the two kinds of dichromatism above mentioned and the typical uniformity of the phenomena, this kind of vision is easy to investigate and can be satisfactorily explained. But there is another kind of anomaly which may be called blue blindness or yellow-blue blindness concerning which our knowledge is still quite scanty. With reference to these disorders, it is worth mentioning that they seem to be seldom congenital, but are more apt to be acquired as the result of sundry diseases of the eye, particularly loosening of the retina. The disturbance in such cases is naturally mostly on one side and frequently confined to certain parts of the visual field.

[1] NAGELS *Handbuch der Physiologie.* III. p. 162. Certain peculiarities in the form of this chart will be mentioned later.

[2] ¶HERING's theory presupposes three visual substances or components that furnish the six fundamental sensations. Different kinds of homogeneous light affect these substances differently. "All coloured lights, except the four primary colours, have three values or valencies, corresponding to their action on the three substances. The physiological value or 'moment' of a light depends upon its physical value and also upon the condition of excitability of the visual mechanism." PARSONS, *Colour Vision*, 253. (J. P. C. S.)

[3] NAGEL's *Handbuch der Physiologie.* III. S. 163.

In five cases investigated and described by KÖNIG[1] the disorder was acquired. These individuals detected no difference between a yellow 566–570$\mu\mu$ or a blue about complementary to it and a colourless mixture. Moreover, special investigations with a series of colour combinations and matches showed that what was involved here was a form of reduction of normal vision. The phenomena are therefore in accord with what might be expected according to the HELMHOLTZ theory, when the blue component is lacking. Consequently, in line with the nomenclature suggested above, this anomaly may be called *tritanopia*.

Some other cases of acquired unilateral anomalies invariably of a similar nature have been described by COLLIN and NAGEL.[2]

In a case described by VINTSCHGAU[3] and HERING[4] the anomaly was on both sides and congenital. Here again yellow and blue lights had the same appearance as composite white light. But not much difference could be noticed for red or green either, so that the condition was to some extent like that of total colour blindness.

Other cases that may be mentioned are the one described by PIPER[5] and another one by LEVY.[6]

With reference to colour matches, it was stated above that certain individual differences are apparent in persons with normal colour vision and also in dichromats of the same type. This fact seems to have been first noted by MAXWELL[7] who also expressed the opinion that possibly the explanation was a physical one and due to the yellow colouration of the fovea, which is different in different individuals. Quite a large mass of experimental work on this subject is available now, which tends to substantiate MAXWELL's assumption for a portion of the phenomena anyhow.

At the instigation of HERING, a direct physical investigation of the macular pigment in the isolated human retina was undertaken by SACHS.[8] He found that the long wave-lengths were not appreciably absorbed. The absorption begins in the yellow-green and continues to increase with decreasing wave-length. As to individual differences in colour matches, v. FREY and the writer[9] showed that for them these differences in a large number of trials with all sorts of combinations

[1] *Sitzungsberichte der Berliner Akademie.* 1897.
[2] *Zeitschrift für Physiologie der Sinnesorgane.* XLI. S. 74. 1906.
[3] PFLÜGERS *Archiv.* LVII. 191. 1894.
[4] Ibid. LVII. S. 308. 1894.
[5] *Zeitschrift für Psychologie.* XXXVIII. S. 155. 1905.
[6] *Archiv f. Ophth.* LXII. 3. S. 464. 1906.
[7] *Philos. Transactions* 1860.
[8] PFLÜGERS *Archiv.* 50. S. 574. 1891.
[9] *Archiv für (Anatomie und) Physiologie,* 1881. S. 336.

were of such nature and so related to one another as to be accounted for by assuming a retinal pigmentation.[1]

The corresponding phenomena occuring with dichromats of the same type are very much simpler. In the case of two protanopes who were carefully investigated by the writer, differences of this sort appeared very distinctly and regularly in "calibrating" a spectrum as described above. The amount of absorption for different kinds of light (or, to be more accurate, the ratio between the absorption for one light and for the other) can be ascertained by comparing the amounts of red in the red-blue mixtures that match. Thus the following table was obtained.

Wave-length	670.8	656	692	568	615	603	591	581	571	561 $\mu\mu$
Calculated ratio	1.1	1.1	1.05	1.00	1.00	1.07	1.07	0.98	0.96	1.02

Wave-length	552	544	536	525	515	505	496	488	480	469 $\mu\mu$
Calculated ratio	0.91	0.91	0.97	0.91	0.65	0.63	0.57	0.42	0.41	0.3

The absorption begins to be evident here at about $525\mu\mu$, in agreement with the results found by SACHS.

The differences observed in the same person between the place of clearest vision itself and the adjacent paracentral places likewise indicate the influence of colouration of the macula on colour matches. Most people can readily notice that, when they try to get a colourless light by mixing red and complementary blue-green, they have to use quite different amounts of the two colours according as the result is obtained by direct fixation or by paracentral observation. If the mixture is correctly adjusted for the fovea, it looks distinctly green when the eye is turned away a little. The differences between colour matches made for the central and paracentral retina have been carefully studied by BREUER[2], and his results likewise support the assumption of an absorbing pigment.

There are difficulties about measuring the absorptions exactly, because the pigmentation, in many cases at least, is quite dilute, and hence colour matches for places that are distinctly outside the macula cannot be made very accurately. The amount of individual variation may be estimated to a certain extent on the basis of observations made on numerous subjects. This method indicates that the strength of the

[1] ¶ABNEY and FESTING, *Phil. Trans. Roy. Soc.* 183A. 1892, p. 532. — W. ABNEY, *Researches in colour vision and the trichromatic theory* 1913. — L. T. TROLAND, Apparent brightness; its conditions and properties. *Trans. Illum. Engin. Soc.* IX. 1916, 947–966.—HECHT, SELIG and R. E. WILLIAMS, The visibility of monochromatic radiation and the absorption spectrum of visual purple. *Jour. Gen. Physiol.* V. 1922. 1–33. (H .L.)

[2] *Zeitschrift für Psychologie*, etc. XIII. S. 464.

blue light that gets through to the visual substance, after being diminished by absorption, is for persons with least macula pigmentation, and for those with most, about in the ratio of 1 : 0.3.

A general colouration of the ocular media, and particularly perhaps of the vitreous humor and lens, may also be taken into consideration in the same sort of way as in the case of the pigment of the yellow spot. It is well known that the lenses of elderly people are slightly yellowish. We can try to get some idea of the importance and effect of this colouration by testing different people for the *twilight values* of yellow and blue lights, the fovea being out of action under these circumstances. That slight differences do occur here, which are however too great to be attributed to errors of observation, had been already found by NAGEL and STARK.[1] The interpretation of these differences as being due to absorption by the pigment of the lens or vitreous humor is, however, rendered questionable, because the pigment of the macula extends quite far out beyond the central region where there is no twilight vision. The twilight equations too, as above stated, are not entirely constant but depend a little on the degree of adaptation.

HESS[2] has recently made twilight equations between yellow and blue (the wave-lengths not being stated) on a large number of persons, obtaining considerable difference in not a few instances, which he attributes to the colourations of the vitreous humor and lens.

This view of macula pigmentation, and its bearing, which has been pretty generally accepted for a long time, has recently been called in question by GULLSTRAND,[3] as stated on p. 304. Thus far the writer has not been convinced that the objections urged against the hitherto accepted opinion were conclusive and that it would be necessary to adopt GULLSTRAND's proposed explanation in place of it.[4]

Another question is whether there may not be other individual differences besides the physical factors alluded to here, as Miss v. MALTZEW[5] is disposed to think. However, as this has to do with observational methods which will be discussed later, it seems best to defer their consideration.

Since the appearance of the first edition of this treatise, an extensive and important addition to our knowledge has resulted from the more exact investigation of those forms of colour vision which KÖNIG speaks of as *anomalous trichromatic systems*. A fact discovered by RAYLEIGH[6] was the starting point of these researches. If a large number of people are required to make colour mixtures with homogeneous red and

[1] STARK, Beitrag zur Lehre von der Farbenblindheit. *Diss.* Freiburg 1897.
[2] *Archiv f. Augenheilkunde* LXIII. S. 164. 1909.
[3] *Arch. f. Ophth.* LXII. S. 1. 1905.
[4] The writer hopes to return to this subject briefly in another place.
[5] *Ztschr. für Physiologie der Sinnesorgane* XLIII. S. 76. 1909.
[6] *Nature* XXV. S. 64. 1881.

yellow-green (671 and 536$\mu\mu$, corresponding to the lines of lithium and thallium) that will match a homogeneous (sodium) yellow (589$\mu\mu$), it is found that the red and green are not always mixed in the same proportion. It is true that in the large majority of cases the differences are only comparatively slight. Moreover, the settings of these persons are always close to one another, so that they are naturally classified in a group of individuals who agree among themselves, not absolutely, of course, but approximately. Anticipating things a little, let us say at once that these are persons whose colour vision may be considered as perfectly normal. We shall have to consider later the differences that occur also in this class with regard to making the RAYLEIGH colour match. At the same time there is found to be a smaller number of persons whose settings are very considerably different from the others. Some of them will use far more green, and others far more red, in the mixture than the large majority do. Thus red-green mixtures which look to these people like sodium yellow are either distinctly green or distinctly red to most people. It is natural to suppose that the way most eyes are found to behave is a better and more perfect way, and to regard deviations in the nature of anomalies; and hence we are warranted here in speaking of *anomalous trichromatic systems.* It should be mentioned at once that the test described above is not the only way, although it is the best and most reliable one, for detecting this peculiar kind of vision, and for that reason it has gained a special importance. Hence, a short name has been invented for colour matches between sodium yellow and lithium-thallium-mixtures, namely, the RAYLEIGH-*equation* (or RAYLEIGH test).

Moreover, for reasons that will be evident as soon as certain facts are stated, as will be done immediately, NAGEL suggested calling that class of these people *protanomalous,* who use more than the normal amount of red in the RAYLEIGH-test; and the other class *deuteranomalous,* who use too much green.

A more exact examination of the anomalous trichromats shows that the deviation from the normal cannot be due to absorption of light by a pigment in the ocular media or in the retina. Were such the case, so that, for example, the amounts of green light reaching the visual substance for a normal and for an anomalous eye were in the ratio of $1:\alpha$, then (other properties of the visual organ being alike in the two cases) the anomalous individual would always have to use more green than the normal person, in the ratio of $1:\alpha$, in trying to match a red-green mixture with any homogeneous light. But this is by no means the case. The fact is rather that the two red-green ratios (for the normal eye and for the anomalous eye) are found to be very different according to the kind of homogeneous light that is matched, and that the quotient of the two ratios varies in a regular way with the wave-length of the homogeneous light. An illustration is given in the following tables.[1]

Homogeneous light	Quotient of the ratio (red : green) for the two observers
628 $\mu\mu$	4.51
615 "	3.74
603 "	3.15
591 "	3.14
581 "	2.68
571 "	2.48
561 "	2.15
552 "	2.12

Homogeneous light	Quotient of the ratio (red : green) for the two observers
625 $\mu\mu$	0.019
613 "	0.123
601 "	0.230
589 "	0.278
579 "	0.262
569 "	0.249
559 "	0.176
550 "	0.080

As stated above, it may be inferred from this fact that these differences are not due to physical causes, but rather to modifications of some kind in the parts of the visual organ that can be affected by light. In order to express this idea and at the same time to bring out the distinction between these forms of colour vision and the reduction systems, which is an important one so far as theory is concerned, the former may be referred to as *alteration systems*.[2] While all colour equations valid for normal vision are also valid for the reduction systems, a characteristic of the alteration system is that this is not the case. This fact is brought out particularly clearly in the case of the RAYLEIGH-equation. Lights that look alike to the normal eye have to be changed considerably to be made to match for vision of this kind.

[1] See v. KRIES *Zeitschrift für Psychologie* etc. XIX. S. 65.

[2] This distinction between "reduced" and "altered" systems is essentially the same as that made by HERING (PFLÜGERS *Arch.*LVII. 308), although the terms he employs (qualitative and quantitative modifications of colour vision) are not very apt and may easily lead to misunderstanding. The subject is discussed by the writer in his paper: "Über Farbensystem, *Zft. f. Psychol. u. Physiol. der Sinnesorgane*, XIII. pp. 246 and foll.

So far a complete "calibration" of a spectrum has not been made for anomalous trichromats any more than for persons with normal vision. Only the less refrangible half has been "calibrated" just as in the case of normal vision. These "calibrations" have been made by two observers, one of whom had one form of the anomaly, and the other the other form.

If these results are used for making calculations similar to those that were mentioned above as having been made for the normal eye, it is found that the colour matches of one group agree very closely with the results for deuteranopes, and those of the other group with the results for protanopes. Thus, in terms of the HELMHOLTZ theory, the presumption is that in each case only one of the three hypothetical component substances of the visual organ is modified as compared with the normal condition. For instance, if colour matches made by an anomalous trichromat are found to be valid for a deuteranope, it may be inferred that the two are alike so far as the red component substance is concerned, and that vision is normal to this extent. If the anomaly has to do with the green component, the colour matches will not indeed be valid for a normal subject, but will be valid, on the other hand, for a deuteranope who has not got this component substance at all. Now, as a matter of fact, as stated above, this is what seems to be the case. Consequently, we can assume that in the protanomalous an alteration occurs in the red component; and in the deuteranomalous an alteration in the green component.

There is some difficulty about giving a theoretical explanation of these anomalies, because neither of them represents a unique type. Both forms include individuals with minor differences amongst themselves, although these differences may perhaps be regarded as different degrees of the same kind of anomaly. They do not come out so clearly in the RAYLEIGH test as might be expected, perhaps because even with normal persons not altogether slight variations are found in making the RAYLEIGH test which are due to physical causes. On the other hand, in connection with a series of other phenomena, the different degrees of the anomaly are manifested very distinctly.

Aside from differences in making colour matches and "calibrations," a number of important peculiarities are to be found among anomalous trichromats. We are indebted especially to NAGEL's investigations[1] for light on this subject. DONDERS had long ago expressed the view that such persons had a "dim sense of colour." This indeed has now been thoroughly established; that is, the anomaly of the colour system (in the sense of having to make a different setting for the RAYLEIGH-

[1] *Klinische Monatsblätter f. Augenheilk.* XLII. S. 356. 1904.—*Zeitschr. für Psychologie* etc. XLI. S. 239, 319, 455. 1906. — Also GUTTMANN, ibid. XLII. S. 24 and 250. 1907.

equation) can be considered as an important and sure sign of "colour infirmity."

No doubt, to some extent, loss of ability for discrimination is a factor here, but the main thing is an impairment of the faculty of recognizing colours.

So far as the first point is concerned, most anomalous trichromats make the RAYLEIGH-equations, and many of them make matches also between other homogeneous lights and red-green mixtures, almost as accurately as persons with normal vision.

However, even in these cases power of discrimination for variations of wave-length is apparently on the wane. For instance, GUTTMANN made errors at 589$\mu\mu$ averaging from 12 to 13$\mu\mu$; whereas in the case of normal persons working under the same conditions the error amounted to only one or two units.

But along with this there is another factor that is more complicated. For example, in order to make a colour match with sodium light or with one of longer wave-length, the mixture of red and green has to be pretty nearly in a fixed ratio. On the other hand, if the experiment is made with a homogeneous light whose wave-length is decidedly less than 589$\mu\mu$, the settings become entirely uncertain and variable. Moderate amounts of red not only do not upset the match, but the red can even be reduced to nothing. In other words, homogeneous light of 536$\mu\mu$ can be displayed alongside the yellow without the two fields ceasing to look at least approximately the same. Evidently, therefore, power of discriminating variations of wave-length in the region from 570 to 535$\mu\mu$ is very much diminished here. These are extreme cases of so-called deuteranomaly. Extreme cases of protanomaly also occur, in which homogeneous yellow, and particularly lights of somewhat longer wave-lengths, appear to be almost the same as spectral red. Here too the setting for colour matches between red-green mixtures and certain homogeneous lights may be entirely wrong.

A still simpler indication of infirmity of colour sense is more or less inability with respect to absolute recognition of colour. The coloured object has to be much larger for anomalous persons than for people with normal vision. Mere reduction of the visual angle is sufficient to make the colour of the object unrecognizable, even when it can easily be perceived by a normal eye.

Another regular difficulty that anomalous persons have is that they take a longer time to decide what the colour is. If the coloured object is exposed to view for very brief, measurable intervals of time, the minimum time required by an anomalous person to recognize the colour will be found to be much longer than that required by a normal person.

Perhaps connected with the difficulty of recognizing the right colour, is another peculiarity that NAGEL found to be very noticeable in all anomalous persons. This was what appeared to be an accentuation of certain contrast phenomena. For example, a yellow field viewed by itself is described correctly as yellow; but if it is displayed alongside a red field, it is then called green. This is the explanation of what is practically a very important fact, namely, that in observing two adjacent objects of different colours, anomalous persons are liable to make big mistakes that would be entirely out of the question in normal vision. Thus a gaslight or electric light of the usual pale yellow colour is taken for green if it is put by the side of a red light.

Note by v. Kries

(Prepared specially for insertion here, January 1924)

In recent years extensive studies have been made on the connections between the various forms of anomalous colour vision. HESS[1], in particular, has carried out very thorough investigations on this subject. The writer will limit himself here to a brief discussion of the matter so far as it has to do with typical dichromats (protanopes and deuteranopes).

The old view, as represented by SEEBECK and HELMHOLTZ, namely, that the cases of congenital partial colour blindness were arranged in two typically different groups not connected by any transitions, had been, as stated above, fully established by systematic testing of colour-mixture equations. HESS has again challenged this fact. The particular support on which he relies is that even within each group there are considerable differences in the mode of vision. However, HESS's observations were not made under conditions that insure pure daylight vision, and for this reason alone decisive importance cannot be attached to them. For the point that is of theoretical interest is, whether or not there is a distinct difference between the *mechanisms of daylight vision* in the protanopic and deuteranopic visual organs. This question can only be decided by observations where precautions have been taken to insure practically pure cone vision, by having the eye light-adapted and the observation confined to a small field. If, for

[1] v. HESS, Die Rot-Grün-Blindheiten. PFLÜGERS *Archiv*, CLXXXV. 1920.—Idem, Die angeborenen Farbensinnstörungen und das Farbengesichtsfeld. *Archiv. f. Augenheilkunde*, LXXXVI. 1920. — v. HESS gave a very thorough account of the subject in his article on "Farbenlehre" in the *Ergebnissen der Physiologie*, Bd. XX. p. 1. 1922. See v. KRIES's review of this article, Zur physiologischen Farbenlehre, *Klinische Monatsblätter f. Augenheilkunde*, LXX. p. 577. 1923.

instance, it was found that different protanopes disagree more or less about what is the brightest place in the spectrum, this might easily be due to the fact that under the conditions in which the observations were made the relation between rod-activity and cone-activity was very unequally adjusted for different individuals. Now it is a well-known fact, without any particular further significance, that the capacity for twilight vision, and hence also the ratio in which the two component parts of *vision* are concerned in the resultant sensations, differs very much in different individuals. There is another point that is more important. HESS finds not only that protanopes as well as deuteranopes are without the red-green sense, but that in the former the yellow-blue sense also is lowered as compared with that of persons with normal vision or with that of deuteranopes. For this reason he is disposed to regard protanopia as being a transition-stage of that dichromatic condition, which is due to lack of the red-green sense on its way to total colour blindness. If this fact should be established, it would constitute a behaviour of great theoretical interest. On the assumption that the normal visual organ (the only question here is as to the daylight mechanism that has colour vision) is converted into a protanopic eye by the lack of *one* component and into a deuteranopic eye by the lack of the *other* component, it may be the case that both of these organs are indeed adapted for producing the sensation of yellow and the sensation of blue, and yet that the yellow-blue sensation, like the entire daylight vision, has an altogether different basis in the two different visual organs. A whole series of *facts* testifies to the fact that in normal daylight vision it is precisely the red component, the very part that the protanope lacks, that has a specially important, indeed a principal, significance.[1] This being the case, the lack of this component, that is, protanopia, must mean a far more serious injury than the lack of the other component, that is, deuteranopia. According to this, we can understand how it is that both protanopes and deuteranopes have indeed the faculty of seeing yellow and blue, but not equally well, and why the deuteranope is as a rule, or at any rate on the average, superior to the protanope in this respect. Nor would it be

[1] KÖNIG long ago called attention incidentally to the curious fact that the luminosity values of the colours, no matter whether they are determined by the direct impression they produce or are measured as peripheral values or flicker values, turn out to be very nearly the same function of the wave-length for deuteranopes and persons with normal vision. Since in the case of deuteranopes light of long wave-lengths has no effect except on the red component, we must suppose also that this action depends on the wave-length in the same way. But then the further result is that the distribution of luminosity is not affected, or at least almost inappreciably affected, by the addition of the green component by which the deuteranopic visual organ is converted into a normal organ; in other words, that in the case of normal vision also the luminosity goes practically hand in hand with the action of the red component.

strange if under some circumstances, dependent on a vision in which rod-function and cone-function were intermingled, this vision should on the average gain the upper hand more strongly with protanopes than with normal persons and deuteranopes. Then indeed we could speak of an approximation to total colour blindness, of course only in the sense in which that term is understood in the duplicity theory. Unfortunately, the method which Hess used in his observations does not enable us to come to any sure decision about these points. Thus, he ascertained the limits of the visual field for seeing yellow and blue with suitable objects, that is, the extent of the so-called colour-fields of vision. But his own observations show that even with persons whose visual organs are the same in other ways, and particularly in the case of persons with normal colour vision, there are very considerable differences in the extent of these fields among different individuals. And hence in the first place it seems hardly permissible to draw any definite conclusions from the differences in this respect that were found between protanopes and deuteranopes.

2. The Phenomena of Daylight Vision Under Conditions that make it Difficult or Impossible to Recognize Colours

The above description of normal and abnormal vision has been confined principally to perfectly definite conditions, those in fact that are most conducive to distinguishing and recognizing colours. Even in the normal organ of vision there are conditions for which the sensations are modified as to their colour specifications or they may be entirely lacking. Most important of all is the case when the light does not fall on the fovea but on more or less eccentric parts of the retina; and another case of importance is when the time of exposure or the apparent size of the object is reduced.

The facts concerning the first case have already been partly considered in the main text (p. 154). As was there stated, the coloured character of the sensation produced by a certain kind of light acting on the eccentric part of the retina depends to a very great extent on the absolute intensity of the light and on the luminous area. Along with the fact that the pigments used were not exactly defined, this serves to explain the many conflicting statements of the earlier workers in this field. In numerous details the conditions have been clarified by Hess's investigations.[1] The appearance of most colours is found to alter more and more towards the periphery of the retina. There are just four hues that tend to fade out without losing their quality. These "invariable colours"[2] are a green of 495$\mu\mu$, a nearly complementary red (spectral

[1] *Archiv f. Ophth.* XXXV. (4) S. 1.

[2] ¶See Parsons, *Colour vision*, p. 70. (J. P. C. S.)

red with a little addition of blue), a yellow of 574.5$\mu\mu$, and a blue of 471$\mu\mu$. The first two lose their colour at a comparatively short distance out from the centre; the last two very much farther out. When the red and green objects are of the same size, saturation and luminosity[1], the limits are found to be the same for both of these colours. The same thing is true with respect to yellow and blue. The limits for yellow and blue being farther out than for red and green, in the case of objects of given size, luminosity and saturation, we can speak of a dichromatic zone; in which the invariable red and green look colourless, and the invariable yellow and blue are seen in their ordinary hues. A simple rule for the qualitative changes of all colours that are comprised between these invariables is that the red (or green) sensation disappears more quickly than the blue (or yellow), and so it approaches the latter. Thus violet and blue-green blend into blue; orange and green-yellow, into yellow.

Vision in the dichromatic zone is, at all events, different from that of the protanopic eye. It might be something like that of the deuteranopic eye; although it is not certain as to whether this agreement is perfectly exact. This matter will be discussed later.

An outermost zone of the retina in which even yellow and blue objects have no colour[2] may be said to be *totally colour-blind* or "monochromatic,"[3] although it must be remembered that the extent of this zone depends very much on the luminosity and size of the object. The

[1] What is meant by the "same saturation" is, for instance, when the two equal sectors on the colour top produce a colourless mixture.

As to "same luminosity," there is some uncertainty about this point in HESS's work. His condition is that the two colours must have the same "white valence." This amounts to defining white valences in two ways, (1) as being the same as what we now call twilight value, and (2) as being the same as what we now call peripheral value (see page 415). At that time it was not understood that these two things are entirely different. Presumably, the condition involved here (so far as the light-adapted eye is concerned) means that the peripheral values of the two colours must be the same.

[2] ¶C. E. FERREE and G. RAND, The absolute limits of colour sensitivity and the effect of intensity of light on the apparent limits. *Psychol. Rev.* XXVII. 1921, 1–23.—The extent and shape of the zones of color sensitivity in relation to the intensity of the stimulus light. *Amer. Jour. Physiol. Optics.* I. 1921, 185–213. — The limits of color sensitivity; effects of brightness of preexposure and surrounding field. *Psychol. Rev.* XXVII. 1921, 377–398.—Some contributions to the science and practice of ophthalmology. *Trans. Internat. Cong. Ophthalmol.* Washington, 1922, pp. 1–36—Perimetry; variable factors influencing breadth of color fields. *Amer. Jour. Ophthalm.* V. 1922, p. 886.—A new laboratory and clinical perimeter. *Jour. Exp. Psychol.*, V. 1922, 46–67. (H. L.)

[3] ¶This use of the word "monochromatic," to describe the type of vision of those "people who apparently see all parts of the spectrum of one hue" (PARSONS) is, of course, entirely different from the way it is used in Physics. The physicist is accustomed to speak of "monochromatic" light, meaning light of a definite wave-length, and of "monochromatic aberrations" (as used by HELMHOLTZ and GULLSTRAND in Volume I). The word "achromatic" is in the same predicament. This is a good instance of the "confusion of tongues" among different groups of scientists that often makes it exceedingly difficult for one to understand the terminology of the other even when both have the same thing in mind. Total colour blindness or monochromatic vision is the same thing as "achromatopia." (J. P. C. S.)

disappearance of colour can be noticed best on the nasal edge of the visual field. When bits of coloured paper on a grey background are carefully inserted in the visual field from this side, they always show up at first as brighter or darker spots without any colour. Under these circumstances any pair of lights can be made to look exactly alike by simply adjusting their luminosities in the right proportion. And hence the mode of vision can be determined exactly as in the case of a totally colour-blind eye, by finding out the requisite luminosities of the various parts of the spectrum or the distribution of the spectral stimulus values that have to be taken into consideration here. The writer has found the distribution of these stimulus values, or *peripheral values* (as he terms them), in the dispersion spectrum of gaslight. In order to show at once the theoretically important fact that the peripheral values depend on the wave-length in an entirely different way from the twilight values, both the peripheral values and the twilight values are given in the following table (the values for sodium light being put equal to 100).[1]

Wave-length	680	651	629	608	589	573	558	530	513 $\mu\mu$
Peripheral value	9.6	37.5	77.5	101	100	79.1	52.2	28.5	14.6
Twilight value	?	3.4	14.0	35.5	100	256	351	321	198

The results are exhibited in Fig. 78 in the usual way.[2]

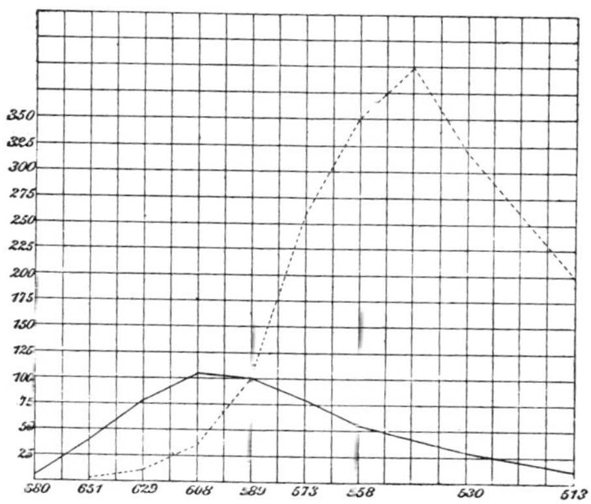

Fig. 78.

Distribution of the peripheral values——— and of the twilight values...... in the prismatic spectrum of gaslight.

As was mentioned above, the sensation can also be deprived of its coloured character by reducing the time of exposure or the apparent size of the stimulating light. The factors in the first case have not yet been accurately determined. On the other hand, SIEBECK[3] has made a systematic study of the other case. He finds that, with spectral lights of average luminosity, colours cannot be made

[1] ¶See PARSONS, *Colour vision*, p. 71. (J. P. C. S.)

[2] ¶Compare PARSONS, *Colour vision*, Fig. 25, where these curves are called "photopic luminosity curve for the totally colour blind peripheral zone of the retina," and "scotopic luminosity curve." (J. P. C. S.)

[3] *Zeitschrift für Physiologie der Sinnesorgane* XLI. S. 71.

to disappear entirely at the fovea by reducing the size of the field; although it is easy to do so at very small distances from the centre (1.5°). Accordingly, it is possible to determine for what luminosities different lights are visible under these conditions; an investigation that is perfectly analogous to that which has just been described, where the colours are made to disappear by being observed far out from the centre. The stimulus values found in this way may be termed *minimum-field luminosities*. The results obtained by SIEBECK are given in the following table, which shows that the minimum-field luminosities are likewise entirely different from the twilight values, but are in approximate agreement with the peripheral values (the maximum being at 601.3$\mu\mu$).

Wave-length	658.0	650.1	642.4	635.0	627.8	620.8	604.2
Minimum-field luminosity	44.9	56.7	66.6	72.1	84.2	91.5	104.3
Wave-length	607.8	601.3	595.2	598.3	584.2	579.1	574.1
Minimum-field luminosity	112.8	128	110.9	100	90	81.6	79.5
Wave-length	569.3	564.8	560.1	556.4	551.4	547.2	542.9
Minimum-field luminosity	72.9	68.6	62.6	58.5	52	47.3	42.8

In finding the minimum-field luminosities and the peripheral values, it is necessary to compare the luminosities of lights of different colours. It is possible to do this just because under the given conditions the sensations do not reveal differences of colour. This question of comparison of the luminosity of lights of different colours is not only one of general interest but has been the subject of much controversy. In these observations, therefore, we may speak of methods of *heterochromatic photometry*.[1]

A series of other observations will, therefore, be included here, because from this point of view they are connected with those that have just been described. The following fact, discovered first by ROOD, constitutes the starting point of these observations. If, by means of a colour disc, the eye is exposed alternately and regularly to two lights, one coloured and the other not, and if the luminosity of the colourless light is systematically varied, there is a certain definite value for which a minimum number of alternations suffices to make the sensation

[1] In the use of this term undoubtedly one has to be on his guard against making a mistake or at least against being misled by very questionable assumptions. It does not mean to imply at all that the luminosity-relations of different lights must necessarily be the same when the colour is made to disappear in some other way; nor does it mean that equality of brightness of two sensations of different colour is any precisely defined relation. See the writer's work on this subject in NAGELS *Handbuch der Physiol.* III. pages 28 and 259.

¶Consult also GLAZEBROOK's *Dictionary of Applied Physics*, IV (1923), pages 453, foll. (J. P. C. S.)

uniform and the flicker disappear. When this luminosity of the colourless light has been found and the disc made to rotate just fast enough for the flicker to disappear, the flicker appears again as soon as the luminosity of the colourless light is increased or decreased. Without implying anything as to the theory of the phenomenon, let us speak of two lights whose intensities are adjusted for the minimum "fusion frequency" as being "equivalent as to flicker"; and call the luminosity of the colourless light which is "equivalent as to flicker" with a given coloured light its *flicker value*. A determination of the flicker values for the prismatic spectrum of gaslight has been made by POLIMANTI.[1] The results are given in the following table.[2]

Wave-length	687	664	642	624	606	588	565	543	526	500 $\mu\mu$
Flicker value	28.3	31.2	56.6	75.2	87.9	100	83.1	48.1	25.0	21.9

Evidently, the distribution of the flicker values agrees, approximately at least, with those of the peripheral values and minimum-field values; but it is entirely different from that of the twilight values.

Quite a number of other methods have been used for comparing the luminosities of lights of different colours; and some have been proposed without ever having been employed.

VIERORDT[3] considered two coloured lights as being equally bright when their luminosities were such that a certain white light added to them could just be perceived. According to another principle, likewise based on flicker phenomena but not to be confused with that of ROOD's mentioned above, lights of different colours may be regarded as being equally bright which, acting intermittently, have to have the same frequency to produce a steady sensation without any flicker in it (HAYCRAFT[4], RIVERS[5]).

Moreover, there is an observation reported by BRÜCKE that indicates something analogous in space-discrimination to ROOD's observation for time-discrimination alluded to above. When a coloured object is displayed against a colourless background, there is a certain perfectly definite luminosity of the background for which it is hardest to recognize the object, no matter what angle it subtends; and the luminosity of the coloured object may then be said to be equal to that of the background.

[1] *Zeitschrift für Psychologie etc.* XIX. S. 263.

[2] ¶See PARSONS, *Colour vision*, Fig. 26. (J. P. C. S.)

[3] POGGENDORFFS *Annalen*. CXXXVII. — *Die Anwendung des Spektralapparates*. Tübingen 1871.

[4] *Journal of Physiology* XXI. S. 126.

[5] Ibid., XXII. S. 137.

One of the methods used by BRÜCKE[1] consisted in mixing small amounts of different greys with the colour to be tested. If a small piece of grey paper is placed on a coloured disc so as to make a ring of the right width when the disc is set in rotation, it can be decided with a comparative degree of certainty whether the ring is brighter or darker than the rest of the disc; that is, whether the grey is brighter or darker than the coloured paper. Fundamentally, this procedure is also a direct comparison of the luminosity of unlike colours (which differ also in saturation). There is the advantage, however, of being able to compare two colours of the same hue that do not differ much in saturation.

Lastly it will be recalled that FRAUNHOFER tried to compare the luminosity of light of different colours directly by their subjective impression, the colours being visible. This method was subsequently undertaken again by KÖNIG[2] on a rather bigger scale. His results on five normal subjects are given in the following table.

Wave-length	Luminosity for Observer				
$\mu\mu$	I	II	III	IV	V
670	0.855	1.120	—	—	—
650	2.381	2.137	1.15	—	0.64
625	3.460	3.413	2.06	1.10	1.24
605	3.650	3.247	2.56	1.66	1.56
590	3.030	2.645	2.38	2.05	1.58
575	2.358	1.923	2.00	2.08	1.56
555	1.695	1.389	1.50	1.65	1.36
535	1	1	1	1	1
520	0.554	0.553	—	—	—
505	0.224	0.250	—	—	—
490	0.0994	0.092	—	—	—

Accordingly, in approximate agreement with the peripheral values and flicker values and also with the minimum-field luminosities, the maximum luminosity, on the average anyhow, is on the red side of the sodium line. The results for the different subjects, it is true, vary a good deal, especially too in the position of this maximum. Incidentally, these tests have been shown to be quite uncertain. The setting of two unlike colours at equal luminosity is very variable and often also seems to be somewhat arbitrary. And hence it is difficult too to decide whether the individual differences exceed the degrees of uncertainties and fluctuations of this nature.

Some of the investigations referred to above have been extended to a larger number of persons, normal and anomalous. These results must now be briefly considered, as they are of some theoretical interest.

[1] PFLÜGERS *Archiv* XCVIII S. 90.

[2] HELMHOLTZ *Festschrift*. 1891. S. 350.

In the first place it seems that even normal persons are subject to not altogether inconsiderable variations in this matter. This is apparent in some degree in the above measurements of KÖNIG. But even in those methods where the uncertainty due to difference of colour is eliminated, variations are manifested. Thus POLIMANTI's results for the distribution of the peripheral values in his own case are distinctly, although not considerably, different from those of the writer.

Miss v. MALTZEW's extensive experiments[1] were concerned mainly with flicker values, and showed that even among normal trichromats there are very considerable differences in this respect. Apparently, these variations are regularly connected in some way with the differences alluded to above in making the setting for the RAYLEIGH-equation. Persons who use comparatively much green in the mixture likewise require a high luminosity of green to give it a flicker value equal to that of a certain red.

As to the phenomena in anomalous vision, the results indicate, as might have been expected that the most favourable conditions for such colour sensations as dichromats have are in the fovea. At a certain distance from the centre, assuming that the conditions are suitable in every way, colour determinations are found to be entirely impossible. Accordingly, peripheral values can be determined in their case in the same way as for normal persons. The result then is found to be that the distribution of the peripheral values for *protanopes* is entirely different from that for persons with normal colour vision, as shown in the following table.[2]

Wave-length	680	651	629	608	587	573	558	530	513
Peripheral value for protanopes	4.1?	10.7	34.0		100		110		36.4
Peripheral value for normal persons	9.6	37.5	77.5	101	100	79.6	52.2	28.5	14.6

A similar difference is found also in the case of the flicker values[3], not for the protanopic eye only, but for the protanomalous eye as well. Thus, generally, in order to get the requisite effect here, the luminosities of the red light have to be very much greater for both of these anomalous kinds of vision than for normal vision.

The differences, between normal colour vision, on the one hand, and protanopic and protanomalous vision, on the other hand, are in fact so large as to be apparent without any question, even when lights of different colours are compared simply by the subjective impression

[1] *Zeitschrift für Physiologie der Sinnesorgane.* XLIII. S. 76. 1909.

[2] v. KRIES, *Zeitschrift für Psychologie.* XV. S. 266. Similar results were subsequently obtained by v. d. WIJDE, *Onderzoekingen*, Utrecht. 4. III. 2.

[3] POLIMANTI, *loc. cit.* S. 274. — LEVY, *Zft. Psychologie, etc.* XXXVI. S. 83. 1904.

of their brightnesses. This fact has been established by a number of observers.[1]

In the writer's first observations a trifling difference, hardly beyond the limit of error, was found between the peripheral values of a deuteranopic eye and those of a normal eye. But ANGIER[2] believed that he could prove that there certainly was a difference in this respect. However, some doubt has been raised as to the correctness of this conclusion by Miss v. MALTZEW's results as to the individual differences of flicker values that are found to occur among normal trichromats. The differences which she got were very large, as has been stated. The deuteranopic and deuteranomalous persons investigated by her arranged themselves in between the normal trichromats. It is at least conceivable that the individual differences in the ratios of the peripheral values are approximately similar to those of the flicker values; and if this is the case, it would be expected that also the differences in this respect between normal vision and deuteranopic or deuteranomalous vision would anyhow not be so large that they could not be accounted for by individual differences within the same type.[3]

On the assumption that trichromats and deuteranopes agree as to peripheral values, and, further, that lights of equal peripheral values, even when they are seen centrally and therefore in different colours, give the impression of being equally bright, it follows that green and bluish-red lights which look alike to the deuteranope will look about equally bright to a person with normal colour vision. As a matter of fact this is the case, as far as we can tell with the great uncertainty that exists in heterochromatic comparisons of brightness. Certain peculiarities of the colour chart mentioned by SCHENCK are connected with this fact. It has been already shown (§20) that the colours that look alike to dichromats lie on straight lines that must all intersect at one point (in case the dichromatic system is a reduction form of normal vision). This point, which generally is outside of the real part of the colour chart, represents the stimuli which act only on the component that is not present in the dichromatic eye.

The construction of the colour chart, and hence also the position of this point, is arbitrary to the extent that unit quantities of the three calibration lights can be selected in any way we like. It is always possible to choose them in such a way that the point is infinitely distant, that is, so that colours that a dichromat cannot tell apart will fall on parallel lines. In order for this to be the case for a deuteranope, according to the assumptions made here, the unit quantities of the three calibration lights would have to be chosen so as to be about equally bright for the normal eye.

We have had to limit this treatment to a discussion of those anomalies of colour vision that are of particular interest from physiological standpoints. Of course, the subject is of much interest in other direc-

[1] LEVY, *loc. cit.* — SCHENCK. PFLÜGERS *Archiv* CXVIII. S. 174. 1907.

[2] ANGIER, Ibid. XXXVII. S. 401. 1905.

[3] ¶PARSONS, *Colour vision.* p. 179. (J. P. C. S.)

tions. As to the general biological relations (inheritance), statistical data and practical significance, HOLMGREN's work is still the standard book of reference on this subject.[1]

From the practical standpoint, the thing of most importance, as has been shown by NAGEL's recent researches, is that not only dichromats but persons with anomalous vision also are below par in a number of respects, and they are particularly unsuitable for railway or marine service. This, therefore, to a certain extent is the chief reason for improving the methods of investigation. The examination of these people is the right way of going at the problem of finding out all persons with abnormal colour vision and eliminating them.

As to the method of investigation, it is of very little value, as has been stated, to conduct tests which consist in finding out the names that are used in describing objects of different colours. In the writer's judgment this is true likewise concerning deductions by colour-contrasts, although a high estimate is placed on these tests in many quarters. The only methods that have proved to be of real value are those that depend on finding out what colours look alike to the patient. However, it is not easy to obtain absolutely definite colours that look exactly alike to all deuteranopes, say; partly because there are individual variations and partly because the technical production of colours is more or less uncertain and the illumination is different, etc. And therefore the best way of making the test amounts to finding out whether certain colours seem to the patient to be *very nearly* alike. This was the principle of HOLMGREN's method in which the patient was required to pick out of a large assortment of coloured wools those hues that matched certain samples. NAGEL's charts, based also on the same principle, have during the last few years been proved to be the most useful and reliable test, having been officially adopted for the merchant marine and the Prussian railway system after very thorough comparison with other methods. No system can entirely supersede the spectroscopic test of the RAYLEIGH-equation, and sometimes it has to be resorted to still. For making this test easily with comparatively inexpensive appliances, SCHMIDT and HAENSCH have constructed an instrument called an "anomaloscope" designed according to NAGEL's specifications. It is used to test the RAYLEIGH-equation by comparing, therefore, a field illuminated by homogeneous yellow with a red-green mixture.[2]

---

[1] *Die Farbenblindheit in ihren Beziehungen zum Eisenbahn- und Marinedienst.* German edition. Leipzig 1878.

[2] As to the various methods of investigation, see especially NAGEL, *Die Diagnose der praktisch wichtigen angeborenen Störungen des Farbensinnes*, 1899 and also a large number of other works by the same author, some of the most important of which are as follows:

Note by v. Kries

(specially prepared for insertion here, January 1924)

During the past decade some new facts have been ascertained which can be utilized for comparing the luminosity of light of different colours. Consider a small spot on a uniformly white ground. If for a very brief period the white light at this place is replaced by coloured light, the colour cannot be recognized any longer. But what happens is that at the moment when the coloured light acts, that particular place appears brighter or darker than the surroundings, depending on the intensity at that point. For a certain definite ratio between the intensity of the two kinds of light the spot looks neither brighter nor darker than the surroundings, and then the difference between them will not be noticed at all. Thus, just as we speak of "minimum-field luminosity" we can speak here of a *minimum-time luminosity*. Thus, when a coloured light substituted for a very brief time on a colourless ground appears neither bright nor dark it is said to be "equivalent" to that colourless light in "minimum time" (*minimal zeitgleich*). ZAHN's observations have shown that the distribution of minimum-time luminosity in the spectrum is almost, although not absolutely exactly, in agreement with the distribution of minimum-field luminosity and with the distribution of the peripheral values and flicker values.[1]

Moreover, very lately PULFRICH[2] has described a method of comparing the luminosity of light of different colours depending on binocular perception of depth, which therefore is called a *stereo-photometric* process. It is based on the following fact. If a dark vertical rod is moved back and forth from right to left and from left to right in a plane in front of a bright background, under ordinary circumstances we get the correct impression of a rod being moved in a frontal plane. But now if a smoked glass is interposed in front of one eye, so as to cut down the light that comes to that eye, we get the impression that each

---

Die Diagnose der anomalen trichromatischen Systeme, *Klin. Monatsblätter für Augenheilkunde* XLII. S. 369, 1904. Zwei Apparate für die augenärztliche Funktionsprüfung (Adaptometer und Anomaloskop), *Zeitschrift für Augenheilkunde* XVII. Fortgesetzte Untersuchungen zur Symptomatologie und Diagnostik der angeborenen Störungen des Farbensinnes, *Zeitschrift für Psychologie und Physiologie der Sinnesorgane*, XLI. S. 237 and 319. 1906. Versuche mit Eisenbahn-Signallichtern an Personen mit normalen und abnormem Farbensinn, Ibid. XLI. S. 455.

[1] ZAHN, Über die Helligkeitswerte reinen Lichter bei kurzen Wirkungszeiten. *Zft. f. Psychologie für Sinnesphysiologie* XLVI. p. 287. 1912.

[2] PULFRICH, Die Stereoskopie im Dienste der isochromen und heterochromen Photometrie. *Die Naturwissenschaften*, 1922. pp. 553, 569, 714, 735 and 751. In book form with the title, *Die Stereoskopie im Dienste der Photometrie und Pyrometrie*. Berlin, 1923.

point of the rod traverses a horizontal ellipse. In the middle of its path it looks therefore farther off when it traverses it one way than when it traverses it the other way. PULFRICH speaks of this phenomenon as that of the "revolving mark." The explanation he gives is as follows. Between the arrival of a stimulus and the production of the sensation in every case there is a certain amount of time lost, which varies however with the intensity of the stimulus, decreasing as the intensity of the stimulus is increased. Thus, for instance, if the right eye is stimulated more highly than the left eye, the image of a moving object in the right eye will be a little ahead of that in the left eye. Hence, in that phase where the rod is moving to the right, the image in the right eye must be a little more to the right and that in the left eye more to the left. On the other hand when the motion is reversed, the case is just opposite; the image in the right eye will be more to the left, and that in the left eye more to the right. Accordingly, we have here the general condition that is necessary for binocular perception of depth, namely, a disparity between what is seen by one eye and what is seen by the other eye, in this case a right-and-left displacement (transversal inequality, binocular parallax). According to the general laws of binocular perception of depth, what is to be expected is that in the case first mentioned the rod would be perceived as being farther away, and in the other case as being closer. And this is exactly how it looks. The rod invariably appears to move in such a way that in the remoter part of the path it proceeds from the eye that is shaded towards the eye that sees brighter. Viewed from above the apparent motion is clockwise when the right eye gets brighter light than the left eye, and counter-clockwise when it is just the other way. Now if a *coloured* glass is inserted in front of one eye instead of the smoked glass, the same effect is produced. This is not surprising, because the coloured glass also absorbs some portions of the light concerned and so tends on the whole to darken it. If, finally, with a coloured glass in front of one eye, colourless glasses of regular gradations of darkness are inserted in front of the other eye, a certain one of them can be found for which the rod will appear to move in a plane; whereas when a brighter or darker glass than this is used, the "revolving" movement will be seen, clockwise in one case and counter-clockwise in the other. Now here we have *one* basis for comparing the luminosity of light of different colours. In PULFRICH's own words, "We get the following definition of equal luminosities: two colours are said to have the same luminosity when the time between stimulation and sensation is the same for both of them, and we can tell when their luminosities are equal by the fact that at the instant the difference vanishes between the times of the two sensations as shown by the revolving mark the circular motion becomes rectilinear" (*loc. cit.*, p. 37).[1]

We must notice that here we are dealing with a definition of equality of luminosity which, however valuable and convenient it may prove to be, still has a conventional significance always, exactly in the same way as if, for example, we should stipulate that a coloured light should be said to be of the same brightness as a light without colour, when under certain special conditions (very eccentric vision, acting on an extremely small part of the field or for a very brief time) it looks precisely the same as the colourless light. In order to express this, the writer has suggested using the term "stereo-equal" when two lights are found to be equal by PULFRICH's method. Thus two lights would be "stereo-equal" when in observing a moving object with both eyes, one eye getting one kind of light and the other eye getting the other kind, no stereoscopic effect of any sort is obtained, that is, the object which actually does move in a frontal plane also appears to move in this plane.

Physiological investigation of the phenomenon[2] has shown that lights, which are equal as to minimum-field and which therefore also have approximately the same peripheral value and minimum-time luminosity, are approximately "stereo-equal" also. However, this is only true with certain limitations, that is, with high light intensities and for the photopic eye. On the contrary, when we change over to reduced intensity of light and more or less dark-adapted eye, there is a very decided shifting of the "stereo-equal" relation. However, it is a remarkable fact that the ratio between the stereo-values for a light of long wave-length and one of short wave-length is not shifted in the sense of the PURKINJE phenomenon, but in the opposite sense. The explanation of this behaviour can readily be found in the fact that both components of the visual organ, that is, the cones and the rods, are concerned to a different extent in the production of the visual impression, and that the reaction of the rods is decidedly more inert, as is shown also by numerous other phenomena (see pages 445, ff). In the transition from daylight-vision to twilight-vision the short waves of light gain in luminosity as compared with the long waves, because they act far more powerfully on the rods. It is exactly the same thing that comes out in the PURKINJE phenomenon. But the rapidity of the reaction, and therefore the stereo-value, is not increased, but diminished, by the greater action of the rods that is in evidence

[1] ¶This striking and beautiful phenomenon can be readily observed. PULFRICH's method is both ingenious and interesting, and certainly of much theoretical value. However, it would hardly seem to be capable of sufficient accuracy for purposes of practical heterochromatic photometry. (J.P.C.S.)

[2] v. KRIES, Über das stereophotometrische Verfahren zur Helligkeitsvergleichung ungleichfarbiger Lichter. *Die Naturwissenschaften*. 1923, p. 461.

here. And so it may happen that by lowering the light and by dark adaptation the ratio between the stereo-values of light of long waves and light of short waves is shifted in the opposite sense of the PURKINJE phenomenon, that is, in favour of the light of long waves. The consequence is that in any case the stereo-photometric process cannot give unequivocal and useful results unless these complications are avoided, that is, unless the operation of the twilight mechanism is practically eliminated by using light of high intensity and making the observations with the light-adapted eye.

The *threshold values* supply another set of facts that can be made to bear on heterochromatic photometry. The luminosity of a light may be defined as being inversely proportional to its threshold value, and accordingly another definition will be obtained as to what is intended when we speak of the luminosity of a light or of two lights of different colours as having the same luminosity. In the first place there are known facts here which show that, provided the observation is made with the scotopic eye, and no special precaution is taken to confine it to the central region where there are no rods, the luminosities of lights of different colours are measured by their twilight values. This case can be excluded here, because under these conditions there is no sensation of colour anyhow, and hence also there is no difficulty about making a comparison between the luminosities of lights of different colours. But the results obtained in the case of purely foveal vision are of more interest. Suppose we use the term *threshold luminosity* to denote the luminosity as measured by the reciprocal of the threshold value; then the question arises again whether it is proportional to the minimum-field luminosity, peripheral value, etc. According to BOSWELL's results, apparently it is not. His observations rather indicated that in the middle part of the spectrum, indeed (where the lights that are on the verge of visibility are seen without colour even in *foveal* vision), the threshold luminosities proceed almost side by side with the minimum-field luminosities. But at the two ends of the spectrum, that is, for light of both long waves and short waves, a light that is just over the threshold will immediately begin to appear coloured. Here the threshold luminosities are different from the minimum-field luminosities; the threshold, in fact, being lower, that is, the sensitivity being greater than it would be if there were a proportionality with the minimum-field luminosities. BOSWELL's conclusion from this was that when a light was just on the verge of visibility something more was involved than the colourless luminosity alone, and that the colour-value promotes the visibility, that is, lowers the threshold. This has been confirmed by KOHLRAUSCH's more recent observations, which, however, have shown also that these phenomena are likewise dependent on the state of adaptation of the fovea. The matter requires further investigation.

II. Theories of Vision

1. The Young-Helmholtz Theory

Helmholtz endeavoured to elucidate the whole nature of the visual sensations, and especially the way in which they depend on the effective stimuli, in terms of the theory which is known by his name and which is based on a conjecture of Thomas Young. In giving a description of this theory and of the numerous subsequent hypotheses that have been conceived along the same line, the writer would like to indicate beforehand the principal point of view which is the motive for doing this.

Even at the present time the theory of Helmholtz is thoroughly justified as to its fundamental conceptions, it is in close agreement with the facts, and as an hypothesis it is qualified to explain a very large mass of actual phenomena. The writer has briefly expounded this elsewhere.[1] The main thing that has to be done now, therefore, is, first, to mention a series of new facts recently discovered which substantiate and support the theory, and then to speak of the limitations, modifications and additions which it certainly must undergo on account of a number of other established results. Incidentally, the opportunity will be afforded at the same time of correcting some misunderstandings under which the theory has laboured in many ways.

It will be necessary also, in the second place, to discuss here the other theories of visions, which, as above stated, are very numerous. However, this will be confined to a general survey and brief outline of the fundamental conceptions, without going into a detailed criticism and weighing all probabilities and dubieties. The writer's views have been stated elsewhere as to the futility of making hypotheses and dissecting and weighing all sorts of possibilities beyond what seems expedient in the light of our positive knowledge.

A perfectly general conclusion should certainly be stated here prominently first of all: Fundamentally, the Helmholtz theory was simply the expression of a direct fact of observation, namely, that *the resultant of all the various light stimuli so far as sensations are concerned can be completely represented as a function of three variables.*[2] It is idle to try to explain this fact except on the assumption that the *result of the stimulation* also can be represented completely as a function of three variables. It is this assumption that is the fundamental conception of the Helmholtz theory. The justification for it has

[1] Nagels *Handbuch der Physiologie* III. S. 266 f.

[2] ¶Some would maintain perhaps that all we are justified in saying here is that an immediate photo-chemical initial process is a function of three variables. (J. P. C. S.)

been shown most of all by the huge difficulties that are encountered by every theory that has been developed on the assumption of a greater number of independent valences.

The next thing to be noticed here is that so far as *dichromatic vision* is concerned and its relation to normal colour vision, the researches described above (page 398) have completely verified HELMHOLTZ's conjecture, that there are two distinct types of this kind, both being reduction forms of normal vision. In fact, therefore, (just as HELMHOLTZ guessed from less accurate data) the conditions occurring here can be explained most satisfactorily as being *deficiency-effects*, that is, on the assumption that one component of the normal organ of vision is lacking in one case and another component in the other case.

Observations have shown too that any other explanation of the difference between the two types of dichromats (especially as being due to physical conditions of the ocular media) cannot be thought of at all. Undoubtedly, therefore, the simple way in which the theory is able to account correctly for these facts is a strong confirmation of it.

Matters are not quite so simple when we come to consider *anomalous trichromatic vision*.

In order to give a theoretical explanation of the facts which have been assembled above, we must keep steadily in mind, first of all, that the question here, as was pointed out then, is one of *alterations*, that is, variations, the effect of which is that matches that are valid for the normal eye are not valid for the anomalous eye. This immediately excludes the idea that one of the normal components, let us say, instead of being lacking entirely as in case of the dichromats, has simply been reduced to a very small amount without being changed in its character. We might indeed suppose that in this way the ability of discrimination in certain respects was lowered more and more; but matches that were valid for the normal eye would still have to be valid for the anomalous eye. A better plan is to adopt an idea which was developed first by FICK[1], in a different connection, it is true. The different appearance of colours is due to their acting unequally on the various component substances of the eye. Imagine the visual organ modified in such a way that the valence curves of two component substances instead of being considerably different, as they usually are, are almost identical; then no matter how these two substances may combine, there will not be much difference in the effects produced. Thus the result of certain differences in the nature of the stimulating light will be smaller differences in the corresponding physiological

[1] *Verhandlungen der physik-mediz. Gesellschaft zu Würzburg.* V 1873. S. 129. —PFLÜGERS *Archiv* XLVII S. 274.

processes, and hence with respect to certain discriminations the response of the organ of vision will be at fault. If the valence curves of two components are identical, the result would be a typically dichromatic organ. Accordingly, we are bound to suppose that the red component substance in the case of one class of anomalous persons, and the green component substance in the case of the other class, has undergone some kind of change by which its valence curve has been so modified that there is less difference between it and the valence curve of the other (normal) component substance than there is in the case of an eye with good colour vision. In this way we can easily conceive of a series of stages of transition into typical dichromatic vision of one kind or the other.

At present we still have no way of calculating these changed valence curves. But if light of all wave-lengths between 589 and 536$\mu\mu$ looks almost exactly alike to the extreme deuteranomalous eye, this means that the valence curves of both components are very nearly in the same proportion in this part of the spectrum. And in the most pronounced cases of protanomalous vision also the behaviour seems to be the same in the region of the long waves about up to the sodium line.

However, as to the connection between dichromatism and normal vision, a somewhat different view appears to be probable. Taking everything into consideration, it is quite likely that anomalous trichromatic colour systems are intermediate forms between normal trichromatism and typical dichromatism; so that, either by separate steps or continuously, the protanomalous eye represents a transition between the normal eye and the protanopic eye, and the deuteranomalous eye a transition between the normal eye and the deuteranopic eye. As a matter of fact, the observations described above show that as soon as the ability of the anomalous eye for discriminating colour is lowered by reducing the size of the object, it behaves nearly in the same way as the dichromatic eye. Whether this agreement is perfect, has not yet been decided in the milder forms of anomalous vision. On the contrary, indeed, it is found, for instance, that a person with pronounced protanomalous vision will make a match between red (671$\mu\mu$) and yellow (589) that agrees with that of the protanopic eye; explaining, however, at the same time that, while the fields match mighty nearly, they are not exactly the same. For deuteranopic vision and deuteranomalous vision the relation is still more sharply established. When the investigation is made with the small fields of the spectrum apparatus, persons may behave like typical deuteranopes, and yet, when tested with large objects, never make the RAYLEIGH-match perfectly correctly, showing themselves to be, therefore, like very anomalous trichromats.[1] Another thing to be noted here is that with reference also to luminosities (peripheral values, flicker values, etc.) protanomalous vision and protanopic vision behave in the same way or at least very much alike, and therefore very differently from normal vision; but the differences in deuteranomalous persons (just as in deuteranopes) are no greater than the individual variations from the normal.

Thus if dichromatic vision is regarded as being the highest manifestation of a modification, which in its lower degree is characteristic of so-called anomalous vision, the conclusion is reached that in such cases there is something more than just simply a lack of one of the component substances of the visual organ. Take the case of protanopic vision, for example. The idea is that, in its capability of being acted on by different kinds of light, the first component substance here, instead of having its own normal quality, has that which belongs normally to the second component substance. So also in the case of deuteranopic vision the valence curves of these two component substances would both be the curve that normally belongs to the first, or red, component. It is obvious that an organ of vision of this nature, in which two of the component substances are bound to act in the way described, must behave as a dichromatic eye.

Undoubtedly, alterations are not so easy to comprehend as pure deficiency effects; and the entire subject no longer seems to be quite so clear and lucid as it looked when there was an absolutely typical and fixed organization that could not be modified except by being lacking in some constituent part. But there cannot be any doubt whatever as to the occurrence of such alterations. (No physical explanation can account for the big discrepancy in some colour matches.) A certain variability of the organ of vision is therefore a directly established fact, and every theory must take it into account.

Moreover, from this point of view typical dichromatic vision would be a remarkable case, because the two valence curves are perfectly identical. And in a rather broad sense this case might perhaps be considered as being a deficiency effect. Thus, suppose, for instance, that A and B are two component substances that normally are provided with photo-sensitive materials a and b, respectively. Now if, instead of this, both A and B have the same material b, it would mean simply that the quality of the light-receptor belonging to A was abnormal; or, of course, it can be considered as being an absence of a. Thus, the importance of the fact that perfectly dichromatic systems are reduction forms of normal vision does not seem to be diminished,

[1] Nagel, *Zeitschrift für Psychologie* etc., XXXIX. p. 96 1905.

even if in their case it is not a question of deficiency effect in the very simplest meaning of the expression.

Incidentally also, according to this view, the answer is obtained to the question, as to what is the nature of the sensations of dichromats; whether, for example, the protanope sees mixed white light as colourless or perhaps blue-green. Although it is impossible to get any certain information as to the sensation of other persons, this latter assumption always did seem very unlikely, particularly in reference to twilight vision, as well as to the phenomena of acquired and unilateral colour blindness. On the theory of dichromatic vision adopted here, it seems clear that colourless light looks exactly the same way to dichromats as it does to normal persons; and it may be conjectured that long waves of light look yellow and short waves blue to both protanopes and deuteranopes.

An extensive array of facts, mainly phenomena connected with adaptation, has been accumulated since HELMHOLTZ's theory of colour vision was formulated. The phenomena of adaptation are at the basis of the duplicity theory (twilight vision, total colour blindness, the PURKINJE phenomenon, etc.). We shall have to see how HELMHOLTZ's theory can be connected with these facts and with the assumptions of the duplicity theory so far as the explanation of these facts is concerned.

At first glance it might indeed seem that the duplicity theory was diametrically opposed to HELMHOLTZ's theory of the sensations of vision. For according to the latter, the sensation of colourless brightness ought to be produced by a combination of processes which of themselves arouse a red-green and a violet sensation; which cannot be considered at all in the case of the activity of the twilight mechanism. However, by imposing some limitations on the HELMHOLTZ theory, as has to be done also for other reasons, this difficulty will be removed. In thus modifying the original theory it is still a question whether this implies something essentially different from what its author had in mind or whether it amounts merely to introducing definite assumptions as to some matters which HELMHOLTZ himself left open to discussion.

In the first place it must be kept in mind here that when HELMHOLTZ made the assumption that the organ of vision was a structure composed of three parts, he did not mean that the *sensation* itself was a combination of three elements (like the three notes of a musical triad). What he meant rather was that, in spite of the composite nature of the physiological process as made up of three independent constituents, the sensation may very well be something perfectly unitary and not capable of psychological sub-division. He regarded the outstanding position of certain sensations (for absence of colour and the so-called pure colours) as the result of psychological relations, connected with the naming of colours, etc. It is this view that stands in such sharp contrast to the one so often entertained nowadays, that the simple investigation of sensations without any auxiliary appliances is sufficient to enable us to find out their "simple elements." In the writer's opinion, from analogy with what is known about other senses, this way of looking at the matter is certainly much less accurate; and HELMHOLTZ's conception is undoubtedly justified to the extent that sensations of complex physiological origin can sometimes convey the impression of being absolutely unitary and typically steadfast.

Thus, while from this point of view there does not seem to be any positive necessity of modifying the HELMHOLTZ theory as proposed above, there are other facts that point in that direction with greater force. We know by experience that the sensation of absence of colour must certainly be exceptionally significant, because in a great many cases colour perceptions cease, and there is nothing left but a colourless sensation. And this is true, indeed, not only under the conditions of twilight vision (where there is a simple explanation for it on the assumptions of the duplicity theory), but also for daylight vision. The variations of vision in passing from the central to the more and more eccentric parts of the retina belong here. Still more important is the fact that by decreasing the size of an object the colour can be made to disappear; as a person with normal vision can very easily verify by using the eccentric parts of his retina, and as can be shown anywhere in the visual field of an anomalous trichromat. Another point in this same connection is that by limiting the time of exposure it can be made impossible for an anomalous trichromat to recognize colour. And, finally, let us allude also to acquired abnormalities of colour vision due to pathological causes, in which likewise colour discrimination is lost. Perhaps, in some of these cases absence of colour can be attributed, as above, to variations of the valence curves; but this explanation is ruled out for the cases of areal and temporal limitations. We are almost bound to make the assumption here that, even when the degrees of activity of the three hypothetical components of the organ of vision are adjusted to correspond to a colour, still in order for the sensation to be really that of a colour, or at least in order for the colour to be recognized as such, some other conditions besides have to be fulfilled; conditions, which by their very nature are in a certain way analogous to the ascent above a threshold value.

On this basis it may be considered as extremely probable that the organization in three components assumed in the HELMHOLTZ theory does not apply to the organ of vision as a whole, but only to those parts that are directly exposed to the action of light and á more or less extended series of parts connected with them; and that, on the other hand, the final results, the immediate substrata of the sensations, are themselves of a different nature; and hence that somewhere along the route the three independent results of stimulus are transformed into processes of a different kind and composition. As to these processes, nothing can be said with certainty, in the writer's opinion, except that in them the colourless sensation has some outstanding physiological significance.

In order to have some short way of referring to this assumption, let us speak of it as a *zonal theory*. From this point of view there seems to be no particular difficulty about supposing that the sensations of vision may be aroused by two different mechanisms more or less independent of each other, only one of which has the tripartite structure in question, whereas the other, being unitary, reacts to its stimulus in a simple monotone.[1]

Colour blindness of the eccentric parts of the retina is a subject about which it is hard to form a more positive opinion at present. Undoubtedly there are many points of resemblance between the defective colour vision of the eccentric parts of the retina and the colour infirmity of anomalous trichromats, and it would seem natural to try to explain them on the same basis, that is, as variations of the valence curves. This idea was developed and explained by Fick. Formerly, the writer was not able to accept this view, but now that it has been positively ascertained that there are alterations of this sort in anomalous trichromats, it seems to him to be more worthy of consideration. However, according to this explanation we should expect to find deviations in the Rayleigh-equation in the eccentric zones of the retina; and so far they have not been found. Should they occur, hardly any doubt would remain as to the correctness of that conception. On the other hand, if these deviations are not found, and if (leaving out of account the influence of the macula pigment) matches that are valid for the central retina are good for the periphery also,[2] we could perhaps do as the writer formerly suggested and fall back on assumptions as to modifications of those relations which exist when the peripheral processes are transformed into the substrata of sensation. The experimental data on the subject are not yet so complete as they should be; so that it would be somewhat premature to discuss the possibilities in detail.

[1] ¶See L. T. Troland, Brilliance and chroma in relation to zone theories of vision. *Jour. Opt. Soc. Amer.* VI. 1922, 3–26. (H. L.)

[2] That this is the case has heretofore generally been tacitly assumed as self-evident, but it can hardly be regarded as having been really proved. Incidentally, it is hard to decide the matter on account of the intermingling of rod functions.

Another point which has not been cleared up, and which incidentally is closely related to that just discussed, is the mode of distribution of luminosity. König called attention to the fact that the distribution of luminosity in the spectrum is approximately, though not very exactly (the great individual differences would prevent this from being the case), in agreement with the stimulus values as calculated for the red component. The same thing is true, as later investigations have shown, for peripheral values and flicker values and also minimum-field luminosities. Clearly associated with this is the fact that these relations for deuteranopic vision and deuteranomalous vision are not greatly different from those for normal vision, although their green components are different in their capacity for being stimulated; whereas in protanopic vision and protanomalous vision, in both of which it is the red component that is deficient in this capacity, the peripheral values, etc., are found to be entirely different from those in normal vision.

If we adopt the above suggestion and assume that the operations taking place in the three components are transformed towards the centre into processes of another kind, then in order to explain these phenomena, some hypothesis will have to be made as to the nature of these new processes and how they are related to the former. However, in the writer's opinion, it is idle to try to do this at present because we do not yet know enough about the facts along these lines. In the first place it would have to be definitely decided whether it is really a fact, as it seems to be, that even in the case of persons with normal colour vision there are individual differences in these respects, and how they are connected with each other or with other differences of the organ of vision, etc.[1] As the writer sees it, it is important to bring out clearly that the idea of each of the components having to contribute something to the total luminosity, which at first glance is the most natural supposition, is by no means the only possible conjecture, and it is more likely that the collective effect is the result of other very different modalities.

Briefly, by way of conclusion, it may be said that as regards the group of facts, that is, the so-called laws of colour mixture, which constituted the original primary basis of the Helmholtz theory, the theory has been confirmed in a remarkable manner, even in the light of the present very extensive state of our knowledge. It accounts for the behaviour of the normal organ of vision and its relations to typically dichromatic vision in a simple and accurately satisfactory fashion.

[1] Indications in this direction that are worth considering are to be found in Miss v. Maltzew's work mentioned above, but they need to be elaborated and confirmed.

The somewhat more complicated conditions of anomalous trichromatic vision can be made to fit in it in an intelligible manner, although not quite so simply. And it affords us at least the simplest way of expressing the observed facts in a comprehensive system. If, on the other hand, it has to be stated that it certainly is not a definitive theory of the whole of vision, as HELMHOLTZ supposed, there can hardly be any doubt as to the correctness of its fundamental conceptions. A rather more general discussion of theoretical questions will be deferred until other theories and other groups of facts have been described.

2. Other Theories of the Sensations of Light and Colour

While HELMHOLTZ deemed it illegitimate or at least untrustworthy to draw conclusions as to physiological processes from the direct psychological character of the sensations, most subsequent theories of vision use this very group of facts as their starting point. The results of this mode of treatment are in harmony as to their main features at least, and so all these theories have a common stamp impressed on them. In the first place, they assign a very special importance to the series of *colourless sensations* ranging in all degrees from black to white.[1] A similar importance is also assigned to certain colours, which are accordingly referred to as pure, simple or primary colours, namely, red, green, yellow and blue.[2] Another thing to be mentioned about it is that in sensation each quality of one pair seems capable of being combined with both qualities of the other pair (that is, red with blue as well as with yellow and yellow with red as well as with green); but the two determinations of each pair are mutually destructive, that is, red and green cannot be combined, nor yellow and blue.[3] This conception is a very old one fundamentally. Of late years it has been developed and advocated chiefly by AUBERT, and may be called the *four-colour theory*.

Unfortunately, this theory has been very inadequately tested as to the very points that are capable of being easily investigated. Obviously, it would be important to find out what objective lights can, under proper conditions, give the impression of "pure" colours, and whether in this respect various observers would agree or disagree. Investigations of this kind are very scarce; and among the advocates

[1] ¶Sometimes called "toneless" or "untoned" or "grey" sensations as distinguished from the "toned" or coloured sensations, namely, red, yellow, green and blue; as mentioned in the text. (J. P. C. S.)

[2] ¶The so-called "psychological primaries." This is another instance of the confusion of colour terminology in physics and psychology. (J.P.C.S.)

[3] ¶These latter combinations are what Mrs. LADD-FRANKLIN terms *colour-fusions* or *colour-extinctions*, to which attention was called elsewhere. (J. P. C. S.)

of the four-colour theory opinions are very much divided concerning a point of fundamental importance. Thus, AUBERT was of the opinion, and HERING has also espoused this view (as will be seen in the discussion of his theory below), that pure red and pure green mixed in proper proportions lose their coloured qualities and produce white; that is, that pure red and pure green are complementary colours. In opposition to this view, Mrs. LADD-FRANKLIN[1] (in agreement, by the way, with an idea expressed long ago by FICK[2]) affirms that pure red and pure green acting together are not colourless, but look yellow. Of course, there are also great differences of opinion in the further development of the four-colour theory, which will be mentioned below.

In the past decade much attention has been bestowed on this theory mainly in the special form which has been given to it by E. HERING.[3] According to him the visual organ is composed of three constituent parts, one of which (the "black-white substance") furnishes the colourless luminosity sensations, the other two visual substances (the red-green and yellow-blue) being responsible for the sensations of colour. In harmony with HERING's general biological ideas, the theory assumes moreover that in each of these substances processes of an opposite kind take place simultaneously; which HERING calls assimilative and dissimilative (usually spoken of as A and D processes). The *ratio* between these opposite processes is what determines the sensation. When they are in equilibrium in one of the coloured visual substances, this means that the sensation does not have the quality of either colour of that particular pair (that is, is neither yellow nor blue, or neither red nor green). When there is a state of equilibrium in *both* coloured visual substances, the sensation is devoid of colour. Equilibrium in the black-white visual substance corresponds to a particular "mean grey." White, red, and yellow, in increasing luminosity or saturation, is the result of the increasing preponderance of the three D processes; similarly, black, green, and blue result from the preponderance of the A processes. This entire conception was called by HERING a *theory of "Gegenfarben"* or *the theory of opponent colours.*

The hypothetical mode of action of different kinds of light on the separate visual substances is evident at once. All kinds of light would have to act on the black-white substance in the dissimilative (white) sense, but in different degree depending on the wave-length. On the other hand, the long waves as far as pure green (extending neither into the yellow nor into the blue) have a dissimilative (yellow) action on

[1] *Zeitschrift für Psychologie* etc. IV. p. 211. 1893.—Article on Vision in BALDWIN's *Dictionary of Philosophy and Psychology.*

[2] PFLÜGERS *Archiv* XLVII. S. 285. 1890.

[3] *Sitzungsberichte der Wiener Akademie*, Mathem.-naturw. Kl. LXIX. S. 131. 1874.

the yellow-blue substance; whereas the short waves from green on have an assimilative (blue) action on it. In the case of the red-green substance there is *D* action for the less refrangible end of the spectrum as far as pure yellow, and *A* action from there on up to pure blue, but beyond this point there is *D* action again in the violet. Pure yellow and blue would have no effect at all on the red-green substance, and pure red or green would have no effect on the yellow-blue substance.

Of particular interest also is the manner in which the effect of the stimulus on the visual organ is considered to take place here, and its dependence on the processes that determine the sensation. If the dissimilative action preponderates, the material substance is diminished and hence the "key" (*Stimmung*) varied in such a way that the effect of the *D* stimuli is reduced while that of the *A* stimuli is augmented. The reverse happens when the *A* processes are preponderant. The conditions are thus regulated automatically, as it were. The only possible permanent condition is that of an equilibrium between *D* processes and *A* processes, that is, a colourless grey of definite luminosity.

Now as to the anomalies that we have in colour vision, the four-colour theory assumes that in dichromatic vision there is an absence of the red-green sense, and both forms therefore are included under the name *red-green blindness*. HERING also accepted this way of regarding it, on the assumption, that the red-green visual substance is lacking in all cases. The differences in the two types that are now called protanopes and deuteranopes, which even at that time were well understood to a certain extent, HERING tried to account for by attributing them to physical causes (absorption of light in the coloured media of the eye, especially in the pigment of the macula). This assumption, hardly compatible with what was known about the matter then, has been shown by subsequent investigations to be thoroughly untenable. The difference between the protanopic and the deuteranopic organ of vision in the case of the long waves is so great and so typically fixed thereby, being also just as pronounced in the periphery as in the centre of the retina, that such an explanation cannot be seriously considered.[1]

We have no more right to place side by side the differences in the forms of dichromatic vision and those that are found in normal vision, as TSCHERMAK[2] has recently tried to do. He supposed that there are two types of persons with normal colour vision, just as there are two types of red-green-blind; one of which would be characterized by a relatively high, and the other by a relatively

[1] Concerning the particulars of the investigation by which HERING was led into this error, see v. KRIES, *Zft. f. Psychol.*, etc., XIII. S. 301. 1897.

[2] *Ergebnisse der Physiologie*. I. 2.

low, power of responding to long waves of light. But a normal eye never responds poorly to the long waves in the way the protanopic eye does. Persons with trichromatic vision who have trouble in connection with the long waves are protanomalous. Similarly, there is a kind of trichromatic vision which differs from normal vision in another way, but only in a limited sense, and which is not characterized by particular sensitivity to long waves, or by low sensitivity to the short. This difference however, is associated with deterioration in colour vision and leads by continuous transition to the deuteranopic system.

There are therefore not two kinds of normal colour vision between which there is the same kind of difference as between the two kinds of dichromats. But there is only one normal colour vision which is capable of two modifications. One of these is characterized by "poor use of long waves," to borrow this expression once; whereas this is not the case with the other. But both involve derangements of colour vision and may lead to dichromatic vision.

At present, therefore, there can be no doubt as to the failure of the four-colour theory at this point, and that, in particular, the explanation which HERING tried to give for the two kinds of colour blindness is not compatible with the facts. Of course, this does not imply that the assumptions of the theory, especially the original fundamental propositions of the theory, may not be right in a sense. But it is certain that the anomalies of colour vision point unavoidably to some set of relations that are not considered in the theory, and that thereby necessitate a limitation or extension of it.

The theory of opponent colours has been just as unsuccessful in explaining total colour blindness. Originally, HERING believed that in these cases the organ of vision was without both coloured visual substances, and was therefore limited to the black-white substance only. Thus he believed that by ascertaining the stimulus values for such an eye the *white valences* would be determined; and that a confirmation of this assumption can be found in the fact that the normal organ of vision in the state of dark adaptation in dim illumination (below the colour threshold) has very nearly just this same kind of vision. Apart from the fact that numerous peculiarities of the totally colour-blind eye would be completely incomprehensible from this point of view, this conception is shattered by the fact that these white valences may be totally different for two kinds of light which look alike to dichromats under the conditions of daylight vision; and also by the fact that, although in strong light the extreme periphery of the light-adapted retina has no colour vision anywhere, it can discern differences of luminosity; and chiefly by the whole series of facts which force us to regard twilight vision as a function of some special constituent of the organ of vision.[1]

[1] The assumption that the effects on the black-white visual substance corresponds to the twilight values under all circumstances was found to conflict in some ways also with the luminosity distribution in the coloured spectrum. HERING tried to get around it in his

The theory has come here into still more direct conflict with the actual facts, but it can indeed escape from this difficulty by a simple and unimportant modification. It would simply have to be completed by the assumptions of the duplicity theory; and the peripheral values, not the twilight values, would then have to be regarded as being the measure of the action on the black-white substance, that is, the peripheral values would be the "white valences."

An amplification of the theory in these directions has been attempted by G. E. MÜLLER.[1] Starting with the fundamental concepts of the four-colour theory, he has developed a theory of the organ of vision intended to take account not only of these facts but also of adaptation. He began with the perfectly sound view that, while it is possible to consider the last substrata of sensation in a manner corresponding to that proposed in the four-colour theory, it is also possible that the arrangements in the peripheral sense organ may be of another character. With respect to these latter, MÜLLER then developed a series of ideas which might seem adapted for explaining the facts known at that time (1897), although since then they have become much more complicated. A detailed account of MÜLLER's theory (which would certainly exceed the space allotted to this article) is not so necessary at present, because the theory would perhaps require to be modified still further to be made to agree with the facts which have been discovered in the last ten years.

TSCHERMAK also has tried to develop further the theory of opponent colours, mainly without much success. As was mentioned above, he abandons HERING's physical explanation of the difference between protanopes and deuteranopes, and speaks of a relatively better or poorer "utilization of the long waves." He has not made any attempt to explain the ground for this unequal utilization or the special connections between the different kinds of colour vision on this basis.[2] Incidentally, in this latter respect, as had been stated, TSCHERMAK starts out with certain notions that are contrary to the facts.

---

theory of the *specific brightness of colours*. There is no need to go into this matter because, from what has been said above, it can no longer be considered seriously. The fact that the distribution of luminosity in the coloured spectrum is approximately the same as it is in case of colourless vision with the eccentric parts of the retina in state of light adaptation, shows that the colours have but slight influence on the luminosity.

[1] *Zeitschrift für Psychologie* etc. X. pp. 1 and 321; XIV. pp. 1 and 161.

[2] TSCHERMAK simply alludes to the possibility that the matter may have something to do with peculiarities of structures which are in front of the visual substances proper. Here he adopts a view occasionally advanced by HERING himself, but never really seriously; which DONDERS on the other hand, had actually tried to apply, and which has been the writer's view ever since his earliest publications (*Die Gesichtsempfindungen und ihre Analyse*, 1882, pages 163, foll.)

On the other hand, SCHENCK[1] tried to develop the assumptions of the HELMHOLTZ theory along certain lines, and particularly to explain the hypothetical separation of the organ of vision into three components by means of conceptions as to its evolution. He supposed that not only the rods but the cones also originally contained a single substance not very different from that in the rods, the decomposition of which by light gave a colourless sensation of brightness in the same way as in the rods. This substance, like that in the rods, was originally not very sensitive to long waves, but at first it underwent a change called "pan-chromatization," by virtue of which (just as in photographic plates) it was made relatively very sensitive to long waves also. Next it differentiated or divided into blue and yellow components, and finally the yellow component was subdivided again into red and green components. For the anomalies of colour vision there are various possibilities. Entire absence of the cone mechanism involves the ordinary form of total colour blindness. Other forms are due partly to failure of pan-chromatization, and especially in some cases to failure of splitting up of the yellow process. This latter is responsible for dichromatic vision, which is deuteranopic or protanopic, according as pan-chromatization has, or has not, taken place, respectively.

The theory certainly is calculated to give an attractive explanation of the occurrence of many forms of colour vision like those which are actually found (although it has always been a curious fact that some of the theoretically possible forms have never been observed). But it seems to the writer that the manner in which the special relations between the different kinds of colour vision are explained is not so smooth and simple as to compel conviction, lacking, as it does, objective observations in support of it. Accordingly, it does not seem necessary to give a detailed description of it here.

LADD-FRANKLIN's theory[2], developed long before, is very similar to that of SCHENCK. The writer has mentioned the latter rather more particularly, because it discusses in special detail just the anomalies of colour vision that concern us here. Some other theories of the visual sensations, whose interest is more from another side, will be briefly alluded to presently.

3. Modulations of the Organ of Vision[3]

Although the manifold anomalies of colour vision are certainly best suited for giving us an insight in the normal structure of the organ of vision, and have been studied and discussed chiefly from this point

[1] PFLÜGERS *Archiv* CXVIII. S. 129. 1907.

[2] *Zeitschrift für Psychologie* etc. IV. S. 211. 1893.

[3] ¶It is difficult to find a precise English equivalent to the German *Umstimmungen* as employed here to include both "fatigue" (see PARSONS, *Colour Vision*, 112) and "re-

of view, there are some other groups of phenomena that are valuable for the same reason. These will be only briefly touched on here in order to record facts which have been made known recently. The phenomena of *the modulation* (*Umstimmung*[1]) of the organ of vision will be mentioned here first. While the fundamental facts of these phenomena have been known for a long time, and, incidentally, can be satisfactorily explained too by all theories, certain questions of special theoretical interest have resulted from researches in these fields. First, there is the question as to the "*persistence of optical equations*," that is, as to whether two mixtures of light, which are not equal objectively, but which appear to be the same for a certain condition of the organ of vision, will continue to look alike for every other condition, or whether they can be made to look unlike by changes in the state of the eye. Facts above stated show that in all parts of the retina where there are both rods and cones the latter is the case, that is, the match does not persist. It will be recalled that mixtures of light which look alike under the conditions of daylight vision may have totally unequal twilight values. Thus, matches of this kind can be destroyed by adaptation, and changed often to an enormous extent. On the other hand, so far as we know, the probability is that in the place where vision is most distinct, and where there are no rods, changes of this nature do not occur at all, or to such a slight extent that there is no sure proof of them at present.[2]

Moreover, as to the parts of the organ of vision that are used in twilight vision, STEGMANN's experiments (page 389) and observations on total colour blindness show that the luminosity relations of different kinds of light are but little affected by changing adaptation. We can therefore assume that for the organ of vision as a whole there is no persistence of optical equations whatever, but that for each of its two mechanisms matches do persist, approximately anyhow.

As to the way in which the response to the stimulus is influenced by the state of the organ of vision at the time and of its component parts, HELMHOLTZ, adopting a suggestion of FECHNER, had conjectured that "the fatigue of the nervous substance of vision has about the same effect on the sensation of fresh incident light as if the objective

---

covery" (or adaptation). The word "modulation" is certainly not an exact rendering of the original, and, doubtless, exception will be taken to it. Still it seems to convey the meaning better than such terms as "mutation," "conversion," etc. (J.P.C.S.)

[1] This term "*Umstimmung*" is employed here instead of HELMHOLTZ's expressions "fatigue" and "recovery," because it is more general and is not prejudiced by any theoretical view.

[2] Concerning researches and controversies on this subject, see NAGELS *Handbuch der Physiologie*. III. pp. 210, foll., and the literature given there.

intensity of this light were diminished by a definite fraction of its amount" (see p. 235).

Accordingly, therefore, if one of the mechanisms of the organ of vision were "tuned" differently, the result would be the same as if all stimuli acting on it were multiplied by a certain coefficient. Incidentally, HERING also made this a fundamental part of his earlier ideas at least, and assumed that the product of the intensity of the light and a coefficient depending on the condition of the organ of vision was a measure of the effect of the stimulus.

The writer calls this the *coefficient law.* So far as its proof and validity are concerned, it must be borne in mind in the first place that the assumption in regard to the functional importance of the eye's being "attuned," which is at the basis of the law, applies merely to the behaviour of a unitary system, that is, of a single component mechanism of the organ of vision. Accordingly, it is to be expected that the law will *not* apply where twilight and daylight vision are in operation together; because, under these conditions, we have to do with changes of condition of both mechanisms, and these changes are generally different. As a matter of fact, BÜHLER[1] found that the effects of dim lights were increased much more by dark adaptation than those of bright lights. Thus, in order to illuminate two adjacent parts of the retina that are "attuned" differently, so as to get the same impression of brightness from them both, the ratio between the two objective brightnesses will have to be reduced (that is, will have to be made nearer unity) as the absolute intensities are increased. And if a match is made between two dim lights, and then both are increased in the same proportion, the stronger light (acting on the fatigued place) will appear too bright. The same result has been obtained recently by DITTLER and ORBELLI.[2]

In apparent conflict with the results of the authors just mentioned, WIRTH[3] found that the coefficient law was valid within fairly wide limits; but the explanation perhaps is that he worked with light-adapted eye and possibly also with stronger lights. Thus an unequivocal answer cannot be given at present to questions that are of peculiar theoretical interest, namely, as to how the ratios turn out for the separate mechanisms of the eye; whether the coefficient law is obeyed here approximately, and if not, in what direction the deviations from it are. To obtain the answers, the test would have to be made with very small fields, directly fixated, and with very bright lights.

[1] Beiträge zur Lehre von der Umstimmung des Sehorgans. *Diss.* Freiburg 1903.

[2] PFLÜGERS *Archiv,* CXXXII. p. 338. 1910. — From the description given by these authors it is not clear whether the conditions were such that the coöperation of the twilight mechanism can be assumed; but they certainly do not indicate that this was not the case.

[3] WUNDT, *Philosoph. Studien* XVI. (4), XVII. (3) and XVIII. (4).

As to more special phenomena, there is a fact connected with the action of white light that should be mentioned here. If in the case of the photopic eye a part of the retina that has been fatigued with white light is tested with coloured light, we find that in order to get the same impression (that is, equal colour saturation) from the adjacent unfatigued places as from this spot, we have to use lights of different intensity but of approximately the same spectral composition. Thus, for instance, if in order to make the match, three times as much white light is required at the fatigued place as on the adjacent parts, this place must be stimulated with approximately three times the amount of coloured light also. Fatiguing the eye by white light appears therefore to modify the effects that stimuli have on the colour discriminations of the sensation.[1]

There are also a number of new results concerning the phenomena of fatigue with coloured lights. If places on the retina that are fatigued for one colour are tested with lights of other colours, the general effect is an appearance which is approximately complementary to that of the colour causing fatigue. We are indebted to HESS[2] for a series of quantitative determinations along this line indicating that these changes are very considerable. A green light ($517\mu\mu$) acting on a portion of the retina fatigued with blue looked the same to him as a slightly greenish yellow ($565\mu\mu$) which he used to make the comparison. The writer[3] found that after red fatigue a yellow looked like green yellow ($556\mu\mu$), and that after green fatigue it looked like orange ($605\mu\mu$). HELMHOLTZ stated that all pure spectral colours seem to be very much more saturated when the eye has been previously fatigued by the complementary colours; and this has since been repeatedly verified.[4] It is especially true with respect to the colours of the less refrangible half of the spectrum and for high intensities.

By making the tests with the same light used for fatiguing the eye, information is obtained concerning the apparent variations that take place in looking at a single colour (or monochromatic light) for a long time. It is a familiar fact that there is loss of saturation under these circumstances; and most colours undergo a change of hue at the same time. According to VOESTE's observations,[5] after steady fixation for a long time, light of long wave-lengths (up to $560\mu\mu$) was found to

[1] v. KRIES, *Berichte der Freiburger Naturf. Gesellschaft.* 1894. — WIRTH, (*Archiv. f. Psychologie.* I. page 49) obtained the same result. On the other hand, opposite results have been found very lately by DITTLER and RICHTER in similar experiments, the explanation of which is not clear at the present time. *Zft. f. Sinnesphysiologie* XLV. page 1, 1910.

[2] HESS, *Archiv f. Ophth.* XXXIX (2), S. 45.

[3] v. KRIES, NAGELS *Handbuch der Physiologie* III S. 215.

[4] HESS, *loc. cit.*—v. KRIES, NAGELS *Handbuch der Physiol.* III. S. 214. 220.

[5] VOESTE, *Zeitschrift f. Psychologie* etc., XVIII. S. 257.

look like a test light of shorter wave-length; and light of wave-length between 560 and 500$\mu\mu$ like a light of longer wave-length. Thus in each case they approach yellow in appearance. Light comprised between 500 and 460$\mu\mu$ again begins to look like light of shorter wave-length, that is, tends to look more like blue.

As previously stated, there is a great variety of satisfactory theoretical explanations of the essential qualitative aspects of these phenomena; but no one has yet succeeded in giving a satisfactory explanation of the details and quantitative relations. And hence, with respect to this whole group of phenomena, we should be cautious about drawing conclusions as to fundamental questions concerning the organization of the eye or how it is composed of separate parts.

HELMHOLTZ's assumption of a *fatigue*, which would take possession of the separate components independently, affecting each of them in proportion to its activity, encounters difficulties with respect to the great increase of saturation which can be produced even with spectral lights by fatigue with complementary colours. At any rate this is so if we adhere to the theory of the components as derived from the anomalies of colour vision. And it is doubtful, to say the least, whether these difficulties can be surmounted by making other assumptions as to the components.

As has been mentioned, HERING originally regarded the modulations, conformably to the coefficient law, as being of such nature that the result of the stimulus was measured by the product of stimulus-intensity and a factor depending on the modulation. Later, as HESS had done before, he substituted for this the notion that every light by its colour modulation (*Farbenumstimmung*) must receive a certain value in complementary valence (*Betrag an gegensinniger Valenz*); a conception which cannot be said to be entirely satisfactory. In all cases additions of some sort have to be made to the theory in order to explain the independence between optical matches and the modulation of the organ of vision.[1] And the theory encounters positive difficulties in connection with the facts given above, which show that by fatigue with white light the excitability of even the underlying mechanism of colour determinations is lowered.

4. Temporal Effects of Stimulations

The *temporal relations* of the processes of stimulation are much more complicated than was formerly supposed. These have theoretical bearings in many ways, and hence a review of the more recent studies may be included here also.

[1] v. KRIES, *Arch. f. (Anatomie und) Physiol.* The validity of this view has been questioned by HERING, without justification. See HERING, PFLÜGERS *Arch.*, XLII. 497. — v. KRIES *Arch. f. Physiol.* 1887, page 113; 1888, page 381.

As to the effect of a single short-lived stimulus, peculiar conditions were found to arise from the so-called PURKINJE *after-image* (positive complementary image), which is a phenomenon that has been known for a long time. Thereupon, systematic investigation of the action of short-lived stimuli revealed that this constitutes only a part of a long series of processes.

As to methods, these researches have been conducted along two lines. One way is to let a short-lived stimulus act on a part of the retina and to observe the successive stages of sensation that follow. The other way consists in letting a spot of light glide over the retina of the immobile eye. Moreover, in the latter case each point of the retina is illuminated for a short time (determined by the size and speed of the image), but one after the other. Thus the various phases of the process are watched at the same time, but separate from each other in space, which makes it much easier to analyze and appreciate them.

The phenomena that are perceived under the most favourable conditions for distinguishing their various stages will now be enumerated.[1]

Suppose a bright object is made to move about in the field of view which is otherwise perfectly dark; then under proper conditions the following stages of the entire process connected with brief stimulation may be distinguished.[2]

1. The primary image, the preliminary appearance of light, which is the first and greatest effect of the stimulus. It comes up generally in the same colour that the light has when it acts continuously. As contrasted with the stationary light, it is elongated more or less, depending on the speed of movement.

2. The primary image is followed by a dark stretch.

3. Thereupon follows closely a so-called secondary image indicating a repeated flashing out. When the light itself is coloured, this image is faint and usually indeed complementary to the primary image. It represents what was observed first by PURKINJE, and what has since been called the positive complementary or PURKINJE after-image. Very sharply delineated and as a rule not drawn out much, this image constitutes the most peculiar and most conspicuous phenomenon of the whole region. For when the movement is moderately rapid, it makes the impression of a second luminous object at a fixed distance behind the first, often of considerable brightness, and, as intimated, having a colour about complementary to the primary image. This is

[1] This is a brief résumé of the writer's description in NAGEL's *Handb. d. Physiol.* III. 221.

[2] ¶See PARSONS, *Colour vision*, pages 87, foll. (J. P. C. S.)

the reason why the phenomenon was called "recurrent vision" (YOUNG, DAVIS), "the pursuant image" (v. KRIES), "ghost" (BIDWELL), and "satellite" (HAMAKER). For a revolving blue object it is shown in Fig. 2 of Plate III.*

4. The secondary image is not sharply terminated behind, and is followed by a second dark gap. Close on this is—

5. Another glow still, which we may call the tertiary image, adopting the term used by SNELLEN and BOSCHA. It has no colour or is faintly coloured like the primary image. It is not sharply outlined, but represents a gradual increase and decrease of brightness extending over several seconds. When the object completes its circuit in from 1.5 to 3 seconds (which is the best speed for seeing the secondary image), this part of the phenomenon, if it is well produced, is like a cloud of light filling the entire circle. Therefore, in order to observe the entire course of the phenomenon, it is better to let the object make only one revolution or part of a revolution.

6. The last phase consists of a darkening following close on the preceding, again without being sharply separated from it. It reveals the path traversed by the bright object as a deep black band. The primary image is by far the brightest, and although the secondary image is much fainter, it is, however, considerably brighter than the tertiary image. Accordingly, supposing that there are certain intensities of the objective light that are best adapted for showing the phenomena in the way they have been described above, we should expect that with lower intensities the primary and secondary images alone would be visible, and with lower intensities still the latter too would disappear. Moreover, the extent of the separate images is variable and depends on circumstances; and hence the primary image may reach as far as the secondary, and the latter again may extend out to the tertiary; and so there will be no dark gaps. In this case the appearance is essentially different in character; and this is manifested especially by the fact that the comparative individuality of the various stimulations supposed to occur here is not clearly brought out any more.

Of the many details only those can be mentioned here that have been most positively observed and are of particular theoretical interest. In the first place, the primary image shows variations which indicate that the different parts of the organ of vision that are affected by the continuous action of light do not all come into activity simultaneously when the light is first turned on. Phenomena of this kind, by the way, can be observed under many conditions, and some of them have been known for a long time. Thus, for example, it has already been stated (p. 255) that for certain rates of rotation of a disc with black and white

*Plate III faces page 384.

sectors, the forward edges of the latter appear reddish and the backward edges bluish. Thus it seems as if the red component reacted more quickly than the others. Incidentally, also, KUNKEL's observations[1] are in good accord with this. HESS[2] states that the forward edge of a red object in motion appears to be a more saturated red, which perhaps can be explained in the same way.

These differentiations are especially noticeable in dark adaptation. In this case, vivid colour appears only on the forward edge (the effect is particularly distinct with blue light). Close behind it there is a paler portion which runs out into a white tail. The primary image then has the appearance shown in Fig. 3, Plate III.* This shows that *the primary stimulation of twilight vision follows that of daylight vision with a little retardation.* The old phenomenon of the so-called "fluttering heart" belongs here also (see p. 258). Bits of coloured paper are fastened on a background of another colour; and when the whole affair is moved slightly to and fro, the pieces of paper seem to dance about on the background, as if they lagged behind it or ran ahead of it. It is an effect that can be induced by suggestion under very many conditions. By far the best way to get it is with bits of red paper on a dark blue ground (or *vice versa*), and with feeble illumination, so that in case of the blue its rod-action preponderates.

Along a radius of a black disc attach a narrow red strip and a blue strip so that one forms the continuation of the other; and observe it in dim illumination with dark adaptation. When the disc is turned slowly, it is easy to see how the red strip runs ahead and the blue strip lags behind. (Of course, the eye must not try to follow the movement of the disc, but must look fixedly in the same direction.) McDOUGALL[3] estimated the amount of this retardation at about 1/18 second.[4]

The *secondary image* appears from a quarter to a sixth of a second after the primary. The most remarkable thing about this image is that it is absent in the fovea. Thus if it is observed in the way described above as an image coming behind, and if then, without changing the fixation of the eye, the primary image is made to glide over the

[1] PFLÜGERS *Archiv.* VI. S. 197. 1872.

[2] PFLÜGERS *Archiv.* CI. S. 226. 1904.

[3] McDOUGALL, *Brit. Journal of Psychology.* I. 1904.

[4] ¶H. E. IVES, Visual diffusivity. *Phil. Mag.* XXXIII 1917, 18–33.—Idem, Resolution of mixed colors by differential visual diffusivity. *Phil. Mag.* XXXV. 1918. 413–421.—F. W. FRÖHLICH, *Grundzüge einer Lehre vom Licht- und Farbensinn. Ein Beitrag zur allgemeinen Physiologie der Sinne.* 1920. — Idem, Über oscillierende Erregungsvorgänge im Sehfeld. Untersuchungen über periodischer Nachbilder. Zur Analyse des Licht u. Farbencontrastes. *Zft. f. Sinnesphysiol.* LII. 52–59; 60–88; 89–103. — Idem, Über den Einfluss der Hell- u. Dunkeladaptation auf den Verlauf der periodischen Nachbilder. *Zft. f. Sinnesphysiol.* LIII. 79–107.—Idem, Über die Abhängigkeit der periodischen Nachbilder von der Dauer der Belichtung. *Zft. f. Sinnesphysiol.* LIII. 108–121. (H. L.)

*Plate III faces page 384.

fovea, the trailing image follows behind it up to a short distance from the fovea, and then disappears there in very singular fashion, just as if it went into a tunnel, to reappear a little the other side of the fixation point.[1] The inference is that the secondary stimulation is a function of the twilight mechanism, which is confirmed by the fact that the luminosity of the trailing image corresponds to the twilight values of the stimulating lights. When light of very long wave-length is used, the secondary image is not observed, at any rate not until the intensity of the light is very considerable. Thus, so far as the luminosity of the secondary image is concerned, we must consider it as being a special temporal characteristic of rod function; and, on the other hand, the reason for its complementary colour is connected with the modulation of the organ of vision that exists during this phase.[2] The way in which the secondary image depends on the adaptation is remarkable and has not yet been explained. Starting with the eye light-adapted and observing the changes that take place in the secondary image, during otherwise constant conditions, as dark adaptation sets in and continues, we notice that for some minutes it is invariably clearer and more beautiful, increasing in brightness and extent. But then it becomes harder to make out, and after very long dark adaptation (two hours or more) the writer cannot see it any longer at all.[3] The tertiary image, being of the same colour as the first, proves to be a second stimulation of the trichromatic mechanism similar to the first, but there seems to be superadded still another, that is, a third luminosity-sensation mediated by the rods.

Thus, the details of the phenomena enable us to keep separate the part in them that is probably taken by each of the two mechanisms of the organ of vision. But at present we are not in the position to give an explanation of the complex temporal configuration of the processes of stimulation in each of these parts.

With respect to the stimulus effects in case of long-continued illuminations, HESS[4] reports that at both the beginning and the end of the stimulus similar oscillatory proceedings can be observed. Long

[1] v. KRIES, *Zeitschrift für Psychologie* XII. S. 83. 1896.

[2] Certain deviations from the rule of complementary colouration are due to the fact that the sensation corresponding to twilight vision is often more or less bluish. This is why the blue colouration is particularly clear and easy to see when yellow lights are used. On the other hand, with blue lights, frequently the secondary image seems to be of the same colour provided the stimulating light is of low saturation; not being yellow, as it should be, unless the saturation is high.

[3] NAGEL and various other observers working with him and the writer had the same experience. HESS states that for him the phenomenon is not essentially different whether the dark adaptation is very long or brief.

[4] PFLÜGERS *Archiv*. CI. S. 226. 1904.

ago EXNER[1] carried out experiments and measurements as to the manner in which the process of stimulation increases at the beginning of the stimulus, and especially with respect to the time taken to reach the maximum.

For this purpose, he stimulated two contiguous portions of the retina at an interval apart of from one-fiftieth to one-sixtieth of a second, and then simultaneously extinguished both lights. The exact moment of extinction can be varied, and so the two illuminations can be made to last longer or shorter, however with the same interval always between their instants of beginning. Now if the periods of illumination are so brief that the maximum is not yet reached, then at the moment the lights are extinguished the stimulation due to the first will be nearer the maximum, that is, higher in value. But if the maximum has already been reached, then the stimulation due to the first will already be beyond that, that is, lower in value. EXNER found that the times thus ascertained for the rise of the sensation became shorter with increased intensity. With the intensities which he used, these times were between 0.150 and 0.287 second. Also when the time of action of one stimulus was diminished, it could be determined how its intensity had to be increased in order for it to remain equal to the other, which was kept constant in period and strength. Hence, the mode of rise of the stimulation could be derived.

Similar experiments were made by KUNKEL[2] with coloured (spectral) lights.

As the comparison between the adjacent fields is made mainly during the after-image phase when the illuminations are of such short duration, the principle is correct only on the assumption that, if the stimulations are equal at the moment the stimuli cease to act, they will also die down together. Owing to the complicated temporal relations of stimulation by short-lived stimuli, this supposition is open to some doubt; and so there is some uncertainty about experiments based on it.[3] Still greater difficulties are encountered in comparing the luminosity of a very brief stimulus with that of one that is continuously visible; as was done by MARTIUS[4] in trying also to find how the

[1] *Sitzungsberichte der Wiener Akademie* Math. naturw. Kl. (2) LVIII. S. 601.

[2] PFLÜGERS *Archiv.* IX. S. 197.

[3] On the other hand, the writer cannot see how these results would conflict with the above mentioned observation of HESS concerning the oscillatory way the stimulations begin or how they would be invalidated thereby. EXNER worked with stationary objects and HESS with moving objects, which is a difference that may be worth considering in this matter. But aside from this, there is nothing whatever against the idea that EXNER's determinations relate to the first rise and first maximum of the process of stimulation, as was rightly pointed out by EXNER himself in reply to HESS's criticisms. (PFLÜGERS *Archiv.* CIII. p. 167, 1904).

[4] *Beiträge zur Philosophie und Psychologie.* I. S. 3.

sensation rises. The conditions here for comparing the luminosity are very complex, and the results are variable; as is shown directly by WATT's observations.[1]

The phenomena are far more diversified still when time and space are both factors in the combined stimulus; as, for example, when images more or less complicated in character are made to pass over the retina. These phenomena are as yet only partially understood, and cannot be classified or positively explained. It may suffice to cite here one example which is especially remarkable because very brilliant colour effects are produced by objects without any colour as a result simply of special space and time relations. When a disc like that represented in Fig. 79 is made to execute single rotations of about 180° in the direction of the arrow (the best way to do it is by hand, because then the right speed can be easily found), what we see is that a metallic luminous yellow red tail attached to the black portion comes on behind it. The tail is deep red when the disc is viewed through a yellow glass. If the disc is made to revolve continuously, at about 4 or 5 revolutions per second, the colour spreads over the whole periphery, that is, we see a yellow-red or red ring, in which incidentally a periodic alternation of bright and dark places is noticeable.

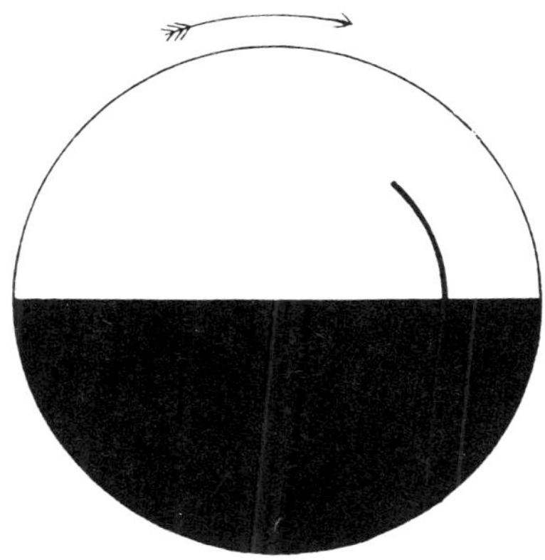

Fig. 79.

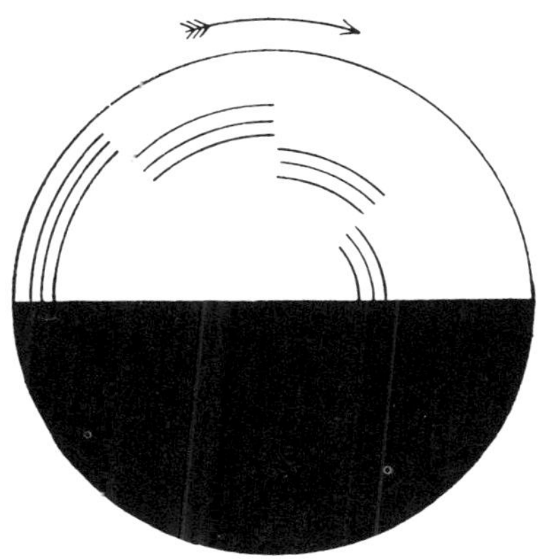

Fig. 80.

The yellow-red colouration occurs on portions of the retina, where the (colourless) illumination begins a little later than on the adjacent parts.

In an experiment described by BIDWELL[2] the same thing was ob-

[1] WATT, PFLÜGERS *Archiv.* CVII. S. 531. 1905.

[2] *Proceedings of the Royal Society London.* LXI. S. 268.

served in a little different form. A disc, half white and half black, has a 45° sector cut out of the white half. It is set to spinning rapidly, 180° of black being followed by 135° of white. Black print on a white background seen intermittently through the gap in the rotating disc looks red. This phenomenon, combined with others, can be observed with the so-called BENHAM top[1] (Fig. 80). Rotated slowly, the inner rings here, which are arranged as described above, look yellow-red, and the outer ones bluish. No distinct colour can be discerned in the two intermediate sets of rings.

5. Review of the Status of the Theoretical Questions

Reviewing the facts here mentioned, we can see that perhaps of all theoretical conceptions as to the structure of the organ of vision the assumptions that constitute the *duplicity theory* may be regarded as comparatively the most secure. It is not necessary to take up again in detail the relations that have been described. It may be added, however, that they are connected with perfectly definite anatomical and physical facts, and therefore have the advantage of dealing with something concrete and known rather than with the abstractions of HELMHOLTZ's components or HERING's visual substances.

Connected with this is the fact that we are on less sure ground the moment we try to form more exact pictures of the trichromatic mechanism. As to these matters, it may be said in the first place that the anomalies of colour vision and their relations to normal vision represent the field in which there has been on the whole most success in obtaining a complete survey of a large number of facts in terms of relatively simple general laws. This field affords therefore what appears to be the most solid foundation for conjectures as to the structure of the organ of vision. Moreover, in a comparatively simple way these facts can be made to dovetail with the HELMHOLTZ theory, but not with any other theory.

It must be remembered also that this theory does not pretend to explain the conditions of vision in the thorough and simple manner it seemed to do fifty years ago. But from an unprejudiced view of the facts there can hardly be any doubt that in its assumptions it lays hold of a fundamental characteristic in the structure of the visual mechanism and is very much to the point. There is nothing inconsistent with this in imagining that the hypothetical articulation (*Gliederung*) of the mechanism of vision is true only for one portion of it; the question still being left open, however, as to what those component parts really are. And we may go on using the ideas of com-

[1] Ibid.

ponents, valence curves, deficiency effect, etc., without being deceived as to their merely provisional meanings and without leaving out of account the fact that subsequent investigations will determine their ultimate significance, probably substituting something else for them equivalent in effect.[1]

As to the conceptions of the four-colour theory, the writer thinks that all that can be said with comparative certainty is that in some way the colourless sensations occupy an important position. But whether a similar statement can be made with respect to what are called pure colour sensations is very much more doubtful. But even on the first assumption by itself, we may still regard the three components of the HELMHOLTZ theory as being a satisfactory representation simply of an outer structure of the organ of vision, the processes occurring within them being transformed into others in the interior, in which the colourless luminosity sensation comes to occupy a position of special importance, and which probably may conform in other ways also to the assumptions of the four-colour theory. It is rather curious to note that most theoretical speculations have led, from different points of view, to be sure, to these notions of a different division of the organ of vision in different zones. This is a view which the author advanced in his earliest work. Starting from the theory of opponent colours, G. E. MÜLLER likewise has had to assume other mechanisms for the places where the light acts directly. TSCHERMAK also, by his distinction between stimulus-receptors and sensation-stimulators, takes practically the same position. And SCHENCK has assumed a distinction of this kind and regards it as indispensable.

So far the conception may be said to be on a pretty solid basis of facts. But when we try to designate physiologically the immediate substrata of sensation, we are landed in the midst of all kinds of uncertainty. Certainly, in this respect HERING's theory is the most interesting by reason of its general biological considerations. But it is precisely from this point of view that there are also weighty objections to it.[2] Moreover, it never seemed plausible to suppose that the state of the sensation both in brightness and darkness has to be adjusted on the average to the same value (neutral grey). And at present it can be regarded as extremely probable that at any rate the greatest changes that go on in the organ of vision are those produced by adapta-

[1] ¶C. LADD-FRANKLIN, On color theories and chromatic sensations. *Psychol. Rev.* XXIII. 1916, 237–249. — Eadem, Practical logic and color theories. *Physiol. Rev.* XXIX. 1922, 180–200. Eadem, Tetrachromatic vision and the development theory of color. *Science*, LV. 1922, 1–6. (H.L.)

[2] See, for example, what FICK says in *Sitzungsberichte der Physik. Med. Gesellschaft Würzburg*. 1900.

tion and are connected with the formation of visual purple; in other words, they have an external basis in the sense that they are not directly dependent on the sensation process.—Among hypotheses connected with HERING's fundamental ideas, the proposals made by PAULI[1] and by BRUNNER[2] may be referred to. They are likewise concerned with the nature of the processes underlying sensations, especially with the mode of antagonism that has to be hypothecated between processes combined in one pair. With respect to the anomalies of colour vision, these theories certainly need to be amplified exactly in the same way as the original theory of opponent colours; and, according to what has been stated already, this amplification would have to do with the external part of the organ of vision. As to the basis of sensation itself, the writer thinks that the proposals made by PAULI and BRUNNER amount to far more considerable changes in HERING's theory than they themselves realize; changes which enable the theory to get rid of many difficulties, but also involve it again in many others. The interest of the entire conceptions, and especially the way in which they differ from the original theory, is connected chiefly with general biological questions; and therefore a thorough discussion has to be omitted here and reserved for another occasion.

An idea that is apparently at the basis of many of these attempted explanations was first expressed by DONDERS. He supposed that the sensations are associated throughout by cleavage-processes. The colourless sensations are due to total cleavages (symmetrical or unsymmetrical) of highly complex molecules, whereas the coloured sensations are due to partial cleavages. SCHENCK's theory mentioned above and LADD-FRANKLIN's theory, which is very similar to it, are closely related to this idea.

On the other hand, BERNSTEIN[3] has developed a theory of visual sensations on an entirely different basis. This theory makes use of special conceptions concerning the processes in the central nervous system, including inhibitory effects.

The very diversity of the results of all these speculations, in spite of the fact that they all start from very similar premises, merely serves to show what wide room there is here for hypotheses. The writer has expressed the opinion over and over again that it is idle at present to tackle these questions; and recent experiments (especially one of SCHENCK's) tend to confirm this opinion. Of course, it is a matter of individual scientific disposition and personal taste as to how far one can go in this direction.

[1] PAULI. *Der kolloidale Zustand und die Vorgänge in der lebendigen Substanz.* 1902.
[2] PFLÜGERS *Archiv* CXXI. S. 370. 1908.
[3] *Naturwissenschaftliche Rundschau.* XXI. S. 497. 1906.

As to the main result of the more recent investigations, the writer would not venture to state that they are in harmony with any definite theory of the organ of vision or tend to give it support. The chief outcome has been rather to establish the fact that the different mechanisms of vision, which are different also in their efficiency, are distinguished from each other in very characteristic fashion by the way in which they react to different kinds of light, that is, by the peculiarity of the lights that appear to be the same. The thing to do, therefore, is to find out *what lights look alike to each mechanism*, because, as far as we know at present, this is the only perfectly unobjectionable method of obtaining information in comparatively simple fashion as to the efficiency of an eye in recognizing and discriminating colours.

The reason why KÖNIG's researches and the vast amount of work that has been carried on in methodical connection with it have had such valuable results is just because, without any theoretical presumption whatever, a certain group of facts was investigated that could be observed directly. And in this way results were obtained which may be regarded, also without any theoretical bias, as general facts as to the mode of behaviour of the different mechanisms of vision, as found by direct observation. It is worth emphasizing this chiefly on account of the great practical value of these tests of colour vision for railway employees, etc. Just such conditions constitute the basis of methods of research, as has been stated before. But these facts must be further emphasized because, although it was formerly supposed that colour matches were particularly adapted for determining the character of vision, for a long time the method was viewed with some doubt, the tendency being to relegate it to a position of subordinate importance. As a matter of fact, HERING's theory of the relative blueness or yellowness of vision as a peculiarity that may belong to the normal eye and the "red-green-blind" eye entirely in the same way, necessarily led to the conception, which is certainly very prevalent, that we are concerned here with peculiarities of comparatively secondary importance, that are of no great scientific interest and certainly without any practical value.[1] On the contrary, we must emphasize that this whole theory has proved to be actually incorrect. The differences in the normal eye are not of a similar kind to those between

[1] HERING's vacillating attitude is responsible for this opinion in great measure. In his work on the subject published in 1885 he left it undecided whether "the division of the red-green-blind into red-blind and green-blind, which is considered more accurate by many persons, can be said to be correct, inasmuch as the higher degrees of blueness or yellowness of vision are more frequent than the medium degrees." Thus he did not think it worth the trouble to come to a definite conclusion on this question. Subsequently, so far as the writer is aware, neither HERING himself nor any of his pupils made any observations of this kind on dichromatic eyes or eyes with poor colour vision.

protanopes and deuteranopes. Normal eyes, colour-blind eyes, and eyes with poor colour vision can be checked in a perfectly regular way by the kind of colour matches they make.

Colour matches are a correct criterion for determining the character of the organ of vision. The differentiation between the different forms of the organ of vision, based on these matches, so as to include normal vision, the two kinds of dichromatic vision (colour blindness), and the two kinds of poor (anomalous) colour vision; the possibility of deciding on the mode of vision of each of these persons and distinguishing them exactly—these are the things in the writer's opinion that constitute the most important outcome of recent investigations. These facts by themselves are not sufficient to constitute a basis of a theory of vision. And as long as there is a lack of objective observations as to the action of light, and as to the morphological or chemical subdivision of the organ of vision, etc., every theory of this kind will float in the air, so to speak. Undoubtedly, however, the chief problem of a future theory supported by determinations of this kind will be to explain these functional facts; and its ability to do this will be the touchstone by which it must be tested.[1]

[1] ¶L. Koeppe, Lässt sich das retinale Sehen neu physikalisch erklären? *Münchner Medizinische Wochenschrift.* No. 16, 1921; and Idem, Die Rolle stehender Lichtwellen im optischen Lamellaraufbaue der lebenden Augenmedien. *Deutsche optische Wochenschrift.* No. 12, 1921. (J. P. C. S.)

The Nature of the Colour Sensations

being a further discussion of this subject

by

CHRISTINE LADD-FRANKLIN[1]

Professor CATTELL, in reviewing for *Science*, in 1898, the second edition of HELMHOLTZ's *Physiologische Optik*, said that this work is "one of the few great classics in the history of science." This very just judgment holds still at the present time, although it is now nearly sixty years since the first edition, which had been some ten years in coming out, was finally issued. Whoever looks over this splendid example of acute scientific thinking and brilliant experimenting will be grateful to the Optical Society of America and to the editor in charge of the translation, Professor SOUTHALL, for having decided to bring out even now (what ought to have been done long ago) an English translation of this great work. Some of the facts here recorded will, it is true, have been superseded by later work, but on the other hand much will be found in it which has been, by accident, simply overlooked in later times. The scientist in the subject of physiological optics will therefore be amply repaid if he reads this translation, and not simply secures it for his bookshelves.

I. The Helmholtz Theory

HELMHOLTZ was a great psychologist as well as a great mathematician, a great physicist, and a great physiologist.[2] If his work were to be brought out now for the first time it would undoubtedly be called Psychological Optics instead of Physiological Optics—there is far more of psychology in it than there is of physiology, and the psychology

[1] ¶As stated in the Preface, this chapter is an addition to the original work, for which the editor assumes the sole responsibility. The writer, as everybody knows, is particularly well qualified to discuss this subject. One of the sessions of the International Congress of Psychology in London in 1892 was devoted to new theories of colour sensation. That evening the papers which had been read in the morning were being discussed privately by a group of scientists of whom HELMHOLTZ happened to be one. He had spoken rather disparagingly of one of the contributions, when somebody asked him what he thought of Mrs. FRANKLIN's colour theory. "Ah," said HELMHOLTZ, "Frau FRANKLIN,—*die* versteht die Sache!" (J.P.C.S.)

[2] It happens that his predecessor in the construction of a theory of the colour-sensations, THOMAS YOUNG, was also a man of the first distinction in half a dozen different branches of learning.

is (for the most part) of an extremely acute, as well as of a highly original kind. Organised (non-philosophical) psychology was not definitely in existence when HELMHOLTZ began issuing this book (1856), and he is, very properly, regarded as one of the first investigators in this field. It is, therefore, one of the most inexplicable of psychological occurrences that so great a scientist paid no attention whatever to the fact that, while the necessary *stimuli* for all the colours in the spectrum (and in the world) can be secured by appropriate mixtures of only *three* wave-lengths, the distinct, different, *sensations* that result are not three in number but five—yellow and white are just as good, just as unitary, light-sensations as are red and green and blue.[1] The things to be accounted for, then, in a theory of the visual sensations are, in the order of their phylogenetic development, a primitive achromatic sensation, the dull whites, and *four* chromatic sensations, first yellow and blue (the bees) and then red and green in addition (normal tetrachromatic vision); it may, therefore, be said that the YOUNG-HELMHOLTZ theory is at most three-fifths of a colour theory—it recognizes the existence of three out of five of the actual sensations. But also it takes no account of why the chromata[2] are developed in this peculiar way in pairs: first *yellow* and blue (although yellow does not exist in this theory)—later red and green. (It is also in this order inverted that the colours are lost in the case of diseases of the eye—tobacco amblyopia, progressive atrophy of the optic nerve—and also restored if the disease is recovered from.) Still less has it occurred to the adherents of this theory to pay attention to the extraordinary fact (absolutely unique in the whole range of the sensations) that these very colours constitute "disappearing" colour-pairs[3]—that no human being has ever seen a red-green or a yellow-blue—though he would be very much surprised if he failed to see, on the other two sides of the colour-triangle, the blue-greens and the red-blues, or, in the taste sensations, *all six* of the possible blends, two and two—the bitter-

[1] Black, as HELMHOLTZ recognized perfectly, is a definite sensation, but it is a constant permanent, background sensation which becomes evident and forms a dual blend with any of the colours—not with white only—when they are faint. It is a non-light sensation and it belongs in a totally different category from the light-sensations. Its function I have given a theory for (see *Dictionary of Philosophy and Psychology*, II, Art. *Vision*, p. 767, and *Psychological Review*, November, 1924).

[2] I have urged the introduction of this name, *chroma*, in the sense of *colour proper*, *getönte Farbe*, since 1913, in order to obviate the hopeless ambiguity that results from using "colour" in a double sense—now including and now excluding the achromatic sensations.

[3] The term "disappearing" colour pair is very necessary in order not to prejudice the mind of the reader, at first, as between the HERING conception (antagonistic chemical processes) and mine (that of *constituent* chemical processes). Still less is "cancellation" (TROLAND) a permissible term for this, unless one has definitely adopted the HERING theory.

sweets, the sweet-acids, etc. The theory is therefore in absolute contradiction to the first of the admirable "axioms" enunciated by Professor G. E. MÜLLER—that in correlation with every distinct sensation some distinct physiological (cortical in the last instance) process must be assumed to exist. To suppose (as v. KRIES does) that for no assignable reason a three-fold process in the retina turns into a four-fold process in the cortex (or conversely, as DONDERS does) is to make admission of a fact, but does not provide a theory to account for it. Not everybody is interested in hypotheses (theories); some are content with a plain diet of fact. But it is well-known that successful theories of complicated occurrences in nature are not only intellectually satisfying but also most important as guides to further investigation.

Our ancestors thought that redness resides in the rose. They had, very naturally, no conception of the fact—now a commonplace of science (save for the physicists)—that there is no redness until specific light-frequencies have passed through the alchemy of the retina (and not always then—some persons, though they can see, are totally chroma-blind). That HELMHOLTZ himself should have used the word "colour" in its primitive, objective, sense is singular in the extreme; two successive sections of his book are called *Die einfachen Farben* and *Die zusammengesetzten Farben*, when what he is discussing is respectively homogeneous (or pure) and non-homogeneous (or mixed) light-rays. A blue-green sensation-blend may come from a homogeneous beam of light, and on the other hand a unitary yellow sensation may come from a mixture of "red" and "green" lights. While this use of the word "Farben" is a mere momentary inadvertence on the part of HELMHOLTZ—he puts the matter in the third volume of this work (§26) in perfectly strict scientific terms—that is far from being the case with most of the physicists who write on colour at the present time, *e. g.*, with Sir OLIVER LODGE, BARTON, JOLY, PEDDIE, etc. This is all the more strange when it is remembered that NEWTON (no one, indeed, before him had made this discovery) says plainly: "The rays, to speak properly, are not coloured. In them is nothing else than a certain power or disposition to stir up a sensation of this or that colour . . . . So colours in the object are nothing but a disposition to reflect this or that sort of rays more copiously than the rest."

Moreover, it is to be noted that in the original theory of THOMAS YOUNG it was the *physiological* difficulty of imagining a sufficient number of tuned retinal fibres for all the rays of light, in accordance with the view of NEWTON—"vibrations running along the aqueous pores or crystalline pith of the capillamenta which pave or face the retina"—that led him to substitute a three-part mechanism, with an *overlapping* in various proportions for the intermediate blue-greens, etc.[1] Unquestionably, THOMAS YOUNG would never have called this hypothesis of his (a limited-number-of-constituents-hypothesis) a *trichromatic* theory, nor (what is no improvement) a "triple nerve-excitation theory."

It is for all these purely psychological reasons that the psychologists have never been able to regard the *colour-theory* of HELMHOLTZ as deserving of serious consideration. (The fact that HELMHOLTZ gave no physiological explanation of what v. KRIES[2] has lately furnished a much-needed name for, namely, the "accessory" visual phenomena, is of far less consequence; the recent finding of FRÖHLICH, *e. g.*, that contrast is simply an after-image of scattered light, would fit into any hypothesis regarding the fundamental process of the colour-sensations.) In fact, Professor CATTELL has said of this theory (in the review already quoted from) that "if it were to be proposed at this time [1898] it would not have a single adherent." WILLIAM JAMES said that HELMHOLTZ is, in the science of colour, more eminent for his experimental work than for his theoretical contributions; but it has been pointed out, on the other hand, that the *Physiologische Optik* was the work of his younger days. In any case the physicists ought surely to take notice of the fact that the psychologists, who are experts in questions of *sensation*, find that the HELMHOLTZ theory is wholly inadequate.

II. The Helmholtz-König Facts of Colour Sensation

But however inadequate the HELMHOLTZ theory may be for explaining the characters of the chromatic sensations—however certain it may be that (1) vision is tetrachromatic, (2) that it has undergone a remarkable and a perfectly well made out course of development,

[1] It would seem to be impossible for the physicists to realize that when once light has struck the retina, wave-lengths cease to exist — that their place is taken by *three* initial chemical products and *mixtures* of them. Thus no interest attaches to a pair of complementary *wave-lengths*. When red and blue-green, or green and blue-red, or blue and yellow (that is, red-green) mixed in the required proportions, make white, what we have is the fundamental YOUNG-HELMHOLTZ fact that white can be made out of the three physical constituents, red and blue and green. (Yellow does not need to be put in separately—yellow is a secondary chemical product that forms itself. See my theory, to be given presently.) A simple consideration of similar triangles (in the map of colour sensations in terms of trilinear coördinates) will show that lines drawn through the whiteness-point will meet the spectral line in points such that their severally combined red and green and blue constituents will be in the correct proportion for making white. I have, therefore, to aid this process of thought, proposed the use of the term "transformer mechanism" (plainly adumbrated, as above, by THOMAS YOUNG) to accentuate the immediate change that takes place as soon as light performs its initial (photo-chemical) work upon the photo-aesthetic retina. There is everything in having a name when things are in danger of being overlooked.

[2] *Allgemeine Sinnesphysiologie.*

namely—(a) white, (b) yellow and blue in addition, which however revert to white, and (c) the addition again of red and green, which revert, when mixed, to yellow,—nevertheless the result of the great work carried out in the HELMHOLTZ laboratory by KÖNIG and his assistants is plain matter of fact. It is indeed the most fundamental of all the facts regarding the colour-processes. However, it is not a fact regarding the *sensations* of colour, but only regarding the initial, photo-chemical process which starts up conditions resulting finally in sensation. Colour vision is not *trichromatic*, but it starts up (in the cones) with an initial "tri-receptor" photo-chemical process.

The distribution-curves of the (four) chromatic sensations which represent the theory of HERING are all purely the work of the imagination, and so are the first tentative curves of HELMHOLTZ (still too frequently reproduced in the books). But the situation is very different when it comes to the later curves, which represent what I may call the HELMHOLTZ facts. These curves are drawn in accordance with the results of a vast number of observations in the game of "matching by mixtures"—the demonstration, by the eye, that all the colours of the

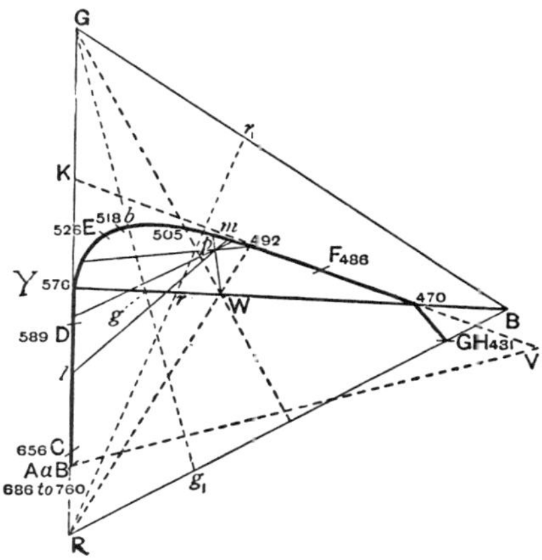

Fig. 81.—The Colour Triangle (KÖNIG).

spectrum can be matched by physical mixtures of red, green and blue lights. When one-half of the field of view of the great HELMHOLTZ instrument for mixing specific light frequencies, the *Farbenmisch-apparat*, is filled with a combination of two different lights and the other half with a homogeneous light, or a different light mixture, or white light, and if the proportions and the character of the several constituents are varied until the two half-fields are indistinguishable, we are said to have before us a colour equation.[1] The results of these measurements are mapped in a *colour-triangle* (what the metallurgists call a triaxial diagram)—a very natural plan for representing (proportionally) by trilinear coördinates functions of three independent variables. The three distribution-curves picture the same facts by means of a different system of representation. But it was only after the incorporation into this work by König of the results of the equations made by the partially chroma-blind that it acquired its present immense significance.

This triangle should always be drawn with the yellow-white-blue line a fundamental (that is, a horizontal) line, as representing the fact that the yellow and blue system of sensations (that of the common form of partial colour blindness, of the normal human mid-periphery, and of the bees) was the first to be developed.

When it came to deciding what wave-lengths to take for the independent variables in this work of matching by mixture, the choice was a difficult one for green (the two ends of the spectrum were naturally chosen, as a first trial, for red and for blue). In the lack of any determining consideration for this choice, a green was taken somewhat at hazard, and tentative curves were determined. (König's names for these curves and for those which later replaced them, "*Elementarempfindungen*" and "*Grundempfindungen,*" are without present significance). Just at this time it happened that it was possible to secure four individuals who were trained observers, and whose vision was dichromatic (two of each type). Would their curves show any coincidence with the curves of the normal eye? It turned out that while the *blue* curves of these defectives coincided with the blue curve of normal vision, the other two curves (both *yellow* in quality, it cannot be too often repeated) were markedly different. But would they perhaps have coincided if some *other* independent variables had been chosen? The question is easily put to the test: it is a simple matter of mathematics (merely a change in the vertices of the triangle of reference) to find out if there are independent variables, that is, unit quantities of lights of particular wave-lengths, such that the entire spectrum as seen by the three classes of individuals (the normal and the two types of defectives) can be built up out of like amounts of two or three of the several constituents—that is, are such that their curves do actually coincide. As a matter of fact König and Dieterici (König, *Gesammelte Abhandlungen*, p. 317) found that it was only necessary

[1] The instrument provides also for the throwing of measured quantities of white light upon either field at pleasure. An improved model has been put on the market quite lately by its makers, Schmidt and Haensch of Berlin. Other equivalent means of securing the same results are now in use in this country.

to make the following substitutions for the colours first chosen to disclose complete coincidence:

$$\mathrm{R} = \frac{R - 0.15G + 0.1V}{0.95}$$

$$\mathrm{G} = \frac{0.25R + G}{1.25}$$

$$\mathrm{B} = V$$

The colours thus fixed upon are (1) a red less yellowish than that of the spectrum, (2) a green of about 505$\mu\mu$, (3) a blue of about 470$\mu\mu$. To repeat constantly that "these stimuli correspond quite closely with three of the fundamental physiological primaries of HERING" (*Colorimetry Report* 1920-21) is to commit a sad error. The *Urfarben* of HERING are complementary colours and therefore cannot be the same as the colours of the KÖNIG curves. KÖNIG and HERING both are perfectly explicit on this point (PFLÜGERS *Arch.*, XLI, 44, 1887 and XLVII, 425, 1890). The first set of KÖNIG curves prove simply that three stimuli *are enough* to reproduce all the colours of the spectrum—they do not show that the actual constituents of normal vision may not be *more* in number. But the extraordinary circumstance that when vision is dichromatic the sets of two constituents (of two very different types) coincide respectively with one or the other pair of the three normal constituents, the blue and the red *or* the blue and the green, is a fact that can only be accounted for by admitting that we have here discovered the actual limited number of constituents of the wilderness of the colour-sensations. (It is this non-occurrence of either the "red" or the "green" *distribution* curve that has made it almost impossible for the physicist to admit the fact that in the case of undeveloped, second-stage, vision it is the more primitive yellow that the defective sees instead of *either* red or green.) In other words, the colour-systems of the two types of chroma-blindness are "reduction systems" (this admirable term is due to v. KRIES). Another way of stating the fact here involved is this: in the colour-triangle all points on lines drawn through the vertices R and G will represent colours which look alike to the defective concerned, and their quality will be that of the whitish yellow (or blue) point in which the line cuts the fundamental Y-W-B line of the triangle. The continued use of the term "confusion colours" shows great ignorance of all these well-established facts of colour-vision. To say that (*Colorimetry Report*, 1920-21, p. 553) "the results cannot be regarded as sufficiently final to justify their adoption in place of a maximally straightforward [!] representation of the facts of colour mixture," and to reproduce KÖNIG's crude, tentative, curves (*loc. cit.*, page 288), is to have missed this point altogether. The curves may be changed by future more exact methods, but the important thing is that they will both (normal and defective) be changed together, so that the *coincidence* will not be lost.

When it was decided to make the first edition of the *Physiologische Optik* instead of the second the basis of the third edition (1909-1910), the editors automatically left out all of this very important work of KÖNIG's in the determination of the distribution-curves. Many of the *theoretical* views of KÖNIG were of such a nature as not to be confirmed by future results—as his belief that the cones have merely a dioptric function and that the photo-chemical process starts (for the chromata) in the epithelium cells. (It is plain that these large cells would have none of the minute space-specificity which is provided for in the cones.) But that is no reason for not recognizing the fact that his *experimental* work is fundamental in the highest degree. His great paper giving the complete account of this work did not appear until after the death of HELMHOLTZ (1896; reproduced in *Abhandlungen*, pp. 214-321), but his final results had been very fully published, and they were expressly incorporated by HELMHOLTZ himself in the *second* edition.[1] To have published an edition of HELMHOLTZ with all this left out was very much like issuing the play of SHAKESPEARE without the part of Hamlet. I therefore give on the opposite page the original diagram of KÖNIG (*loc. cit.*, p. 310), never before reproduced as it happens, except schematically (HELMHOLTZ); which exhibits the definitive coincidence between the normal distribution curves (KÖNIG and DIETERICI) and those of two each of the two common types of dichromatic vision (*yellow* and blue, in both cases, as sensations).

III. The Development Theory of the Colour Sensations[2]

It can never be known beforehand what ones of a highly complicated (and apparently contradictory) collection of facts will be the ones to throw light upon the whole bewildering subject—to suggest a theory which will reconcile the facts in question and fuse them into one all-embracing conception. In 1891-92, when I had the good fortune to have successive semesters in the laboratories of G. E. MÜLLER and KÖNIG, and as a consequence to have the HELMHOLTZ and the HERING points of view both very "warm" in my consciousness, I found the antagonistic states of mind produced by these two absolutely

[1] Up to page 640 this edition is the work of HELMHOLTZ solely (as KÖNIG expressly states in his preface); the account of the work of KÖNIG is given in pages 357–370.

[2] This has been variously called, by its adherents, the genetic theory, the evolution theory, and the development theory.

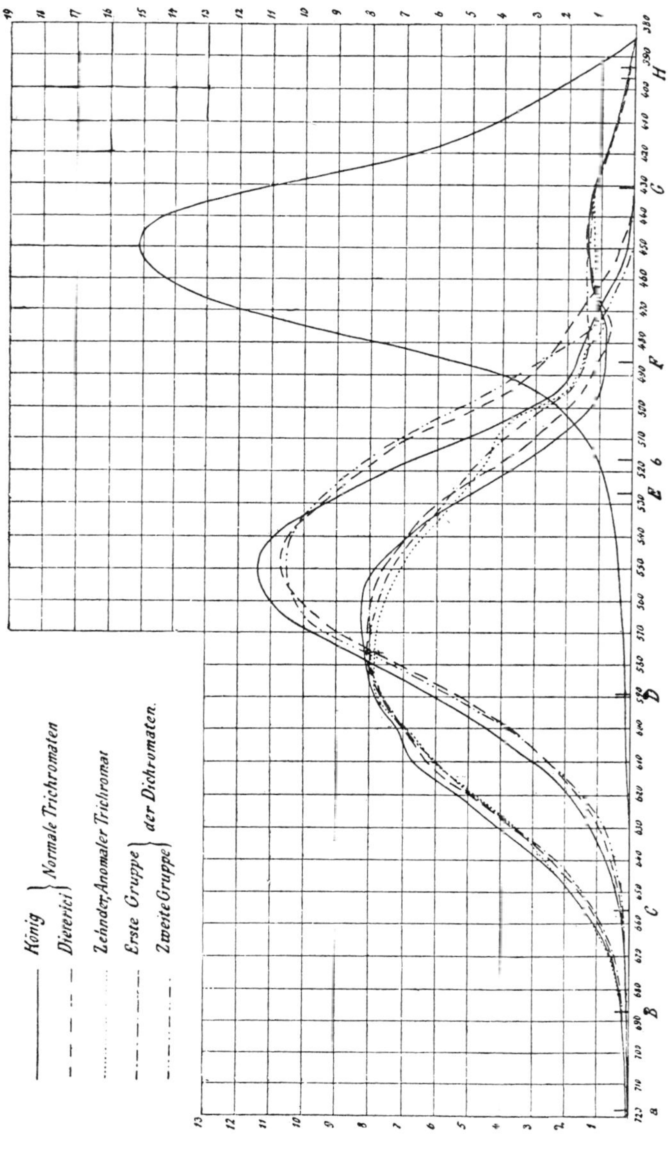

Fig. 82.—The HELMHOLTZ-KÖNIG Distribution Curves (final, corrected, form)

incompatible arrays of facts to be very irksome. It was (as I state in my first paper on the subject, *Mind*, N. S. II, 1893) while I was engaged in writing an article to show that a certain theory by DONDERS was better than that of HELMHOLTZ or of HERING, that it suddenly dawned upon me that a far better theory still was possible. The theory of DONDERS does nothing towards reconciling the views of HELMHOLTZ and of HERING; but (the converse of the position of v. KRIES) it does at least recognize the fact that if vision is tetrachromatic in the retina, it must, to take account of the HELMHOLTZ argument, become in some way "trichromatic" in the cortex. But this admission does not constitute a *theory* of the fact. All theories of colour (with the exception of the one which I have proposed) fall into one or the other of two classes —they accept (and explain) *either* the facts of HERING *or* the facts of HELMHOLTZ. Thus we have, as theories of "trichromatism"[1] and tetrachromatism, respectively: The HELMHOLTZ School—LODGE, JOLY, BARTON, etc.; and the HERING School—DONDERS, G. E., MÜLLER, FRÖHLICH, SCHJELDERUP, etc. Not one of these theorists does more than to shut his eyes to the facts which the rival theories explain. This applies also to the ardent exposition by FRÖBES of the theory of MÜLLER, although he comes nearer than is customary to understanding the adverse facts.

But in the light of the order of development of the colour-sense, the question became less insoluble. The remarkable fact of *the double structure and the double function* of the retina (rods and cones—MAX SCHÜLTZE, PARINAUD) was already sufficiently well established to be made the foundation stone of a new theory of colour. And not only was it known that rod-vision is white vision, and that chromatic vision is cone-vision; it was also plain that the chroma-pairs did not occur both at the same time, that the yellow-blue pair preceded the red-green pair. With this it became possible, by what HEGEL might have called a "higher synthesis," to reconcile the YOUNG-HELMHOLTZ tri-receptor process in a cone with tetrachromatism for sensation (LEONARDO DA VINCI, BRÜCKE, AUBERT, HERING), and to explain at the same time those singular phenomena, the reversion of the red-greens to yellow and of the yellow-blues to white.

The theory thus indicated may be described in the following terms. It is assumed that there is a light-sensitive substance in the rods which gives off, under the influence of light, a reaction-product which is the basis of the primitive sensation of whiteness. In the cones, in

[1] This word is a misnomer, if we take *chroma* in its actual significance. The vision of the totally chroma-blind is not monochromatic but achromatic. The initial three-fold photo-chemical process which actually takes place in a cone should be called a *tri-receptor* process.

the next higher stage of development of the colour-sense (the yellow and blue vision of the bees, and of our own mid-periphery), this same light-sensitive substance has become, by a simple molecular re-arrangement, *more specific* in its response to light, and in such a way that the two ends of the spectrum act separately to produce nerve-excitant substances which, however, when they are produced both at once, unite chemically to form the "white" nerve-excitant out of which they were developed. In the third and final stage the "yellow" nerve-excitant has again undergone a development in the direction of greater specificity, and red and green vision are acquired. These reaction-products are, however, the constituents of the more primitive "yellow" nerve-excitant, and hence when they both occur at once—when red and green light fall together on the retina—they revert to the "yellow" nerve-excitant. If blue light is now added, the white sensation is again produced. Thus "yellow" and "white" are, in tetrachromatic vision, secondary products; at the same time *they are the identical nerve-excitants* which produced the more primitive forms of vision. In other words, (1) a light-sensitive "mother substance" in the rods which, on dissociation by light, gives off a cleavage-product, W, resulting (in the cortex) in the dull white sensations, becomes (2) capable of giving off two subsidiary cleavage-products, Y and B; Y is split off by light of low frequency and B by light of high frequency; Y is the nerve-excitant for the sensation of yellow, B for the sensation of blue. But suppose that, chemically $Y+B \equiv W$; then if Y and B are both split off at the same time in the same cone they (being by hypothesis the chemical constituents of W) immediately unite to form W, and the *sensation* produced will be the primitive white. So in the third stage we shall have, in brief:

$$R + G \equiv Y\,, \qquad Y + B\ (\equiv R + G + B) \equiv W\,.$$

The accessory phenomena of colour are also given a perfectly simple and satisfactory explanation in this theory.

It is not necessary to discuss here the theory of Professor HERING; in addition to all its other difficulties it is absolutely incompatible with the HELMHOLTZ fact of the tri-receptor process and consequently it has, naturally, never appealed to the physicists. There is no occasion for considering (as does PARSONS) at great length all the minor merits and demerits of the HELMHOLTZ and HERING theories. The situation is simply that HERING confutes HELMHOLTZ and HELMHOLTZ confutes HERING.

Before the time of LAVOISIER, when chemistry was not yet in existence, the alchemists might by chance have discovered that on putting hydrogen and chlorine together in a test-tube, under certain conditions both of these substances disappear and hydrochloric acid takes their place. Being not yet chemists, they might have explained this experiment in this way: they might have said, "Hydrogen and chlorine are a naturally antagonistic pair of elements—when they are put together in a test-tube they both vanish, and a hydrochloric acid which was there all the time takes their place." This would be analogous to the HERING explanation of the fundamental event in colour. But since chemistry *does* now exist, it would be a pity not to take advantage of its effectiveness for explaining disappearances and appearances.

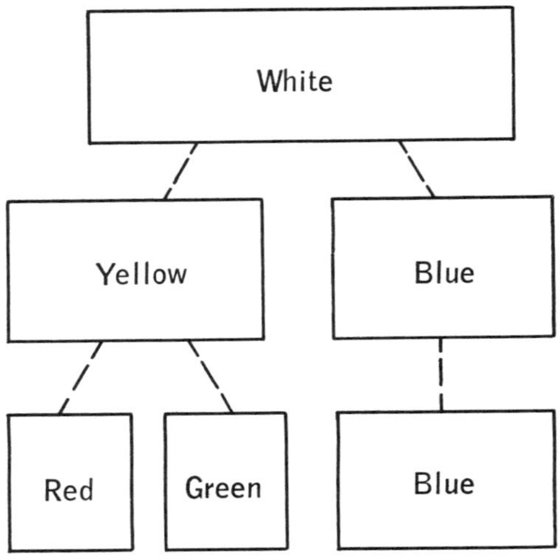

Fig. 83.—Stages 1, 2 and 3 of the actual development of the colour sense

The assumptions which I make (representing the psychological actualities) have been confirmed in a remarkable manner. (1) That the cones are *anatomically* more highly developed rods has now been put beyond question by RAMON Y CAJAL. It is natural therefore to think that the light-sensitive substance which they contain has also undergone development, and that too in the direction of greater specificity. (2) But following upon the discovery of WEIGERT (the photochlorides) that a specific light-sensitive substance need not show colour to the human eye, HECHT has proved that that in the cones actually *is the same substance* as that in the rods, save that it has undergone a "molecular rearrangement"—the very phrase that I am in the habit of using to characterize the change in the colour-molecule made necessary by the *psychological* considerations (the same final whiteness-sensation in the cones as in the rods, though due to a three-part mechanism). (3) Moreover nothing could be simpler, chemically, than this situation. In fact (as Dr. ACREE has pointed out to me) there is a perfect analogy for it in a certain dye-stuff, a rosaniline carboxylate (no longer in practical use because it has been superseded by other less labile dyes). This is a substance such that (under proper conditions of light, heat and moisture), (*a*) hydrogen, chlorine and ethyl alcohol can either one of them be given off separately; but (*b*) when hydrogen and chlorine are given off together, they unite to

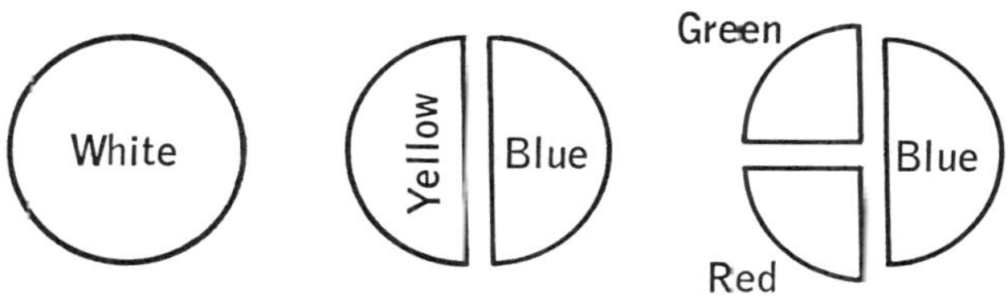

Fig. 84.*—The cleavage products in the three stages of the colour sense. This diagram does not represent the entire light-sensitive molecule, but only the specific cleavage products which, according to the LADD-FRANKLIN theory, constitute the several nerve-excitants for the colour sensations (See WOODWORTH'S *Psychology*). For other diagrams, see *Psychological Review*, 23:247, 1916; *Zeitschrift f. Psychologie*, Bd. 6, etc.

form hydrochloric acid (analogue of the yellows); (*c*) when ethyl alcohol and *either* hydrogen or chlorine are given off, they do not unite—they persist as mixtures (analogue of the blue-greens and the blue-reds); (*d*) when all three of these substances are given off at once they unite to form ethyl chloride (analogue of the three-part leucogenic nerve-excitant in the cones). In other words, the ethyl alcohol set free does not unite with either the hydrogen or the chlorine until after they have first united with each other, exactly as a "blue" constituent in the retina does not chemically unite with either a "red" or a "green" constituent *unless* they have first united with each other to make yellow. Nothing could be more perfectly analogous to what is required for the phenomena of colour-vision.

In conclusion it must be kept in mind that no theory of colour-sensation is deserving of consideration which is not built upon, at once, (1) the fact discovered by THOMAS YOUNG (and magnificently confirmed in the laboratory of HELMHOLTZ)—that three light-stimuli are sufficient, as a physical cause, to start up the retinal photo-chemical processes: (2) the apparently contradictory fact that nevertheless the

*Figure 83 and 84 are reproduced in color facing page 385.

sensations are *five* in number—yellow and white have been somehow added; (3) the very illuminating fact that the *order of development* of the colour sense can be made to account fully for this anomaly and also for (4) the disappearance of the red-greens (and of the yellow-blues) and the appearance in their stead of yellow (and of white). Any proposed theory should be subjected to the test: does it meet all of these "minimal requirements"? (See *Journ. Opt. Soc. Amer.*, etc., 7, pp. 66–68.)

A fuller account of the Development Theory of Colour Sensation will be found in: *Zft. f. Psychologie*, 4, 1892; *Dictionary of Philosophy and Psychology*; *American Cyclopedia of Ophthalmology*, 1913; the *Psychological Review*, XXIII, 1916 and XXIX, 1922; *Mind*, 1892, 1893; and *Science*, XXII, pp. 18, 19. In the last two places I have discussed its fundamental difference from the theory of DONDERS. My theory has been taken over by SCHENCK without due acknowledgment, as has been pointed out for me by v. BRÜCKE (*Zentralbl. f. Physiologie*, 1905). SCHENCK, however, failed completely to see all that my theory accounts for: he felt, himself, no necessity for explaining (he was not a psychologist) what I explain very simply in the light of development: namely, the reversion of what we should see as the red-greens and the yellow-blues to the primitive sensations, yellow and white.

Partial Bibliography 1911-1924 for Volume II

The following list of some articles and books that have appeared since the publication of the third German edition of the *Handbuch der physiologischen Optik* has been compiled by the editor with the idea that, very imperfect and far from complete as it necessarily is, it would still be of much use to the student who is interested in the manifold subjects treated in Volume II. The list does not even include some of the more recent works that are cited in the new footnotes. It should be borne in mind that this manuscript was in the hands of the printer in June 1924 and of course contains nothing later than that date.

1911. W. de W. Abney, Colour-blindness and trichromatic theory of colour vision. II. Incomplete colour-blindness. *Proc. Roy. Soc.*, Ser A. **86**. pp. 43–56.

F. Allen, A new method of measuring the luminosity of the spectrum. *Phil. Mag.*, 604–607.

E. Estanave, Synthesis of complementary colours by ruled gratings. *C. R.*, **153**, pp. 1464–1466.

E. Stigler, A new binocular photometer. *Akad. Wiss Wien*, Ber. 120, 2a. pp. 1069–1073.

C. Zakrzewski, The half-shadow interferometer as photometer *Acad.Sci. Cracovie*, Bull. 8a. pp. 545–547.

1912. W. de W. Abney, Trichromatic theory of colour vision—the measurement of retinal fatigue. *Proc. Roy. Soc.*, Ser. A. **87**, pp. 415–427.

L. Arons, A chromoscope. *Ann. d. Phys.* **39**, 3. pp. 545–568.

W. E. Barrows, *Light, photometry and illumination.*

L. Bell, Factors in heterochromatic photometry. *Electr. World*, **59**, pp. 201–203.

A. Brückner and R. Kirsch, Untersuchungen über die Farbenzeitschwelle. *Zft. f. Sinnesphysiol.* **46**, 229–287.

W. B. Croft, Contrast colours in the use of zone-plates. *Nature*, **89**, p. 581.

E. P. Hyde, Spectro-photometry. *Astrophys. Journ* , **35**, 237–267.

H. E. Ives, Studies in the photometry of light of different colours. *Phil. Mag.*, **24**, pp. 149–188, 352–370, 845–853.

Idem, Primary standard of light. *Astrophys. Journ.*, **36**, pp. 321–329.

A. C. Jolley and A. J. Bull, Measurement of colour. *Phys. Soc., Proc.* **24**, pp. 417–421.

H. Kruess, Spectrophotometer with colour-mixing apparatus *Zft. Instrumentenk.*, **32**, pp. 6–13.

P. G. Nutting, New precision colorimeter. *Washington Acad. Sci., Journ.*, **2**, pp. 183–185.

T. C. Porter, Study of flicker. *Proc. Roy. Soc.*, Ser. A, **86**, pp. 495–513.

H. S. Ryland, Optical experiments. *Nature*, **89**, pp. 554, 555.

J. Thovert, Heterochromatic photometry. *Jour. de Phys.*, **2**, Ser. 5, pp. 34–40.

M. Wertheimer, Experimentellen Studien über das Sehen von Bewegungen. *Zft. f. Psychol.*, **61**, pp. 162-266.

1913. W. de W. Abney, *Researches in colour vision and the trichromatic theory.*
W. W. Coblentz, Radiometer attachment for monochromatic illuminator. *Journ. Franklin Inst.*, **175**, pp. 151, 152.
C. Fabry and H. Buisson, Absorption of ultra-violet light by ozone and limit of solar spectrum. *C. R.*, **156**, pp. 782–785.
R. A. Houstoun, Luminous efficiency throughout the visible spectrum. *Phil. Mag.*, **25**, pp. 715–722.
O. Lummer, Rod vision. *Phys. Zft.*, **14**, pp. 97–102.
J. R. Milne, New types of spectrophotometer on the "flicker" principle. *Proc. Roy. Soc. Edinburgh*, **33**, pp. 257–263.
T. A. Neelin, The sensitiveness of the eye to light and color. *Trans. Roy. Soc. Canada*, **2**, pp. 195–208.
W. E. and R. Pauli, Objective photometry, *Ann. der Phys.*, **41**, 4, pp. 812–828.
J. P. C. Southall, *The Principles and Methods of Geometrical Optics*. Rev. ed. New York.
H. C. Stevens, Flicker photometer and the iris. *Phil. Mag.*, **26**, pp. 180–183.
W. Watson, Luminosity curves of persons having normal and abnormal colour visions. *Proc. Roy. Soc.*, Ser. A, **88**, pp. 404–428.
Idem, On the luminosity curve of a colour-blind observer. *Proc. Roy. Soc.*, Ser. A, **89**, pp. 36–38.
1914. L. Bell, Types of abnormal colour vision. *Proc. Amer. Acad.*, **50**, pp. 3–13.
H. E. Ives and E. J. Brady, Apparatus for the spectroscopic synthesis of colour. *Journ. Franklin Inst.*, **178**, pp. 89-96.
Iidem, New form of flicker photometer *Phys. Rev.*, **4**, pp. 222-227.
H. E. Ives and E. F. Kingsbury, The theory of the flicker photometer. *Phil. Mag.*, **28**, pp. 708–728.
L. A. Jones, Colour analyses of two component mixtures. *Phys. Rev.*, **4**, pp. 454–466.
M. Luckiesh, Growth and decay of colour sensations. *Phys. Rev.*, **4**, pp. 1–11.
A. Mallock, Intermittent vision. *Proc. Roy. Soc.*, Ser. A, pp. 407–410.
P. G. Nutting, Visibility of faintly illuminated objects. *Nature*, **93**, p. 480.
Idem, Visibility of radiation. *Am. Illum. Engin. Soc., Trans.*, **9**, 7, pp. 633–642.
A. H. Pfund, Contrast photometer. *Phys. Rev.*, **4**, pp. 477–481.
S. P. Thompson, Intermittent vision. *Proc. Roy. Soc.*, Ser. A, **90**, pp. 448, 449.
L. T. Troland, Adaptation and the chemical theory of sensory response. *Am. Journ. of Psychol.*, **25**, pp. 500–527.
F. H. Verhoeff and L. Bell, Effect of ultra-violet radiation on the eye. *Science*, **40**, pp. 452–455.
W. Watson, On anomalous trichromatic colour vision. *Proc. Roy. Soc.*, Ser. A, **90**, pp. 443–448.
1915. W. de W. Abney and W. Watson, Threshold of vision for different coloured lights. *Roy. Soc., Phil. Trans.*, **216**, pp. 91–128.
W. E. Burge, Effect of ultra-violet light on the eye. *Electr. World*, **65**, pp. 912–914.
E. P. Hyde and W. E. Forsythe, Visibility of radiation in the red end of the visible spectrum. *Phys. Rev.*, **6**, pp. 68-70.
H. E. Ives, Establishment of photometry on a physical basis. *Journ. Franklin Inst.*, **180**, pp. 409–436.
Idem, The transformation of colour-mixture equations from one system to another. Ibid., **180**, pp. 673–701.
Idem, Precision artificial eye. *Phys. Rev.*, **6**, pp. 334–344.
Idem, Physical photometry. *Am. Illum. Engin. Soc., Trans.*, **10**, pp. 101–125.
Idem, On the choice of a group of observers for heterochromatic measurements. *Am. Illum. Eng. Soc., Trans.*, **10**, pp. 203–209.
H. E. Ives and E. F. Kingsbury, Heterochromatic photometry by Crova's method. *Am. Illum. Eng. Soc., Trans.*, **10**, pp. 716–723.
Iidem, Flicker photometer measurements with a monochromatic green solution. *Phys. Rev.*, **5**, pp. 220–235.

1915. E. F. KINGSBURY, Flicker photometer attachment for the LUMMER-BRODHUN photometer. *Journ. Franklin Inst.*, **180**, pp. 215–223.
P. G. NUTTING, The visibility of radiation. *Phil. Mag.*, **29**, pp. 301–309.
W. OSTWALD, Theory of pigments. I. Fundamental properties of pigments and size of grain. *Kolloid. Zft.*, **16**, 1, pp. 1–4.
J. H. PARSONS, *An introduction to the study of colour vision.*
I. G. PRIEST, New method of heterochromatic photometry. *Phys. Rev.*, **6**, pp. 64–66.
C. V. RAMAN, Intermittent vision. *Phil. Mag.*, **30**, pp. 701, 702.
M. O. SALTMARSH, Brightness of intermittent illumination. *Phil. Mag.*, **29**, 646–650.
1916. P. W. COBB, Photometric considerations pertaining to visual stimuli. *Psychol. Rev.*, **23**, pp. 71–88.
Idem, The effect on foveal vision of bright surroundings. *Journ. Exper. Psychol.*, **1**, pp. 419–425.
E. C. CRITTENDEN and F. K. RICHTMYER, An "average eye" for heterochromatic photometry, and a comparison of a flicker and an equality-of-brightness photometer. *Am. Illum. Eng. Soc., Trans.*, **11**, pp. 331–367.
F. DITTMERS, Über die Abhängigkeit der Unterschiedschwelle für Helligkeiten von der antagonistischen Induktion. *Zft. f. Sinnesphysiol.*, **51**, pp. 214–232.
CHRISTINE LADD-FRANKLIN, On colour theories and chromatic sensations. *Psychol. Rev.*, **23**, pp. 237–249.
R. A. HOUSTOUN, Theory of colour vision. *Proc. Roy. Soc.*, **92**, pp. 424–432.
H. E. IVES, Minimum radiation visually perceptible. *Astrophys. Journ.*, **44**, pp. 124–127.
H. E. IVES and E. F. KINGSBURY, Mechanical equivalent of light. *Phys. Rev.*, **8**, pp. 177–190, 254–258.
Iidem, The theory of the flicker photometer. *Phil. Mag.*, **31**, pp. 290–321.
W. OSTWALD, Theory of colour. *Phys. Zft.*, **17**, pp. 322–332.
Idem, Absolute system of colours. *Zft. phys. Chem.*, **91**, 2, pp. 129–142.
P. G. NUTTING, The retinal sensibilities related to illuminating engineering. *Am. Illum. Eng. Soc., Trans.*, **11**, pp. 1–21; 131–136.
L. T. TROLAND, Flicker frequency as affected by colour. *Journ. Franklin Inst.*, **181**, pp. 853–855.
Idem, Absence of PURKINJE-effect in the fovea of the eye. Ibid., **182**, pp. 111, 112.
Idem, The heterochromatic brightness discrimination threshold. Ibid., **182**, pp. 112–114.
Idem, Notes on flicker photometry: flicker photometer frequency as a function of light intensity. Ibid., **182**, pp. 261–262.
F. H. VERHOEFF, L. BELL and C. B. WALKER, The pathological effects of radiant energy on the eye. *Proc. Amer. Acad.*, **51**, pp. 630–793.
1917. W. DE W. ABNEY, The fourth colourless sensation in the three-sensation spectrum, measured on the centre of the retina. *Proc. Roy. Soc.*, **94**, pp. 1–13.
F. BEST, Untersuchungen über die Dunkelanpassung des Auges mit Leuchtfarben. *Zft. f. Biol.*, **68**, pp. 111–146.
H. BUISSON, Minimum radiation visually perceptible. *Astrophys. Journ.*, **45**, pp. 296, 297.
Idem, The minimum energy necessary to produce the sensation of light. *Journ. de phys.*, **7**, pp. 68–74.
W. W. COBLENTZ and W. B. EMERSON, Sensibility of the average eye to light of different colours. *Bureau of Stand., Bull.* **14**, pp. 167–236.
Iidem, Mechanical equivalent of light. Ibid., pp. 255–265.
J. GUILD, Mechanism of colour vision. *Proc. Roy. Soc.*, **29**, pp. 354–361.
G. F. GÖTHLIN, Studien über die Energieschwelle für die Empfindung Rot in ihrer Abhängigkeit von der Wellenlänge der Lichtstrahlung. *Kungl. Vetenskapsakadem. Handlingar*, **58**, No. 1, 89 pages.

1917. C. HESS, Der Farbensinn der Vögel und die Lehre von den Schmuckfarben. PFLÜGERS *Arch.*, **166**, pp. 381–426.

A. V. HÜBL, Colour apparatus. *Phys. Zft.*, **18**, pp. 270–275.

H. E. IVES, Visual diffusivity. *Phil. Mag.*, **33**, pp. 18–33.

Idem, Polarization flicker photometer. Ibid., pp. 360–380.

Idem, Hue difference and flicker-photometer speed. Ibid., **34**, pp. 99–112.

Idem, Explanation of persistence of vision as a physical conduction phenomenon. *Journ. Franklin Inst.*, **183**, pp. 779, 780.

Idem, Units of radiation and illumination. *Astrophys. Journ.*, **45**, pp. 39-49.

L. A. JONES, The fundamental scale of pure hue and retinal sensibility to hue differences. *Journ. Opt. Soc. Amer.*, **1**, pp. 63–77.

R. I. LLOYD, Hemianopsia. *Ophthalmic Record*, Oct. 1917.

M. LUCKIESH, Equality point in spectral energy and luminosity distribution curves. *Journ. Franklin Inst.*, **183**, pp. 633, 634.

Idem, The physical basis of color technology. *Journ. Franklin Inst.*, **183**, pp. 73–93 and **184**, pp. 227–250.

H. E. MERVIN, Optical properties and theory of colour of pigments and paints. *Proc. Amer. Soc. for Testing Materials*, **17**, pp. 1–34.

P. REEVES, Effect of light on pupillary area and retina of the eye. *Eastman Kodak Co.'s Research Lab.*, Comm. No. 52.—*Journ. Franklin Inst.*, **184**, pp. 717–719.

Idem, The minimum radiation visually perceptible. *Astrophys. Journ.*, **46**, pp. 167–174.

H. N. RUSSELL, Minimum radiation visually perceptible. Ibid., **45**, pp. 60–64.

L. T. TROLAND, On the measurement of visual stimulation intensities. *Journ. Exp. Psychol.*, **2**, pp. 1–33.

1918. J. BLANCHARD, The brightness sensibility of the retina. *Phys. Rev.*, **11**, pp. 81–99.

F. EXNER, Some experiments and remarks on colour theory. *Akad. Wiss. Wien*, Ber. **127**. 2a. pp. 1829–1864.

L. W. HARTMAN, Visibility of radiation in the blue end of the visible spectrum. *Astrophys. Journ.* **47**, pp. 83–95.

G. HOLST and J. S. DE VISSER, Brightness of the black body and the mechanical equivalent of light. *K. Akad. Amsterdam Proc.*, **20**, 7, pp. 1036–1042.

R. A. HOUSTOUN, A statistical survey of colour vision. *Proc. Roy. Soc.*, **94A**, pp. 576–586.

E. P. HYDE, W. E. FORSYTHE and F. E. CADY, Visibility of radiation. *Journ Franklin Inst.*, **185**, pp. 829–831.

H. E. IVES, Resolution of mixed colours by differential visual diffusivity. *Phil. Mag.*, **35**, pp. 413–421.

L. LUMIERE, Persistence of luminous impressions on the retina. *C. R.*, **166**, pp. 654–656.

P. METZNER, Adaptation of the eye for low illuminations. *Deutsch. phys. Gesell.*, Verh. **20**, pp. 183–186.

I. G. PRIEST, The law of symmetry of the visibility function. *Phys. Rev.*, **11**, pp. 498–502.

Idem, A precision method for producing artificial daylight. Ibid., 502–504.

P REEVES, The rate of pupillary dilation and contraction. *Physiol. Rev.*, **25**, pp. 330–340.

Idem, Visibility of radiation. *Phil. Mag.*, **35**, pp. 174–181.

J. P. C. SOUTHALL, *Mirrors, Prisms and Lenses*, New York.

1919. F. ALLEN, On the discovery of four transition points in the spectrum and the primary colour sensations. *Phil. Mag.*, **38**, pp. 55–81.

Idem, The persistence of vision for colours of varying intensity. Ibid., pp. 81–89.

E. H. BARTON and H. M. BROWNING, Syntonic hypothesis of colour vision with mechanical illustrations. *Phil. Mag.*, **38**, pp. 338–348.

1919. A. Blondel, A physical method of heterochromatic photometry. *C. R.*, **169**, pp. 830–835.

P. W. Cobb, Dark-adaptation, with especial reference to the problems of night-flying. *Psychol. Rev.*, **26** pp. 428–453.

M. Dresbach, J. E. Sutton, Jr., and S. R. Burbage, Dark adaptation of the peripheral retina. *Proc. Amer. Physiol. Soc.;Amer. J. Physiol.*, **51**, p. 188.

C. E. Ferree and G. Rand, Effect of colour of light on acuteness of vision. *Am. Illum. Eng. Soc., Trans.*, **14**, pp. 107–116 (discussion, pp. 117–132).

Iidem, Chromatic thresholds of sensation from center to periphery of the retina and their bearing on color theory, *Psychol. Rev.*, **26**, pp. 16–41 and 150–163.

J. W. French, The unaided eye. *Trans. Opt. Soc.*, **20**, pp. 209–232 and **21**, pp. 1–33 and 127–156.

R. A. Houstoun, Theory of colour vision. *Phil. Mag.* **38**, pp. 402–417.

E. P. Hyde, W. E. Forsythe and F. E. Cady, Brightness of a black body and mechanical equivalent of light. *Phys. Rev.*, **13**, pp. 45–58.

H. E. Ives, Photometric scale. *Journ. Franklin Inst.*, **188**, pp. 217–235.

M. Luckiesh, Infra-red radiant energy and the eye. *Amer. J. Physiol.*, **50**, pp. 383–398.

O. Meissner, Colorimetric measurements according to Ostwald's scale. *Phys. Zft.*, **20**, pp. 83–85.

Idem, Ostwald's colorimetry. Ibid., pp. 344–346.

S. G. Rich, The Purkinje phenomenon in subtropical moonlight. *Psychol. Bull.*, **16**, pp. 388, 389.

F. Woog, Variation in persistence of vision in different regions of the retina. *C. R.*, **168**, pp. 1222–1224.

1920. M. Becke, The nature of colour perception. *Färber-Ztg.*, **31**, pp. 50–52 and 61–64.

U. Ebbecke, Über das Augenblicksehen. Über das Sehen in Flimmerlicht. Pflügers *Arch.*, pp. 181–195, 196–223.

F. W. Edridge-Green, *The physiology of vision, with special reference to colour blindness.*

Idem, Colour vision and colour blindness; new spectrometer. *Journ. Soc. Arts*, **69**, pp. 40–47.

F. Exner, v. Helmholtz's three-colour scheme. *Akad. Wiss. Wien*, Ber. **129**, 2a, pp. 27–46.

C. E. Ferree and G. Rand, The extent and shape of the zones of color sensitivity in relation to the intensity of the stimulus light. *Am. J. Physiolog. Optics*, **1**, pp. 185–213.

C. E. Ferree and G. Rand, An illuminated perimeter with campimeter features. *Trans. Amer. Ophthalm. Soc.* **18**, pp. 164–173.

J. Fison, Position of macula lutea and optic disc. *J. Anat.*, **54**, pp. 184–187.

H. Hartridge, Colourimeter design. *Cambr. Phil. Soc. Proc*, **19**, pp. 271–282.

Idem, Colour blindness and Young's Hypothesis. *Brit. J. Ophth.*, **4**, pp. 318–322

S. Hecht, The photochemistry of the visual purple. *Journ. of Gen. Physiol.*, **3**. pp. 1–13.

S. Hecht, The dark adaptation of the human eye. *J. Gen. Physiol.*, **2**, pp. 499–517.

H. Henning, Optische Versuche an Vögeln und Schildkroten übe die Bedeutung der roten Ölkugeln im Auge. Plfügers *Arch.*, **178**, pp. 91–123.

E. Hering, *Grundzüge der Lehre vom Lichtsinn.*

C. v. Hess, Die angeborenen Farbensinnstörungen und das Farbengesichtfeld. *Arch. f. Augenheilk.*, **86**, pp. 317–335.

Idem, Die Rotgrünblindheiten. *Arch. f. d. ges. Physiol.*, **185**, pp. 147–164.

Idem, Untersuchungen zur Lehre von der Wechselwirkung der Sehfeldstellen. Ibid., **179**, pp. 50–72.

Idem, Die Grenzen der Sichtbarkeit des Spektrums in der Tierreihe. *Naturwissenschaften*, **8**, pp. 197–200.

1920. E. P. HYDE and W. E. FORSYTHE, Colour match and spectral distribution. *Journ. Franklin Inst.*, **189**, pp. 663, 664.

E. R. JAENSCH, Parallelgesetz über der Verhalten der Reizschwellen bei Kontrast und Transformation. *Zft. f. Psychol.*, **85**, pp. 342–352.

L. A. JONES, Subtractive colorimeter. *Journ. Opt. Soc. Amer.*, **4**, pp. 420–431.

Idem, An instrument for the measurement of the visibility of objects. *Phil. Mag.*, **39**, pp. 96–134.

A. JUHÁSZ, Über die komplementär gefärbten Nachbilder. *Zft. f. Psychol. u. Physiol. d. Sinnesorgane II. Zft. f. Sinnesphysiol.*, **51**, pp. 233–263.

K. W. F. KOHLRAUSCH, Contributions to the theory of colour. *Phys. Zft.*, **21**, pp. 396–403.

Idem, Brightness of pigment colours. Ibid., pp. 423–426.

Idem, Remarks on OSTWALD's theory of pigment colours. Ibid., pp. 473–477.

J. KUNZ, Threshold phenomena in radiation. *Phys. Rev.*, **15**, pp. 129, 130.

R. I. LLOYD, The stereoscopic campimeter slate. *N. Y. Med. Journ.*, Dec. 1920.

P. G. NUTTING, Sensitiveness of the eye to light and colour. *Journ. Opt. Soc. Amer.*, **4**, pp. 55–79.

W. OSTWALD, The foundations of colorimetry. *Zft. tech. Phys.*, **1**, 9, pp. 173–175 and **1**, 12, pp. 261–271.

A. H. PFUND, Colorimetry of nearly-white surfaces. *Journ. Franklin Inst.*, **189**, pp. 371–379.

I. G. PRIEST, Note on the relation between the frequencies of complementary hues. *Journ. Opt. Soc. Amer.*, **4**, pp. 402–404.

Idem, Preliminary note on the relations between the quality of color and the spectral distribution of light in the stimulus. Ibid., **4**, pp. 388–401.

P. REEVES, The response of the average pupil to various intensities of light. *Journ. Opt. Soc. Amer.*, **4**, pp. 35–43.

F. K. RICHTMYER and E. C. CRITTENDEN, The precision of photometric measurements. *Journ. Opt. Soc. Amer.*, **4**, pp. 371–387.

F. E. ROSS, Photographic photometry and the PURKINJE effect. *Astrophys. Journ.*, **52**, pp. 86–97.

E. SCHRÖDINGER, Colorimetrics. *Zft. f. Physik*, **1**, 5, pp. 459–466.

M. SO, The visibility of radiation throughout the spectrum. *Phys. Math. Soc. Japan, Proc.*, **2**, pp. 177–184.

J. P. C. SOUTHALL, Refraction and visual acuity of the human eye. *Am. J. Physiol. Optics.* **1**, pp. 277-316.

L. T. TROLAND, The enigma of color vision. *Amer. Journ. Physiol. Optics*, **1**, pp. 317–337.

Idem, The nature of the visual receptor process. *Journ. Opt. Soc. Amer.*, **4**, pp. 3–15

Idem, The "all or none" law in visual response. Ibid, **4**, pp. 160–185.

R. L. WATERFIELD, Sensitiveness of different parts of the retina. *Journ. Brit. Astron. Assoc.* also *Journ. Roy. Astronom. Soc. Canada*, **15**, p. 84.

1921. A. AMES, JR., Systems of color standards. *Journ. Opt. Soc. Amer.*, **5**, pp. 160–170.

F. W. EDRIDGE-GREEN, The effect of red fatigue on the white equation. *Proc. Roy. Soc.*, **92B**, pp. 232–234.

W. EINTHOVEN, Über die Beobachtung und Abbildung dünner Faden. PFLÜGERS *Archiv*, **191**, pp. 60–98.

F. EXNER, Determination of brightness in the red-blind colour system. *Akad. Wiss. Wien, Ber.*, **130**, 2a, No. 7–8, pp. 355–361.

H. S. GRADLE, The blind spot. III. The relation of the blind spot to medullated nerve fibres in the retina. *J. Amer. Med. Assoc.*, **77**, pp. 1483–1486.

H. HARTRIDGE, Apparatus for projecting spectra. *Cambr. Phil. Soc. Proc.*, **20**, p. 480.

S. HECHT, Foveal dark adaptation. *J. Gen. Physiol.*, **4**, pp. 113-139.

H. HONIGMANN, Über Lichtempfindlichkeit und adaptierung des Vogelauges. PFLÜGERS *Arch.*, **189**, pp. 1–72.

1921. S. Hecht, Photo-chemistry of visual purple. II. *J. Gen. Physiol.* **3**, pp. 285–290.
Idem, Time and intensity in photosensory stimulation. *J. Gen. Physiol.*, **3**, 367–390.
F. L. Hopwood, An auto-stroboscope and an incandescent colour top. *Trans. Opt. Soc.*, **23,** 2, pp. 93–98.
K. Horovitz, Theory of visual space. *Akad. Wiss. Wien., Ber.* **130**, 2a, No. 7–8, pp. 405–420.
R. A. Houstoun, A new method of investigating colour blindness with a description of 23 cases. *Proc. Roy. Soc. Edinburgh,* **42,** pp. 75–88.
R. A. Houstoun and M. A. Dunlop, A statistical survey of colour vision. *Phil. Mag.*, **41,** pp. 186–200.
H. E. Ives, A proposed standard method of colorimetry. *Journ. Opt. Soc. Amer.*, **5,** pp. 469–478.
J. Joly, A quantum theory of vision. *Phil. Mag.*, **41,** pp. 289–304.
H. Koellner, Über theorien des Farbensinnes. *Münch. med. Woch.*, **68,** pp. 1045–1047.
L. Koeppe, *Die ultra- und polarisationsmikroskopische Erforschung des lebenden Auges.* Published by E. Bircher, Bern, Switzerland.
R. I. Lloyd, The stereocampimeter and its uses. *Scientific and Technical Publications of Bausch & Lomb Optical Co.*, Rochester, N. Y., No. 12.
Idem, The stereo-campimeter and its use; presentation of a new instrument for the study of central field defects. *The Journal of Ophthalm., Otology and Laryngology,* Oct., 1921.
T. R. Merton, Spectrophotometry in the visible and ultra-violet spectrum. *Proc. Roy. Soc.*, **99,** pp. 78–84.
G. Michailesco, Sur les variations de la sensibilité autour de la tache de Mariotte. *C. R.*, **173,** pp. 604–606.
P. L. du Noüy, Energy and vision. *J. Gen. Physiol.*, **3,** pp. 743–764.
J. H. J. Poole, The photoelectric theory of vision. *Phil. Mag.*, **41,** pp. 347–357.
Idem, Minimum time necessary to excite the human retina. Ibid., **43,** pp. 345–348.
I. G. Priest, The spectral distribution of energy required to evoke the gray sensation. *Bureau of Stand. Sci. Papers,* No. 417; *Journ. Opt. Soc. Amer.*, **5,** pp. 205–209.
P. Reeves, Reaction of the eye to light; energy required to produce visual sensation. *Trans. Opt. Soc.*, **22,** pp. 1–14.
F. Schanz, Das Sehen. *Munch. med. Woch.*, **68,** pp. 1390–1392.
L. T. Troland, The colors produced by equilibrium photopic adaptation. *Journ. Exp. Psychol.*, **4,** pp. 344–390.
F. Weigert, Ein photochemisches Modell der Retina. Pflügers *Arch.*, **190,** pp. 177–197.
R. S. Woodworth, *Psychology, A study of mental life.*
1922. F. Aigner, Resonance theory of colour vision. *Akad. Wiss. Wien., Ber.* **131,** 2a. No. 4–5, pp. 299–320.
E. H. Barton, Colour vision and syntony. *Nature,* **110,** pp. 357–359.
A. Contino, Measurement of visual acuity. *Riv. Ottica e Mecan.*, **2,** pp. 40–59.
A. Duschek-Frankfurt, Perception of brightness in relation to colour. *Akad. Wiss. Wien.*, Ber. **131**. 2a. No. 3. pp. 171–197.
F. W. Edridge-Green, Colour-vision theories and colour-blindness. *Phil. Mag.*, **44,** pp. 916–920.
F. W. Ellis, The study of visual processes by means of momentary retinal shadows. *Journ. Opt. Soc. Amer.*, etc., **6,** pp. 922–931.
F. Exner, Colour vision. *Akad. Wiss. Wien., Ber.* **131,** 2a. No. 10, pp. 615–641.
C. Fabry and H. Buisson, A universal photometer without diffusing screen. *Rev. d'optique,* **1,** pp. 1–12.
C. E. Ferree and G. Rand, A new laboratory and clinic perimeter. *J. Experim. Psychol.*, **5,** pp. 46-67.

1922. C. E. FERREE and G. RAND, An illuminated perimeter with campimeter features. *Amer. Jour. Ophthal.*, **5**, pp. 455-465.

C. E. FERREE and G. RAND, Perimetry: Variable factors influencing the breadth of color fields. *Am. J. Ophthalm.*, **5**, pp. 886-895.

C. E. FERREE and G. RAND, Some contributions to the science and practice of ophthalmology. *Trans. Internat. Congress Ophthalmol.*, Washington, D. C.

W. E. FORSYTHE, Accuracy in color matching. *Journ. Opt. Soc. Amer.*, etc., **6**, pp. 476–482.

W. P. GRAHAM, The absorption of the eye for ultra-violet radiation. *Journ. Opt. Soc. Amer.*, etc. **6**, pp 605–614.

C. v. HESS, Farbenlehre. *Ergebn. d. Physiol.*, **20**, pp. 1–107.

R. A. HOUSTOUN, An investigation of the colour vision of 527 students by the RAYLEIGH test. *Proc. Roy. Soc.*, **102**, pp. 353–360.

K. HOROVITZ, Heteromorphia due to variation of effective aperture and visual acuity. *Journ. Opt. Soc. Amer.*, etc., **6**, pp. 597–604.

E. P. HYDE and F. E. CADY, New principle in contrast photometry. *Journ. Opt. Soc. Amer.*, etc., **6**, pp. 615–619.

H. E. IVES, A theory of intermittent vision. *Journ. Opt. Soc. Amer.*, etc. **6**, pp. 343–361.

L. KOEPPE, *Die Bedeutung der Gitterstruktur in den lebenden Augenmedien für die Theorie der subjektiven Farbenerscheinungen.* (E. BIRCHER, Bern, Switzerland.)

Idem, Über das Problem der Krümmungen der elastischen Membranen und Lamellen in den lebenden Augenmedien. *Deutsche optische Wochenschrift,* Nos. 36, 37.

Idem, *Die Mikroskopie des lebenden Auges in natürlichen Lichte.* Vol. I (1920) and Vol. II (1922); published by JULIUS SPRINGER, Berlin.

CHRISTINE LADD-FRANKLIN, Tetrachromatic vision and the development theory of color. *Science,* **55**, No. 1430.

Eadem, Practical logic and color theories. *Psychol. Rev.*, **29**, pp. 180–200.

M. LUCKIESH, *Visual illusions, their causes, characteristics and applications.*

W. OSTWALD, What advances has the new colour theory made? *Zft. Elektrochem.* **28**, pp. 398–404.

W. PEDDIE, *Colour vision.*

F. SCHANZ, Theory of vision. *Zft. f. Physik,* **12**, 1 and 2, pp. 28–37.

H. SCHULZ, *Das Sehen.*

M. C. SHIELDS, The visibility function and visibility thresholds for color-defectives. *Journ. Opt. Soc. Amer.*, etc., **6**, pp. 362–368.

L. T. TROLAND, Psychophysics as the key to the mysteries of physics and metaphysics. *Journ. Washington Acad. Sci.*, **12**, pp. 141–162.

Idem, The facts and theories of color vision. *Trans. Internat. Congress of Ophthalm.*

Idem, The present status of visual science. Bulletin No. 27, *National Research Council.*

Idem, Brilliance and chroma in relation to zone theories of vision. *Journ. Opt. Soc. Amer.*, etc., **6**, pp. 3–26.

Idem, Report of the nomenclature and stadards sub-committee on colorimetry for 1920–1921. *Journ. Opt. Soc. Amer.*, etc., **6**, pp. 527–604.

F. WEIGERT, The physical chemistry of colour vision. *Zft. Phys. Chem.*, **100**, pp. 537–565.

1923. F. ALLEN, On reflex visual sensations. *Journ. Opt. Soc. Amer.* etc., **7**, pp. 583–626.

Idem, On reflex visual sensations and color contrast. Ibid., **7**, pp. 913–942.

T. Y. BAKER, Limits of visual acuity. *Phil. Mag.*, **46**, pp. 640–642.

R. T. BEATTY, Monochromator for the ultra-violet, visible, and near infra-red. *Journ. Sci. Instr.*, **1**, pp. 33–42.

L. BLOCH, Colour measurement. *Zft. tech. Phys.*, **4**, 4, pp. 175–182.

E. L. CHAFFEE, W. T. BOVIE and ALICE HAMPSON, The electrical response of the retina under stimulation by light. *Journ. Opt. Soc. Amer.*, etc., **7**, pp. 1–43.

1923. P. W. Cobb, The relation between field brightness and speed of retinal impression. *Journ. Franklin Inst.*, **195**, pp. 855–857.

C. C. Colby, Jr. and C. M. Doolittle, A distribution photometer of new design. *Trans. Amer. Illum. Eng. Soc.*, **18**, pp. 273–289.

E. C. Crittenden, The measurement of light. *Journ. Washington Acad. Sci.*, **13**, pp. 69–90.

H. Dember, An entoptic phenomenon. *Zft. tech. Physik*, **4**, 1, p. 7.

R. Dietzius, The ultra-violet end of the solar spectrum. *Meteorlog. Zft.*, **40**, pp. 297–301.

G. M. B. Dobson, A flicker type of photoelectric photometer giving high precision. *Proc. Roy. Soc.*, **104**, pp. 302–315.

F. W. Edridge-Green, Some curious phenomena of vision. *Journ. Roy. Soc. Arts*, **71**, pp. 469–478.

O. Emersleben, Theory of vision. *Zft. f. Physik*, **15**, 2 and 3, pp. 180–183.

C. E. Ferree and G. Rand, Theory of flicker photometry. *Amer. Illum. Eng. Soc., Trans.*, **18**, pp. 151–199.

W. E. Forsythe, Color match and spectral distribution. *Journ. Opt. Soc. Amer.*, etc., **7**, pp. 1115–1122.

K. S. Gibson and E. P. T. Tyndall, Visibility of radiant energy. *Bureau Stand., Sci. Papers*, No. 475, pp. 131–191.

E. Goldberg, The grey wedge and its use in sensitometry. Faraday *Soc., Trans.* **19**. pp. 349–354.

H. Hartridge, Physiological limits to accuracy of visual observation and measurement. *Phil. Mag.*, **46**, pp. 49–79.

F. L. Hopwood, Time lag in vision. *Proc. Phys. Soc.*, **35**, p. 214.

R. A. Houstoun, *Light and colour.*

R. A. Houstoun and E. Dow, The evolution of the colours of the spectrum in terms of the three primary colours. *Phil. Mag.*, **45**, 2, pp. 169–176.

R. A. Houstoun and W. M. Heddle, Visibility of red light. *Phil. Mag.*, **46**, pp. 699–706.

R. A. Houstoun and W. H. Manson, A new method of investigating colour-blindness. *Roy. Soc. Edinb.*, Proc. **43**. 2. pp. 216–218. (1922–1923).

H. E. Ives, A colour-match photometer. *Journ. Opt. Soc. Amer.*, etc., **7**, pp. 243–261.

Idem, Simultaneous matches by trichromatic and monochromatic analysis. *Journ. Opt. Soc. Amer.*, etc., **7**, pp. 287–298.

Idem, A chart of the flicker photometer. Ibid., **7**, pp. 363–373.

Idem, Transformation of colour-mixture equations from one system to another. *Journ. Franklin Inst.*, **195**, pp. 23–44.

Idem, A variable aperture rotating sector-disc. *Journ. Opt. Soc. Amer.*, etc. **7**, pp. 683–687.

E. Karrer, A photometric sectored disc. *Journ. Opt. Soc. Amer.*, etc. **7**, pp. 893–899.

M. Luckiesh, Demonstrating color-mixture. *Journ. Opt. Soc. Amer.*, etc., **7**, pp. 657–660.

Martin and Gamble, *Color and methods of color reproduction.* New York, 1923.

L. C. Martin, The photometric matching field. *Proc. Roy. Soc.*, **104**, pp. 302–315.

T. Otashiro, The visibility curve. *Proc. Phys Math. Soc. Japan*, **5**, pp. 57–59.

W. Peddie, Colour vision. *Phil. Mag.*, **45**, pp. 1058–1062.

I. G. Priest, The colorimetry and photometry of daylight and incandescent illuminants by the method of rotary dispersion. *Journ. Opt. Soc. Amer.*, etc., **7**, pp. 1175–1209.

Idem, Rotatory dispersion method of colorimetry and heterochromatic photometry. *Am. Illum. Eng. Soc., Trans.*, **18**, pp. 861–864.

H. M. Randall, Infra-red spectra. *Am. Phil Soc., Proc.*, **62**. 6 pp. 326–340.

R. H. Sinden, Studies based on the spectral complementaries. *Journ. Opt. Soc. Amer.*, etc., **7**, pp. 1123–1153.

1923. J. P. C. SOUTHALL, *Mirrors, Prisms and Lenses*. Rev. ed. New York.

W. WENIGER, Summary of investigations in the infra-red spectrum of long wave-lengths. *Journ. Opt. Soc. Amer.*, etc. **7**, pp. 517–527.

1924. F. ALLEN, On the reflex origin of the self light of the retina. *Journ. Opt. Soc. Amer.*, etc., **8**, pp. 275–286.

Idem, A new tri-colour mixing spectrometer. *Journ. Opt. Soc. Amer.*, etc., **8**, pp. 339–341.

Idem, On reflex visual sensations—I. Experimental. *Amer. Journ. Physiol. Optics*, **5**, pp. 341–375.

Idem, On reflex visual sensations. II. Discussion of results. *Am. J. Physiolog. Optics*, **5**, pp. 420-437.

Idem, The reflex origin of color contrast. *J. Opt. Soc. Am.*, etc., **9**, pp. 375-388.

W. A. ANDERSON, Reflex visual sensations and anomalous trichromatism. *Journ. Opt. Soc. Amer.*, etc., **8**, pp. 731–765.

A. BLONDEL and J. REY, New verification of the law of perception of brief illuminations at the limit of their range.. *C. R.*, **178**, pp. 276–280.

A. BLONDEL and J. REY. The law of the limiting perception of luminous flashing signals. *Comptes Rendus*, 178. pp. 1245-1247.

E. L. CHAFFEE and ALICE HAMPSON, Effects of varying the wave-length of the stimulating light upon the electrical response of the retina. *Jour. Opt. Soc. Amer.*, etc., **9**, pp. 1–25.

J. H. CLARK, The Ishihara test for color blindness. *Amer. Journ. Physiolog. Optics*, **5**, pp. 269–276.

P. W. COBB, The speed of vision. *Am. Illum. Eng. Soc., Trans.* **19**, pp. 150–163; discussion, pp. 163–175.

H. G. DORSEY, A visual method of measuring short time intervals. *Journ. Opt. Soc. Amer.*, etc., **9**, pp. 351, 352.

H. ERFLE and H. BOEGEHOLD, *Grundzüge der Theorie der optischen Instrumente nach* ABBE, *von* S. CZAPSKI *und* O. EPPENSTEIN. Dritte Auflage bearbeitet von den wissenschaftlichen Mitarbeitern der ZEISSischen Werkstätte. Leipzig.

C. E. FERREE and G. RAND, Further studies on the extent and shape of the color fields in relation to the intensity of the stimulus light. *Am. J. Physiolog. Optics*, **5**, pp. 409-419.

C. E. FERREE and G. RAND. Effect of composition of light on important ocular functions. *Am. Illum. Eng. Soc., Trans.* **19**, pp. 424-447.

E. F. FINCHAM, A new form of Corneal Microscope with Combined Slit Lamp Illuminating Device. *Opt. Soc., Trans.*, **25**. 3. pp. 113–118; Disc., 118–122. (1923–1924.)

H. GEORGE and E. BAYLE, Spectrophotometric Definition of Colours of Fluorescence. *Comptes Rendus*, **178**, pp. 1895–1897.

K. S. GIBSON, Special characteristics of test solutions used in heterochromatic photometry. *Journ. Opt. Soc. Amer.*, etc., **9**, pp. 113–121.

J. GUILD, New flicker photometer for heterochromatic photometry. *Journ. Sci. Instruments*, **1**, pp. 182–189.

A. GULLSTRAND, Optische Systemgesetze zweiter und dritter Ordnung. (175 pages, 4to.) *Kungl. Svenska Vetenskapsakademiens Handligar*. Bd. 63. No. 13 (Supplement).

S. HECHT, Visibility of the spectrum. *Journ. Opt. Soc. Amer.*, etc., **9**, pp. 211–222.

OTTO HENKER, *Introduction to the theory of spectacles*. English translation by R. KANTHACK. Jena School of Optics.

A. HOLLENBERG, Visual sensory reflexes and color blindness. *J. Opt. Soc. Am.*, etc., **9**, pp. 389-401.

H. S. HOLLENBERG, On the verification of the principle of reflex visual sensations. *Journ. Opt. Soc. Amer.*, etc., **8**, pp. 713–730.

H. E. IVES, A primary standard of light. *Journ. Franklin Inst.*, **197**, pp. 147–182 and 359–400.

1924. G. JAECKEL, The faintest visible. *Phys. Zft.*, **25**, pp. 13–17.

D. B. JUDD, The spectral energy distributions produced by rotary mixing of complementary papers. *Journ. Opt. Soc. Amer.* etc., **9**, pp. 95–111.

E. KARRER, An interesting phenomenon in photopsia. *Journ. Opt. Soc. Amer.*, etc., **8**, pp. 319, 320.

Idem, New type of photometric disc. *Journ. Opt. Soc. Amer.*, etc., **8**, pp. 541–543.

E. KARRER and A. PORITSKY, Low brightness photometer. *Opt. Soc. Amer.*, etc., **8**, pp. 355–359.

J. v. KRIES, Über einige Aubgaben der Farbenlehre. *Zft. f. techn. Physik*, 6 Nr. 8, pp. 327-349.

P. LASAREFF, Application of the WEBER-FECHNER law to photometry. *Rev. d'optique*, **3**, pp. 65–70.

C. LEISS, Monochromator for 200$\mu\mu$ to 800$\mu\mu$. *Zft. f. Physik*, **21**, 1. pp. 60–62.

O. MEISSNER, The OSTWALD double-colour cone. *Zft. f. Physik*, **21**, **1**, pp. 68–72.

J. H. PARSONS, *An introduction to the study of colour vision.* Second edition. Cambridge.

W. PEDDIE, The development and present position of colour vision theory. *Science Progress*, No. 72, pp. 635–639.

H. PIÉRON, The minimum of energy in the luminous excitation of the retina by brief illumination. *Comptes rendus*, **178**, pp. 966–968.

I. G. PRIEST, Apparatus for the determination of color in terms of dominant wavelength, purity and brightness. *Jour. Opt. Soc. Amer.*, etc. **8**, pp 173–200.

I. G. PRIEST, H. J. MCNICHOLAS and M. K. FREHAFER, Some tests of the precision and reliability of measurements of spectral transmission by the KOENIG-MARTENS spectrophotometer. Ibid., **8**, pp. 201–212.

I. G. PRIEST (in collaboration with L. B. TUCKERMAN, H. E. IVES, and F. K. HARRIS), The Computation of Colorimetric Purity. *J. Opt. Soc. Amer.*, etc., **9**, pp. 503–520.

C. PULFRICH, The Colours Produced by Rotation of the Plane of Polarisation in a Quartz Plate Gound Perpendicular to the Axis. *Zft. Instrumentenk.*, **44**. p. 261–270.

O. SCHNEIDER. Effect of colour of light on visual fatigue. *Zft. techn. Physik*, **5**, 8. pp. 355-357.

C. SHEARD, On the effects of quantity and quality of illumination upon the human eye and vision. *Am. J. Physiolog. Optics.*, **5**, pp. 463–485.

W. J. SMITH, The law of recurrences and decay of after images. *Proc. Roy. Soc. Edinburgh* (1923-24), **44**, Part III., pp. 211-217.

L. T. TROLAND, The optics of the nervous system. *Journ. Opt. Soc. Amer.*, etc., **8**, 389–409.

Idem, The interrelations of modern physics and modern psychology. *Journ. Franklin Inst.*, **197**, pp. 479–504.

E. P. T. TYNDALL and K. S. GIBSON, Visibility of radiant energy equation. *J. Opt. Soc. Am.*, etc., **9**, pp. 403, 404.

J. W. T. WALSH, *The elementary principles of lighting and photometry.*

DOVER PHOENIX EDITIONS

A series of hardcover reprints of major works in mathematics, science and engineering.
All editions are $5\frac{5}{8} \times 8\frac{1}{2}$ unless otherwise noted.

Mathematics

Theory of Approximation, N. I Achieser. Unabridged republication of the 1956 edition. 320pp. 49543-4

The Origins of the Infinitesimal Calculus, Margaret E. Baron. Unabridged republication of the 1969 edition. 320pp. 49544-2

A Treatise on the Calculus of Finite Differences, George Boole. Unabridged republication of the 2nd and last revised edition. 352pp. 49523-X

Space and Time, Emile Borel. Unabridged republication of the 1926 edition. 15 figures. 256pp. 49545-0

An Elementary Treatise on Fourier's Series, William Elwood Byerly. Unabridged republication of the 1893 edition. 304pp. 49546-9

Substance and Function & Einstein's Theory of Relativity, Ernst Cassirer. Unabridged republication of the 1923 double volume. 480pp. 49547-7

A History of Geometrical Methods, Julian Lowell Coolidge. Unabridged republication of the 1940 first edition. 13 figures. 480pp. 49524-8

Linear Groups with an Exposition of Galois Field Theory, Leonard Eugene Dickson. Unabridged republication of the 1901 edition. 336pp. 49548-5

Continuous Groups of Transformations, Luther Pfahler Eisenhart. Unabridged republication of the 1933 first edition. 320pp. 49525-6

Transcendental and Algebraic Numbers, A. O. Gelfond. Unabridged republication of the 1960 edition. 208pp. 49526-4

Lectures on Cauchy's Problem in Linear Partial Differential Equations, Jacques Hadamard. Unabridged reprint of the 1923 edition. 320pp. 49549-3

The Theory of Branching Processes, Theodore E. Harris. Unabridged, corrected republication of the 1963 edition. xiv+230pp. 49508-6

The Continuum, Edward V. Huntington. Unabridged republication of the 1917 edition. 4 figures. 96pp. 49550-7

Lectures on Ordinary Differential Equations, Witold Hurewicz. Unabridged republication of the 1958 edition. xvii+122pp. 49510-8

Mathematical Methods and Theory in Games, Programming, and Economics: Two Volumes Bound as One, Samuel Karlin. Unabridged republication of the 1959 edition. 848pp. 49527-2

Famous Problems of Elementary Geometry, Felix Klein. Unabridged reprint of the 1930 second edition, revised and enlarged. 112pp. 49551-5

Lectures on the Icosahedron, Felix Klein. Unabridged republication of the 2nd revised edition, 1913. 304pp. 49528-0

On Riemann's Theory of Algebraic Functions, Felix Klein. Unabridged republication of the 1893 edition. 43 figures. 96pp. 49552-3

A Treatise on the Theory of Determinants, Thomas Muir. Unabridged republication of the revised 1933 edition. 784pp. 49553-1

A Survey of Minimal Surfaces, Robert Osserman. Corrected and enlarged republication of the work first published in 1969. 224pp. 49514-0

The Variational Theory of Geodesics, M. M. Postnikov. Unabridged republication of the 1967 edition. 208pp. 49529-9

DOVER PHOENIX EDITIONS

An Introduction to the Approximation of Functions, Theodore J. Rivlin. Unabridged republication of the 1969 edition. 160pp. 49554-X

An Essay on the Foundations of Geometry, Bertrand Russell. Unabridged republication of the 1897 edition. 224pp. 49555-8

Elements of Number Theory, I. M. Vinogradov. Unabridged republication of the first edition, 1954. 240pp. 49530-2

Asymptotic Expansions for Ordinary Differential Equations, Wolfgang Wasow. Unabridged republication of the 1976 corrected, slightly enlarged reprint of the original 1965 edition. 384pp. 49518-3

Physics

Semiconductor Statistics, J. S. Blakemore. Unabridged, corrected, and slightly enlarged republication of the 1962 edition. 141 illustrations. xviii+318pp. 49502-7

Wave Propagation in Periodic Structures, L. Brillouin. Unabridged republication of the 1946 edition. 131 illustrations. 272pp. 49556-6

The Conceptual Foundations of the Statistical Approach in Mechanics, Paul and Tatiana Ehrenfest. Unabridged republication of the 1959 edition. 128pp. 49504-3

The Analytical Theory of Heat, Joseph Fourier. Unabridged republication of the 1878 edition. 20 figures. 496pp. 49531-0

States of Matter, David L. Goodstein. Unabridged republication of the 1975 edition. 154 figures. 4 tables. 512pp. 49506-X

The Principles of Mechanics, Heinrich Hertz. Unabridged republication of the 1900 edition. 320pp. 49557-4

Thermodynamics of Small Systems, Terrell L. Hill. Unabridged and corrected republication in one volume of the two-volume edition published in 1963–1964. 32 illustrations. 408pp. 6½ x 9¼. 49509-4

Theoretical Physics, A. S. Kompaneyets. Unabridged republication of the 1961 edition. 56 figures. 592pp. 49532-9

Quantum Mechanics, H. A. Kramers. Unabridged republication of the 1957 edition. 14 figures. 512pp. 49533-7

The Theory of Electrons, H. A. Lorentz. Unabridged reproduction of the 1915 edition. 9 figures. 352pp. 49558-2

The Principles of Physical Optics, Ernst Mach. Unabridged republication of the 1926 edition. 279 figures. 10 portraits. 336pp. 49559-0

The Scientific Papers of James Clerk Maxwell, James Clerk Maxwell. Unabridged republication of the 1890 edition. 197 figures. 39 tables. Total of 1,456pp.
Volume I (640pp.) 49560-4; *Volume II* (816pp.) 49561-2

Vectors and Tensors in Crystallography, Donald E. Sands. Unabridged and corrected republication of the 1982 edition. xviii+228pp. 49516-7

Principles of Mechanics and Dynamics, Sir William (Lord Kelvin) Thompson and Peter Guthrie Tait. Unabridged republication of the 1912 edition. 168 diagrams. Total of 1,088pp. *Volume I* (528pp.) 49562-0; *Volume II* (560pp.) 49563-9

Treatise on Irreversible and Statistical Thermophysics: An Introduction to Nonclassical Thermodynamics, Wolfgang Yourgrau, Alwyn van der Merwe, and Gough Raw. Unabridged, corrected republication of the 1966 edition. xx+268pp. 49519-1

Engineering

Principles of Aeroelasticity, Raymond L. Bisplinghoff and Holt Ashley. Unabridged, corrected republication of the original 1962 edition. xi+527pp. 49500-0

Statics of Deformable Solids, Raymond L. Bisplinghoff, James W. Mar, and Theodore H. H. Pian. Unabridged and corrected Dover republication of the edition published in 1965. 376 illustrations. xii+322pp. 6½ x 9¼. 49501-9